T0282620

CAMBRIDGE LIBRARY COLLECTION

Books of enduring scholarly value

Mathematics

From its pre-historic roots in simple counting to the algorithms powering modern desktop computers, from the genius of Archimedes to the genius of Einstein, advances in mathematical understanding and numerical techniques have been directly responsible for creating the modern world as we know it. This series will provide a library of the most influential publications and writers on mathematics in its broadest sense. As such, it will show not only the deep roots from which modern science and technology have grown, but also the astonishing breadth of application of mathematical techniques in the humanities and social sciences, and in everyday life.

A Treatise on Dynamics of a Particle

As senior wrangler in 1854, Edward John Routh (1831–1907) was the man who beat James Clerk Maxwell in the Cambridge mathematics tripos. He went on to become a highly successful coach in mathematics at Cambridge, producing a total of twenty-seven senior wranglers during his career – an unrivalled achievement. In addition to his considerable teaching commitments, Routh was also a very able and productive researcher who contributed to the foundations of control theory and to the modern treatment of mechanics. This textbook, first published in 1898, offers extensive coverage of dynamics, providing formulae and examples throughout. While the growth of modern physics and mathematics may have forced out the problem-based mechanics of Routh's textbooks from the undergraduate syllabus, the utility and importance of his work is undiminished.

Cambridge University Press has long been a pioneer in the reissuing of out-of-print titles from its own backlist, producing digital reprints of books that are still sought after by scholars and students but could not be reprinted economically using traditional technology. The Cambridge Library Collection extends this activity to a wider range of books which are still of importance to researchers and professionals, either for the source material they contain, or as landmarks in the history of their academic discipline.

Drawing from the world-renowned collections in the Cambridge University Library and other partner libraries, and guided by the advice of experts in each subject area, Cambridge University Press is using state-of-the-art scanning machines in its own Printing House to capture the content of each book selected for inclusion. The files are processed to give a consistently clear, crisp image, and the books finished to the high quality standard for which the Press is recognised around the world. The latest print-on-demand technology ensures that the books will remain available indefinitely, and that orders for single or multiple copies can quickly be supplied.

The Cambridge Library Collection brings back to life books of enduring scholarly value (including out-of-copyright works originally issued by other publishers) across a wide range of disciplines in the humanities and social sciences and in science and technology.

A Treatise on Dynamics of a Particle

With Numerous Examples

EDWARD JOHN ROUTH

CAMBRIDGE UNIVERSITY PRESS

Cambridge, New York, Melbourne, Madrid, Cape Town,
Singapore, São Paolo, Delhi, Mexico City

Published in the United States of America by Cambridge University Press, New York

www.cambridge.org
Information on this title: www.cambridge.org/9781108050340

This edition first published 1898
This digitally printed version 2013

ISBN 978-1-108-05034-0 Paperback

A TREATISE ON

DYNAMICS OF A PARTICLE.

London: C. J. CLAY AND SONS,
CAMBRIDGE UNIVERSITY PRESS WAREHOUSE,
AVE MARIA LANE.
Glasgow: 263, ARGYLE STREET.

Leipzig: F. A. BROCKHAUS.
New York: THE MACMILLAN COMPANY.
Bombay: E. SEYMOUR HALE.

A TREATISE ON

DYNAMICS OF A PARTICLE

WITH NUMEROUS EXAMPLES

BY

EDWARD JOHN ROUTH, Sc.D., LL.D., M.A., F.R.S., &c.,

HON. FELLOW OF PETERHOUSE, CAMBRIDGE;
FELLOW OF THE UNIVERSITY OF LONDON.

CAMBRIDGE
AT THE UNIVERSITY PRESS
1898

Cambridge:
PRINTED BY J. AND C. F. CLAY,
AT THE UNIVERSITY PRESS.

PREFACE.

SO many questions which necessarily excite our interest and curiosity are discussed in the dynamics of a particle that this subject has always been a favourite one with students. How, for example, is it that by observing the motion of a pendulum we can tell the time of the rotation of the earth, or knowing this, how is it that we can deduce the latitude of the place? Why does our earth travel round the sun in an ellipse and what would be the path if the law of gravitation were different? Would any other law give a closed orbit so that our planet might (if undisturbed) repeat the same path continually? Is there a resisting medium which is slowly but continually bringing our orbit nearer to the sun? What would be the path of a particle in a system of two centres of force? When a comet passes close to a planet does it carry with it in its new orbit some tokens to prove its identity?

Such problems as these (which are merely examples) excite our curiosity at the very beginning of the subject. When we study the replies we find new objects of interest. Beginning at the elementary resolutions of the forces we are led on from one generalization to another. We presently arrive at Lagrange's general method, by which when a single function (worthily called after his great name) has been found we can write down, in any kind of coordinates, all the equations of motion cleared of unknown reactions. A little further on we find Jacobi's method by which the whole solution of a dynamical problem can be made to depend on a single integral.

The last word has not yet been said on these problems. The student finds as he proceeds much left to discover and many new questions to ask.

When we extend our studies so as to include the planetary perturbations and to take account of the finite size of the bodies the mathematical difficulties are much increased. In the dynamics of a particle we confine ourselves to simpler problems and easier mathematics.

As the subject of dynamics is usually read early in the mathematical course, the student cannot be expected to master all its difficulties at once. In this treatise the parts intended for a first reading are printed in large type and the student is advised to pass over the other parts until they are referred to later on.

The same problem may be attacked on many sides and we therefore have several different ways of finding a solution. In what follows the most elementary method has in general been put first, other solutions being given later on. For the sake of simplicity they have also generally been treated first in two dimensions. In these ways the difficulties of dynamics are separated from those of pure geometry and it is hoped that both difficulties may thus be more easily overcome.

Some of the examples have been fully worked out, on others hints have been given. Many of these have been selected from the Tripos and College papers in order that they may the better indicate the recent directions of dynamical thought.

I cannot conclude without thanking Mr Dickson of Peterhouse. He has kindly assisted me in correcting most of the proofs and has given material aid by his verifications and suggestions.

EDWARD J. ROUTH.

PETERHOUSE,
July, 1898.

CONTENTS.

CHAPTER I.

ELEMENTARY CONSIDERATIONS.

CHAPTER II.

RECTILINEAR MOTION.

CHAPTER III.

MOTION OF PROJECTILES.

CHAPTER IV.

CONSTRAINED MOTION IN TWO DIMENSIONS.

CHAPTER V.

MOTION IN TWO DIMENSIONS.

CHAPTER VI.

CENTRAL FORCES.

CHAPTER VII.

MOTION IN THREE DIMENSIONS.

CHAPTER VIII.

SOME SPECIAL PROBLEMS.

ERRATUM.

page 155, line 15.

for $-2v \cdot np$ *read* $+2v \cdot np$.

CHAPTER I.

Velocity and Acceleration.

1. THE science of dynamics is divided into two parts. In one the geometrical circumstances of the motion are considered apart from the physical causes of that motion. In the other the mode in which the motion is produced by the action of forces is investigated. The first is usually called *kinematics*, the second is called sometimes *kinetics* and sometimes *dynamics*.

2. Let us consider the geometrical motion of a point on a given curve. *The motion is said to be uniform when equal spaces are described in any two equal times. The space described in any unit of time measures the velocity.*

The word "any" in this definition is important. If all the spaces described in successive units of time were equal, the motion need not be uniform. For example, the hands of a clock move over equal spaces in successive seconds, but in some clocks each space is described by a jump at the end of each second.

In discussing the geometry of the motion, the time is regarded as the independent variable. It is merely some continually increasing quantity. So far as our present purpose is concerned, we may suppose that the time is measured by the space described by some standard point moving in a straight line always in the same direction.

Let s be the distance at the time t of a point P moving uniformly on a curve measured along the arc from some fixed point on the curve. Let s_0 be the arc-distance at the time t_0. Since v is the space described in a unit of time, the arc $s-s_0$

described in $t-t_0$ units of time is given by $s-s_0=v(t-t_0)$. This leads to the converse equation, *in uniform motion the velocity is equal to the space described in any time divided by that time.*

3. When all the arcs described in equal times are not equal, the velocity is variable. By the principles of the differential calculus we consider the arcs described in infinitely short times. The point being in any position P at the time t, let δs be the arc described in a following interval of time δt. If this arc were described uniformly the velocity would be $\delta s/\delta t$. The limiting value when δt is indefinitely small is $v=\dfrac{ds}{dt}$. This may be defined to be the velocity in the position P. This equation is usually expressed in the following words.

The velocity of a point when variable is measured by the space or arc which would be described in a unit of time if the point were to move uniformly with the velocity it had at the moment under consideration.

It is worth while to give a more formal proof of the important equation $v=ds/dt$. Let, as before, δs be the arc described in the next interval δt. Let v_1, v_2 be the greatest and least velocities of the point in that interval. The space δs must lie between $v_1\delta t$ and $v_2\delta t$, and therefore $\delta s/\delta t$ must lie between v_1 and v_2. In the limit v_1 and v_2 become equal to each other and therefore each is equal to ds/dt. This therefore must be the value of v.

4. Parallelogram of velocities. *Velocities may be compounded by the parallelogram law.* Let a point P move with a uniform velocity u along a finite straight line OA and arrive at A at the end of a given time, then $OP=ut$. Let the straight line OA move, always remaining parallel to itself, with a uniform velocity v and come into the position BC in the same time. It is evident, from the properties of similar figures, that the point P has described the diagonal OC of the parallelogram, two adjacent sides of which are OA and OB. The two velocities u, v are proportional to the lengths of the straight lines OA and OB, and are evidently represented by those lines in direction and magnitude. *When therefore a particle moves with two simultaneous velocities represented in direction and magnitude by the straight*

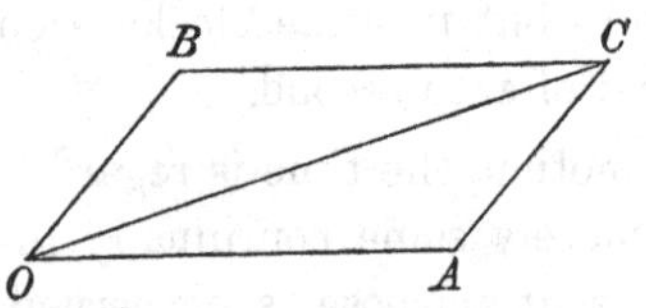

lines OA, OB, its motion is the same as if it were moved with a single velocity represented in direction and magnitude by the diagonal OC of the parallelogram constructed on OA, OB as sides.

5. This rule is the same as that given in Statics for compounding forces which act at the same point. Hence all the rules of Statics, which are derived from the parallelogram of forces, will also apply to velocities.

We may therefore infer the triangle of velocities, and all the various rules for resolving and compounding velocities, both by rectangular and oblique resolutions.

6. *Moment of a velocity.* The moment of a velocity about a point may be defined in the same way as the moment of a force. Let a point P be moving with a velocity v in a direction represented by the straight line APB. Let $CN = p$ be the perpendicular drawn from any point C on the straight line APB. *The moment of the velocity v about C is then defined to be equal to vp.*

Using the same proof as that adopted in Statics, we infer that *the moment of the velocity of a point about any straight line is equal to the sum of the moments of its components.*

7. This theorem enables us to express the moment of the velocity about the origin in several different forms, all of which are in common use.

Let a point P move along a curve. It is proved in Art. 12 that the polar components of the velocity are dr/dt and $rd\theta/dt$; the moments of these about the origin are respectively zero and $r^2d\theta/dt$. The moment of the velocity is therefore $r^2\dfrac{d\theta}{dt}$.

In the same way, the Cartesian components being dx/dt and dy/dt, the moment of the resultant velocity is $x\dfrac{dy}{dt} - y\dfrac{dx}{dt}$.

Lastly let A be the polar area bounded by the path, the moving radius vector r and any fixed radius vector. It is clear that pds is twice the area dA traced out by the radius vector. The moment of the velocity about the origin is $pv = 2\dfrac{dA}{dt}$.

8. The definition given above is strictly the moment of the velocity about a straight line drawn through C perpendicular to the plane containing C and the straight line APB. When we require the moment of the velocity of a point moving along AB about any straight line CD which is inclined to the plane CAB, we use the same extended definition as in Statics.

Let MN be the shortest distance between AB and CD; resolve the velocity v along AB into two components, one along Nz parallel to CD and the other along Ny perpendicular to CD. The former is $v \cos \theta$, the latter $v \sin \theta$, where θ is the angle contained by AB and CD. The moment of the former is defined to be zero, the moment of the latter is $v \sin \theta . p$ where $p = MN$.

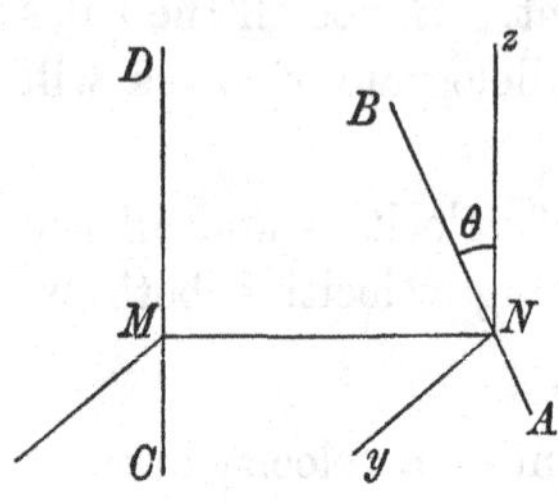

If a point move along AB with a velocity v, the moment of that velocity about CD is $vp \sin \theta$, *where p is the shortest distance between AB and CD and* θ *is the angle contained by those lines.*

The symmetry of this result shows that the moment about AB of a velocity along CD is the same as that about CD of an equal velocity along AB.

9. *Ex.* Given the two straight lines $\frac{x-f}{\lambda} = \frac{y-g}{\mu} = \frac{z-h}{\nu}$; $\frac{x-f'}{\lambda'} =$ &c., where λ, μ, ν; λ', &c. are the direction cosines of the two lines. A particle is moving along one of them; prove that the moment of the velocity about the other is vi, where i is the determinant in the margin.

$$\begin{vmatrix} f-f', & g-g', & h-h' \\ \lambda & \mu & \nu \\ \lambda' & \mu' & \nu' \end{vmatrix}$$

10. Relative velocity. Two points P, Q are moving along two straight lines AB, CD with velocities u, v. It is required to find their relative velocity.

Let any number of bodies be situated within a space and let that space be moved carrying the bodies with it (as in a railway carriage); it is evident that the relative positions of the bodies are unchanged. If then we impress on both the points P, Q a velocity equal and opposite to that of one of them, say P, the relative positions and motions are unaltered. The point P is now at rest and the velocity of Q is the resultant of its own velocity, viz. v, and the reversed velocity, viz. $-u$, of P.

To find the relative velocity of Q with regard to P, we compound the actual velocity of Q with the reversed velocity of P according to the parallelogram law.

Ex. A circle is rotated in its own plane about a point in its circumference with an angular velocity ω and a point P moves on the circle in the opposite direction with angular velocity 2ω relative to the circle. Prove that P moves in a straight line and find its velocity. [Coll. Exam. 1896.]

11. Coordinate velocities. Let P, P' be the positions of a point moving on a curve APP' at the times t and $t+dt$ respectively. Let

$$OM=x, \quad MP=y$$

be the coordinates of P; OM', $M'P'$ those of P'. Let PL, drawn parallel to Ox, cut $P'M'$ in L, then

$$PL=dx, \quad LP'=dy.$$

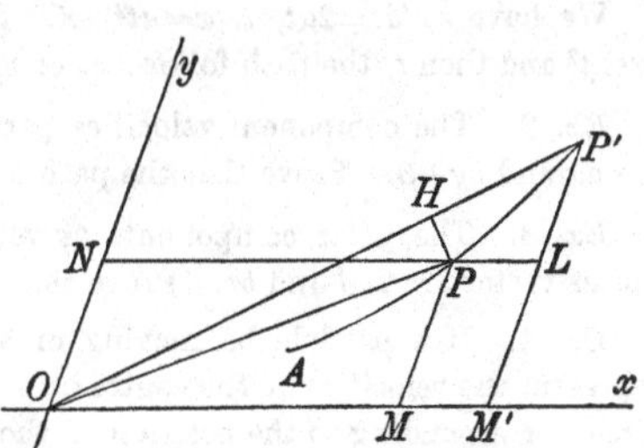

By the triangle of velocities the sides PL, LP' of the triangle PLP' represent the oblique components of velocity on the same scale that PP' represents the resultant velocity. The components of velocity are therefore PL/dt and $P'L/dt$. *If then a point move on a curve, and its coordinates are x, y, the Cartesian components of its velocity are equal to* $\dfrac{dx}{dt}$ *and* $\dfrac{dy}{dt}$.

12. Let PH be a perpendicular drawn from P on OP'. The sides of the triangle PHP' will ultimately represent on the same scale the component velocities perpendicular and parallel to the radius vector OP. These components are therefore PH/dt and HP'/dt. If $OP=r$ and the angle $POx=\theta$ we know by the elementary principles of the differential calculus that $PH=rd\theta$ and $HP'=dr$ ultimately. *The components of velocity along and perpendicular to the radius vector are therefore* $\dfrac{dr}{dt}$ *and* $r\dfrac{d\theta}{dt}$.

13. Let Q be another point whose coordinates are x', y'. The components of its velocity are dx'/dt and dy'/dt. To find the component velocities of Q relative to P we follow the rule of Art. 10. Reversing the component velocities of P and adding the results to those of Q, it is clear that the component relative velocities of Q are $\dfrac{dx'}{dt}-\dfrac{dx}{dt}$ and $\dfrac{dy'}{dt}-\dfrac{dy}{dt}$.

We may put the argument in another form. Let ξ, η, be the coordinates of Q referred to axes having their origin at the moving point P, their directions remaining parallel to the original axes. The component relative velocities are then $d\xi/dt$ and $d\eta/dt$. But since $\xi=x'-x$, $\eta=y'-y$, we arrive by differentiation at the same results as before.

14. *Ex.* 1. The component velocities of a point in the directions of two axes are $2at$ and $2bt+\beta$. Prove that the path is a parabola whose axis is parallel to

$$ay=bx.$$

We have $dx/dt=2at$, $\therefore x=at^2+A$. Similarly y may be found. Eliminating first t^2 and then t, the path follows at once.

Ex. 2. The component velocities parallel to the axes of x and y respectively are ax and $by+\beta$. Prove that the path is $(by+\beta)^a=Ax^b$.

Ex. 3. The polar components of velocity parallel and perpendicular to the radius vector are $2a\theta$ and br. Prove that the path is $br=a\theta^2+A$.

Ex. 4. If a particle be moving in a hypocycloid with velocity u, and v, V represent the velocities of the centre of curvature and the centre of the generating circle corresponding to the position of the particle, prove that

$$\frac{u^2}{(c-b)^2}+\frac{v^2}{(c+b)^2}=\frac{4V^2}{(c-b)^2},$$

c being the distance between the centres of the generating circles, and b the radius of the moving circle. [Math. Tripos.]

15. Acceleration. This word is used to express the rate at which the velocity is increasing. It may be either uniform or variable.

If a point move in such a manner that the increments of velocity gained in any equal times are the same in direction and equal in magnitude, the acceleration is said to be uniform. The increment of velocity in each unit of time measures the magnitude of the acceleration.

16. *First, let the point move in a straight line.* Let v_0 be the velocity at any time t_0; after a unit of time has elapsed, let v_0+f be the velocity. After a second unit of time the velocity must be v_0+2f, because equal increments are gained in equal times. Hence after $t-t_0$ units of time the velocity has increased by $f(t-t_0)$. If v be the velocity at the time t, we have

$$v=v_0+f(t-t_0).$$

The quantity f is the acceleration.

17. *If the point does not move in a straight line* the explanation is only slightly altered. Let Oy represent the direction in which the constant increments of velocity are given to the point, and let Ox be the direction of motion at the time $t=t_0$. Let u_0, v_0 be the components of the velocity in the directions of the axes Ox and Oy respectively at the time t_0. After a unit of time has elapsed the component of velocity parallel to Oy is v_0+f, but that parallel to Ox is unchanged because no velocity has been added in that direction. After $t-t_0$ units of time, the component of velocity parallel to Oy is $v_0+f(t-t_0)$, while that parallel to Ox is still u_0. If u, v are the components of velocity at the time t, we have

$$u=u_0, \qquad v=v_0+f(t-t_0).$$

The magnitude of the acceleration is f, and its direction is Oy.

18. When the increments of velocity in equal times are unequal in magnitude, or not the same in direction, the acceleration is said to be variable. To obtain a measure we follow the method adopted to measure variable velocity.

Acceleration when uniform is measured by the velocity generated in any unit of time. When variable, the acceleration at any instant is measured by the velocity which would be generated in the next unit of time if the acceleration had remained constant in magnitude during that interval and fixed in direction.

19. *To find the equations of motion of a point moving in a straight line with a variable acceleration f.*

Let v and $v+dv$ be the velocities at the times t and $t+dt$. Assuming the principles of the differential calculus, dv being the increment in the time dt, it follows by a simple proportion that dv/dt is the velocity which would be added in a unit of time, if the acceleration had remained constant. Hence, by Art. 16,

$$f=dv/dt.$$

The argument is usually put into a more elementary form. Let δv be the velocity generated in the time δt. Let f_1, f_2 be the greatest and least accelerations of the particle during the interval δt. Then since the actual rate at which the velocity is increasing is always less than the one and greater than the other, the velocity added is less than $f_1\delta t$ and greater than $f_2\delta t$. In the limit f_1 and f_2 coincide and we have $f=dv/dt$.

20. Let the geometrical position of the point at the time t be determined by its distance s from a fixed point in the path. Let v be the velocity, f the acceleration, then

$$v=\frac{ds}{dt}, \qquad f=\frac{dv}{dt}=\frac{d^2s}{dt^2}=v\frac{dv}{ds}.$$

All these expressions for the acceleration are of great importance.

21. We notice that velocity and acceleration are dynamical names for the first and second differential coefficients of s with regard to the independent variable t. If the third differential coefficient were required, we should use some such name as the hyper-acceleration, but this extension is not necessary to dynamics.

22. It appears that acceleration bears the same general relation to velocity that velocity bears to space. When a point moves in a straight line the velocity is the rate of increase of the space, the acceleration is the rate of increase of the velocity.

23. Just as velocity is positive or negative according as the space measured in the positive direction is increasing or decreasing, so acceleration is positive or negative according as the velocity is increasing or decreasing. A negative acceleration is sometimes called a retardation.

24. *To find the motion of a point P moving in a straight line with a uniform acceleration f.*

Let the position of the point at the time $t = t_0$ be given by $s = s_0$, and let v_0 be the velocity. Since $f = d^2s/dt^2$, we have

$$v = ds/dt = ft + A.$$

Hence
$$v_0 = ft_0 + A,$$
and
$$v = f(t - t_0) + v_0.$$

Integrating again, since $v = ds/dt$,

$$s = \tfrac{1}{2}f(t - t_0)^2 + v_0 t + B.$$

Hence $s_0 = v_0 t_0 + B$, and therefore

$$s = \tfrac{1}{2}f(t - t_0)^2 + v_0(t - t_0) + s_0.$$

25. The three fundamental formulæ of elementary kinematics follow from this result. If the point start from the position $s = 0$ at the time $t = 0$,

$$s = \tfrac{1}{2}ft^2 + v_0 t,$$
$$v = ft + v_0,$$
$$v^2 = 2fs + v_0^2.$$

26. *Ex.* 1. A particle describes a space s in time t with a uniform acceleration, the velocities at the beginning and end of this period being v_0 and v. Prove that $s = \frac{1}{2}(v_0 + v)t$. Notice that the coefficient of t is the mean of the two velocities.

Ex. 2. A particle moves from rest with a uniform acceleration. Prove that the average velocity is half or two-thirds of the final velocity, according as the time or the space is divided into an infinite number of equal portions and the average taken with regard to these. [St John's Coll., 1895.]

Ex. 3. Two points P, Q move on a straight line AB. The point P starts from A in the direction AB with velocity u and acceleration f, and at the same time Q starts from B in the direction BA with velocity u' and acceleration f'; if they pass one another at the middle point of AB and arrive at the other ends of AB with equal velocities, prove that $(u+u')(f-f')=8(fu'-f'u)$. [Coll. Exam. 1896.]

Ex. 4. A heavy particle, projected horizontally on a smooth table with velocity v, is reduced to rest by the resistance of the air after describing a space s. Supposing the resistance of the air to be a uniform force, prove that, when the particle is projected vertically upwards with any velocity, the squares of the times of ascent and descent to the point of projection are in the ratio $2gs-v^2$ to $2gs+v^2$.

Ex. 5. A particle is projected vertically upwards from a point A. If the resistance of the air were constant and equal to ng, where n is less than unity, prove that the times of ascent and descent are as $\sqrt{(1-n)} : \sqrt{(1+n)}$.

Ex. 6. A particle is projected vertically upwards in vacuo from a given point P. Prove that the product of the times of passing through another given point Q is independent of the velocity of projection from P.

Ex. 7. Two particles P, P' starting simultaneously from the points A, A' with initial velocities u, u', move in the straight line AA' with accelerations f, f'. If v, v' are their velocities when the distance PP' exceeds the initial distance AA' by s, then

$$(v'-v)^2=(u'-u)^2+2(f'-f)s.$$

See Arts. 10 and 39.

27. *Ex.* A point P, at any given moment, is in the position O moving in the direction Ox with a velocity u. A uniform acceleration f is given to it in the direction Oy. It is required to exhibit geometrically the position and direction of motion after t seconds.

To find the direction of motion we measure lengths OA, OB along Ox, Oy to represent on any scale the velocities u and ft respectively. The direction of motion after t seconds is parallel to the diagonal OD of the parallelogram AOB.

To find the position of the point we measure lengths equal to the spaces, viz. $OE=ut$, $OF=\frac{1}{2}ft^2$. If OG is the diagonal of EOF, the point is at G moving in a direction parallel to OD.

To find the direction of motion we compound the velocities, to find the position we compound the spaces.

28. The parallelogram of accelerations. This theorem follows at once from the parallelogram of velocities. Let a point be moving in any direction at the time t with any velocity. Referring to the figure of Art. 4, let OA, OB represent in direction and magnitude two uniform accelerations given to the point. Then by definition OA, OB represent the two velocities given to the point per unit of time. By the parallelogram of velocities the diagonal OC of the parallelogram constructed on OA, OB represents the resultant increment of velocity per unit of time. The point is therefore uniformly accelerated, and the acceleration is represented in direction and magnitude by OC.

The actual velocity at the time $t+t'$ (if required) could be found by compounding the velocity at the time t, either with both the components OA, OB, each multiplied by t', or with their resultant after multiplication by t'.

29. Hodograph. Let a point move in a curve and let P be its position at any time t. From the origin O draw a straight line OH to represent in direction and magnitude the velocity v at P. *Then OH is parallel to the tangent at P and its length is equal to κv*, where κ is an arbitrary constant introduced to show the scale on which OH represents the velocity.

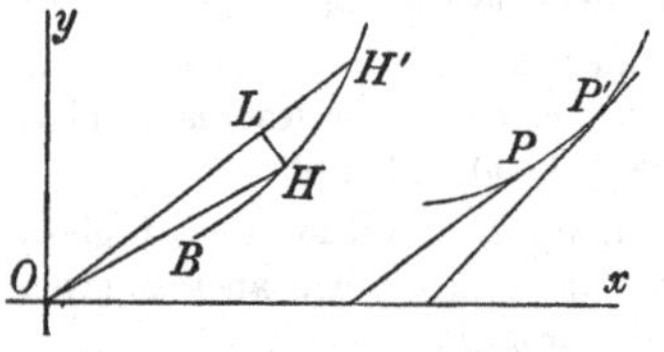

As the point travels from P along its path, *the point H describes a second curve which is called the hodograph of the first.*

Let P, P' be two positions of the point at the times t, $t+dt$; H, H' the corresponding points on the hodograph. Since OH, OH' represent the velocities at P, P' in direction and magnitude, the third side HH' of the triangle HOH' must represent in direction and magnitude the velocity given to the particle in the time dt. It follows by a simple proportion that HH'/dt represents the velocity which would have been added to the velocity at P if the acceleration had remained constant for a unit of time.

The tangent at H therefore represents the acceleration in direction and the ratio of an elementary arc HH' to the time dt of describing it measures the magnitude of the acceleration on the same scale that the radius vector OH represents the velocity.

In this way *the hodograph represents to the eye the motion of a point on a curve.* In general language, the radius vector represents the velocity, the arc gives the acceleration. If r is the radius vector and σ the arc BH, then $r=\kappa v$ and $d\sigma/dt=\kappa f$, where f is the acceleration.

30. *To find the hodograph* when both the curve described by P and the velocity of P are given. If ψ be the angle the tangent at P makes with some fixed straight line taken as the axis of x, we notice that κv and ψ are the polar coordinates of H. From the conditions of the question we first find v and ψ in terms of some one quantity. Then eliminating that quantity we obtain the polar equation of the hodograph. Several examples will be given in the chapter on central forces.

31. *To find the equations of motion of a point moving in a curve with variable acceleration.*

We may deduce the components of acceleration parallel to the axes of coordinates from the acceleration of a point moving in a straight line. Referring to the figure of Art. 11, let $OM=x$,

$ON = y$. The components of the velocity of P have been shown to be the actual velocities of M and N as they move along the axes of x and y respectively. This being true for all positions of P, the acceleration of P is the resultant of the accelerations of M and N. If then X, Y are the component accelerations of P, we have

$$X = \frac{d^2x}{dt^2}, \quad Y = \frac{d^2y}{dt^2}.$$

32. *Ex.* 1. When a point Q describes a circle with a uniform velocity, its projection P on any diameter $x'Ox$ oscillates on each side of the centre O through a length equal to the radius. Prove that the acceleration of P tends towards O and varies as the distance from O.

Let the arc described by Q per unit of time subtend an angle n at the centre, let the angle QOx be α when $t=0$. Then at the time t, the angle $QOx = nt + \alpha$. If a be the radius, the length $OP = a \cos(nt + \alpha)$, hence the acceleration

$$d^2x/dt^2 = -an^2 \cos(nt + \alpha) = -n^2x.$$

The minus sign shows that the acceleration tends towards O.

An oscillatory motion represented by $x = a \cos(nt + \alpha)$ is usually called *a simple harmonic oscillation.*

Ex. 2. A point P moves *towards* a fixed point O so that its velocity varies as x^n, where $x = OP$. Prove that the acceleration varies as x^{2n-1}. Is the acceleration to or from O?

33. The Cartesian components of acceleration are not the only ones which are required in dynamics. The components in polar coordinates and those along the tangent and normal are continually used. Besides these there are the components for moving axes and the extension of all these formulæ to three dimensions. In order to avoid raising unnecessary difficulties at the beginning of the subject we shall confine our attention in the present chapter to the simpler cases. The others will be taken up in the sections on resolved velocities and accelerations.

34. The general principle on which the component of velocity or acceleration in any fixed direction has been defined may be summed up in the following manner.

Since the component of acceleration is the rate at which the component of velocity in that direction is increasing, we have by the definition of a differential coefficient

$$\left.\begin{array}{r}\text{resolved}\\ \text{acceleration}\end{array}\right\} = \text{Limit}\,\frac{(\text{res. vel. at time } t + dt) - (\text{res. vel. at time } t)}{dt}.$$

In the same way if the fixed direction is called the axis of x,

$$\left.\begin{array}{r}\text{resolved}\\ \text{velocity}\end{array}\right\} = \text{Limit}\,\frac{(\text{abscissa at time } t + dt) - (\text{absc. at time } t)}{dt}.$$

35. *To find the resolved accelerations of a point in polar coordinates.*

Let $OP = r$, $POx = \theta$ be the polar coordinates of P. By

Art. 12 the components of velocity at P along and perpendicular to OP are $u = dr/dt$ and $v = rd\theta/dt$. At the time $t + dt$ let the particle be at P', the components of velocity along and perpendicular to the radius vector of P', viz. OP', are $u_1 = u + du$ and $v_1 = v + dv$. Since the angle $POP' = d\theta$, the component of velocity at the time $t + dt$ in the direction OP is

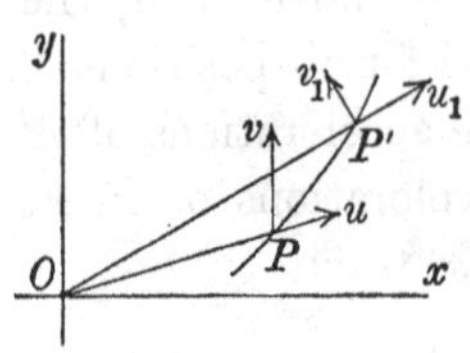

$$u_1 \cos d\theta - v_1 \sin d\theta.$$

This direction being fixed in space for the time dt, the acceleration along the radius vector OP is

$$\text{Limit}\,\frac{(u + du)\cos d\theta - (v + dv)\sin d\theta - u}{dt} = \frac{du}{dt} - v\frac{d\theta}{dt}.$$

Similarly the acceleration perpendicular to the radius vector OP is

$$\text{Limit}\,\frac{(u + du)\sin d\theta + (v + dv)\cos d\theta - v}{dt} = u\frac{d\theta}{dt} + \frac{dv}{dt}.$$

Substituting for u, v their values given above, the accelerations R and S along and perpendicular to the radius vector at the time t are respectively

$$R = \frac{d^2r}{dt^2} - r\left(\frac{d\theta}{dt}\right)^2,$$

$$S = \frac{dr}{dt}\frac{d\theta}{dt} + \frac{d}{dt}\left(r\frac{d\theta}{dt}\right) = \frac{1}{r}\frac{d}{dt}\left(r^2\frac{d\theta}{dt}\right).$$

36. *To find the resolved accelerations along the tangent and normal.*

Let the arc $AP = s$. By Art. 3, the velocity v at P is along the tangent and $v = ds/dt$. At the time $t + dt$ the point is at P', its velocity v_1 is in the direction of the tangent at P' and $v_1 = v + dv$. The components of v_1 in the directions of the tangent and normal at P are therefore $v_1 \cos d\psi$ and $v_1 \sin d\psi$, where $d\psi$ is the angle the tangents at P and P' make with each other. The acceleration along the tangent at P is therefore

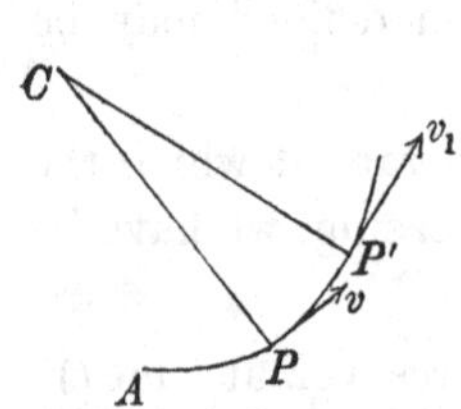

$$T = \text{Limit}\,\frac{(v + dv)\cos d\psi - v}{dt} = \frac{dv}{dt}.$$

Similarly that along the normal in the direction in which the radius of curvature ρ is measured positively, is

$$N = \text{Limit}\,\frac{(v + dv)\sin d\psi}{dt} = v\frac{d\psi}{dt} = \frac{v^2}{\rho}.$$

37. We have now obtained three different sets of components for the accelerations of a moving point. These are the components X, Y along the axes, the components R, S along and transverse to the radius vector, and the components T, N along the tangent and normal. *Any one set can be deduced from any other set by a simple resolution.*

The components R, S are evidently connected with X, Y by the equations

$$R = X\cos\theta + Y\sin\theta, \quad S = -X\sin\theta + Y\cos\theta.$$

Writing $X = d^2x/dt^2$, $Y = d^2y/dt^2$ and substituting $x = r\cos\theta$, $y = r\sin\theta$ we arrive by a simple but rather long differentiation at the values of R and S given in Art. 35. In the same way we have

$$T = X\cos\psi + Y\sin\psi, \quad N = -X\sin\psi + Y\cos\psi.$$

The process of deducing the polar and the tangential-normal components of acceleration from the Cartesian components may be shortened by the following artifice. If $x = r\cos\theta$ we have by differentiation

$$X = \frac{d^2x}{dt^2} = \left\{\frac{d^2r}{dt^2} - r\left(\frac{d\theta}{dt}\right)^2\right\}\cos\theta - \left\{2\frac{dr}{dt}\frac{d\theta}{dt} + r\frac{d^2\theta}{dt^2}\right\}\sin\theta.$$

Since the axis of x is arbitrary in position, let it be so taken that the radius vector r as it turns round the origin is passing through x at the time t. We then have $\theta = 0$, and X becomes R; hence $R = \dfrac{d^2r}{dt^2} - r\left(\dfrac{d\theta}{dt}\right)^2$. To find the acceleration S perpendicular to the radius vector, we take the positive side of the axis of x parallel to the direction in which S is to be measured; that is, the axis of x must be one right angle in advance of the radius vector. Putting therefore $\theta = -\frac{1}{2}\pi$, we find $S = \dfrac{1}{r}\dfrac{d}{dt}\left(r^2\dfrac{d\theta}{dt}\right)$.

38. The three elementary sets of components may be summed up in the following table. They are to be measured positively in the direction in which the length named in the fourth column is measured positively.

	velocity	acceleration	positively
axis of x	$\dfrac{dx}{dt}$	$\dfrac{d^2x}{dt^2}$	x
axis of y	$\dfrac{dy}{dt}$	$\dfrac{d^2y}{dt^2}$	y
along rad. vect.	$\dfrac{dr}{dt}$	$\dfrac{d^2r}{dt^2} - r\left(\dfrac{d\theta}{dt}\right)^2$	r
perpendic. rad. vect.	$r\dfrac{d\theta}{dt}$	$\dfrac{1}{r}\dfrac{d}{dt}\left(r^2\dfrac{d\theta}{dt}\right)$	θ
tangent	$\dfrac{ds}{dt}$	$\dfrac{d^2s}{dt^2}$	s
normal	0	$\dfrac{v^2}{\rho}$	ρ

39. Relative accelerations. Two points P, Q are moving along two curves, it is required to find the acceleration of Q relative to P. By the same reasoning as in Art. 10, it follows that if we impress on both points an acceleration equal and opposite to that of one of them, say P, their relative motions and accelerations are unaltered. This leads at once to the following rule; *the acceleration of Q in space is the resultant of its acceleration relative to P and of the acceleration of P.* As we generally require the components of acceleration, we say that *the component of the acceleration of Q in any direction is equal to its component relative to P plus the component of the acceleration of P.*

40. *Ex.* 1. The position of a point P is given by its polar coordinates r, θ, referred to a fixed origin O and the axis of x. The position of Q is given by its polar coordinates r_1, θ_1, referred to P as origin with the axis of x_1 parallel to x. It is required to find the component accelerations of Q in space.

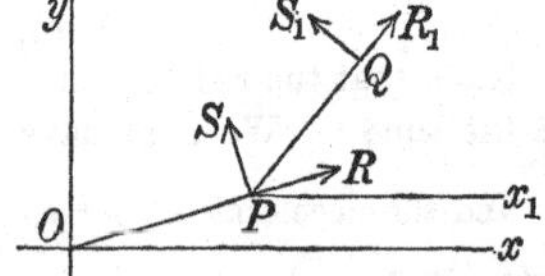

The polar accelerations of P are

$$R=\frac{d^2r}{dt^2}-r\left(\frac{d\theta}{dt}\right)^2, \quad S=\frac{1}{r}\frac{d}{dt}\left(r^2\frac{d\theta}{dt}\right).$$

If R_1, S_1, represent similar quantities when r_1, θ_1, are written for r, θ, these are the accelerations of Q relatively to P. If $\phi=\theta_1-\theta$, we see by a simple resolution, that the resolved part of the space acceleration of Q in the direction PQ is

$$=R_1+R\cos\phi+S\sin\phi.$$

The resolved part perpendicular to PQ is

$$=S_1-R\sin\phi+S\cos\phi.$$

In the same way the resolved parts along and perpendicular to OP are

$$R+R_1\cos\phi-S_1\sin\phi,$$
$$S+R_1\sin\phi+S_1\cos\phi.$$

Ex. 2. The point P describes a circle of radius a with a uniform velocity u. The point Q describes a circle of radius b relatively to P with a uniform velocity v. Prove that the components of the space acceleration of Q along and perpendicular to PQ are respectively $v^2/b+\cos\phi\,.\,u^2/a$ and $\sin\phi\,.\,u^2/a$, where $\phi=(v/b-u/a)\,t$.

Ex. 3. A point P describes a circle of 1 foot radius in 1 hour, and a point Q describes a concentric circle of 4 feet radius in 14 hours, both points move in the counter-clockwise direction; show that the line joining them rotates in the counter-clockwise direction for a period of $43\frac{1}{13}$ minutes followed by a period of $21\frac{7}{13}$ minutes in the clockwise direction. [Coll. Exam. 1896.]

Ex. 4. A circular wire of radius a moves in its own plane without rotation so that its centre has a simple harmonic motion of amplitude a (Art. 32): a bead moves on the wire uniformly, completing a circuit in the period of the simple harmonic motion, and being in the line of the motion of the centre when the centre is in its mean position and is moving in the direction towards the bead; prove that the acceleration of the bead is towards the centre of the simple harmonic motion and that its path is an ellipse of eccentricity $(\frac{3}{2}\sqrt{5}-\frac{5}{2})^{\frac{1}{2}}$. [Coll. Exam. 1897.]

Ex. 5. A railway passenger seated in one corner of the carriage looks out of the windows at the further end and observes that a star near the horizon is traversing these windows in the direction of the train's motion and that it is obscured by the partition between the corner windows on his own side of the carriage and the middle window while the train is moving through the seventh part of a mile. Prove that the train is on a curve the concavity of which is directed towards the star and which, if it be circular, has a radius of nearly three miles, the breadth of the carriage being seven feet and the breadth of the partition four inches.

[Math. Tripos, 1860.]

41. Angular velocity and acceleration. A rigid body is said to be turning round an axis OA when each point is describing a circle whose plane is perpendicular to OA and whose centre lies in OA. Let ϕ be the angle which the plane containing any point P of the body and the axis OA makes with some plane fixed in space and passing through OA. The rate at which the angle ϕ is increasing is called the angular velocity of the body. Following the same line of argument as in the case of linear velocities, the angular velocity is measured by $d\phi/dt$ and the angular acceleration by $d^2\phi/dt^2$.

We notice that if P_1 be any other point in the body and ϕ_1 the angle the plane P_1OA makes with the plane of reference, the angle $\phi-\phi_1$ is independent of the time, so that $d\phi/dt=d\phi_1/dt$.

If Q be any point of the body, r its distance from the axis OA and $\omega=d\phi/dt$ be the angular velocity, the point Q is moving perpendicularly to the plane QOA with a velocity equal to ωr.

If the rotation continue only for a time dt the axis OA (by rotation about which the motion in that time can be constructed) is called the *instantaneous axis* and ω is the *instantaneous angular velocity*.

42. An angular velocity ω about an axis is geometrically represented by a length OA proportional to ω measured along the axis. The direction of the rotation is determined by the convention used in Statics to indicate the direction of rotation of a couple. If OA be the direction in which the length is measured the rotation when positive, should appear to be in some standard direction to a spectator placed with his feet at A and head at B. This standard direction is often taken to be the direction of rotation of the hands of a clock.

43. Parallelogram of angular velocities. *If two instantaneous angular velocities of a body are represented in magnitude and direction by two lengths OA, OB, the diagonal OC of the parallelogram constructed on OA, OB as sides is the resultant instantaneous axis of rotation, and its length represents the magnitude of the resultant angular velocity.*

Let Q be any point which at the time t lies in the plane AOB; r_1, r_2, the distances of Q from OA, OB, p its distance from OC. Let $\omega_1=OA$, $\omega_2=OB$, $\Omega=OC$. The velocity of Q due to the two rotations ω_1, ω_2 is $\omega_1r_1+\omega_2r_2$, while that due to the single rotation Ω is Ωp. To prove that these are equal it is sufficient to notice that if OA, OB represented forces and OC the resultant, the equality merely asserts that the sum of the moments of OA, OB about Q is equal to that of the resultant OC.

Let v be the velocity of Q, then

$$v = \omega_1 r_1 + \omega_2 r_2 = \Omega\rho.$$

If Q lie on OC, $\rho = 0$, and therefore every point of OC is at rest. Hence OC is the resultant axis of rotation. Also since $\Omega = v/\rho$ the angular velocity about the axis is Ω.

44. The theorem of *the parallelogram of angular accelerations* follows from that of angular velocities, just as the parallelogram of linear accelerations follows from that of linear velocities.

45. The rule for compounding angular velocities being the same as that used in Statics to compound forces, we may interpret the limiting case when the intersection O is at infinity as we do the corresponding case in Statics. It is however simpler to deduce the result independently.

Let the body have instantaneous angular velocities ω, ω', about two parallel axes OA, $O'B$ distant a from each other. The resultant velocity of any point Q in the plane of OA, $O'B$ and distant y and $y+a$ from them respectively is $\omega y + \omega'(y+a)$.

Firstly, let $\omega+\omega'$ not be zero. Equating the velocity of Q to zero, we see that every point on a straight line $O''C$ determined by $y = \dfrac{-\omega' a}{\omega+\omega'}$, is at rest. The resultant axis of rotation is therefore parallel to OA, $O'B$ and at a distance y from the former. To find the resultant angular velocity Ω we notice that the velocity of a point Q situated on OA is represented both by $\Omega(-y)$ and $\omega' a$. Hence substituting for y, $\Omega = \omega + \omega'$.

Secondly, let $\omega + \omega' = 0$. The resultant velocity of Q is independent of y and is equal to $\omega' a$. Hence every point in the plane of the axes (and therefore every point of the rigid body) is moving with the same velocity in the same direction. We infer, that *two equal and opposite instantaneous angular velocities about parallel axes are together equivalent to a translation in a direction perpendicular to the plane containing the axes.*

46. Units of space and time. The ordinary unit of time is the second of mean solar time. Space is measured either in feet or centimetres. The metre is 39·37 inches nearly, while the centimetre is the hundredth part of the metre. The unit velocity is then either one foot or one centimetre per second, and the unit of acceleration is a gain of one unit of velocity per second.

47. We are not however restricted to use these units. Let the unit of space be σ feet and the unit of time τ seconds. The unit of velocity is then σ feet per τ seconds, i.e. σ/τ of the feet-seconds units of velocity. The unit of acceleration is a gain of σ/τ feet per second, to be added on every τ seconds, i.e. σ/τ^2 feet per second added on every second. The unit of acceleration is therefore σ/τ^2 feet-seconds units of acceleration.

Let F be the measure of an acceleration when the units are σ, τ; and f the measure of the *same* acceleration when feet and seconds are used. Then since the measure of the same thing varies inversely as the length of the units employed, we have $F = f\frac{\tau^2}{\sigma}$.

48. *Ex.* 1. If the acceleration of a falling body due to gravity is $g = 32\cdot19$ when a foot and a second are the units, show that the acceleration is 981·17 when a centimetre and a second are the units.

Ex. 2. A point moving with uniform acceleration describes 20 feet in the half second which elapses after the first second of its motion. Prove that the acceleration is to that of gravity as 32 to 32·18. Prove also that if a minute be the unit of time and a mile that of space the acceleration will be measured by 240/11.

[Math. Tripos, 1860.]

Ex. 3. If the area of a field of ten acres is represented by 100 and the acceleration of a heavy falling body by $58\frac{2}{3}$, find the unit of time. [Coll. Ex.]

Since an acre is 4840 square yards, 100 new square units is equal to $4840 \times 9 \times 10$ square feet. The new measure of length is therefore 66 feet. Let τ be the required unit of time, then $58\frac{2}{3} = \frac{\tau^2}{66} . 32$. This gives $\tau = 11$ seconds.

Laws of Motion.

49. If one portion of matter, say A, act on another, B, the mutual action is in dynamics called *force.* If we are examining the motion of A only, disregarding B, this force is said to be *external* to A, but if we are taking both portions into consideration, the action is an *internal force.* An external force is usually called *an impressed force.* The mutual actions and reactions between the molecules or parts of a body are internal forces. These forces have different names according to the circumstances of the case. When the bodies are apparently in contact, their mutual action is called *pressure,* when at a distance, the action is called *attraction.*

Nothing has been said of the size of the body, but it is convenient to divide bodies into small portions. A body so small that its position in space when free is determined by the coordinates of one point may be called *a particle.* This division into indefinitely small particles is not necessary for our present purpose. All that we require is that there shall be no rotation. A particle may be said to have no rotation; the rotation of finite bodies is usually regarded as a part of Rigid Dynamics.

50. Our object in dynamics is to investigate the motion of a body. We have then to consider (1) how a body A moves when left to itself; (2) how the motion is affected by the action of an external force, say, due to the presence of another body B; (3) how the action of B on A is related to the reaction of A on B. The answers to these questions are given in Newton's Laws of Motion.

The strict definition of the meaning of the word force as used in dynamics is determined by these laws. We do not consider all the actions which one body can exert on another but those only which tend to alter the instantaneous motion of the body. The following definition or explanation is commonly given. *The word force is used to express any cause which produces or tends to produce a change in the existing state of rest or motion of the body.*

The velocity of a body has both direction and magnitude, we must therefore suppose that the cause of this motion also has both direction and magnitude. To determine a force we require to know (1) its point of application, (2) its direction, and (3) its magnitude. The unit of magnitude will be considered presently.

51. Newton's Laws of Motion are as follows*:

Law 1. *Every body continues in its state of rest or of uniform motion in a straight line, except in so far as it may be compelled to change that state by impressed forces.*

Law 2. *Change of motion is proportional to the impressed force, and takes place in the direction of the straight line in which the force acts.*

Law 3. *To every action there is always an equal and contrary reaction; or, the mutual actions of any two bodies are always equal and oppositely directed.*

52. The first law of motion asserts that the internal forces of a body do not alter the uniform motion. This law is not a repetition of the explanation of the word force given in Art. 50. The law asserts that the causes of motion must be *external.*

* The reader who desires something more than the slight sketch here given of the laws of motion may refer to Newton's *Principia*, to a treatise on Matter and Motion by the late J. Clerk Maxwell and to the *Elements of Natural Philosophy* by Thomson and Tait. There are also Maxwell's two reviews of the latter book in *Nature*, vol. VII. and vol. XX. Several points of controversy are discussed in an essay by R. F. Muirhead to which a Smith's Prize was awarded in 1886, see the *Phil. Mag.* 1887.

The law is sometimes expressed by saying that *the body has inertia.* The body has no power of itself to change its state of rest or motion, but goes on moving in the same direction with the same velocity when not acted on by an impressed force.

53. To define a uniform state of motion we require the measurements of space and time. If we assume the truth of the first law for some particular body, we can measure time by the space passed over by that body. The first law then asserts that the spaces described by any other body (not acted on by any external force) are equal when the spaces simultaneously described by the clock-body are equal. There remains the practical difficulty of obtaining a body free from external forces, which could be used as a clock. For this purpose we have recourse to some other dynamical result.

Applying the principles of dynamics, as developed from the laws of motion, to a rotating body, it can be proved that the motion of rotation about a certain axis is uniform if the external forces have no moment about that axis. The rotation of such a body may be used very conveniently as a clock.

The rotating body actually chosen is the earth. The forces which tend to alter the period of rotation are so small as to be only scientifically perceptible. This period, scientifically amended where necessary, is used as a unit of time. The practical methods of adapting our clocks to the rotation of the earth are described in treatises on astronomy.

We have specially mentioned the rotation of the earth because that supplies the measure of time in common use. Other phenomena may also be used, for example, the velocities of the different kinds of light and their wave lengths in vacuo are constants. Their numerical values have been calculated, and from these we could deduce unalterable units of space and time. The numerical values connected with any perpetual phenomenon would enable future observers to discover our present units from their determinations of the same periods and lengths.

54. The words "change of motion" in the second law mean "change of momentum."

The quantity of matter in a body is called its mass. This may be measured by taking any given lump of matter as the unit of mass. Confining our attention, for the moment, to any the same kind of matter, the mass of any other lump may be deduced by taking the ratio of the volumes.

The momentum of a body, all the points of which are moving in parallel straight lines with equal velocities, is the product of the mass by the velocity.

We notice that the momentum of a body has direction and magnitude. It may be compounded and resolved by the parallelogram law. Let m be the mass, v the velocity, and let θ be the angle the direction of motion makes with some fixed straight

2—2

line; then $v\cos\theta$ is the component of velocity, and $mv\cos\theta$ the component of momentum in the direction of that straight line.

55. The force spoken of in the second law is an external force. It includes the ideas of the magnitude of the force and the time during which its action is considered. During this time the direction and magnitude of the force continue unchanged. We may also regard it as an impulse by which the whole momentum is instantaneously communicated to the body.

Consider the case in which a uniform force F acts on a moving particle in the direction of its motion, and in the time $t'-t$ let the velocity be increased from v to v'. The second law asserts that the change of momentum produced in a unit of time, viz. $m(v'-v)$, divided by the time $t'-t$, is proportional to the magnitude of F.

If the force F is not uniform, the time $t'-t$ must be replaced by dt and the velocity $v'-v$ by dv, Art. 19. The law then asserts that the product of the mass and the acceleration is proportional to the instantaneous magnitude of the force F. We then have, F varies as mf.

56. The arguments for the truth of Newton's laws may be classed under three heads.

First, we can make an appeal to common experience; this is considered to suggest the laws in a general way. We then try some simple experiments so arranged that they can be conducted with considerable accuracy. These test the laws only within the limits of error of the experiments, but, by taking care, these can be reduced to a small amount.

Secondly, we can show that having granted some portions of the laws as being truly founded on an experimental basis other portions follow by pure reasoning.

Lastly, we can assume the laws as a working hypothesis and deduce from them the proper motions of a variety of bodies. If these are found to agree with the observed motions, the laws are tested within the limits of error of the observations. Let us consider these latter tests a little more fully.

57. The position of a planet, the times of the beginning and end of an eclipse and some other phenomena can be observed with great accuracy and are therefore severe tests of the truth of the results of dynamics. The calculations by which these predictions are obtained are very complicated, depending on the combination of many forces acting diversely. There are therefore many causes of error. The predictions in the Nautical Almanac are made some years beforehand, so that any small error, say in a velocity, might be expected by accumulation to produce a sensible effect. Yet notwithstanding both these opportunities of detecting errors, the predicted places agree with the observations. In many of the astronomical calculations the truth of the law of gravitation is assumed. The comparison of the predictions with observations is a test of the truth of that law, as well as of the principles of dynamics.

The solutions of the equations of motion have also in some cases led to unexpected results, which had never been discovered until they were suggested by theory. For example, no one had noticed the slow rotation of the plane of vibration of a pendulum due to the rotation of the earth until Foucault deduced it from dynamical principles.

Our belief in the truth of the laws of motion may be made to rest on these latter considerations. We may regard these laws as the axioms on which the science of dynamics is founded. All its predictions have as yet been verified. It is only when we arrive at a result contradicted by experience, after due allowance has been made for the necessary errors of observation, that we can be called on to amend so much of the laws as has led to the error.

Still such a course would be felt to leave something wanting. We require to know how the laws were discovered, or at least what considerations would make them probable. For this reason a very brief summary of the arguments has been given in the following articles.

58. *First law.* Let a body be set in motion by any cause, say it is projected along a horizontal plane. We notice that when the cause ceases to act, the body continues in motion. It has therefore some power of retaining the motion given to it. There is only a question of degree; does it retain the whole or only some portion of the velocity given to it? The body gradually comes to rest, but we also observe that there are forces tending to stop the body, such as friction and the resistance of the air. We observe that when these resistances are small the body continues in motion for a long time. This suggests that the diminution of the velocity may be entirely due to the resistances, though it does not prove that fact. We improve the argument by having recourse to some experiments sufficiently accurate to allow measurements to be made. Any of the ordinary problems given in treatises on elementary dynamics may be utilized for this purpose, but the one most commonly used is Atwood's machine.

59. Before proceeding to that experiment let us consider some points connected with the second law. How is the action of a force affected by the previously existing motion of the body? We must show that both in direction and in magnitude the action is independent of the velocity. Let us take gravity as the force to be experimented on. We find that a stone dropped from a moving support, say, the ceiling of a railway carriage in rapid uniform motion, hits the same point of the floor that it would have hit had the carriage been at rest. Since, by the first law, the stone retains the horizontal velocity of the carriage, gravity must have acted vertically on the moving particle, that is *in the same direction as if the particle were at rest.*

If a number of balls are simultaneously projected horizontally from a platform with different velocities, they reach the ground at the same time; only one knock is heard. Gravity has therefore pulled all the balls through equal vertical spaces in the same time. This experiment suggests that *the magnitude of gravity* is not altered by the existing motion of the particle attracted.

These experiments cannot be made with great accuracy. They are first attempts to answer the question placed at the beginning of this article.

60. In Atwood's machine two heavy particles are attached together by a string which passes over a pulley. If w, w' are the weights of the particles, the moving force is $w-w'$ while the weight of the mass moved is $w+w'$. By choosing nearly equal values of w and w' the motions produced by gravity can be made as slow as we please. The spaces described and the velocities generated can therefore be measured with some degree of accuracy, and the results compared with the laws of falling bodies. The machine being carefully constructed, some allowance may be made for the inertia of the pulley, the friction, &c. Even the resistance of the air, owing to simplicity of the motion, could be allowed for; but it is almost imperceptible in such slow motions. By an arrangement of platforms small weights can be added or subtracted so that the moving force can be suddenly increased or decreased at pleasure. By making this force balance the resistances we can test the first law. By other changes we can determine whether the effect of a force is modified by a previously existing velocity.

61. *The equation $F=mf$.* Let F be the force which will just support a body when attracted by the earth. Then reversing this force we can imagine the body to be acted on in the same direction by two forces each equal to gravity. Each of these forces can act only by producing motion in the body and we have just seen that this action is not modified by any existing motion. Assuming this, each force will generate the same velocity in the body in the same time. Thus twice the velocity is produced by twice the force, and generally the velocity produced varies as the force when the mass is constant.

Again, if we suppose that two equal volumes of the *same material* are placed side by side, and each acted on by equal forces, equal velocities are generated in the same time. If the initial velocities are equal, the bodies will continue to move side by side, without pressing on each other, and we may suppose them to be united into one mass. Thus twice the force will produce in twice the mass the same velocity, and generally the force varies as the mass when the velocity produced is constant. Varying both the velocity and the mass, we conclude that the magnitude of the force varies as the mass multiplied by the velocity generated. This product is called momentum, Art. 54.

62. Lastly let us consider how far the equality of action and reaction is suggested by elementary considerations. If we press a stone with the finger, the finger is pressed back by the stone. If a horse pull a body by a rope, the tension of the rope impedes the progress of the horse. To determine if these actions are equal, we shall examine separately the conditions when the bodies are in contact and when they act at a distance.

We have to prove that when two bodies in contact press on each other, the momentum lost by one is equal to that acquired by the other. In our test experiment, we arrange the circumstances so that these changes of momenta can be readily observed. Let us suspend two spherical balls by strings and allow them to impinge on each other. The initial positions being given we can find the velocities just before impact. By observing their subsequent motion we can deduce the velocities just after the impulse is concluded. In this way Newton showed that the changes of velocity were such that the momentum lost by one was equal to that gained by the other.

Let us next compare the forces exerted by two mutually attracting bodies. It was a well-known fact that a magnet attracts iron, but Newton showed experimentally that the iron attracts the magnet with an equal force. This he effected by floating both in separate vessels in standing water. The vessels being placed in contact, neither was able to propel the other. The resultant force on each body was therefore zero. Admitting that the mutual action and reaction of the vessels in contact are equal and opposite, it follows that the attraction of each of the distant bodies on the other was equal to the pressure between the vessels and therefore equal to each other.

63. Units of mass. The second law of motion enables us to extend our measurement of mass to bodies of different materials. We first select some quantity of a standard substance and define that to be the unit of mass. Such a quantity of the same or another substance is then said to be of the same mass when two forces, known to be equal, acting on the two masses generate equal velocities in equal times. The second law then asserts that, with this definition of mass, the momentum generated by every force is proportional to that force. Art. 55.

The British unit of mass is defined by Act of Parliament. It is a quantity of platinum preserved in the office of the Exchequer and called the *Imperial standard pound Avoirdupois.* One seven-thousandth part of it is declared to be a grain, and 5760 grains to be a pound Troy. The French standard of mass is called the gramme. This is the one-thousandth part of a certain mass of platinum preserved in the Archives and called a *kilogramme.* The English pound is very nearly equal to 453·59 grammes, and a kilogramme to 2·2 pounds. The system of units derived from the centimetre, gramme and second is usually called the C.G.S. system. That founded on the foot, pound and second may be called the F.P.S. system. It should be noticed that the pound and the gramme are measures of mass, not weight.

A very full account of the history of the English standards of weight and of their comparison with the French standards was given by the late Prof. W. H. Miller in the *Phil. Trans.* for 1856.

64. Units of Force. *The unit of force is that force which, acting on the unit of mass for a unit of time, generates a unit of velocity.* This is usually called Gauss' absolute unit of force.

When the unit of mass is the Imperial pound and the units of space and time are a foot and a mean solar second, the unit of force is called *a poundal.* When the unit of mass is the gramme, and the units of space and time are a centimetre and a second, the unit of force is called *a dyne.*

Since the pound is 453·59 grammes and a metre is 39·37 inches, it is clear that the poundal generates a velocity of 1200/39·37 centimetres in 453·59 grammes. By the second law the magnitude of a force is proportional to the product of the mass by the velocity generated; the poundal is therefore equal to

$$\frac{1200 \times 453\cdot59}{39\cdot37} \text{ dynes.}$$

This makes the poundal equal to 13825 dynes nearly.

When a force F, constant in magnitude and fixed in direction, generates in a mass m a velocity v in a unit of time, we know by the second law that $F = \lambda mv$ where λ is some constant depending on the units of m, v and F. Since F is a unit when m and v are units, $\lambda = 1$. Hence $F = mv$.

When the force F is not constant in magnitude for any finite time, we have recourse to the principles of the differential calculus. Let f be the acceleration, then f is equal to the velocity which would be generated in a unit of time if the force F continued constant in magnitude for that time. Hence $F = mf$, see Art. 55.

65. The determination of the magnitude of a force by experiments on the velocity generated is an inconvenient method of proceeding. We have recourse to the attraction of the earth on them. The law of gravitation asserts that the forces of attraction of the earth on different bodies at the same place are proportional to the masses of those bodies. This is true whatever be the materials of which the body is made, provided only they may be regarded as particles when compared with the size of the earth.

This is an experimental fact which is independent of the laws of motion, and is referred to here as a practical method of comparing forces. Forces therefore may be compared by measuring the weights which they would support at any the same place on the surface of the earth.

Let W be the force of attraction of the earth on a mass m at any given place, let g be the acceleration, then the equation $F = mf$ becomes $W = mg$.

The law of gravitation asserts that g is a constant at the same place on the surface of the earth. It is sometimes called the constant of gravitation.

The average value of g for the area of Great Britain is about 32·18 when the units of space and time are a foot and a second. When the unit of space is changed to centimetres, the numerical value of g becomes 981.

The equation $W = mg$ shows that the weight of a unit of mass is g. The poundal, or unit of force, is therefore $1/g$th part of the weight of the unit piece of platinum, Art. 63. Since 16 oz. make the pound, the poundal is roughly equal to the weight of half an ounce. The dyne is consequently equal to 1/13800th part of half an ounce, Art. 64, roughly a 64th part of a grain.

66. There are two elementary experiments by which it may be shown that g is a constant at the same place and from which the numerical value may be deduced.

In Atwood's machine, let m_1, m_2 be the masses suspended by a string over the pulley, Art. 60. If the law of gravitation is true, the weights are $m_1 g$ and $m_2 g$. The mass moved being $m_1 + m_2$ and the moving force $(m_1 - m_2) g$, the equation $W = mf$ shows that

$$(m_1 + m_2) f = (m_1 - m_2) g$$

where f is the acceleration. By measuring the initial and terminal velocities we can find the value of f and therefore of g for any assumed masses m_1, m_2. Repeating the experiment with other masses, we find that the constancy of g is verified as far as the imperfections of the machine allow.

67. The method adopted by Newton is more accurate. He measured the times of oscillation of hollow wooden balls which he filled with substances of different kinds. Whatever the matter placed inside might be, the time of oscillation (under similar circumstances) was found to be the same. The forces of attraction, measured dynamically by the motion communicated, must therefore have been proportional to the masses moved.

The theory of the oscillation of a particle suspended by a string is given in the chapter on constrained motion. Many experiments have been made since Newton's time for the purpose of determining the numerical value of g. In these the oscillations of bodies of finite size have been observed. An account of some of these experiments is given in the author's *Rigid Dynamics*, vol. I.

68. **Accelerating Force.** The quantity f in the equation $F = mf$ is the acceleration measured, as already explained, by the velocity generated per unit of time. *The quotient F/m is called the accelerating force.* It is equal to the acceleration and the word "force" appears to have been added merely to show from which side of the equation the quantity is derived. It is a convenient phrase to use when we wish to call attention to the fact that the impressed forces under discussion are proportional to the masses acted on.

The product of the mass and the acceleration is called the effective force. Thus md^2x/dt^2 and md^2y/dt^2 are the Cartesian components of the effective force on the particle m. The utility of this name will be better understood when we come to the discussion of the motion of several connected particles.

69. **The vis viva** of a particle whose mass is m and velocity v is mv^2. The half of this quantity has also been called the vis viva, but in England it is more usual to call this latter quantity, viz. $\frac{1}{2}mv^2$, *the kinetic energy.*

70. **The work of a force.** The theory of work is so much used in statics that only a very brief account is necessary here.

Let the point of application A of a force F be moved to a point B, where $AB = ds$. Let θ be the angle made by the direction of motion of A with the direction of the force. Then *$F \cos \theta ds$ is the work of F for the indefinitely small displacement ds.* It is also called *the virtual moment* of F. The work may also be defined to be the product of the force by the resolved displacement of the point of application in the direction of the force.

If the point continue to move and describe any curve, the integral $\int F \cos \theta ds$ is defined to be the work.

If a weight W descend a space dz, the work done is Wdz.

If the space is finite and equal to h, the work is $\int_0^h Wdz$. The work is therefore Wh.

71. The theoretical unit of work is the work done by a dynamical unit of force acting through a unit of space. As explained in Art. 64, this unit of force might be the poundal and the unit of space the foot.

The work required to raise a given weight a given height is taken as a practical unit of work. The unit adopted by English engineers is that required to overcome a force equal to the gravity of a pound through a space of a foot. This unit is called a *foot-pound.*

In the C.G.S. system the theoretical unit is the work done by a dyne in acting through one centimetre. This unit is called the *erg.*

The work done when a kilogramme (Art. 63) is raised one metre is the practical unit and is written kilogramme-metre. A kilogramme-metre is 7·23 foot-pounds very nearly.

72. The rate of doing work is measured by the work done per unit of time. Thus, if the particle describe a space ds in the time dt, the rate of doing work is $F \cos\theta \, ds/dt$. The rate is therefore $Fv \cos\theta$.

The term *horse-power* is used to express the work done per unit of time in practical measure. The unit of horse-power is usually taken to be 550 foot-pounds per second.

The term *force de cheval* corresponds to horse-power, but with different units. The unit of force de cheval is 75 kilogramme-metres per second. A force de cheval is therefore 541 foot-pounds per second; *i.e.* ·98 of one horse-power.

Ex. 1. If the unit of space is σ feet, the unit of time τ seconds, and the unit of mass μ pounds, prove that the unit of force is $\mu\sigma/\tau^2$, the unit of energy is $\mu\sigma^2/\tau^2$, the unit of horse-power $\mu\sigma^2/\tau^3$; see Art. 47.

If F, E, H represent the force, energy and horse-power with these units, find their measures where feet, seconds and pounds are the units.

Ex. 2. Prove that a foot-pound is ·138, and an inch-ton is 25·8 kilogramme-metres.

The Equations of Motion.

73. Equations of Motion. When the resolved part F of the impressed force in any direction and the mass m are given, the corresponding equation of motion is found by equating F/m to the resolved acceleration in that direction. For example in Cartesian coordinates, if X_1, Y_1, be the components of the impressed force, we unite X_1/m, Y_1/m for X, Y in Art. 31. We thus have

$$\frac{d^2x}{dt^2} = \frac{X_1}{m}, \quad \frac{d^2y}{dt^2} = \frac{Y_1}{m}.$$

The polar and other resolutions may be treated in the same way.

74. To make the meaning of these equations clear, let us consider the case of a particle moving in a straight line under the action of several forces, F_1, F_2, &c. The corresponding theorems when there are no restrictions on the motion of the particle will be considered later on.

If m be the mass in motion, the equation of motion takes the form

$$mv\frac{dv}{ds} = m\frac{dv}{dt} = F_1 + F_2 + \ldots \;\ldots\ldots\ldots\ldots\ldots\ldots(1),$$

where s is the space described, and v the velocity at the time t.

This equation may be integrated in two ways. Taking the time t as the independent variable, we have

$$mv - mv_0 = \int F_1 dt + \int F_2 dt + \ldots \;\ldots\ldots\ldots\ldots\ldots\ldots(2),$$

where v_0 is the velocity at the time t_0, and the limits of integration are t_0 to t. The forces F_1, F_2, &c. may not act during the whole time, thus F_1 might act from t_1 to $t_1 + \alpha$, F_2 might act from t_2 to $t_2 + \beta$ and so on. In such cases the limits of each integral should be from the time of beginning to the time of ending of the force. For the sake of conveniently using the equation we notice (what really follows at once from the second law) that *each force F adds to the moving mass a momentum equal to $\int F dt$, where the integration extends over the time of action of the force.* This is called *the time-integral* of the force. The equation (2) is called *the equation of momentum.*

75. Taking the space s as the independent variable, we have

$$\tfrac{1}{2}mv^2 - \tfrac{1}{2}mv_0^2 = \int F_1 ds + \int F_2 ds + \ldots \;\ldots\ldots\ldots\ldots\ldots\ldots(3).$$

It follows that the increase of the kinetic energy of the mass moved is equal to the sum of the works of the several forces. *Each force F communicates to the moving mass an amount of kinetic energy equal to ∫Fds where the integration extends over the space described while F acts on the mass.* This is called *the space-integral* of the force. The equation (3) is called sometimes *the equation of vis viva* and sometimes *the equation of energy.*

If the velocity of the mass is the same at any two *times,* the momentum added on by some of the forces must be equal to that removed by other forces.

If again the velocity is the same in any two *positions,* the work added on by some of the forces must be equal to that subtracted by other forces.

In this way we obtain two equations to find the one quantity v. If the forces F_1, F_2, &c. are constant both the space and time-integrals can be at once found. We therefore use either or both the equations (2) and (3). If the forces are functions of either t or s, only one of the integrations can be immediately effected. We use the equations (2) or (3) according as the forces depend on the time or on the position of the particle.

76. When the system contains more than one particle, their mutual actions may have to be taken into consideration. Suppose, for example, that two particles P, P', whose masses are m, m', are constrained to slide on the straight lines Ox, Ox', and are acted on by the forces F, F' in these directions. Let these be connected by a string of given length which passes over a smooth pulley C. The two equations of energy are

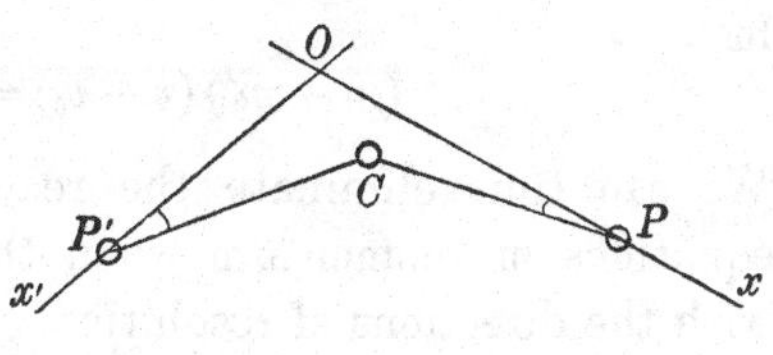

$$\tfrac{1}{2}m\,(v^2 - v_0^{\,2}) = \int F ds \;- \int T\cos\theta ds,$$
$$\tfrac{1}{2}m'\,(v'^2 - v_0'^2) = \int F' ds' - \int T\cos\theta' ds'$$

where θ, θ' are the angles the two portions of the string make with Ox, Ox'. To use these equations we must eliminate the unknown tension T.

We notice that the string is in equilibrium under the action of the tensions at its extremities P, P'; hence, by the principles of statics, their total virtual moment or work is zero. We have therefore

$$T\cos\theta ds + T\cos\theta' ds' = 0.$$

Adding therefore the two equations of energy together

$$\tfrac{1}{2}m(v^2 - v_0^2) + \tfrac{1}{2}m'(v'^2 - v_0'^2) = \int F ds + \int F' ds'.$$

The tension therefore may be omitted in forming the equation of energy, when both the particles are brought into the equation.

77. Consider next the two equations of momenta

$$m(v - v_0) = \int F\,dt - \int T\cos\theta\,dt.$$
$$m'(v' - v_0') = \int F' dt - \int T\cos\theta' dt.$$

The tension T measures the whole momentum transferred per unit of time from one particle to the other along the string. The components transferred are respectively $T\cos\theta$, $T\cos\theta'$, and these are not equal. The transverse components $T\sin\theta$, $T\sin\theta'$ are destroyed by the reactions of the rods Ox, Ox'. If however the pulley C is situated at the intersection O of the rods, θ and θ' are always zero, and the component momentum added to one particle is equal to that taken from the other.

Since the particles must now move with equal velocities, we have $v' = -v$. Eliminating T from the equations of momenta, we have

$$(m + m')(v - v_0) = \int F dt - \int F' dt.$$

We can thus eliminate the reaction T by combining the two equations of momentum when the reaction makes equal angles with the directions of resolution.

78. Examples*. *Ex.* 1. Two heavy rings P, P', of unequal mass, slide on two smooth rods Ox, Ox' at right angles and equally inclined to the horizon at an angle $\alpha = \frac{1}{4}\pi$. The rings are connected by a straight string of given length l and start from rest at distances a, a' from O. Find the motion.

Let s, s' be the distances of P, P' from O at the time t. Since the particles start from rest the equation of vis viva becomes

$$\tfrac{1}{2}(mv^2 + m'v'^2) = \int mg\sin\alpha ds + \int m'g\sin\alpha ds' = g\sin\alpha\,\{m(s-a) + m'(s'-a')\}\ldots\ldots(1),$$

the limits of integration being $s = a$ to s and $s' = a'$ to s'. The length of the string being given we have the geometrical equation

$$s^2 + s'^2 = l^2 = a^2 + a'^2 \ldots\ldots(2).$$

Differentiating (2) we have $\quad sv + s'v' = 0 \ldots\ldots(3).$

* Most of these examples are taken from the examination papers for the entrance and minor scholarships in the several colleges.

The equations (1) and (3) give v and v'. When the particles again come to rest, $v=0$, $v'=0$. Substituting in (1) and using (2) we find, besides the initial solution $s=a$, $s=a'$,

$$s=\frac{2mm'a'+(m^2-m'^2)a}{m^2+m'^2}, \qquad s'=\frac{2mm'a-(m^2-m'^2)a'}{m^2+'m^2}.$$

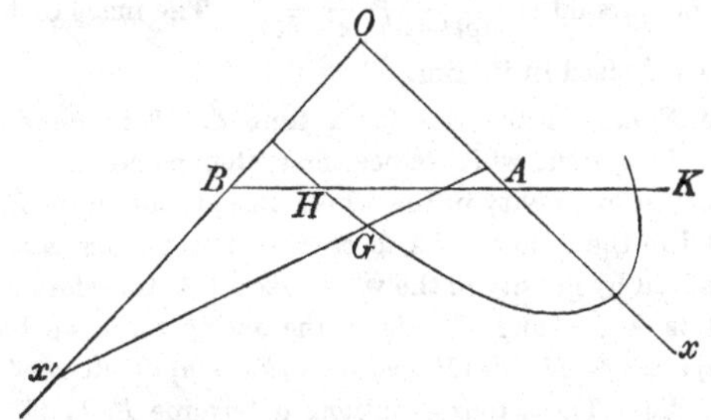

Let y_0 be the initial depth of the centre of gravity of the particles below the horizontal line through O, y the depth at the time t. The equation (1) then gives

$$\tfrac{1}{2}(mv^2+m'v'^2)=g(m+m')(y-y_0)\text{..........................}(4).$$

The centre of gravity G cannot therefore rise above the horizontal line AB drawn through the initial position H, for if it could, the right-hand side of (4) would be negative while the left-hand side is essentially positive. Since the distances of the centre of gravity from Ox, Ox' are respectively $\eta=s'm'/M$ and $\xi=sm/M$, where $M=m+m'$, we see from (2) that the path of the centre of gravity is the ellipse

$$\frac{\xi^2}{m^2}+\frac{\eta^2}{m'^2}=\frac{l^2}{M^2}.$$

This conic cuts the straight line AB in two points H, K. If both these points lie between the rods the centre of gravity continually oscillates in the elliptic arc having H, K for the extreme points. If either H or K lies outside the rods, one particle will pass through the intersection O.

If the string instead of being straight were bent by passing through a small pulley at the intersection of the rods, we could eliminate T from the two equations of momentum. We then have

$$(m+m')v=\int mg\sin\alpha dt-\int m'g\sin\alpha dt=g\sin\alpha(m-m')t.$$

The equation of vis viva is the same as before, but since $v'=-v$ and $s'-a'=a-s$, it takes the simpler form

$$\tfrac{1}{2}(m+m')v^2=g\sin\alpha(m-m')(s-a).$$

These equations give s and v in terms of the time t. We notice that if $m>m'$, the particle P descends along the rod Ox and finally draws P' up to O.

Ex. 2. Two small rings of masses m, m' are moving on a smooth circular wire which is fixed with its plane vertical. They are connected by a straight weightless inextensible string. Prove that, as long as the string remains tight, its tension is $\dfrac{2mm'g\tan\alpha\cos\theta}{m+m'}$, where 2α is the angle which the string when tight subtends at the centre and θ is the inclination of the string to the horizon. [Pemb. Coll. 1897.]

Equate the tangential accelerations of the two particles.

Ex. 3. A bucket of mass M lbs. is raised from the bottom of a shaft of depth h feet by means of a light cord which is wound on a wheel of mass m lbs. The wheel is driven by a constant force which is applied tangentially at its rim for a certain time and then ceases. Prove that if the bucket just comes to rest at the top of the shaft, t seconds after the beginning of the motion, the greatest rate of working in foot-poundals per second is $\dfrac{2M^2g^2ht}{Mgt^2 - 2h(m+M)}$. The mass of the wheel may be considered to be condensed in its rim. [Coll. Ex. 1896.]

Let the force F act on the rim for a time t'. This force communicates a momentum Ft' to the system, which (since the system comes to rest after a time t) is equal to that removed by gravity in the whole ascent, therefore $Ft' = Mgt$. If s' is the space ascended in the time t', the force F communicates a work Fs', which is equal to that removed by gravity in the whole ascent h, therefore $Fs' = Mgh$. Since the mass moved is $M+m$ and $F - Mg$ is the acting force we have also the two equations $(M+m)v' = (F - Mg)t'$, $(M+m)s' = \frac{1}{2}(F - Mg)t'^2$ where v' is the velocity at the time t' (Art. 25). These four equations determine F, t', v', s'. The rate of adding work to the system is Fv (Art. 72), and this is greatest when v is greatest, i.e. when $v = v'$. The result follows without difficulty.

Ex. 4. A train of mass m runs from rest at one station to stop at the next at a distance l. The full speed is V and the average speed is v. The resistance at the rails when the brake is not applied is uV/lg of the weight of the train and when the brake is applied it is $u'V/lg$ of the weight of the train. The pull of the engine has one constant value when the train is starting and another when it runs at full speed. Prove that the average rate at which the engine works in starting the train is $\frac{1}{2}mV^2(u+U)/l$, where $\dfrac{1}{U} = \dfrac{2}{v} - \dfrac{2}{V} - \dfrac{1}{u'}$. [Coll. Ex. 1895.]

There are three stages of the journey. During the first the engine pulls with force F, the acceleration is $F/m - uV/l$, and the velocity increases from zero to V. During the second stage the velocity is uniform and equal to V, the pull F' of the engine just balancing the resistance. During the third the engine stops working, the brake is applied and the acceleration is $-u'V/l$. Using the formulæ of Art. 25, and remembering that the sum of the spaces in the three stages is l, while the average velocity is l divided by the sum of the times, we deduce F. The average rate of working is the quotient "work by time," Art. 72; during the first stage this is $Fs_1/t_1 = \frac{1}{2}FV$.

Ex. 5. The cage of a coal-pit is lowered for the first third of the shaft with a constant acceleration, for the next third it descends with uniform velocity, and then a constant retarding force just brings it to rest as it reaches the bottom of the shaft. If the time of descent is equal to that taken by a particle in falling four times the whole depth, prove that the pressure of the man inside on the bottom of the cage was at the beginning 23/48ths of his weight. [Coll. Ex. 1897.]

The initial acceleration f is found to be $25g/48$. If R be the pressure required the equation of motion of the man is $mf = mg - R$. This leads to the value of R.

Ex. 6. One engine A starting from rest generates in two minutes in a train a velocity of 45 miles per hour while it passes over a distance of 1 mile on the level. Another engine B of equal weight can pull the same train up an incline of $\sin^{-1} 1/80$ at a full speed of 20 miles per hour. Assuming that the resistance due to friction, &c. is constant and equal to the weight of 12 lbs. per ton, prove that the time average of the horse-power at which A works for the two minutes is 1·52... times the horse-power of B. [Math. Tripos, 1893.]

Ex. 7. A window is supported by two cords passing over pulleys in the framework of the window (which it loosely fits) and is connected with counterpoises each equal to half the weight of the window. One cord breaks, and the window descends with acceleration f. Prove that the coefficient of friction between the window and the framework is $\dfrac{(g-3f)\,a}{(g+f)\,b}$, where a is the height and b the breadth of the window. [Coll. Ex. 1896.]

Let the pressures of the window against the framework on one side at the bottom, on the other at the top, be R, R'. Since the window does not move sideways or turn round, we have the statical conditions $R=R'$, $Tb=2Ra$. Considering the vertical motion for the weight alone and for both bodies respectively, we have

$$\tfrac{1}{2}Mf=T-\tfrac{1}{2}Mg, \quad \tfrac{3}{2}Mf=\tfrac{1}{2}Mg-\mu 2R.$$

These determine μ.

Ex. 8. A two-wheeled vehicle is being drawn along a level road with velocity v: the wheels (radius c) are connected by an axle (radius r) fixed to them and the weight of the vehicle exclusive of the wheels and axle is W, and its centre of gravity is vertically above the middle point of the axle. Prove that if the shafts are in a horizontal plane with the tops of the wheels, the horse is working at the rate $\dfrac{Wvr\sin\lambda}{\sqrt{(c^2-r^2\sin^2\lambda)}}$, where λ is the angle of friction between the axle and its bearings. [Coll. Ex. 1895.]

The vehicle, being in *uniform* motion, is in equilibrium under the action of the pull F of the horse, the reaction R of the axle acting at some angle θ to the vertical and the friction $R\tan\lambda$. The equations of Statics give F, R, and θ, and the required rate of working is Fv.

Ex. 9. A particle of mass m is suspended from a fixed point O by a string of length a, and from m is suspended another particle of mass m' by a string of length b. If a horizontal velocity be suddenly communicated to m, show that the tensions of the strings are immediately increased by amounts which are in the ratio

$$1+\frac{mb}{m'(a+b)}:1. \qquad \text{[Coll. Ex. 1895.]}$$

Let T, T' be the tensions of the strings above and below m. Since m describes a circle whose centre is O, its vertical acceleration is v^2/a, hence $\dfrac{mv^2}{a}=T-T'-mg$. The vertical acceleration of m' is equal to that of m plus that due to the relative motion. Relatively to m it begins to describe a circle of radius b with a velocity v, the relative vertical acceleration is therefore v^2/b, see Art. 39. Hence

$$m'\left(\frac{v^2}{a}+\frac{v^2}{b}\right)=T'-m'g.$$

Solving these equations the result follows at once.

Ex. 10. In the system of pulleys in which the string, passing round each pulley, has one end attached to a fixed beam and the other to the pulley next above, there is no "power" and no "weight." The n moveable pulleys are all of equal weight, they are smooth, and can all be treated as particles in calculating their motions. The string is without mass. Prove that the acceleration of the lowest pulley is $3g/(2^n+1)$. [Coll. Ex. 1896.]

The equation of momentum for the rth pulley counting downwards is

$$md^2y_r/dt^2 = mg - 2T_r + T_{r+1},$$

where T_1 and T_{n+1}, being the power and weight, are zero. Also the velocity of each pulley is half that of the one just above. Multiplying these equations by 1, 2, $2^2 \ldots 2^{n-1}$ beginning at the lowest and adding the results the tensions disappear.

Ex. 11. In the system of pulleys in which each string is attached to the weight, there are two pulleys, the weight of the moveable pulley being w, the power P and the weight W. Prove that the acceleration of W is $g\dfrac{3P+w-W}{9P+w+W}$. [Coll. Ex. 1897.]

Ex. 12. A prism with axis horizontal and whose section by a plane perpendicular to it is a regular polygon $ABCD\ldots$ of $4n$ sides is fixed with the uppermost face AB horizontal, and n equal particles are placed at the middle points of AB, BC, &c. These are connected by a continuous string which passes over smooth pulleys at the corners B, C, &c. Assuming that the faces are smooth, prove that the initial acceleration is $\dfrac{g}{2n}\left(\cot\dfrac{\pi}{4n}-1\right)$. [Coll. Ex. 1897.]

Ex. 13. Two equal particles are connected by a string one point of which is fixed and the particles are describing circles of radii a and b about this point with the same angular velocity so that the string is always straight. The string is suddenly released, prove that the tensions of the two portions are altered in the ratios $(a+b) : 2a$ and $(a+b) : 2b$. [Coll. Ex. 1895.]

Before the release the tensions are mv_1^2/a and mv_2^2/b, where $v_1/a = v_2/b = \omega$. After the release the relative space velocity is $v = v_1 + v_2$. The acceleration of each particle being T/m, the relative acceleration is $2T/m$. Since the relative path of either is a circle of radius $r = a+b$, the relative acceleration is v^2/r. Equating these, the tension is $mv^2/2r$. The result follows.

Ex. 14. A cubical box slides down a rough inclined plane, whose coefficient of friction is μ, two sides of the base being horizontal. If the box contain sufficient water just to cover the base of the vessel, prove that the volume of the water is $\frac{1}{2}\mu$ times the internal volume of the vessel. [Coll. Ex. 1897.]

The relative acceleration of a particle of water and the box must be perpendicular to the surface.

79. Linear and Angular Momentum. Let the momentum mv of any particle P of a system be represented in direction and magnitude (Art. 54) by a straight line PP'. Since velocities obey the parallelogram law, we may proceed as in Statics and replace the momentum PP' by three linear momenta at any assumed origin in the directions of the axes, and three couple momenta.

Let the coordinates of the particle be x, y, z and the direction cosines be λ, μ, ν. The three linear momenta being the resolved parts of mv are $mv\lambda$, $mv\mu$, $mv\nu$ respectively. These are often called linear momenta. The three couple momenta are the moments of the momentum mv about the axes. We know by the corresponding theorem in Statics that these moments are

$$mv\begin{vmatrix} y & z \\ \mu & \nu \end{vmatrix}, \quad mv\begin{vmatrix} z & x \\ \nu & \lambda \end{vmatrix}, \quad mv\begin{vmatrix} x & y \\ \lambda & \mu \end{vmatrix}.$$

These are called the angular momenta about the axes.

The linear momentum of a particle in any direction is the resolved part of the momentum in that direction. The angular momentum about a straight line is the moment of the momentum about that straight line.

Impulsive Forces.

80. Impulsive forces. In some cases the forces act only for a very short time, yet, being of great magnitude, produce perceptible effects. Let a force F act on a particle of mass m for a time T. Let v be the velocity at any time t less than T, and let V, V' be the velocities at the beginning and end of the interval T. We have

$$m\frac{dv}{dt}=F, \quad \therefore\ m(V'-V)=\int_0^T F\,dt \text{ ....(1).}$$

Let the force F increase without limit while the duration T decreases without limit. The integral may have a finite limit, say P. The equation then becomes

$$m(V'-V)=P \text{ ....(2).}$$

If v_1, v_2 are the greatest and least velocities during the impact, the space described lies between v_1T and v_2T, and both these are zero in the limit. *The particle therefore has not had time to move, but its velocity has been changed from V to V'.* This sudden change of velocity is the distinguishing characteristic of an impulse.

We may consider that a proper measure has been found for a force when from that measure we can deduce all the effects of the force. Since in the case of the limiting force the change of velocity is the only element to be determined we may measure such a force by the quantity P. When P is known, the change of velocity is given by (2).

81. An impulse or blow is the limit of a force whose magnitude is infinitely great and time of action infinitely small. A finite force F is measured by *the momentum generated per unit of time.* An impulse P is measured by *the whole momentum generated* during the whole time of action, that is, $P=\int F dt$.

When the direction of the force F remains fixed in space during its time of action, the resolved part of P in any direction is also the limit of the resolved part of F. When the direction of F is not fixed in space, we resolve F into its components X, Y. The integrals of these, viz. $X_1=\int X\,dt$, $Y_1=\int Y dt$, are defined to be the components of the limiting impulse.

Strictly speaking, there are no impulsive forces in nature, but there are some forces which are very great and which act only for a short time. The blow of a hammer is a force of this kind. Such forces should be treated as finite forces if the small displacements during the time of action cannot be neglected, and as impulses when these are imperceptible.

82. The general equations of impulsive motion follow from those of finite forces. If (u_1, v_1) are the Cartesian components of velocity we have, by Art. 73,

$$m\frac{du_1}{dt}=X,\quad m\frac{dv_1}{dt}=Y,$$

where X, Y are the components of a finite force F. Let (u, v), (u', v') be the components of the velocity just before and just after the action of any impulse. Let $X_1=\int X\,dt$, $Y_1=\int Y\,dt$ be the components of the impulse, Art. 81. We then have by integration,

$$m(u'-u)=X_1,\quad m(v'-v)=Y_1.$$

These equations may be summed up in the following working rule,

$$\begin{pmatrix}\text{Res. Mom.}\\ \text{after impulse}\end{pmatrix}-\begin{pmatrix}\text{Res. Mom.}\\ \text{before impulse}\end{pmatrix}=\begin{pmatrix}\text{Resolved}\\ \text{impulse}\end{pmatrix}.$$

83. Elastic smooth bodies. When two spheres of any hard material impinge on each other they appear to separate almost immediately and a finite change of velocity is generated in each by the mutual action. Let the centres of gravity of the spheres be moving before impact in the same straight line with velocities u, v. After impact they will continue to move in the same straight line; let u', v' be their velocities. Let m, m' be the masses, R the action between them. The equations of motion are

$$m(u'-u)=-R,\quad m'(v'-v)=R\ \ldots\ldots\ldots\ldots(1).$$

These equations are not sufficient to determine the three quantities u', v' and R. To obtain a third equation we must consider what takes place during the impact.

Each of the balls is slightly compressed by the other, so that they are no longer perfect spheres. Each also in general tends to return to its original shape, so that there is a rebound. The period of impact may therefore be divided into two parts. Firstly, the period of compression, during which the distance between the centres of gravity of the two bodies is diminishing and secondly, the period of restitution in which the distance is increasing. The first period terminates when the two centres of gravity have the same instantaneous velocity, the second when the bodies separate.

The ratio of the magnitude of the action between the bodies during the period of restitution to that during compression is found to be different for bodies of different materials. If the bodies regain their original shapes very slowly the separation may take place before this occurs and then the action during restitution is less than that during compression.

In some cases the force of restitution may be neglected, and the bodies are then said to be *inelastic*. In this case we have just after the impact $u' = v'$. This gives

$$R = \frac{mm'}{m+m'}(u-v), \quad \therefore\ u' = \frac{mu + m'v}{m+m'} \quad \text{.........(2).}$$

If the force of restitution cannot be neglected, let R be the whole action between the balls, R_0 the action up to the moment of greatest compression. The magnitude of R can be found by experiment. This may be done by observing the values of u' and v' and thus determining R by means of the equations (1). Such experiments were made in the first instance by Newton and led to the result that R/R_0 is a constant ratio which depends on the materials of which the balls are made. Let this constant ratio be called $1 + e$. The quantity e is never greater than unity; in the limiting case when $e = 1$ the bodies are said to be perfectly elastic.

The Newtonian law $R/R_0 = 1 + e$ gives only *a first approximation* to the motion, and is not to be regarded as strictly true under all circumstances.

The value of e being supposed to be known the velocities after impact may be easily found. The action R_0 must be first calculated as if the bodies were inelastic, the value of R may then be deduced by multiplying by $1 + e$. This gives

$$R = \frac{mm'}{m+m'}(u-v)(1+e) \quad \text{.................(3).}$$

The three equations comprised in (1) and (3) give the whole motion. Substituting from (3) in (1), we have

$$\left.\begin{aligned} u' &= \frac{mu + m'v}{m + m'} - \frac{m'e}{m + m'}(u - v) \\ v' &= \frac{mu + m'v}{m + m'} + \frac{me}{m + m'}(u - v) \end{aligned}\right\} \quad \text{..............(3).}$$

84. We notice as a useful corollary that

$$v' - u' = -e(v - u) \quad \text{......................(4).}$$

The relative velocity after impact bears to the relative velocity before impact the ratio of $-e$ *to* 1.

By the third law of motion the momentum gained by one ball is equal to that lost by the other; *the whole momentum being unaltered by the impact.* Hence

$$mu' + m'v' = mu + m'v \quad \text{....................(5).}$$

This result follows also by eliminating R between the equations (1).

The equations (4) and (5) may be used to determine u', v', when the impulse R is not required.

85. *When two perfectly elastic spheres of equal mass impinge on each other the bodies exchange velocities.* In this case, by (3),

$$R = m(u - v)$$

and the equations (1) then show that $u' = v$, $v' = u$. Conversely we may show in the same way that if the spheres exchange velocities their masses are equal and the elasticity is perfect.

86. When a sphere impinges on a fixed plane, we regard the plane as an infinitely large mass. Putting m' infinite, we find

$$R = mu(1 + e), \quad u' = -eu, \quad v' = 0,$$

the velocity of the sphere is therefore reversed in direction and its magnitude is multiplied by e.

Ex. If the plane be in motion with a velocity V, prove that the velocity of the sphere after the rebound is $-eu + V(1+e)$.

87. If one sphere of mass m impinge directly on another of mass m' which is at rest and if $m = m'e$, the equation (3) gives $R = mu$. The impinging sphere therefore loses its whole momentum and is reduced to rest.

In the same way, let n spheres be placed in a row at rest and let their masses form a geometrical progression of ratio $1/e$.

If any velocity is given to the first, it will strike the next in order and be reduced to rest. The second will strike the third and remain at rest and so on. Finally the last sphere will proceed onwards with the whole momentum communicated to the first.

If the spheres are perfectly elastic, $e=1$ and the same things happen when the masses are equal.

If the spheres are placed close together, they are only in apparent contact; and each impact will still be concluded before the next begins. Each ball transfers the momentum to the next in order and remains in apparent rest, the last ball moving onwards with the whole momentum communicated to the first.

This may partly explain why, in some cases when blows have been given by the wind or sea to masses of masonry, the stones to leeward have been more disturbed than those exposed to the blows.

88. *Ex.* A series of perfectly elastic balls are arranged in the same straight line, one of them impinges directly on the next and so on; prove that if their masses form a geometrical progression of which the common ratio is 2, their velocities after impact will form a geometrical progression of which the common ratio is 2/3. [Math. Tripos, 1860.]

89. *Two smooth homogeneous spheres A and B impinge obliquely on each other. To find the subsequent motion.*

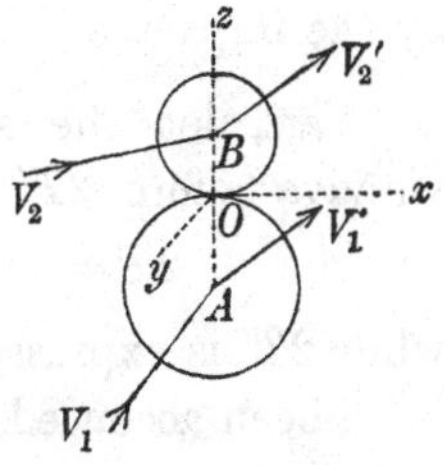

Let the common tangent plane at the point of contact O be the plane of xy, and let the common normal be the axis of z. The spheres being smooth the mutual impulse acts along the axis of z.

Let V_1, V_2 be the velocities of the two spheres, before impact, V_1', V_2' the velocities after. Let

$$(u_1, v_1, w_1), \quad (u_2, v_2, w_2)$$

be the components of the velocities V_1, V_2, and let the same letters, when accented, represent the components of V_1', V_2'. Let m, m' be the masses.

Since the impulse has no components parallel to the axes of x and y, we have

$$u_1' = u_1, \quad v_1' = v_1; \quad u_2' = u_2, \quad v_2' = v_2.$$

Considering next the normal impulse, we find as before

$$R=\frac{mm'}{m+m'}(w_1-w_2)(1+e),\quad w_1'-w_1=-\frac{R}{m},\quad w_2'-w_2=\frac{R}{m'}.$$

These equations determine the components of the velocities after the impact.

When the bodies are rough, the mutual impulse does not necessarily act along the common normal. The problem then becomes more complicated. The reader will find this case discussed in the author's *Rigid Dynamics*.

90. *When two imperfectly elastic spheres impinge on each other, vis viva is always lost.*

First, let the spheres impinge directly on each other. We have, as in Art. 83,

$$R=\frac{mm'}{m+m'}(u-v)(1+e),\quad u'=u-\frac{R}{m},\quad v'=v+\frac{R}{m'}.$$

$$\therefore\ mu'^2+m'v'^2=mu^2+m'v^2+\left\{2(v-u)+R\,\frac{m+m'}{mm'}\right\}R$$

$$=mu^2+m'v^2-\frac{mm'}{m+m'}(u-v)^2(1-e^2).$$

The last term being essentially negative, the vis viva is decreased by the impact.

Next, let the spheres impinge obliquely. Let $2T$ be the vis viva before, $2T'$ that after the impulse. Then, as in Art. 69

$$2T=m(u_1^2+v_1^2+w_1^2)+m'(u_2^2+v_2^2+w_2^2),$$

while $2T'$ is expressed by the same formula after the letters u, v, w have been accented. Hence

$$2T'-2T=-\frac{mm'}{m+m'}(w_1-w_2)^2(1-e^2).$$

It follows that vis viva is always lost.

If V is the relative normal velocity before impact, the vis viva lost is $\dfrac{mm'}{m+m'}V^2(1-e^2)$.

The vis viva after impact is equal to the vis viva before only when $e=1$, *that is, when the bodies are perfectly elastic.* It is evident that w_1 cannot be equal to w_2 or $e=-1$.

91. *Ex.* 1. Particles are projected from a given point A in all directions and obliquely impinge on a fixed plane of elasticity e. Prove that after reflexion the directions of motion diverge from a point B, where AB intersects the fixed plane at right angles in some point M, and $BM = e \,.\, AM$.

Let AP be the path of a particle before impact, PQ that after Let QP produced intersect the perpendicular AM produced in some point B. The com-

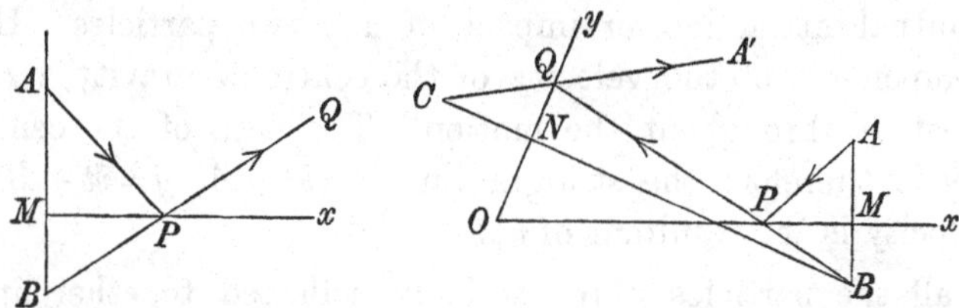

ponent of velocity, u, along MP is unchanged by the impact, while that perpendicular, viz. v, becomes ev and is reversed in direction,

$$\therefore \tan QPx = ev/u = e \tan APM.$$

It immediately follows that $MB = e \,.\, AM$, so that every reflected path intersects the perpendicular from A in the same point.

By using this theorem we can trace the course of a particle after successive reflexions from any number of fixed planes. To take a simple case, let it be required to find how a particle should be horizontally projected from a given point A on the floor, that after reflexion at two vertical walls Ox, Oy, it may pass through another given point A'. We draw a perpendicular AB to the first wall and take $MB = eAM$. A perpendicular is drawn from B to the second wall, and C is taken so that $CN = e \,.\, BN$. Then, since all the paths after the first and second reflexions pass through B and C respectively, the required path $AQPA'$ is found by joining A' to C, Q to B and P to A.

Ex. 2. A particle of elasticity e is projected along a horizontal plane from the middle point of one of the sides of an isosceles right-angled triangle so as after reflexion at the hypothenuse and remaining side to return to the same point; prove that the cotangents of the angles of reflexion are $e+1$ and $e+2$ respectively.

[Math. Tripos, 1851.]

92. *A free system of mutually attracting particles is in motion. Prove* (1) *that the centre of gravity moves in a straight line with uniform velocity, and* (2) *that the motion of the centre of gravity is not affected by any impacts between the particles.*

The mutual attraction between any two particles is measured by the momentum transferred from one to the other per unit of time; the mutual impulse is measured by the whole momentum transferred. In either case it follows by the third law of motion that the whole momentum of the two particles and the components in any directions, are unaltered by their mutual action.

Let (x_1, y_1), (x_2, y_2), &c. be the Cartesian coordinates and (u_1, v_1), (u_2, v_2), &c. the components of velocity at any time t. Since

$$\bar{x}\Sigma m = \Sigma mx, \quad \bar{y}\Sigma m = \Sigma my,$$

we have by differentiation $\bar{u}\Sigma m = \Sigma mu$, $\bar{v}\Sigma m = \Sigma mv$. It has just been shown that the components Σmu, Σmv are unaltered by the mutual attraction or impact of any two particles. Hence the components of the velocity of the centre of gravity, viz. $\bar{u}$, $\bar{v}$, are constant throughout the motion. The path of the centre of gravity is therefore the straight line $\bar{x} = \bar{u}t + A$, $\bar{y} = \bar{v}t + B$, and the velocity is the resultant of $\bar{u}$, $\bar{v}$.

If all the particles were suddenly collected together at the centre of gravity, each particle having its momentum unaltered in direction and magnitude, the momentum of the collected mass would be the resultant of the transferred momenta. The equations $\bar{u}\Sigma m = \Sigma mu$, $\bar{v}\Sigma m = \Sigma mv$ assert that the centre of gravity of the particles before collection moves exactly as the collected mass does.

93. The effect of the mutual action of two particles (whether attracting or impinging on each other) is to transfer a momentum from one to the other whose direction is the straight line joining the particles. Hence the moment of the momentum about any straight line is unaltered by the transference. The moment of the momentum of the whole system (that is, its angular momentum, Art. 79), about any straight line is unaltered by the mutual actions of the particles.

In a system of mutually attracting or impinging particles, the components of its linear momentum along, and the angular momenta about, any fixed straight lines are constant, except so far as they may be altered by the action of external forces. This is only the third law of motion more fully explained.

94. Examples*. *Ex.* 1. If a system of mutually attracting particles were suddenly to become rigidly connected together, determine the conditions that the rigid body should be at rest.

The rigid body will possess the same momenta as the system but differently distributed. If the momenta of all the particles are in equilibrium, the rigid body has no component of momentum in any direction and no moment of momentum

* Many of these examples are taken from the examination papers for the entrance and minor scholarships in the several colleges.

about any straight line. It is therefore at rest. By the rules of Statics the necessary and sufficient conditions for the equilibrium are (1) the whole linear momentum along each axis of coordinates is zero, (2) the angular momentum about each axis is zero.

Ex. 2. Particles of equal mass travel round the sides of a closed skew polygon in the same direction, one starting from each corner and the velocity of each is proportional to the side along which it moves. Prove that their centre of gravity is at rest and that it coincides with the centre of gravity of the sides of the polygon supposing the masses of the sides to be equal. Prove also that if one particle be removed, the centre of gravity of the remaining particles describes a polygon whose sides are parallel and proportional to those of the original polygon.

Since the sides exert no pressures on the particles the centre of gravity moves in a straight line with uniform velocity whatever the momenta of the particles may be. When, as in the problem, the momenta are parallel and proportional to the sides of a closed figure, the components Σmu and Σmv of Art. 92 are zero, and the centre of gravity is therefore at rest. The other parts of the question then follow at once.

Ex. 3. An explosion occurs in a rigid body at rest, and the particles fly off in different directions. If in any subsequent positions they were suddenly connected together, prove that the rigid body thus formed would be at rest.

Ex. 4. A number of particles originally in a straight line fall from rest, and rebound from a partially elastic horizontal plane. Prove that, at any time, the particles which have rebounded once lie in a parabola. [Coll. Ex. 1897.]

Ex. 5. Two small spheres of equal mass can move inside a rough endless horizontal tube of length l. One sphere impinges with velocity v on the other at rest. If the friction of the tube produce a retardation f in either sphere and if after impact the spheres just meet again, prove that $2fl=v^2e$. [Coll. Ex. 1896.]

Ex. 6. Four equal balls of the same material are projected simultaneously with equal velocities from the corners of a square towards its centre, and meet in the neighbourhood of the centre. Show that they return to the corners with velocities reduced in the ratio of the coefficient of restitution to unity.

[Coll. Ex. 1892.]

Ex. 7. Two equal spheres each of mass m are in contact on a smooth horizontal table, a third equal sphere of mass m' impinges symmetrically on them. Prove that this sphere is reduced to rest by the impact if $2m'=3me$, and find the loss of kinetic energy by the impact. [Coll. Ex. 1897.]

Ex. 8. Two equal balls lie in contact on a table. A third equal ball impinges on them, its centre moving along a line nearly coinciding with a horizontal common tangent. Assuming that the periods of the two impacts do not overlap, prove that the ratio of the velocities which either ball will receive according as it is struck first or second is $4:3-e$, where e is the coefficient of restitution.

[Math. Tripos, 1893.]

Ex. 9. A heavy particle tied to a string of length l is projected horizontally with a velocity V from the point to which it is attached. Show that the energy lost by the impulse is a minimum when $V^2=lg/\sqrt{3}$: see Arts. 27, 90.

[Coll. Ex. 1896.]

Ex. 10. A particle of mass m lies at the middle point C of a straight tube AB of mass M and length $2a$, both of whose ends are closed. It is shot along the tube with velocity V. Prove that it will pass the middle point of the tube in the same direction after a time $\frac{a}{V}\left(1+\frac{1}{e}\right)^2$, e being the coefficient of restitution between the particle and either end of the tube; and that in this time the tube will have moved forward a distance $\frac{am}{M+m}\left(1+\frac{1}{e}\right)^2$. [Coll. Ex. 1895.]

The particle traverses the length $CA=a$ in a time a/V and after impact has a relative velocity eV. It therefore traverses the length $AB=2a$ in a time $2a/eV$, and after impact at B has a relative velocity e^2V. It traverses the remaining length $BC=a$ in the time a/e^2V. The whole time T is the sum of these three times. The particle is now at the same point C of the tube as before, the distance traversed by the tube is therefore equal to that traversed by the centre of gravity of the system. Since the initial velocities of the particle and tube are V and zero, the velocity of the centre of gravity is $\bar{v}=mV/(M+m)$. The distance traversed is therefore $\bar{v}T$.

Ex. 11. A particle is projected inside a straight tube of length $2a$, closed at each end, which lies on a smooth horizontal table and whose mass is equal to that of the particle. Prove that, at the moment just before the fourth impact the tube has described a distance $15a$, if the coefficient of restitution is $\frac{1}{2}$, and find the proportion of kinetic energy which has disappeared. [Coll. Ex. 1895.]

Ex. 12. A smooth particle of mass m is at rest in a rectangular box of mass M which is free to move down a smooth plane inclined at an angle α to the horizon, the lowest edge of the box being horizontal, and the particle at its middle point. Suddenly the box is started down the plane with velocity V. Prove that if the coefficient of restitution be unity, the particle will strike the top and bottom of the box after equal successive intervals of time; and that the spaces travelled by the box in the first and second of these intervals are as

$$V^2+gl\sin\alpha : \frac{M-m}{M+m}V^2+3gl\sin\alpha,$$

where $2l$ is the length of the box. [Coll. Ex. 1896.]

Ex. 13. A perfectly elastic ball is projected vertically with velocity v_1, from a point in a rigid horizontal plane, and when its velocity is v_2 an equal ball is projected vertically from the same point also with velocity v_1; show, (1) that the time that elapses between successive impacts of the two balls is v_1/g, (2) that the heights at which they take place are alternately

$$(3v_1-v_2)(v_1+v_2)/8g \text{ and } (3v_1+v_2)(v_1-v_2)/8g,$$

(3) that the velocities of the balls at the impacts are equal and opposite and alternately $\frac{1}{2}(v_1-v_2)$ and $\frac{1}{2}(v_1+v_2)$. [Math. Tripos, 1896.]

Since the balls exchange velocities at each impact, we may suppose that they pass through each other, one ball following the other at an interval $\tau=(v_1-v_2)/g$.

Ex. 14. A weight of mass m and a bucket of mass m' are connected by a light inelastic string which passes over a smooth pulley. These bodies are released from rest when a particle whose mass is p and coefficient of elasticity e falls with vertical velocity V upon the bucket. Prove that a second collision will occur between the particle and bucket after a time $e(m+m')V/mg$ and find the condition that the bodies should then be in their initial positions. [Coll. Ex. 1895.]

Ex. 15. A particle is projected from a point on the inner circumference of a circular hoop, free to move on a horizontal plane. Prove that if the particle return to the position of projection after two impacts, its original direction must make with the radius through the point an angle $\tan^{-1}\{e^3/(1+e+e^2)\}^{\frac{1}{2}}$.

[Coll. Ex. 1897.]

Ex. 16. Two balls of masses M, m (centres A and B), are tied together by a string, and lie on a smooth table with the string straight. A ball of mass m' (centre C) moving on the table with velocity V parallel to the string strikes the ball of mass m, so that the angle ABC is acute and equal to α. Prove that M starts with a velocity $\dfrac{Vmm'\cos^2\alpha\,(1+e)}{Mm'\sin^2\alpha+m\,(M+m+m')}$, e being the coefficient of restitution between m and m'. [Coll. Ex. 1895.]

Let U' be the velocity of m' after impact in the direction CB, v_1' the common velocity of M, m in the direction AB, v_2' the velocity of m perpendicular to AB; then $m'(U'-V\cos\alpha)=-R$. Since $R\cos\alpha$ has to move both M and m, while $R\sin\alpha$ affects m only,

$$(M+m)\,v_1'=R\cos\alpha,\quad mv_2'=R\sin\alpha.$$

At the moment of greatest compression, the velocities of m', m along CB are equal

$$U'=v_1'\cos\alpha+v_2'\sin\alpha.$$

These equations give R. Multiplying the result by $1+e$ the second equation then gives v_1'.

Ex. 17. Three particles A, B, C whose masses are m, m', m'', connected by straight strings, are placed at rest on a smooth table, and the obtuse angle ABC is $\pi-\alpha$. If A receive a blow F parallel to CB prove that C will begin to move with a velocity $\dfrac{m'F\cos^2\alpha}{m'\Sigma m+mm''\sin^2\alpha}$.

Let T, T' be the impulsive tensions of AB, BC. Since A, B must have equal velocities along BA

$$(F\cos\alpha-T)/m=(T-T'\cos\alpha)/m'.$$

Since B, C have equal velocities along BC

$$(T\cos\alpha-T')/m'=T'/m''.$$

These equations determine T and T', and the result required is T'/m''.

Ex. 18. Two smooth spheres whose coefficient of restitution is e are attached by inextensible strings to fixed points. One of them, whose mass is m, describing a circle with velocity v, impinges upon the other whose mass is m' and which is at rest. If the line of centres makes an angle θ with the string attached to m and the strings at that instant cross each other at right angles, then m' begins to describe a circle with velocity $\dfrac{mv\sin\theta\cos\theta\,(1+e)}{m\cos^2\theta+m'\sin^2\theta}$. [Coll. Ex. 1896.]

Let A, B be the centres of m, m', and let the strings be attached to D, E. Let DA intersect EB in C. The force R on m acts along BA and makes an angle θ with AD. Let v', w' be the velocities of m, m' along EC and CD. Then

$$\left.\begin{aligned}0&=-T+R\cos\theta\\ m(v'-v)&=-R\sin\theta\end{aligned}\right\},\qquad \left.\begin{aligned}m'w'&=R\cos\theta\\ 0&=-T'+R\sin\theta\end{aligned}\right\}.$$

At the moment of greatest compression, the velocities of m, m' along AB are equal, $\therefore v'\sin\theta=w'\cos\theta$. This determines the value of R, and the required velocity is $R\,(1+e)\cos\theta/m'$.

Ex. 19. A smooth inelastic sphere of radius r and mass m is suspended by a string above a horizontal table, and another smooth inelastic sphere of radius r' and mass m' is moving on the table; prove that the cotangent of the angle through which the direction of motion of the second sphere is deflected by a collision is $\frac{1}{mb}\frac{m'(r+r')^2+mb^2}{\{(r+r')^2-a^2-b^2\}^{\frac{1}{2}}}$ where a and b are the vertical and horizontal distances of the centre of the first sphere from the path of the second before impact.

[Coll. Ex. 1892.]

We notice that the vertical motion of one sphere is stopped by the reaction of the table, while that of the other is not stopped by the tension of the string.

Ex. 20. Four equal particles are connected by three equal strings AB, BC, CD and lie on a horizontal plane with the strings taut in the form of half a regular hexagon. An impulse is applied at A in the direction DA. Prove that the initial tension of BC is one-fourteenth of the impulse. [Coll. Ex. 1897.]

Ex. 21. If three inelastic particles, m_1, m_2, m_3, moving with velocities v_1, v_2, v_3 making angles α, β, γ, with each other, impinge and coalesce, prove that the loss of energy is $\frac{\Sigma m_1 v_1^2(m_2+m_3)-2\Sigma m_1 m_2 v_1 v_2\cos\gamma}{2\Sigma m}$. [Coll. Ex. 1896.]

Ex. 22. A shot whose mass is m penetrates a thickness s of a fixed plate of mass M, prove that, if M is free to move, the thickness penetrated is $s\Big/\left(1+\frac{m}{M}\right)$.

[Coll. Ex. 1896.]

The mass m strikes M with a velocity v_0 and continues to move inwards until m and M have the same velocity $v_1=mv_0/(M+m)$. If F be the resistance regarded as constant, x and $x+\sigma$ the spaces described by M and m,

$$m(v_1^2-v_0^2)=-2F(x+\sigma),\quad Mv_1^2=2Fx.$$

Eliminating x, we find $2F\sigma=v_0^2Mm/(M+m)$. When M is infinite, $2Fs=v_0^2m$. The ratio σ/s follows. This problem may also be easily solved by considering the relative motion.

Ex. 23. A smooth uniform hemisphere of mass M is sliding with velocity V on an inelastic horizontal plane with which its base is in contact; a sphere of smaller mass m is dropped vertically so as to strike the first on the side towards which it is moving, at an inclination of 45°; prove that if the hemisphere be stopped dead, the sphere must have fallen through a height $\frac{V^2(2M-em)^2}{2g(1+e)^2m^2}$ where e is the coefficient of restitution between them. [Math. Tripos, 1887.]

CHAPTER II.

RECTILINEAR MOTION.

Solution of the Equation of Motion.

95. LET us suppose that a particle of mass m is constrained to move in a straight line, which we may call the axis of x, under the action of forces whose component along x is F. Let $F = mX$. We have seen in the previous chapter that the equation of motion is

$$\frac{d^2x}{dt^2} = \frac{F}{m} = X.$$

Properly this equation gives X when x is a known function of t, and therefore answers the question, *given the motion, what is the force?* Usually we require the solution of the converse problem, *given the accelerating force* X (Art. 68), *find the motion.* To determine this, we must regard the equation of motion as a differential equation and seek for its solution.

96. In the general case X may be a function of x and t and also of the velocity v of the particle. But the equation can only be solved in limited cases. We shall examine these solutions in turn.

Let us suppose that X *is a function of* t *only*, say $X = f(t)$. By integration we have

$$v = \frac{dx}{dt} = f_{,}(t) + A,$$

$$x = f_{,,}(t) + At + B,$$

where suffixes have been used to represent integrations with regard to t.

In this way x has been expressed as a function of t, leaving the constants A and B undetermined. As this value of x satisfies the differential equation, whatever values A and B may have, there is nothing in that equation to help us in finding these two constants. We must have recourse to some other data. These are the initial conditions of the motion. Let us suppose that the particle was projected at a time $t=a$, from a point determined by $x=b$ with a velocity $v=c$. Then remembering that $v=dx/dt$, we have

$$c=f_{,}(a)+A, \quad b=f_{,,}(a)+Aa+B.$$

Solving these, we find A and B. The motion is therefore given by

$$x=f_{,,}(t)+\{c-f_{,}(a)\}t+\{b-ac+af_{,}(a)-f_{,,}(a)\}.$$

97. *Let X be a function of x only*, say $X=f(x)$.

$$\therefore \frac{d^2x}{dt^2}=f(x) \quad\text{.......................(1).}$$

Multiply by $\dfrac{dx}{dt}$, $$\frac{dx}{dt}\frac{d^2x}{dt^2}=f(x)\frac{dx}{dt}.$$

Integrate $$\left(\frac{dx}{dt}\right)^2=2f_{,}(x)+A \quad\text{...................(2),}$$

$$\therefore v=\frac{dx}{dt}=\pm\{2f_{,}(x)+A\}^{\frac{1}{2}} \quad\text{.............(3).}$$

To determine the value of A and the sign of the radical we use the initial conditions. Let us suppose that when $t=a$, $x=b$, and $v=c$. We then have

$$c^2-2f_{,}(b)=A \quad\text{.......................(4),}$$

$$c=\pm\{2f_{,}(b)+A\}^{\frac{1}{2}} \quad\text{.......................(5).}$$

If c is not zero, the radical must have the same sign as c, i.e. the radical is positive or negative according as the direction of the initial velocity makes x increase or decrease. If however $c=0$, we notice that the particle will begin to move in the direction in which the force acts; the radical therefore follows the sign of the initial value of X. Since X is a function of x only, it is obvious that if the initial value of X is also zero, the particle is at rest in a position of equilibrium and that there will be no motion.

We now have

$$\int\frac{dx}{\{2f_{,}(x)+A\}^{\frac{1}{2}}}=t+B \quad\text{...................(6).}$$

Representing the left-hand side of this equation, after the integration has been effected, by $\phi(x)$, we have

$$\phi(x) = t + B \qquad \text{.........................(7)}.$$

To find B we recur again to the given initial conditions, viz. that $x = b$ when $t = a$, hence $B = \phi(b) - a$.

98. The equation (7) determines t when x is known, i.e. it gives the time at which the particle passes over any given point of the straight line along which it moves. If we require the position of the particle at any given time, we must solve the equation and express

$$x = \psi(t) \qquad \text{..........................(8)}.$$

The solution of this algebraical equation may lead to different values of x, thus we may have $x = \psi_1(t)$, $x = \psi_2(t)$, &c. We have yet to determine which of these represents the actual motion. We notice that since the equation (7) is satisfied by $x = b$, $t = a$, *one at least* of these values of x must satisfy this condition. All the others must then be excluded as not agreeing with the given initial conditions. If more than one of these solutions could satisfy this condition, the equation obtained by putting $t = a$ in (7), viz.

$$\phi(x) = a + B,$$

must have equal roots. Hence $\phi'(x) = 0$ when $x = b$. Since $\phi(x)$ represents the left-hand side of (6) it immediately follows that $2f_1(b) + A$ is infinite. But by (5) this cannot happen if the initial velocity c is finite.

99. *Subject of integration infinite.* Other points requiring attention arise when the integrals which occur are such that the subject of integration is infinite at some point B of the path. Since the forces in nature are necessarily finite this cannot happen in the integral (2), for if $f_1(x)$ were infinite its differential coefficient, $f(x)$ for any *finite* value of x, would also be infinite. In the integral (6) the subject of integration is infinite when the velocity is zero.

We can use the integral (6) to find the time of transit from any point A to a point P as near as we please to B on the same side of B as A. If the result is infinite the particle never reaches B. If the time of arrival at B is finite we have to find the subsequent motion.

As the particle approaches B the velocity is numerically decreasing and therefore the accelerating force X has the opposite sign to the velocity. Supposing X not also to vanish at B, *the particle after arriving at B must begin to retrace its steps.* Considering B as a new initial position, the subsequent motion may be deduced from (3) by putting $c = 0$. If $X = 0$ also at B, the particle, as explained above, will remain there in equilibrium.

100. *Ex.* 1. A particle moves in a straight line under a central force tending to the origin and equal to n^2/x^3. Investigate the motion.

We have
$$\frac{d^2x}{dt^2} = -\frac{n^2}{x^3} \quad\text{.................................. (1).}$$

The minus sign is introduced because the left-hand side represents the acceleration in the positive direction of x and the force acts towards the origin. We then find
$$\frac{dx}{dt} = \pm\left\{\frac{n^2}{x^2} + A\right\}^{\frac{1}{2}} \quad\text{.................................. (2).}$$

Let us suppose that the particle starts from rest at a very great or infinite distance from the origin; then when x is infinite, $dx/dt = 0$. Hence $A = 0$, and the equation becomes
$$\frac{dx}{dt} = \pm\frac{n}{x} \quad\text{.................................. (3).}$$

Since the particle begins to move towards the centre of force the velocity is initially negative. We therefore take the negative sign.

Multiplying by x and integrating, we find
$$x^2 = B - 2nt \quad\text{.................................. (4).}$$
Initially when $t=0$, the particle is infinitely distant from the origin, i.e. x is infinite and therefore B is infinite. It follows that the particle does not get within a finite distance of the origin until after the lapse of an infinite time.

If the initial conditions are slightly altered we may obtain a finite result. Let us suppose the particle to be initially projected at a distance $x=b$ (b being positive) with a velocity n/b towards the centre of force. Proceeding as before we find $A=0$, and as it is given that the initial velocity of the particle is negative, the radical has still the negative sign. We thus again arrive at the equation (4). Since $x=b$ when $t=0$, we find $B=b^2$, and
$$x = \pm(b^2 - 2nt)^{\frac{1}{2}} \quad\text{.................................. (5).}$$
Since x is initially positive we must give the radical the positive sign.

As t increases we see that x continually diminishes and when $t=b^2/2n$ the particle arrives at the origin. Its velocity at that moment is found by putting $x=0$ in (3) and is easily seen to be infinite.

Cases in which either the velocity or the force is infinite do not occur in nature. If we construct a central force by placing some attracting matter at the origin there would be an impact before the particle reached the origin and the whole motion would be changed. But as a matter of curiosity we may enquire what would be the subsequent motion if our equations held true for infinite velocities and forces.

In this case the particle arrives at the origin with a negative velocity, we must therefore suppose that the radical in (2) does not change sign when the quantity passes through infinity at the origin. Hence since x now becomes negative, we must take the positive sign in (3) instead of the negative one hitherto used. This gives $x^2 = B + 2nt$, where B need not necessarily have the same value as before. To find B we notice that at the initial stage of this part of the motion, $x=0$ and $t=b^2/2n$; we easily find that $B=-b^2$. The motion after the particle has passed the origin is therefore given by $x = -(2nt - b^2)^{\frac{1}{2}}$.

Ex. 2. If $x = at^n$ we have $\frac{d^2x}{dt^2} = At^{n-2} = A\left(\frac{x}{a}\right)^{\frac{n-2}{n}}$, where $A = an(n-1)$. Let us suppose that $n > 2$.

A particle is placed at rest at the origin. Show that if acted on by $X = At^{n-2}$ the subsequent motion is given by $x = at^n$, but if acted on by $X = A(x/a)^{\frac{n-2}{n}}$ the motion is given by $x = 0$.

Ex. 3. A particle is projected from the origin O with a velocity $\mu p^{\frac{3}{4}}$ under the action of an accelerating force $X = -\frac{3}{4}\mu^2(p-x)^{\frac{1}{2}}$. Prove that the particle comes to rest in the position of equilibrium defined by $x = p$.

101. *Let the acting force X be a function of the velocity only, say* $X = f(v)$. The equation of motion now takes the form

$$\frac{dv}{dt} = f(v) \qquad \qquad (1).$$

Integrating this, we have

$$\int \frac{dv}{f(v)} = t + A \qquad \qquad (2);$$

writing $\phi(v)$ for the integral on the left-hand side, this becomes

$$\phi(v) = t + A \qquad \qquad (3).$$

Supposing as before that the particle is initially projected at a time $t = a$, with a velocity c, we have $A = \phi(c) - a$.

Two rules are given in the theory of differential equations for the solution of the equation (3). The first rule requires us to solve the equation for v and find $v = \psi(t)$, and as already explained that solution is to be chosen which makes $v = c$ when $t = a$. Remembering that $v = dx/dt$ we then obtain x by integration.

If the equation (3) cannot be solved for v, we use the second rule. This requires us to recur to the form (1), eliminating dt by using the equation $v = dx/dt$, we have

$$\frac{v\,dv}{f(v)} = dx;$$

$$\therefore \int \frac{v\,dv}{f(v)} = x + B \qquad \qquad (4).$$

Thus after integration both x and t are expressed by (2) and (4) in terms of a subsidiary quantity v. We notice also that this subsidiary quantity has a dynamical meaning, viz. the velocity of the particle.

102. *Ex.* 1. *A particle is projected with a velocity V in a medium whose resistance is κv^n, where n is a positive quantity.* The equation of motion is then

$$\frac{dv}{dt} = -\kappa v^n \quad\text{.........(1).}$$

$$\therefore \frac{dv}{v^n} = -\kappa dt\,; \quad \therefore \frac{v^{1-n}}{1-n} = -\kappa t + A \quad\text{.........(2).}$$

Measuring t from the moment of projection we have when $t=0$, $v=V$, hence $A = \dfrac{V^{1-n}}{1-n}$. We therefore find

$$v^{1-n} - V^{1-n} = -(1-n)\,\kappa t \quad\text{.........(3).}$$

If $n<1$ the velocity decreases continually from its initial value V, and *vanishes after a finite time*, viz. $t = \dfrac{V^{1-n}}{(1-n)\,\kappa}$. The particle will then remain at rest, since $X=0$.

If $n>1$, writing (3) in the form

$$\frac{1}{v^{n-1}} - \frac{1}{V^{n-1}} = (n-1)\,\kappa t \quad\text{.........(4),}$$

we see that the velocity decreases continually and *vanishes after an infinite time.*

If $n=1$, these equations take an indeterminate form. Returning to the equation (2) we have

$$\log v = -\kappa t + A\,; \quad \therefore v = Ve^{-\kappa t} \quad\text{.........(5).}$$

It follows that the velocity decreases continually and *vanishes after an infinite time.*

In all these cases we can find the space described in any time t. Remembering that $v = dx/dt$, we have from (3),

$$x = \int \{V^{1-n} - (1-n)\,\kappa t\}^{\frac{1}{1-n}}\,dt = \frac{-1}{(2-n)\,\kappa}\{V^{1-n} - (1-n)\,\kappa t\}^{\frac{2-n}{1-n}} + B.$$

Determining B from the condition that $x=0$ when $t=0$ we find

$$-(2-n)\,\kappa x = \{V^{1-n} - (1-n)\,\kappa t\}^{\frac{2-n}{1-n}} - V^{2-n} \quad\text{.........(6).}$$

We may also find the velocity after the particle has described any space x. We begin with

$$v\frac{dv}{dx} = -\kappa v^n.$$

$$\therefore v^{1-n}dv = -\kappa dx\,; \quad \therefore v^{2-n} = V^{2-n} - (2-n)\,\kappa x \quad\text{.........(7).}$$

Let us find the space described by the particle when $v=0$.

If $n<1$, we have $x = \dfrac{V^{2-n}}{(2-n)\,\kappa}$ and $t = \dfrac{V^{1-n}}{(1-n)\,\kappa}$ as shown above; thus the particle comes to rest after describing a finite space in a finite time.

If $n>1$ and <2, we have $x = \dfrac{V^{2-n}}{(2-n)\,\kappa}$ while t is infinite; the particle therefore comes to rest after describing a finite space in an infinite time. If $n>2$, we find that v vanishes when x is infinite and the particle describes an infinite space in an infinite time before it comes to rest.

Ex. 2. If the resistance is κv, show that the particle comes to rest after describing the finite space V/κ in an infinite time.

Ex. 3. If the resistance is κv^2, prove that the particle describes an infinite space in an infinite time before coming to rest.

103. *Ex.* 1. If $X=\phi(v).f(x)$ or $X=\phi(v)f(t)$, prove that the equation of motion can be solved by separating the variables.

In the former case we use $v\,dv/dx=X$, in the latter $dv/dt=X$.

Ex. 2. If $X=f(x)v^n+F(x)v^2$ show that the equation of motion becomes linear by writing $v^{2-n}=y$.

Ex. 3. If $X=f(v^2/x)$ show that the equation of motion becomes homogeneous, and that the variables can be separated by writing $v^2=xy$.

Motion of a heavy particle.

104. *A heavy particle starting from rest slides down a rough straight line which is inclined to the vertical at an angle θ. It is required to find the motion.*

Let O be the initial position of the particle, OV the vertical, Q the particle at any time t. The accelerating force due to

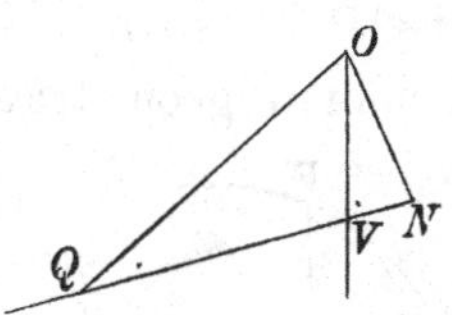

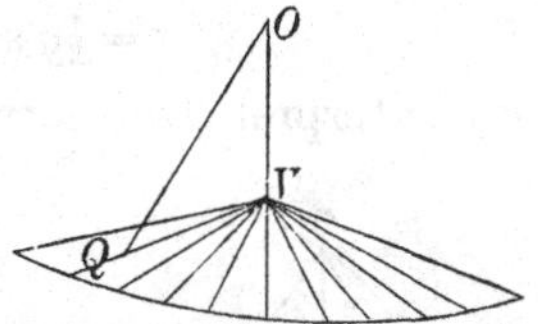

gravity is $g\cos\theta$. The pressure on the straight line being $mg\sin\theta$, the retarding force due to friction is $\mu g\sin\theta$, where μ is the coefficient of friction. The whole accelerating force is therefore

$$f=g(\cos\theta-\mu\sin\theta)=g\sec\epsilon\,.\cos(\theta+\epsilon),$$

where $\mu=\tan\epsilon$. Writing $OQ=s$, the equation of motion is

$$\frac{d^2s}{dt^2}=v\frac{dv}{ds}=g\sec\epsilon\,.\cos(\theta+\epsilon)\ \ldots\ldots\ldots\ldots\ldots(1).$$

Integrating, we find

$$v^2=2gs\sec\epsilon\cos(\theta+\epsilon)+A.$$

Since the particle starts from rest, v and s vanish together. We therefore have $A=0$, and

$$v^2=2gs\sec\epsilon\cos(\theta+\epsilon)\ldots\ldots\ldots\ldots\ldots\ldots\ldots(2).$$

To interpret this formula we make the angle $VON=\epsilon$ and draw any straight line NVQ perpendicular to ON cutting the vertical in V and the straight line along which the particle travels in Q. Then $ON = s\cos(\theta + \epsilon)$. It follows that *the velocity acquired in describing any chord OQ is independent of θ and is equal to that acquired in describing OV.*

If the chord OQ is taken on the same side of the vertical OV as N, the angle θ as above measured becomes negative. Since the friction varies as the pressure taken positively, it must now be represented by $-\mu g \sin\theta$. The theorem therefore only applies to the chords on the side of the vertical opposite to ON.

If we make the figure turn round the vertical OV, the straight line QV will describe a right cone having OV for its axis and $\frac{1}{2}\pi - \epsilon$ for the semi-vertical angle. *The velocity acquired in descending any chord from rest at O to the surface of this cone is equal to that acquired in descending OV.*

105. By integrating (1) twice with regard to t, and remembering that both s and ds/dt vanish when $t = 0$, we find

$$s = \tfrac{1}{2} g \sec\epsilon \cos(\theta + \epsilon)\, t^2 \quad \text{..................(3).}$$

We may interpret this formula by a similar geometrical con-

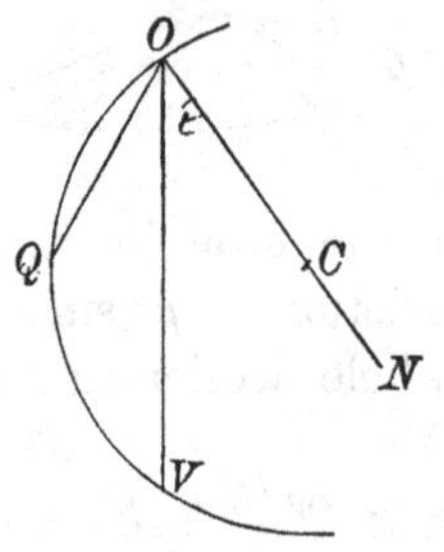

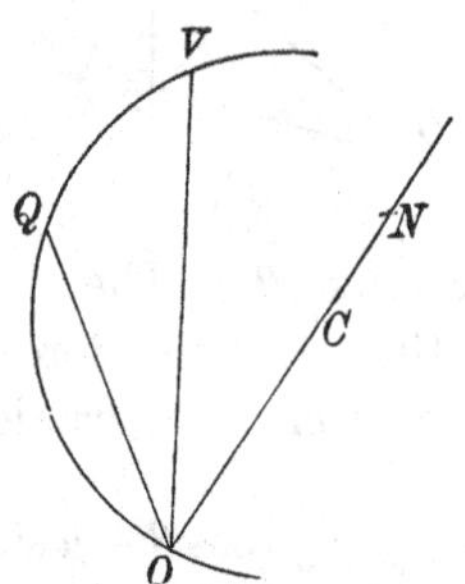

struction. Making as before the angle $VON = \epsilon$, we see that, when t is constant, (3) represents the polar equation of a circle whose radius vector is s and whose centre C is situated on ON. We have therefore the following theorem. Describe any circle passing through O and having its centre on ON, and let it cut the vertical through O in some point V. *The time of descent from rest at O down any chord OQ of this circle is the same as that down OV.* The chord OQ must be on the side of OV remote from the centre.

In the same way if the circle is drawn above O, we can show that the time of descent from rest at any point Q of the circle to O is equal to the time down VO.

106. When the straight line down which the particle slides is smooth ON coincides with the vertical. The cone in Art. 104 becomes a horizontal plane, and the circle in Art. 105 has OV for a diameter. We thus fall back on the well-known theorems (1) that the velocity acquired in descending from rest to a given horizontal plane is the same for all chords, (2) that the time of descending from rest at the highest point of a circle to the circle is the same for all chords.

107. If the motion take place in the air we must make allowance for its resistance. Supposing the resistance to vary as the velocity, the equation of motion is

$$\frac{d^2s}{dt^2}=f-\kappa v \quad\text{.........................(1)},$$

where $f=g\sec\epsilon\cos(\theta+\epsilon)$. Remembering that $v=ds/dt$ we find by integration

$$\frac{ds}{dt}=ft-\kappa s \quad\text{.........................(2)},$$

the constant being omitted because s and v vanish together. Transposing κs, the equation can be integrated again by following the ordinary rule for linear equations. We have

$$\frac{d}{dt}(e^{\kappa t}s)=fte^{\kappa t};$$

$$\therefore\ \kappa e^{\kappa t}s=f\left(te^{\kappa t}-\frac{1}{\kappa}e^{\kappa t}+c\right).$$

Noticing that s should vanish when $t=0$, we have $c=1/\kappa$. Hence, restoring the value of f,

$$s=\frac{g}{\kappa^2}\sec\epsilon\cos(\theta+\epsilon)\{\kappa t-1+e^{-\kappa t}\} \quad\text{...........(3)}.$$

When t is constant and $(\theta+\epsilon)$ is regarded as variable we see that (3) is again the equation of a circle having its centre on ON. *The theorem* of Art. 105 *is therefore also true when the particle slides on a rough chord in a medium resisting as the velocity. The times of descent from rest at O down all chords of the circle are equal.*

108. There is another method of proof by which the solution of the differential equation is evaded. We notice that if we write $s=\sigma\cos(\theta+\epsilon)$, the equation (1) of Art. 107 becomes

$$\frac{d^2\sigma}{dt^2}=g\sec\epsilon-\kappa\frac{d\sigma}{dt},$$

from which the angle θ has disappeared. The initial conditions now become $\sigma=0$ and $d\sigma/dt=0$ when $t=0$; these also are independent of θ. Hence the time of describing any given length σ is independent of θ. But if any value is given to σ, the equation $s=\sigma\cos(\theta+\epsilon)$ is the equation of a circle, s being the radius vector.

109. When a heavy body is immersed in a fluid it is partly supported by the surrounding fluid. Let V be the volume of the body, D its density, ρ that of the fluid. If the body were removed, a mass $V\rho$ of fluid would just fill the vacant place and be supported by the pressures of the surrounding fluid. The apparent weight of the body is therefore $(VD - V\rho)\,g$, and the accelerating force of gravity is

$$g' = g\left(1 - \frac{\rho}{D}\right).$$

This value of g' should properly replace g when the moving body is immersed in a resisting medium. It is sometimes called the *relative acceleration*.

110. *Ex.* 1. Prove that when $\kappa = 0$, the formula for s in Art. 107 reduces to

$$s = \tfrac{1}{2}ft^2.$$

This may be shown by expanding the expression in powers of κ.

Ex. 2. The plane of a circle is inclined to the vertical, prove that the times of descent down all smooth chords from rest at the highest point are equal.

Ex. 3. Two tangents AB, CD are drawn to touch a vertical circle at its highest and lowest points A, B. A variable tangent PQR cuts AB, CD in P, R and touches the circle at Q. Prove that the velocity acquired in descending from rest at P to R under gravity is the same for all positions of the tangent. Prove also that the time of descent from P to R is proportional to the length PR and the time from P to Q is proportional to the distance of P from the centre of the circle.

Ex. 4. If the resistance per unit of mass is κv^2 and the particle slide on a smooth straight wire inclined at an angle θ to the vertical, prove that the space s described in time t from rest is given by $e^{\kappa s} = \frac{1}{2}(e^{bt} + e^{-bt})$ where $b^2 = \kappa g \cos\theta$.

111. Limiting Velocity. When a particle is projected vertically downwards in a medium whose resistance varies as the nth power of the velocity, the equation of motion is

$$\frac{dv}{dt} = g - \kappa v^n,$$

where g is the relative acceleration of gravity.

If the particle is projected downwards with a velocity L such that $\kappa L^n = g$ it is clear that dv/dt is initially zero. There is nothing to change the velocity and the force of gravity continues to be balanced by the resistance. The particle therefore descends with a uniform velocity equal to L. If the particle is projected downwards with a velocity less than L, gravity exceeds the resistance and the velocity of the particle is increased. If the velocity of projection is greater than L, the resistance exceeds gravity and the velocity is decreased. If the particle is projected upwards, the resistance and gravity combine to bring the particle to rest, after which it descends in the manner just described.

In all cases the velocity tends to become more and more nearly equal to the velocity L given by the equation $\kappa L^n = g$. This velocity is called sometimes the *limiting velocity* and sometimes the *terminal velocity*. The latter name is commonly ascribed to Huygens. Other names are given under other circumstances. When the body considered is a ship, the constant g may represent the force of the engine and κv^n the resistances. The ship is said to be *at full speed* when these balance each other.

112. When the body is in the beginning of its fall from rest, the term κv^n is nearly zero and is much smaller than g. The body begins to fall nearly as in a vacuum, and the velocity at first increases rapidly. If the resistance is so great that L is small, the velocity will soon be so nearly equal to the limiting velocity that the motion will be sensibly uniform.

This result has many applications in nature. In a shower of rain, the velocity of a drop is not proportional to the time elapsed since it began to fall. The drops, being observed some little time after the motion has begun, move with a velocity which is sensibly uniform and independent of the height of the cloud.

113. The magnitude of the coefficient κ of the resistance depends on the size and form of the falling body as well as on the nature of the resisting medium. To illustrate this let us suppose that, for similar bodies falling in similar positions in an indefinitely extended fluid, the resistance varies (1) as the surface of the body, (2) as the nth power of its velocity, and (3) as the density ρ of the fluid. If l be the length of any side the surface varies as l^2, while the mass moved varies as $l^3\sigma$ where σ is the density of the body. The accelerating force on the body is therefore

$$f = g - \gamma \frac{l^2 \rho v^n}{l^3 \sigma} = g - \gamma \frac{\rho v^n}{l\sigma},$$

where γ is some constant depending on the form and position of the falling body. Equating f to zero, it follows that the limiting velocity varies as $(l\sigma/\rho)^{\frac{1}{n}}$. We see therefore that *the smaller the size of the body the less is the limiting velocity.* For example, large drops of rain fall with greater velocity than small ones. The particles of a mist are so small and their limiting velocities so slight that the falling drops seem to have no motion.

We have supposed that the falling body is so far symmetrical about a vertical axis that it is not made to rotate by the resistance.

114. *Ex.* 1. A particle falling freely from rest in vacuo acquires a velocity L in β seconds. Show that the same particle, falling in a medium in which the resistance varies as the velocity and the terminal velocity is L, will acquire half its terminal velocity in about $\frac{7}{10}\beta$ seconds and two-thirds of that velocity in $1\frac{1}{10}\beta$ seconds.

To prove this we use the formulae proved in Art. 107 for v. Remembering that $L=g/\kappa$ when the resistance varies as the velocity we have $\kappa=1/\beta$.

Ex. 2. Show that the effect of the resistance of a medium on the motion of a heavy body is less the greater the size and density of the body.

115. Resistance $=\kappa v^2$. A particle is projected vertically upwards with a velocity V in a medium resisting as the square of the velocity. It is required to find the motion.

During the ascending motion the resistance acts downwards and the equation of motion is

$$v\frac{dv}{ds}=\frac{dv}{dt}=-g-g\frac{v^2}{L^2},$$

where L is the limiting velocity. When the particle descends the resistance acts upwards, but *since* v^2 *does not change sign with* v, *the equation of motion must be changed to*

$$v\frac{dv}{ds}=\frac{dv}{dt}=-g+g\frac{v^2}{L^2},$$

where in both equations s and v are measured positively upwards. This discontinuity occurs whenever the power of v in the law of resistance is even.

Following the second rule given in Art. 101 we express both s and t in terms of v. We have for the ascending motion

$$\frac{gt}{L}=-\int\frac{Ldv}{L^2+v^2}=-\tan^{-1}\frac{v}{L}+\tan^{-1}\frac{V}{L}\quad\text{.......................(1)},$$

$$\frac{2gs}{L^2}=-\int\frac{2vdv}{L^2+v^2}=-\log\frac{L^2+v^2}{L^2+V^2}\quad\text{.......................(2)},$$

the constants being determined by the condition that $v=V$ when $t=0$ and $s=0$.

The time T of ascent and the space h ascended are deduced by putting $v=0$. We thus find

$$T=\frac{L}{g}\tan^{-1}\frac{V}{L},\quad h=\frac{L^2}{2g}\log\left(1+\frac{V^2}{L^2}\right)\quad\text{.......................(3)}.$$

The time of ascent and the space ascended are less than in a vacuum, for both gravity and the resistance join in bringing the particle to rest.

For the descending motion we have in the same way

$$\frac{g(t-T)}{L}=-\int\frac{Ldv}{L^2-v^2}=\tfrac{1}{2}\log\frac{L+v}{L-v}\quad\text{.......................(4)},$$

$$\frac{2g(s-h)}{L^2}=-\int\frac{2vdv}{L^2-v^2}=\log\frac{L^2-v^2}{L^2}\quad\text{.......................(5)},$$

the constants being determined from the condition that when $v=0$, $t=T$, $s=h$.

The velocities at which the particle passes upwards and downwards through any given point of space are connected by a simple relation. Taking the given point as the point of projection upwards, let the two velocities be V and V'. Putting $s=0$ in (5) we find

$$-\frac{2gh}{L^2}=\log\left(1-\frac{V'^2}{L^2}\right).$$

Eliminating h between this equation and (3) we arrive at

$$\frac{1}{V'^2}-\frac{1}{V^2}=\frac{1}{L^2}.$$

If σ be the space descended and τ the time, we find by eliminating v

$$e^{g\sigma/L^2}=\tfrac{1}{2}(e^{g\tau/L}+e^{-g\tau/L}).$$

See Art. 110, Ex. 4.

116. Resistance $=\kappa v^n$. A particle is projected vertically upwards with a velocity V in a medium resisting as the nth power of the velocity. It is required to find the motion.

We write the equation for the ascending motion in the form

$$v\frac{dv}{ds}=\frac{dv}{dt}=-g-g\left(\frac{v}{L}\right)^n.$$

It will be convenient to put $v=xL$. Proceeding as in the case when $n=2$, we find for the whole time T and space h of ascent

$$\frac{gT}{L}=\int_0^a\frac{dx}{1+x^n},\quad \frac{gh}{L^2}=\int_0^a\frac{x\,dx}{1+x^n},$$

where the initial upward velocity is $V=aL$.

To find the time and space in which the velocity is decreased from aL to bL we take the limits from b to a.

We can find superior limits to the values of t and h by making the initial velocity V infinitely great. In this case $a=\infty$, and both the integrals are given in the Integral Calculus. We then have

$$\frac{gT}{L}=\frac{\pi}{n\sin\pi/n},\quad \frac{gh}{L^2}=\frac{\pi}{n\sin 2\pi/n},$$

the former requiring $n>1$ and the latter $n>2$. It is remarkable that both these limits are finite, though the upward velocity of projection may be as great as we please.

For the descending motion it is often convenient to *measure s downwards* from the highest point. We thus avoid using a negative velocity. Adopting this plan, the equation of motion is

$$v\frac{dv}{ds}=\frac{dv}{dt}=g-g\left(\frac{v}{L}\right)^n.$$

Putting $v=xL$ as before, we find for the time and space necessary to acquire a velocity aL,

$$\frac{gT'}{L}=\int_0^a\frac{dx}{1-x^n},\quad \frac{gh'}{L^2}=\int_0^a\frac{x\,dx}{1-x^n}.$$

These integrals can be found without difficulty when n is an integer by using the method of partial fractions, see Greenhill's *Differential and Integral Calculus*, Art. 190. Roberts' *Integral Calculus*, Art. 35. The result when n has its general integral value is too complicated to be reproduced here.

117. *Ex.* 1. A heavy particle is projected upwards with a velocity L in a medium resisting as the nth power of the velocity. Prove that the whole space (up and down) described when the velocity downwards is V is equal to LT when L is the limiting velocity and T is the time in which the particle falling from rest in the medium will acquire a velocity V^2/L.

Ex. 2. A particle is projected upwards with velocity L in a medium resisting as the cube of the velocity. Show that the whole time and space of the ascent are connected by the equation $s + LT = \frac{2\pi}{3\sqrt{3}} \frac{L^2}{g}$.

The linear differential equation.

118. The Linear equation. The most important equation of motion which occurs in this part of dynamics is the linear equation with constant coefficients. The simplest form of this equation is

$$\frac{d^2x}{dt^2} + bx = c \qquad\qquad (1),$$

where b and c are two constants.

When $b = 0$ the equation represents the motion of a particle acted on by a constant accelerating force equal to c, and the solution is obviously

$$x = \tfrac{1}{2}ct^2 + At + B \qquad\qquad (2).$$

When b is not zero, we can simplify the equation by putting

$$x = c/b + \xi \qquad\qquad (3),$$

we then have

$$\frac{d^2\xi}{dt^2} + b\xi = 0 \qquad\qquad (4).$$

This can be solved without difficulty by the method already explained in Art. 97. But a simpler solution can be obtained by following the rules for solving equations with constant coefficients given in books on differential equations. We assume as a possible solution

$$\xi = Ae^{\lambda t} \qquad\qquad (5).$$

Substituting we find $A(\lambda^2 + b)e^{\lambda t} = 0$. The equation is therefore satisfied if $\lambda = \pm\sqrt{(-b)}$. If b is negative and equal to $-b'$, we have two *real* values of λ, either of which give a solution. The equation is clearly satisfied by

$$x = \frac{c}{b} + Ae^{t\sqrt{b'}} + Be^{-t\sqrt{b'}} \qquad\qquad (6),$$

and this is the complete integral because it contains the two arbitrary constants A and B.

If b is positive, λ is imaginary; but remembering that an imaginary exponential is a trigonometrical expression, we replace the assumption (5) by

$$\xi = A \sin(\lambda t + B) \quad \text{.......................(7).}$$

Substituting we find $A(-\lambda^2 + b)\sin(\lambda t + B) = 0$. The equation is therefore satisfied by $\lambda = \pm\sqrt{b}$. These two values of λ give the same solution, the effect of changing the sign of λ being merely that of changing the signs of the arbitrary constants A and B. The complete integral is therefore

$$x = c/b + A \sin(t\sqrt{b} + B) \quad \text{.................(8).}$$

It may also be written in either of the forms

$$x = c/b + A' \sin t\sqrt{b} + B' \cos t\sqrt{b} \quad \text{.............(9),}$$

$$x = c/b + A'' \cos(t\sqrt{b} + B'') \quad \text{.................(10).}$$

119. Harmonic Oscillation. The dynamical meaning of the linear equation is important. Consider first the case in which b is positive. Putting $b = n^2$, we have

$$\frac{d^2x}{dt^2} + n^2x = c \quad \text{.........................(1),}$$

$$x = c/n^2 + A \sin(nt + B) \quad \text{.......................(2).}$$

First, we notice that as t continually increases the value of x alternates between the limits $c/n^2 \pm A$. We therefore infer that the differential equation (1) represents an oscillatory motion and that *the arc of oscillation is constant.* The semi-arc of oscillation is A and its magnitude depends on the initial conditions. The semi-arc is called *the amplitude* of the oscillation.

Secondly. The middle point of the arc is determined by $x = c/n^2$, and this point is independent of the initial conditions. If the particle is placed *at rest* in the position defined by this value of x, the equation (1) shows that the accelerating force (viz. d^2x/dt^2) is zero. *The middle point of the arc of oscillation is therefore a position of equilibrium.*

Thirdly. When t is increased by $2\pi/n$, the values of x recur in the same order, but when increased by π/n they recur with opposite signs. *The period of a complete oscillation is therefore*

$2\pi/n$. *This period is independent of the initial conditions.* The quantity n is called the *frequency* of the oscillation.

The time of a complete oscillation is the time occupied by the particle in describing twice the whole arc of oscillation starting from any point and returning finally to the same point again. When the period is independent of the length of the arc, the motion is sometimes called *tautochronous.*

Fourthly. The constant B depends on the instant from which the time t is measured, thus if we write $t+\alpha$ for t, nothing is changed except that B is increased by $n\alpha$.

Fifthly. Let $x=x_0$, $dx/dt=v_0$ be the given values of x and v at the time t_0. Writing the equation (2) in the form (9) of Art. 118 and equating the values of x and dx/dt to x_0 and v_0 when $t=t_0$, we find the values of A' and B'. The solution therefore becomes

$$x=\frac{c}{n^2}+\left(x_0-\frac{c}{n^2}\right)\cos n\,(t-t_0)+\frac{v_0}{n}\sin n\,(t-t_0).$$

Comparing this with the solution (2) we see that

$$A\sin B=x_0-c/n^2,\quad A\cos B=v_0/n.$$

The semi-arc A of oscillation is therefore given by

$$A^2=(x_0-c/n^2)^2+(v_0/n)^2.$$

120. Consider next the case in which b is negative. Writing $b=-n^2$, the differential equation and its solution become

$$\frac{d^2x}{dt^2}-n^2x=c,$$

$$x=-\frac{c}{n^2}+Ae^{nt}+Be^{-nt}.$$

First, we notice that the motion is not oscillatory.

Secondly. If A is not zero the particle travels in an infinite time to an infinite distance from the origin. If $A=0$ the particle after an infinite time arrives at the point determined by $x=-c/n^2$.

Thirdly. The position of equilibrium is given by $x=-c/n^2$.

Fourthly. The particle can change its direction of motion only once. This change occurs when

$$\frac{dx}{dt}=n\,(Ae^{nt}-Be^{-nt})=0.$$

This gives $2nt = \log(B/A)$. This is imaginary if A and B have opposite signs, and gives only one real value of t if A and B have the same sign. The particle can change its direction only if this real value of t is subsequent to the beginning of the motion.

Fifthly. If the values of x and v are respectively x_0 and v_0 at the time $t = t_0$, the value of x at any time t is

$$x = -\frac{c}{n^2} + \frac{1}{2}\left(x_0 + \frac{c}{n^2} + \frac{v_0}{n}\right)e^{n(t-t_0)} + \frac{1}{2}\left(x_0 + \frac{c}{n^2} - \frac{v_0}{n}\right)e^{-n(t-t_0)}.$$

121. When the equation of motion is

$$\frac{d^2x}{dt^2} + 2a\frac{dx}{dt} + bx = c \quad\text{.................................} \quad (1),$$

we take as the trial solution

$$x = \frac{c}{b} + Ae^{\lambda t} \quad\text{.....................................} \quad (2).$$

It is easily seen that this satisfies the differential equation if

$$\lambda^2 + 2a\lambda + b = 0 \quad\text{.......................................} \quad (3).$$

If $a^2 - b$ is positive, the roots of the equation are real. Representing these by λ_1, λ_2, the solution is

$$x = \frac{c}{b} + A_1e^{\lambda_1 t} + A_2e^{\lambda_2 t} \quad\text{................................} \quad (4),$$

where A_1, A_2 are two arbitrary constants.

If $a^2 - b$ is negative, say $= -n^2$, the two roots are $-a \pm n\sqrt{(-1)}$. By an easy reduction the solution (4) becomes

$$x = \frac{c}{b} + e^{-at}B_1 \sin(nt + B_2) \quad\text{............................} \quad (5),$$

where B_1, B_2 are two arbitrary constants.

If $a^2 - b = 0$, the general solution is

$$x = \frac{c}{b} + (At + B)e^{-at} \quad\text{.................................} \quad (6).$$

Considering the solution (5) as the more important of the three, we notice that the trigonometrical term vanishes whenever $nt + B_2$ is a multiple of π, the particle therefore passes through the position defined by $x = c/b$ at intervals each equal to π/n. Since it necessarily passes through this point alternately in opposite directions, the interval between two consecutive passages in the same direction is $2\pi/n$. This is called the time of a complete oscillation. The point defined by $x = c/b$ is evidently the position of equilibrium.

To find the times at which the system comes momentarily to rest we put $dx/dt = 0$. This gives $\tan(nt + B_2) = n/a$. The extent of the oscillations on each side of the position of equilibrium may be found by substituting the values of t given by this equation in the expression for $x - c/b$. Since these occur at a constant interval equal to π/n we see that the amplitude of the oscillation continually decreases and the successive arcs on each side of the position of equilibrium form a geometrical progression whose common ratio is $e^{-a\pi/n}$.

122. The following differential equations occur in dynamics.

(1) Solve $$\frac{d^2x}{dt^2}+n^2x=\phi(t).$$

Multiplying by $\sin nt$, both sides become perfect differentials, hence

$$\frac{dx}{dt}\sin nt-nx\cos nt=\int\phi(t)\sin nt\,dt+A.$$

Multiplying by $\cos nt$, both sides are again perfect differentials,

$$\frac{dx}{dt}\cos nt+nx\sin nt=\int\phi(t)\cos nt\,dt+B.$$

These two simultaneous equations give both x and dx/dt.

To solve $\frac{d^2x}{dt^2}-n^2x=\phi(t)$ we use e^{nt} and e^{-nt} as the two successive multipliers.

(2) When $\phi(t)$ is trigonometrical another method can be used. Let the equation be

$$\frac{d^2x}{dt^2}+n^2x=E\sin(\lambda t+F).$$

Assuming $x=M\sin(\lambda t+F)$ as a trial solution, we see at once that the equation is satisfied if $M(-\lambda^2+n^2)=E$. Adding the solution found in Art. 118 we see that the complete integral is

$$x=A\sin(nt+B)+\frac{E}{-\lambda^2+n^2}\sin(\lambda t+F).$$

This method fails when $\lambda=n$. In this case we take $x=Mt\cos(\lambda t+F)$ as a trial assumption.

We find $-2Mn=E$. The complete integral is therefore

$$x=A\sin(nt+B)-\frac{Et}{2n}\cos(nt+F).$$

Motion under a centre of force.

123. Central force varying as the distance. A particle constrained to move on a smooth straight line is acted on by a central force tending to a fixed point O outside the straight line, whose magnitude varies as the distance of the particle from O.

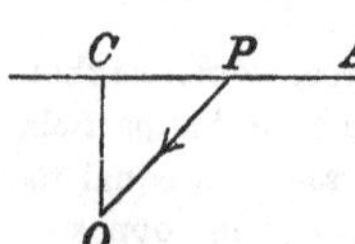

Let $OC=h$ be the perpendicular on the straight line AC. Let P be the particle, $CP=x$. The force on P being $n^2 . OP$, the component along PC is n^2x. Supposing the straight line to be smooth and the motion to take place in vacuo, the equation of motion is

$$\frac{d^2x}{dt^2}=-n^2x.$$

This is the standard form discussed in Art. 119. The particle therefore oscillates about C as the middle point of the arc, and the time of a complete oscillation is $2\pi/n$.

To find the time of oscillation numerically the magnitude of the force must be known at some given distance from the centre O. Suppose that the force is equal to gravity at a distance a, then $n^2a = g$, and the time of a complete oscillation is $2\pi\sqrt{(a/g)}$. If $g = 32{\cdot}18$, the distance a must be measured in feet and the formula gives the time in seconds.

The extent of the arc of oscillation depends on the initial conditions. If the particle start from a point distant x_0 from C with an initial velocity v_0 measured positively from C, the whole subsequent motion is expressed by the fifth result of Art. 119.

124. *Ex.* Any two places on the surface of the earth are joined by a straight tunnel. A particle dropped from one falls towards the other under the sole attraction of the earth. Assuming that the resultant attraction tends to the centre and varies as the distance therefrom, prove that the particle will arrive at the second place after about 42 minutes, the radius of the earth being taken as 4000 miles.

125. *Ex. Effect of friction.* If the straight line in Art. 123 is sensibly rough, it is required to take account of the friction.

Since the normal pressure on the straight line is equal to n^2h and is therefore constant, the limiting friction is also constant. Let us represent this by f. The equation of motion is therefore

$$\frac{d^2x}{dt^2} + n^2x = \pm f.$$

We notice that the frictional accelerating force acts opposite to the direction of motion, so that the sign must be negative or positive according as the particle is

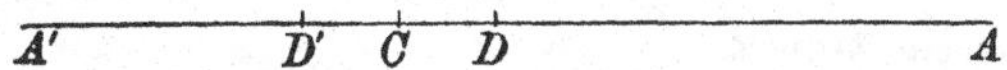

moving in the direction in which x is measured or the opposite. *The equation therefore presents the discontinuity which so frequently occurs whenever friction has to be taken account of.*

Let the particle start from rest at A where $CA = a$. Initially the resolved attraction is n^2a and unless n^2a is greater than the friction f, the particle will not move. Supposing this inequality to hold we write the equation in the form

$$\frac{d^2x}{dt^2} = -n^2\left(x - \frac{f}{n^2}\right).$$

The motion therefore from A towards C is the same as if the centre of force were displaced a distance $CD = f/n^2$ towards A. The particle comes to rest at a point A' on the other side of D where $DA' = AD$. On the return journey we take CD' also equal to f/n^2 and the particle moves as if D' were the centre of force. Thus the centre of force is alternately moved at each oscillation a constant distance, always opposite to the direction of motion. The friction reduces the extent of each successive semi-arc of oscillation by $2f/n^2$. The particle comes finally to rest when the extent of the semi-arc is less than f/n^2.

126. Resistance of the air. If the motion take place in the air its resistance must be allowed for. As a sufficient illustration of the general effects of this force, let us suppose that the resistance varies as the velocity. Excluding friction the equation of motion is then

$$\frac{d^2x}{dt^2} = -n^2x - 2\kappa\frac{dx}{dt} \qquad (1).$$

Assuming $n > \kappa$ the solution is (Art. 121)

$$x = Ae^{-\kappa t}\sin(pt + B) \qquad (2),$$

where $p^2 = n^2 - \kappa^2$. *The constancy of the period of oscillation is therefore unaffected by the resistance of the medium*, Art. 121. The time of oscillation is however longer than in a vacuum.

The successive arcs on each side of the position of equilibrium decrease continually in geometrical progression and vanish only after an infinite time.

In many cases the resistance of the medium is very slight compared with the other forces acting on the particle. The quantity κ is then small, and we see that the period of any one oscillation differs from that in a vacuum by the squares of small quantities. In using the equation (2) we must however remember that when the position of the particle *after a great many oscillations is required we cannot regard pt as the same as nt*; for though p and n differ by a very small quantity, that difference is here multiplied by the time t.

127. By making observations on the lengths of the arcs of oscillation we may test the correctness of the assumed law of resistance. A convenient method of trying the experiment is to use the particle as a pendulum. It may be shown that when the oscillations are small the resolved action of gravity represents the force n^2x while the resistance is $2\kappa\, dx/dt$. The measurements show that the successive arcs do decrease in geometrical progression when the arcs are small, but the decrease follows another law when not small. This, as Poisson remarks, is a justification of the statement that for *small velocities* the resistance varies nearly as the velocity.

The common ratio of the geometrical progression is $e^{-\kappa\pi/p}$. By measuring successive arcs the numerical value of κ can be found.

128. Discontinuity of resistance. When the resistance varies as the velocity the analytical expression $2\kappa v$ changes sign with v. It therefore represents the retardation due to the resisting medium both in sign and magnitude. If the resistance varies as the square (or any even power) of the velocity, the analytical expression $2\kappa v^2$ represents the retardation in magnitude only. Whenever the particle changes its direction of motion it will then be necessary to change the sign of κ. *Thus a discontinuity is introduced into the equations* similar to that which occurs when friction acts on the particle, Arts. 125 and 115.

129. *Ex.* 1. A particle oscillates in a straight line under the action of a central force tending to a fixed point C on the straight line and varying as the distance therefrom. Supposing the motion to take place in a medium *resisting as the square of the velocity*, find the relation between any two successive arcs on each side of C.

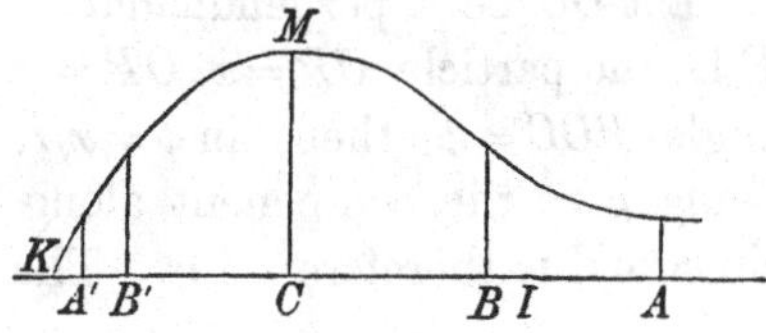

Supposing that the particle is moving in the negative direction (Art. 128) the equation of motion is

$$vdv/dx = -n^2x + \kappa v^2.$$

By Art. 103 this gives $v^2e^{-2\kappa x} = C + \dfrac{n^2}{\kappa}\left(x + \dfrac{1}{2\kappa}\right)e^{-2\kappa x}$. If x_0, x_1 be two successive arcs, x_1 being negative, we have $\left(x_1 + \dfrac{1}{2\kappa}\right)e^{-2\kappa x_1} = \left(x_0 + \dfrac{1}{2\kappa}\right)e^{-2\kappa x_0}$

We notice that this relation is independent of the strength of the attractive force.

To interpret this relation we trace the curve $y = \left(x + \dfrac{1}{2\kappa}\right)e^{-2\kappa x}$. If the particle start from rest at any place A it will come to rest again at A' where the ordinates of A and A' are equal. Taking $CB = CA'$, the third point of rest is at a distance CB' from C on the side of C opposite to A', the ordinates of B, B' being equal, and so on. Thus if the particle start from rest at an *infinite distance* from C it will *first* come to rest at K, where $CK = 1/2\kappa$ numerically.

The general character of the motion is that the successive arcs decrease rapidly at first, but afterwards become more and more nearly equal, the motion never ceasing.

If CI is the abscissa of the point of inflexion, $CI = CM = CK = 1/2\kappa$

Ex. 2. Prove that the times of describing all chords of a circle starting from rest at the same point A under the action of a centre of force situated on the diameter through A and varying as the distance are equal. The chords are to be regarded as smooth and the motion to be in a vacuum.

Ex. 3. A heavy particle whose mass is m is suspended from a fixed point O by an elastic string whose unstretched length is a. If the particle oscillate up and down in a vacuum, prove that the complete period of an oscillation is $2\pi\sqrt{(ma/E)}$, where E is Young's modulus.

Ex. 4. A particle oscillates in a straight line in a medium whose resistance per unit of mass is κ times the square of the velocity. There is a centre of force situated in the straight line whose attraction is μ times the square of the distance from the centre of force. If a and b are the distances from the centre of force of two successive positions of instantaneous rest, and μ is not zero, prove that

$$(\kappa^2b^2-\kappa b+\tfrac{1}{2})\,e^{2\kappa b}+(\kappa^2a^2+\kappa a+\tfrac{1}{2})\,e^{-2\kappa a}=1. \qquad \text{[Art. 135.]}$$

130. The inverse square of the distance. A particle, constrained to move in a straight line, is acted on by a central force tending to a fixed point O external to the line and varying inversely as the square of the distance therefrom. It is required to find the motion.

Let OC be a perpendicular on the straight line, $OC=h$. Let P be the particle, $CP=x$, $OP=r$. See fig. of Art. 123. Let the angle $POC=\phi$, then $\sin\phi=x/r$. The accelerating force on P being μ/r^2, the component along PC is found by multiplying by $\sin\phi$ and is therefore $\mu x/r^3$. The equation of motion is

$$v\frac{dv}{dx}=-\frac{\mu x}{r^3} \quad\text{.........................(1)}$$

Since $r^2=h^2+x^2$, we have $rdr=xdx$. Hence

$$vdv=-\frac{\mu dr}{r^2}, \qquad \therefore\ v^2=C+\frac{2\mu}{r}.$$

If the particle start with a velocity u at some point A distant a from O, we have

$$v^2-u^2=2\mu\left(\frac{1}{r}-\frac{1}{a}\right) \quad\text{.....................(2).}$$

If the particle is projected from C along CA with a velocity u greater than $\sqrt{(2\mu/a)}$, it is clear that the velocity v cannot vanish or change sign. The particle therefore will move continually away from the centre of force.

131. When the centre of force lies on the straight line of motion, the time occupied by the particle in travelling from the initial position A to any point P can be found without difficulty. We put

$$x=b\cos^2\theta, \qquad \therefore\ dx/dt=-2b\sin\theta\cos\theta\, d\theta/dt.$$

The equation of motion is

$$\left(\frac{dx}{dt}\right)^2=2\mu\left(\frac{1}{x}-\frac{1}{b}\right); \qquad \therefore\ 2\cos^2\theta\frac{d\theta}{dt}=\pm\sqrt{\frac{2\mu}{b^3}}.$$

We notice that x begins at $x=a$ with dx/dt initially positive; x then increases until $dx/dt=0$, i.e. until $x=b$. At this point the particle begins to return and dx/dt becomes negative. To represent these changes we make θ begin at $\theta=-\beta$ where $\cos\beta=+\sqrt{(a/b)}$ because this makes dx/dt positive when $x=a$. We then make θ increase through zero and finally become $\frac{1}{2}\pi$ when the particle arrives at the centre of force. Thus the two times at which the particle passes through any point P are distinguished by the sign of θ. Since, according to this arrangement, θ continually increases with the time we give the positive sign to the radical in the expression for $d\theta/dt$. We then find after integration that the time from $\theta=-\beta$ to θ is

$$\sqrt{\frac{b^3}{2\mu}}\,(\theta+\tfrac{1}{2}\sin 2\theta+\beta+\tfrac{1}{2}\sin 2\beta).$$

The time from rest at a distance $x=a$ follows from the preceding or may be found independently. We have

$$\frac{dx}{dt}=\sqrt{2\mu}\sqrt{\left(\frac{1}{x}-\frac{1}{a}\right)};\quad \therefore\ \sqrt{\frac{2\mu}{a}}\,t=-\int x^{\frac{1}{2}}(a-x)^{-\frac{1}{2}}\,dx,$$

the limits being $x=a$ to x. Putting $x=a\cos^2\theta$ we easily find that the time t of moving from $x=a$ to x is

$$t=\sqrt{\frac{a^3}{2\mu}}.(\theta+\tfrac{1}{2}\sin 2\theta).$$

The time of arriving at the centre of force starting from rest at a distance a is found by putting $\theta=\frac{1}{2}\pi$. The result is $\frac{\pi}{2}\sqrt{\frac{a^3}{2\mu}}$.

132. *Ex.* 1. A particle falls from rest at a point A whose altitude above the surface of the earth is equal to the radius. Show that the velocity on arriving at the surface is equal to that acquired by a particle falling from rest through half that space under a constant force equal to g, where g represents gravity at the surface of the earth.

Notice that if μ/r^2 is the attraction of the earth, a the radius, $\mu/a^2=g$.

Ex. 2. If a particle fall from an infinite distance towards the earth, prove that the velocity at the surface is equal to that acquired in falling from rest through a space equal to the radius under a constant force equal to g.

Ex. 3. If any heavenly body were isolated in space, prove that the least velocity with which a particle must be projected from its surface that it may not fall back on the body is $\sqrt{\left(\frac{M}{E}\cdot\frac{2ga^2}{r}\right)}$ feet per second, where M and E are the masses, r and a the radii of the body and the earth. The resistance of the atmosphere is to be neglected.

Show that for the moon this velocity is about one and a half miles per second, taking its mass and radius to be $\frac{1}{81}$th and $\frac{1}{4}$th of the mass and radius of the earth, and the radius of the earth to be 4000 miles.

133. *Ex.* 1. A particle, constrained to move along *a rough straight line* whose coefficient of friction is f, is acted on by a force tending to O and varying as the inverse square. Prove that if the particle start from rest at any point A, it will next come to rest at a point B such that OM bisects the angle AOB, where M is the point on the straight line at which the resolved attraction is balanced by the limiting friction.

Following the same notation as in Art. 130, the equation of motion takes the form

$$v\frac{dv}{dx} = -\frac{\mu x}{r^3} + f\frac{\mu h}{r^3}.$$

Multiply by dx and put $x = h\tan\phi$, where ϕ represents the angle POC. Integrating as before, we find

$$v^2 = \frac{4\mu}{h}(\cos\phi - \cos\phi_0 + f\sin\phi - f\sin\phi_0)$$

$$= -\frac{4\mu}{h}\sec\epsilon\sin\frac{\phi-\phi_0}{2}\sin\left(\frac{\phi+\phi_0}{2}-\epsilon\right),$$

where ϕ_0, ϵ are the angles COA, COM, so that $f = \tan\epsilon$. It is evident that $v=0$ when $\frac{1}{2}(\phi+\phi_0) = \epsilon$.

Ex. 2. If the force to O vary as the inverse fourth power of the distance and the particle starting from rest at A come to rest again at B, prove that the angles COA, COB are complements of each other when $\sin 2(COA) = (4f-2)/(f+1)$. Thus if $f = \frac{1}{2}$ a particle starting from rest at an infinite distance will just reach C.

Ex. 3. A particle is constrained to move in a straight rough tube CA, and is acted on by a central repulsive force λ/r, where r is the distance from the centre of force O and OCA is a right angle. The particle is projected from A away from C with a velocity v; prove that if it come to rest at a point P, the angle COP is a value of θ satisfying the equation $\mu\theta - \log\sec\theta = v^2/2\lambda$, where μ is the coefficient of friction. [Coll. Ex. 1893.]

134. *Ex.* 1. The earth and moon being held at rest, find the least velocity V with which a particle must be projected from the moon to reach the earth.

Let a be the radius of the earth, $b = \frac{3}{11}a$ that of the moon, $60a$ their distance apart from centre to centre. Let E and $\frac{1}{81}E$ be the masses of the earth and moon. If x is the distance of the particle from the centre of the moon, the equation of motion is

$$\frac{d^2x}{dt^2} = \frac{E}{(60a-x)^2} - \frac{1}{81}\frac{E}{x^2} \quad \text{.............................} \quad (1).$$

This equation can be integrated by the rule of Art. 97. The constant of integration can be found in terms of V by remembering that $dx/dt = V$ when $x = b$.

There is evidently a certain point between the earth and moon where the attractions of these bodies balance each other. By equating the right-hand side of (1) to zero, this point is easily seen to be at a distance $6a$ from the centre of the moon. If V is such that dx/dt vanishes when $x = 6a$, it follows that a velocity of projection ever so slightly greater than V will carry the particle to the earth.

Remembering that $E/a^2 = g$ and taking a to be 4000 miles, we find that V is approximately $1\frac{1}{2}$ miles per second.

Ex. 2. If the earth and moon were placed at rest, they would fall towards each other under the influence of their mutual attractions. Supposing the initial distance to be equal to their present distance from each other show that they would meet after about four and a half days.

Consider their relative motion. If E, M be the masses of the earth and moon, the attraction on the earth per unit of mass is M/r^2. By Art. 39 we apply this, reversed in direction, as an acceleration to both bodies. The earth is thus reduced to rest, while the moon is acted on by the two accelerating forces M/r^2 and E/r^2.

The whole accelerating attraction on the moon causing the relative motion is therefore $(E+M)/r^2$. We must also apply to each an initial velocity equal and opposite to that of the earth (Art. 10), but this, in our problem, is zero. The time is then found as in Art. 131.

Ex. 3. Two mutually attracting spheres, each one foot in diameter, and the density of each the same as the mean density of the earth, are placed at rest in a vacuum, the distance between their surfaces being one quarter of an inch. Prove that they will meet in less than 250 seconds. This problem is due to Newton, its history is given in Todhunter's *History of the Theory of Attractions*, &c., Art. 725.

Ex. 4. Two particles A, B, mutually attracting each other according to the Newtonian law, are placed at rest at a given distance a apart. The particle B is now constrained to move away from A along the straight line joining them with a uniform velocity u, show that A will catch B up if $u^2 < 2\mu/a$ where μ is the mass of B. Show also that the time will be $\frac{1}{2}(\pi + 2\beta + \sin 2\beta)\sqrt{b^3/2\mu}$ where $\cos^2\beta = a/b$ and $2\mu/b = 2\mu/a - u^2$. [Reduce B to rest, see Art. 131.]

Ex. 5. A body of mass M is moving in a straight line with velocity U, and is followed at a distance r by a smaller body of mass m, moving in the same line with a smaller velocity u. The two bodies attract each other with a force varying as the inverse square of the distance and equal to κ for two unit masses at unit distance. Prove that the smaller body will overtake the other after a time

$$\left\{\frac{1}{\kappa(M+m)}\right\}^{\frac{1}{2}} \left\{\frac{r}{1+\omega}\right\}^{\frac{3}{2}} \{\pi + \sqrt{(1-\omega^2)} + \cos^{-1}\omega\},$$

where $\kappa(M+m)(1-\omega) = (U-u)^2 r$. [Math. Tripos, 1887.]

135. Discontinuity of a centre of force. A particle constrained to move on a smooth straight line is acted on by a force X tending to a point C situated on the line and varying as the nth power of the distance therefrom. It is required to find the motion.

Let the particle P start from rest at A, $CA = a$, $CP = x$. The equation of motion is

$$\frac{d^2x}{dt^2} = v\frac{dv}{dx} = -\mu x^n \text{ ....(1)},$$

$$\therefore\ v^2 = \frac{2\mu}{n+1}(a^{n+1} - x^{n+1}) \text{..................(2)},$$

the constant of integration being determined by the condition that $v = 0$ when $x = a$. If $n = -1$ the integral takes a logarithmic form.

If n is an odd integer this equation shows that the velocity is again zero at a point A' determined by $x = -a$. The particle therefore oscillates on each side of C, the amplitudes on each side being equal.

If n is an even integer, the expression for v vanishes for no real value of x except $x=a$. Since the particle must obviously oscillate on each side of C through equal arcs, it follows that the equation (2) cannot represent the dynamical facts of the problem.

The reason is that the force X (as given in the question) varies as the nth power of the distance *taken positively* and always acts *towards* C. Now x is the distance of the particle from C taken with its proper sign. We must therefore write

$$X = -\mu x^n \quad \text{or} \quad +\mu(-x)^n \text{.................(3)},$$

according as the particle is on the positive or negative side of the origin C. These are identical if n is odd and in that case the equation (2) holds throughout the motion. If n is even, different equations of motion hold on each side of the origin.

The particle arrives at C with a velocity v_0 obtained by putting $x=0$ in (2). This is a finite velocity if n is positive. After passing C, the equation of motion (1) must be changed to

$$vdv/dx = \mu(-x)^n = \mu x^n \text{...................(4)},$$

since n is even. We then find

$$v^2 = \frac{2\mu}{n+1}(a^{n+1} + x^{n+1}) \text{.....................(5)},$$

the constant of integration being found by the condition that (2) and (5) must agree when $x=0$. The equation (5) shows that v is again zero when $x=-a$, so that the particle in its oscillations describes equal arcs on each side of C.

After the particle has passed through C on its return journey the equation of motion resumes the form (1). The integration is the same as before, but the constant C must now be determined from the condition that the value of v *at the origin* is the same as that given by (5). The resulting value of v^2 is however the same as that given by (2), so that the motion on the positive side of the origin is always that represented by (2), and the motion on the negative side that represented by (5).

136. The time of travelling from A to C is given by

$$\sqrt{\frac{2\mu}{n+1}}\,t = \int_0^a \frac{dx}{\sqrt{(a^{n+1} - x^{n+1})}}.$$

To integrate this in terms of gamma functions we write $x^{n+1} = a^{n+1}\xi$ or $= a^{n+1}/\xi$

according as $n+1$ is positive or negative. We then have

$$\sqrt{2\mu}\,.\,t=\left(\frac{\pi}{(n+1)\,a^{n-1}}\right)^{\frac{1}{2}}\frac{\Gamma\left(\dfrac{1}{1+n}\right)}{\Gamma\left(\dfrac{1}{1+n}+\dfrac{1}{2}\right)}\text{ or }\left(\frac{\pi a^{1-n}}{-n-1}\right)^{\frac{1}{2}}\frac{\Gamma\left(\dfrac{1}{2}-\dfrac{1}{1+n}\right)}{\Gamma\left(1-\dfrac{1}{1+n}\right)}.$$

Ex. 1. A particle starts from rest at a distance a from a centre of force which attracts as the inverse cube of the distance. Show that the time of arriving at the centre is $a^2/\sqrt{\mu}$.

Ex. 2. A particle starts from rest at a distance a from a centre of force which attracts inversely as the distance. Prove that the time of arriving at the centre is

$$a\,(\pi/2\mu)^{\frac{1}{2}}.$$

Small Oscillations and Magnification.

137. Small Oscillations. A particle, constrained to describe a straight line, is under the action of a force tending to a point O external to the straight line and varying as some given function of the distance from O. It is required to discuss the motion when the arc of oscillation decreases without limit.

Let OC be a perpendicular on the straight line, P the particle, $OC=h$, $CP=x$, $OP=r$. Let the accelerating force be $rf(r)$. The equation of motion is therefore

$$\frac{d^2x}{dt^2}=-rf(r)\,.\,\frac{x}{r}\quad\text{.....................(1).}$$

Since $r^2=h^2+x^2$, we can expand $xf(r)$ in powers of x. The equation then takes the form

$$d^2x/dt^2=A_1x+A_2x^2+\dots\quad\text{.................(2),}$$

where A_1, A_2, &c. are known constants. Supposing the series to be convergent when x decreases without limit, we may ultimately omit all the terms after the first which does not vanish. Assuming x to be initially small we proceed to discuss the subsequent motion.

When A_1 is not zero, the equation reduces to

$$d^2x/dt^2=A_1x\quad\text{.........................(3).}$$

The motion represented by this equation has been discussed in Art. 119. If A_1 is negative and equal to $-n^2$, the time of a complete oscillation is $2\pi/n$. It appears therefore that when the arc of oscillation is continually diminished, the displacement and velocity of the particle are ultimately zero, but the limiting time is finite. *This finite time is called the time of a small oscillation*, and the equilibrium position is said to be stable.

If A_1 is positive, we know by Art. 120 that the value of x contains a real exponential and that the motion is not oscillatory. As the displacement x does not remain small we cannot continue to reject the higher terms of the series (2) as compared with the first. The subsequent motion is not represented by equation (3). The equilibrium position is then unstable.

If $A_1 = 0$, let the first power which does not vanish be the nth. The equation is then ultimately

$$d^2x/dt^2 = A_n x^n \quad\quad (4).$$

This equation has been discussed in Art. 135. If A_n is negative the time of oscillation has been found in gamma functions, with a factor $a^{-\frac{1}{2}(n-1)}$, where a is the semi-arc of oscillation. The limiting time of oscillation is therefore infinite if n is positive and greater than unity. If A_n is positive, the value of x becomes great and the higher powers of x cannot be neglected.

138. *Ex.* 1. If Saturn's ring were rigid and held at rest show that the position of Saturn placed at its centre would be one of unstable equilibrium for displacements in the plane of the ring. If the force between the ring and the planet were repulsion instead of attraction that position of Saturn would be stable and the time of a small oscillation would be $2\pi\sqrt{(2a^3/M)}$, where a is the radius of the ring and M its mass.

Show also that the time measured in seconds is $2\pi\sqrt{(2a^3/nb^2g)}$ where n is the ratio of the mass of the ring to that of the earth, b the radius of the earth, and g is gravity at the surface of the earth, a and b being measured in feet.

To prove this, we let x be the distance of Saturn S from the fixed centre C of the ring. Let P be a point on the ring, $PCS = \theta$, $SP = \rho$. The attraction on S in the direction CS is then seen to be $F = \dfrac{M}{2a\pi}\displaystyle\int \frac{ad\theta}{\rho^2}\,\frac{a\cos\theta - x}{\rho}$. Substituting $\rho = a - x\cos\theta$, expanding in powers of x/a and integrating, we find $F = Mx/2a^3$. This force being positive, the equilibrium is unstable. Reversing its sign the time of a complete oscillation follows by Art. 123. The time in seconds is found by using the equation $E/b^2 = g$, see Art. 134.

Ex. 2. If the ring attract Saturn, show that the central position of the planet is stable for displacements perpendicular to the ring, and that the time of a small oscillation is $2\pi\sqrt{(a^3/M)}$.

Ex. 3. A particle is in equilibrium under the influence of two centres A, B of repulsion each varying as the inverse nth power of the distance. Prove that the position of equilibrium is stable for displacements in the straight line AB and that the time of a small oscillation is $2\pi\sqrt{(ab/n\,(a+b)\,F)}$, where a, b are the distances of the particle from A and B, and F is the accelerating repulsion of either force on the particle in the position of equilibrium.

139. Magnification. A particle, oscillating in a straight line under the action of a centre of force whose acceleration is n^2x, is also acted on by the two accelerating forces $X = E \cos \lambda t$, $Y = F \cos \mu t$. It is required to find the motion.

The equation of motion is

$$d^2x/dt^2 = -n^2x + E \cos \lambda t + F \cos \mu t.$$

The solution of this, by Art. 122, is

$$x = A \cos (nt + B) + E' \cos \lambda t + F' \cos \mu t,$$

where $$E' = \frac{E}{n^2 - \lambda^2}, \quad F' = \frac{F}{n^2 - \mu^2}.$$

If the particle start from rest at a distance a from the origin when $t = 0$, we have $A = a - E' - F'$ and $B = 0$.

The motion of the particle is therefore compounded of three oscillations, one has the period $2\pi/n$ due to the central force, while the other two have the same periods, viz. $2\pi/\lambda$ and $2\pi/\mu$, as the forces X and Y.

This example is important because it shows that *the dynamical effects of oscillatory forces are not necessarily in proportion to their magnitudes, but depend also on their periods.* Thus the ratio of E' to F' is a function of λ and μ as well as of E and F.

If the period of the force X is nearly equal to that of the oscillation caused by the central force, $n^2 - \lambda^2$ is small, while, if no such near equality hold for the force Y, $n^2 - \mu^2$ is not small. It follows that if E and F are nearly equal, E' is much greater than F'. If also E and F were so small that the effect of Y on the motion of the particle were insensible, that of X might still be very great. The general result is, that *of two forces X, Y, that one produces* (cæteris paribus) *the greatest oscillation whose period is most nearly equal to the period of the oscillation due to the central force.*

On the other hand we notice that a near equality between the periods of the forces X and Y has no dynamical significance. The reason is that these forces being explicit functions of *the time* do not modify each other, each producing its own effect. But the central force, viz. $-n^2x$, depends on *the abscissa* of the particle and this is more or less altered by the action of the forces X and Y. The solution shows that the alteration is considerable when the period of either X or Y is nearly equal to that due to the central force alone.

If the period of X is exactly equal to that of the oscillation due to the central force the solution of the differential equation takes a different form. By reference to Art. 122 we see that

$$x = a\cos nt + \frac{Et}{2n}\sin nt + F'\cos\mu t,$$

so that the amplitude of the oscillation becomes very great as t increases.

We may also notice that if λ is very great the terms which contain E' as a factor are very small. It follows that an oscillatory force *whose period is very short* produces very little effect on the motion of the particle.

140. As an example of these effects consider how great an oscillation can be generated in a heavy swing by a series of little pushes and pulls if properly timed. If we push when the swing is receding and pull when it is approaching us, the motion is continually increased and the amplitude of the oscillations becomes greater at each succeeding swing. Such a series of alternations of push and pull is practically an oscillatory force, such as X, whose period is exactly equal to that of the swing. If however the alternations of push and pull follow each other at an interval only nearly equal to that of the period of the swing, a time will come when the effects are reversed. The push will be given when the swing is approaching us and the pull when the swing is receding. Thus, though a great oscillation of the swing is at first produced, that oscillation will be presently destroyed only to be again reproduced and so on continually.

141. *Second approximations.* In determining the small oscillations of a particle in Art. 137, it is explained that the terms containing x^2, &c. are usually neglected. These terms are indeed very small in the differential equation, but we know from Art. 139 that their effects may in certain conditions be so magnified that they become perceptible in the value of x. It is therefore sometimes necessary to proceed to a second or a third approximation before we can find a value of x which represents the actual motion. Some examples of this will be given later on, but the reader will find the theory given at length in the Author's *Rigid Dynamics*, vol. II. chap. VII.

142. *Ex.* A heavy particle P is suspended at rest from a point A by an elastic string whose initial and unstretched length is a. The point A at the time $t=0$ begins to oscillate up and down, so that its displacement (measured downwards) at the time t is $c\sin\lambda t$. Prove that the length of the string at the time t is

$$a + \frac{g}{n^2}(1-\cos nt) - \frac{cn\lambda}{n^2-\lambda^2}\sin nt + \frac{cn^2}{n^2-\lambda^2}\sin\lambda t.$$

Discuss the interpretation of this result (1) when λ is nearly equal to n, and (2) when λ is very great.

Notice that if d^2x/dt^2 is to be the acceleration of P, x must be measured from a point *fixed in space*, say the initial position of A.

Chords of quickest descent.

143. *To find the straight lines of quickest and slowest descent from rest at a given point O to a given curve.* The straight line is supposed to be smooth and the motion to be in vacuo.

The solution of this problem depends on the theorem that the curve which possesses the property, that the times of descent

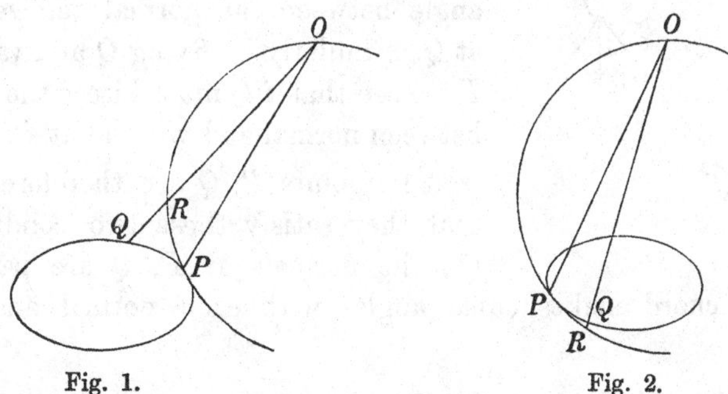

Fig. 1. Fig. 2.

down all radii vectores from rest at O are equal, is a circle having O for the highest or lowest point. See Art. 106.

Describe a circle having its highest point at O and touching the given curve in some point P. There are two cases, according as the circle touches the given curve on one side or the other. These are represented in figures (1) and (2).

If OQ be any chord cutting the circle in R, the time down OP is equal to the time down OR and is therefore less than the time down OQ in fig. (1) and greater than that time in fig. (2). Thus OP is the chord of quickest or slowest descent according to the mode in which the circle of construction touches the given curve.

If C is the centre of the circle, the angles CPO and COP are equal. Since CO is vertical *the chord of quickest or slowest descent from rest at O meets the given curve at a point P such that OP bisects the angle between the vertical and normal at P.*

If the position of a point P on a given curve is required such that the time of descent from P to a given point O is a maximum or minimum, we follow the same construction except that O is to be the *lowest point* of the circle of construction instead of the highest. The result is that the particle must start from a point P such that PO bisects the angle between the vertical and normal at P.

144. *To find the chords of quickest and slowest descent from rest at one given curve to another given curve.*

Let PQ be the required chord. Then since the time down PQ is less than the time down any neighbouring chord drawn from P to the other curve, PQ must bisect the angle between the normal and vertical at Q. Similarly by fixing Q and varying P we see that PQ must bisect the angle between normal and vertical at P.

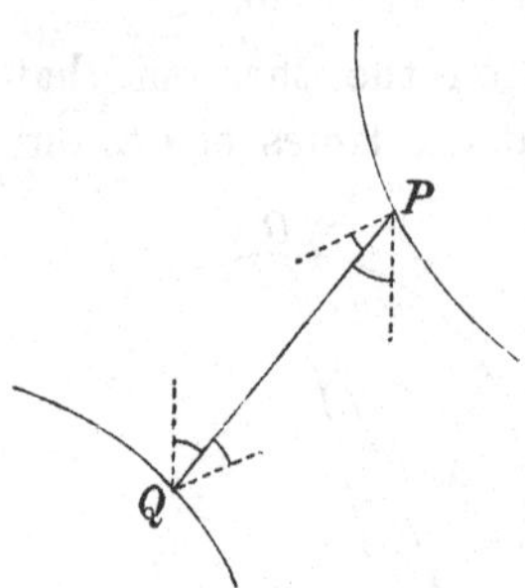

The points P, Q are therefore such that they satisfy these two conditions, (1) the normals at P, Q are parallel, (2) the chord makes equal angles with each normal and the vertical.

145. To find the chord of quickest descent from rest *in a medium whose resistance varies as the velocity* we use the same construction, because the times of descent down all chords of a circle from rest at the highest point are equal. Art. 107.

If the resistance vary as the square of the velocity the curve which possesses the property of equal times for the chords is not a circle; see Art. 110, Ex. 4. The geometrical construction is therefore inapplicable.

146. *If the chords of quickest or slowest descent are rough* we slightly modify the rule. To find the rough chord of quickest descent from O to a given curve we describe a circle to touch the given curve in some point P, but such that the diameter through O makes an angle with the vertical equal to the angle of friction, Art. 105.

The result is that the required chord meets the curve at a point P such that OP makes equal angles with the normal at P and a straight line inclined to the vertical at the angle of friction.

147. *Ex.* 1. A point A and a straight line BC are given in the same vertical plane. Show how to draw (when possible) a straight line from A to BC, so that the time of descent from rest under gravity may be equal to a given time t. When there are two such lines, intersecting BC in P and Q, prove that the radius of the circle described about APQ is $\frac{1}{4}gt^2$.

Ex. 2. Two parabolas are placed in the same vertical plane with their foci coincident, axes vertical and vertices downwards. Prove that the chord of quickest descent from the outer to the inner parabola passes through the focus and makes an angle equal to $\frac{1}{3}\pi$ with the axis.

The normals at the extremities of the chords are parallel and the parabolas are similar. The chord therefore passes through the centre of similitude, i.e. the focus S. If PG be a normal, the second condition of Art. 144 shows that the triangle SPG is equilateral, i.e. each angle is equal to $\frac{1}{3}\pi$.

Ex. 3. Find the smooth chord along which a particle must travel starting from rest at some point on one given curve and ending at another given curve, so that *the velocity acquired* may be a max-min. The force acting on the particle tends to a fixed centre O and varies as some function of the distance from O. The result is that if P be either extremity of the required chord, either the force is zero at P or OP is a normal to the given curve at P.

To prove this, let the central force be $f'(r)$. We then find $v^2=2f(r_1)-2f(r_2)$ where r_1, r_2 are the distances of the extremities P, Q from O. Fixing Q let us vary P along the arc (as in Art. 144), then $dv^2/ds=0$. Hence $f'(r_1)\,dr_1/ds=0$, i.e. the component of the central force along the tangent to the curve is zero.

Ex. 4. Prove that the smooth chord of quickest descent from rest at one given circle to another given circle when produced passes through the highest point of the first circle and the lowest point of the other.

Prove also that the smooth chord of longest descent between the same two circles is either a horizontal straight line or (when produced if necessary) passes through the lowest point of the first circle and the highest of the other.

Ex. 5. Prove that the locus of the points from which the times of descent to three given points in space are the same is a rectangular hyperbola. Prove also that the locus of the points from which the times of shortest descent to three equal spheres, given in position in space, are the same is a rectangular hyperbola.

[Math. Tripos, 1885.]

Ex. 6. Prove that the rough chord of quickest descent from rest at some point on a given straight line to some point on a given circle (not intersecting), (1) when produced passes through a point B on the circle such that a particle placed at B is in equilibrium with limiting friction, (2) bisects the angle between the diameter through B and the perpendicular from B on the given straight line.

Ex. 7. Heavy particles slide down chords of a circle whose plane is vertical starting from rest at the highest point A in a medium resisting as the square of the velocity. Prove that the chords of slowest and quickest descent are the vertical diameter and a chord making an infinitely small angle with the horizon.

These results may be deduced from the formulæ given in Art. 115, but the following line of argument is worth noticing. Let $AB=2a$ be the vertical diameter, AQ a chord making an angle θ with AB, then $AQ=2a\cos\theta$. We have to find the time of describing $2a\cos\theta$ from rest with an acceleration $g\cos\theta-\kappa(dx/dt)^2$. Writing $x=\xi\cos\theta$, this time is equal to that of describing $2a$ from rest with an acceleration $g-\kappa\cos\theta\,(d\xi/dt)^2$. In vacuo, where $\kappa=0$, this is independent of θ and therefore all chords are described in the same time. Also this time is increased by the presence of the resisting medium because the acceleration is thereby diminished. This increase of time is zero when $\cos\theta=0$, i.e. $\theta=\frac{1}{2}\pi$, becomes greater as $\cos\theta$ is greater and is greatest when $\cos\theta=1$, i.e. $\theta=0$. The time of descent therefore increases as θ passes from $\frac{1}{2}\pi$ to 0.

Infinitesimal Impulses.

148. When the effect of an impulse acting on a body is required, we commonly disregard all finite forces which act simultaneously with it. The duration T of the impulse being infinitesimal, Art. 80, a finite force F can generate only a momentum FT which vanishes in the limit when compared with the finite momentum communicated by the impulse. If, however, the impulse is itself very small these may be comparable in magnitude and it will then be necessary to take account of both forces in the same equation of motion*.

This generally happens when the mass of the body changes during the motion.

149. Let a body of mass M whose resolved velocity parallel to x is v be acted on by a finite force X. *Let this body lose a small portion $m=-dM$ of its mass in each element of time dt.* It is required to find the motion.

The momentum at the time t is Mv, and the gain in the time dt is $d(Mv)$. In this time the force increases the linear momentum by $X dt$, while the momentum lost by diminution of mass is mv. Hence

$$d(Mv)=X\,dt+v\,dM, \quad \therefore\ M\frac{dv}{dt}=X \text{..............(1)}.$$

Here there are no impacts; the particles merely separate with their common velocity without mutual action.

If $X=mg$, the equation becomes $dv/dt=g$, and each portion moves parallel to x with an acceleration g.

Next, let us suppose that the body gains a mass $m=dM$ in the time dt and let the resolved velocity of this increment before it is attached to M be v'. The total gain of momentum is now, $X\,dt$ due to the force and mv' due to the impact produced by the sudden junction of the masses M and m with different velocities.

* Problems on infinitesimal impulses were solved in the lecture room of the late Mr Hopkins as long ago as 1850. A problem of this kind was set in the Smith's Prize examination in 1853 by Prof. Challis, and a solution given in Tait and Steele's *Dynamics*. Another was proposed in 1869 by Prof. Cayley who published the solution in the *Mathematical Messenger* in 1871. Two problems were also solved in the *Quarterly Journal* in 1870 by Dr Besant.

The equation of motion is therefore

$$d(Mv) = X\,dt + v'dM \quad\text{..................(2).}$$

If $v' = v$ this reduces to the former result.

150. *Ex.* 1. A uniform chain of mass M_1 and length l, is coiled up on a small horizontal ledge at the top of a plane, inclined at an angle a to the horizon, and has masses M_2, M_3 fastened to its two ends. If M_2 is gently pushed off the ledge, prove that the velocity of M_3 just before it leaves the ledge is v', and just after is v'', where

$$v'^2 = \frac{2gl \sin a\,\{\frac{1}{3}M_1^2 + M_1M_2 + M_2^2\}}{(M_1+M_2+M_3)^2}, \qquad v''^2 = v'^2\left(\frac{M_1+M_2+M_3\cos a}{M_1+M_2+M_3}\right)^2 + (v'\sin a)^2.$$

[Coll. Ex. 1897.]

Let x be the distance of the lowest point of the chain from the edge, m the mass of a unit of length of the chain. The momentum at the time t is $(M_2+mx)\,v$. In the time dt a mass mdx without velocity is taken from the ledge and added to the moving length. Also gravity adds a momentum $(M_2+mx)\,g'dt$, where $g' = g\sin a$.

$$\therefore\ d\{(M_2+mx)\,v\} = (M_2+mx)\,g'dt \quad\text{.................... (1).}$$

To make the formation of this equation more clear, let the coil be at a short distance a from the edge, and let the edge be rounded off in a circular arc of radius b. We here only require the limiting case when both a and b are zero. As each element passes over the edge, the velocity is at first horizontal and the change of direction is effected by the normal pressures at the rounded edge. The momentum generated by the weight of the chain on the rounded edge is ultimately zero since the radius b can be made as small as we please.

To integrate (1) we multiply both sides by $(M_2+mx)\,v$, then remembering that $v = dx/dt$,

$$(M_2+mx)^2\,v^2 = \{(M_2+mx)^3 + C\}\,\frac{2g'}{3m} \quad\text{....................(2).}$$

Since x and v vanish together $C = -M_2^3$. When all the chain has left the ledge $x = l$, and

$$(M_2+ml)^2\,v^2 = \left\{\frac{m^2l^2}{3} + mlM_2 + M_2^2\right\}\,2g'l \quad\text{.................. (3).}$$

At this instant there is an impact, the tension acts on M_3 horizontally, hence if v' be the velocity of M_3 and the chain just before M_3 reaches the edge

$$(M_2+ml+M_3)\,v' = (M_2+ml)\,v \quad\text{........................ (4).}$$

The mass M_3 immediately reaches the edge with a horizontal velocity v', while the chain is moving along the plane with an equal velocity. There is therefore another impact, the component of momentum $M_3v'\sin a$ perpendicular to the chain remains unchanged, while the component $M_3v'\cos a$ is joined to that of the chain. If u be the common velocity of M_3 and the chain parallel to the plane just after M_3 has left the ledge,

$$\left.\begin{array}{c}(M_3+M_2+ml)\,u = (M_2+ml)\,v' + M_3v'\cos a\\ v''^2 = u^2 + (v'\sin a)^2\end{array}\right\} \quad\text{............... (5).}$$

The equations (4) and (5) give the required results.

Ex. 2. A chain of length l is coiled at the edge of a table. One end is fastened to a particle whose mass is the same as that of the whole chain. The other end is put over the edge. Prove that immediately after leaving the table the particle is moving with velocity $\frac{1}{2}\sqrt{(\frac{5}{3}gl)}$. [Coll. Ex. 1896.]

Ex. 3. A mass M is attached to one end of a chain whose mass per unit of length is m. The whole is placed with the chain coiled up on a smooth table and M is projected horizontally with a velocity V. Prove that when a length x of the chain has become straight, the velocity of M is $MV/(M+mx)$.

[Cayley, *Math. Messenger*, 1871.]

Ex. 4. A uniform chain of length l and mass ml is coiled on the floor, and a mass mc is attached to one end and projected vertically upwards with velocity $\sqrt{2gh}$. Prove that, according as the chain does or does not completely leave the floor, the velocity of the mass on finally reaching the floor again is the velocity due to a fall through a height $\frac{1}{3}\{2l-c+a^3/(l+c)^2\}$ or $a-c$; where $a^3=c^2(c+3h)$.

[Coll. Ex. 1896.]

When descending each portion moves with a uniform acceleration g, as explained in Art. 149.

Ex. 5. A chain brake is used at railway depôts for arresting runaway trucks, consisting of a coil of chain between the metals, having a hook at one end so placed as to catch on to the axle of the truck. If the mass of the truck be equal to that of a length l of the chain, less than the whole length, then the truck running on the level with velocity V will be stopped when it has dragged a length x of chain over the rough ground, where $V^2/\mu g=\frac{1}{3}(2x+3l)x^2/l^2$.

[Coll. Ex. 1897.]

Ex. 6. A weight W is connected with a coil of heavy chain by means of a fine weightless thread passing over a smooth peg above the coil which rests on a table; if W be allowed to fall a height h whereupon the thread becomes tight, find the motion, and show that if $w=3W$ then in setting the coil in motion energy to the amount $hWw/(W+w)$ is dissipated. [Coll. Ex. 1887.]

Ex. 7. Rain is falling vertically with a uniform velocity of 20 feet per second at the rate of two inches depth per day on a cart with a cylindrical cover of semi-circular section and horizontal axis. Prove that, if the cover of the cart is 10 feet long and 6 feet in diameter, the resultant pressure on it due to the impact of the rain is about the weight of one-twelfth of a cubic inch of water. [Coll. Ex. 1895.]

Theory of Dimensions.

151. Many theorems follow at once from some simple considerations on the dimensions of the quantities with which we are dealing. Each side of an equation must be of the same dimensions in space, for we could not have, for instance, an area equal to a length. Again one side of an equation could not be the square of a time and the other side a cube, and so on.

In dynamics we are concerned with the four quantities space, time, mass, and force; but the dimensions of these quantities are so related that force is mass . space/(time)^2. Taking into account this relation we have the general principle that both sides of every equation must be of the same dimensions in regard to (1) space, (2) time, (3) mass.

152. As an example let us apply this principle to the following problem already considered in Art. 136.

A particle starts from rest at a distance a from a centre of force O whose accelerating force at a distance x is μx^n. To find the time T of arriving at the centre of force.

It is clear that T is some function of a and μ, n being merely a number without dimensions. Expanding T in powers of a and μ we have

$$T = \Sigma A a^p \mu^q \qquad (1).$$

Now the accelerating force μx^n is of the dimensions space/(time)^2, hence μ is $1-n$ dimensions in space and -2 in time. We also notice that a is one dimension in space and none in time, while T is one in time and none in space.

Considering the equation (1) and counting the dimensions of each side first in space and secondly in time, we have

$$0 = p + (1-n)\,q, \qquad 1 = -2q \qquad (2).$$

Hence $q = -\frac{1}{2}$ and $p = \frac{1}{2}(1-n)$. As these equations give only one set of values to p, q, the equation (1) contains only one term, viz.

$$T = A a^{\frac{1}{2}(1-n)} \mu^{-\frac{1}{2}} \qquad (3).$$

It follows that the time of arriving at the centre of force O varies as the $\frac{1}{2}(1-n)$th power of the initial distance. If the central force vary as the distance, $n=1$ and the time of arrival at O is the same for all initial distances; a theorem which has been proved in Art. 136 by integrating the equation of motion. If the central force vary according to the Newtonian law, $n=-2$ and the square of the time varies as the cube of the initial distance, a result in accordance with one of Kepler's laws.

The symbol A represents a number and as it has no dimensions its magnitude cannot be deduced from the theory of dimensions.

153. *Ex.* 1. A particle moves with an acceleration g, prove that the velocity acquired in describing a space s varies as $\sqrt{(gs)}$, and that the time varies as $\sqrt{(s/g)}$.

Ex. 2. A particle starts from rest at a given distance from a centre of force whose attraction varies as the distance and moves in a medium whose resistance varies as the velocity. Prove that the time of arriving at the centre of force is independent of the initial distance. See Art. 126.

Ex. 3. A particle P moves from rest under the action of a constant accelerating force f and a centre of force whose attraction is μ times the distance, both tending to the same point O and the initial distance $OP = a$. Prove that

$$t\sqrt{\mu} = \phi\,(a\mu/f),$$

where t is the time of arrival at O.

CHAPTER III.

MOTION OF PROJECTILES.

Parabolic Motion.

154. General principle. The particle moves under the action of a force which, being fixed in direction and magnitude, is independent of the position of the particle. It follows that all the circumstances of the motion parallel to any fixed direction are independent of those of the motion parallel to any other direction. These circumstances may therefore be deduced from the formulæ for rectilinear motion by taking account solely of the resolved initial velocity and the resolved force of gravity.

155. Cartesian axes. Let the particle be projected from a point O with an initial velocity V in a direction making an angle α with the horizon. Let v be the velocity at any point P of the path; v_x, v_y its horizontal and vertical components.

Consider the horizontal motion. Since the component of gravity in this direction is zero, *the horizontal velocity is constant throughout the motion and is equal to* $V \cos \alpha$. We therefore have

$$x = V \cos \alpha t, \quad v_x = V \cos \alpha \;\ldots\ldots\ldots\ldots\ldots\ldots(1).$$

This gives an obvious and useful rule to find the time of describing any arc of the trajectory, viz. *the time of transit is equal to the horizontal space divided by the horizontal velocity.*

Consider next the vertical motion. Since the component of gravity is g we infer from the formulæ of rectilinear motion (Art. 25) that

$$y = V \sin \alpha t - \tfrac{1}{2} g t^2, \quad v_y^2 = V^2 \sin^2 \alpha - 2gy \;\ldots\ldots\ldots(2).$$

The Cartesian equation of the path is found by eliminating t between (1) and (2); we have

$$y = x \tan \alpha - \tfrac{1}{2} g x^2 / V^2 \cos^2 \alpha \quad \text{..................(3).}$$

This is the well-known equation of a parabola.

To find the greatest altitude of the particle. We consider only the descending motion; the particle starts downwards with a zero vertical velocity and arrives at the level of the original point of projection with a vertical velocity which, by the theory of rectilinear motion, is equal to that with which it was projected upwards. If h is the greatest altitude we have $V^2 \sin^2 \alpha = 2gh$.

To find the time of flight. We again consider the vertical descending motion, disregarding the horizontal motion. If T be the time of ascent and descent, we have $V \sin \alpha = \tfrac{1}{2} gT$.

To find the range on a horizontal plane. We consider the horizontal motion; the constant horizontal velocity is $V \cos \alpha$, and the time of flight has just been found. The range is therefore $V^2 \sin 2\alpha / g$. The range is greatest for a given velocity when the direction of projection makes an angle of 45° with the horizon, and continually decreases as the angle increases to a right angle or decreases to zero.

156. When the motion with regard to an inclined plane passing through the point of projection is required, it is useful to take the axis of x along the line of greatest slope and the axis of y perpendicular to the inclined plane.

If the direction of projection is not in the plane of xy, let V and W be the components of the velocity in and perpendicular to that plane. The motion perpendicular to the plane of xy is uniform and $z = Wt$.

Turning our attention to the motion in the plane of xy, let γ be the angle the direction of the velocity V makes with Ox and β the inclination of the plane to the horizon. The initial component velocities being $V \cos \gamma$ and $V \sin \gamma$, the formulæ of rectilinear motion (Art. 25) give

$$\left.\begin{aligned} x &= V \cos \gamma t - \tfrac{1}{2} g \sin \beta t^2 \\ y &= V \sin \gamma t - \tfrac{1}{2} g \cos \beta t^2 \end{aligned}\right\}, \quad \left.\begin{aligned} v_x^2 &= (V \cos \gamma)^2 - 2g \sin \beta x \\ v_y^2 &= (V \sin \gamma)^2 - 2g \cos \beta y \end{aligned}\right\} \text{...(4).}$$

To find the time of flight T before reaching the plane, we consider the motion perpendicular to the plane. The descending motion gives $V \sin \gamma = g \cos \beta \,.\, T/2$. It also follows that the time of describing the arc from O to the point where the tangent is parallel to Ox is $T/2$. In the same way by considering the motion parallel to the plane we see that the time from O to the point where the tangent is parallel to Oy is $V\cos\gamma/g\sin\beta$.

To find the range r on the inclined plane, we use the expression for x. We easily find $r = 2V^2\sin\gamma\cos(\gamma+\beta)\,.\,\sec^2\beta/g$.

157. Oblique axes. Let the direction of motion of the particle at any point P of the path be PT and let the velocity be V. The particle being acted on by gravity in the direction PN, let Q be its position after a time t.

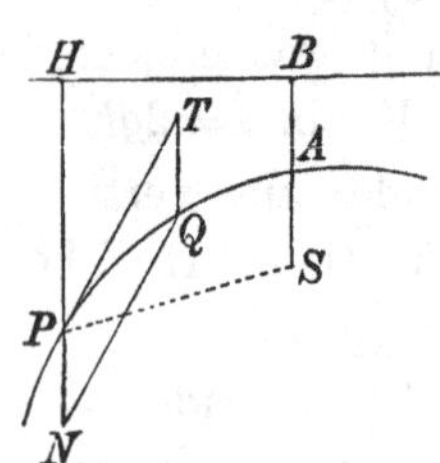

Consider separately the motion in the two directions PT and PN. The oblique components of V in these directions are V and zero, while those of gravity are zero and g. We therefore have $PT = Vt$, and $TQ = \frac{1}{2}gt^2$ ....(5). Draw QN parallel to TP and let $PN = \eta$, $QN = \xi$. The equation of the path is therefore

$$\xi^2 = \frac{2V^2}{g}\eta \quad \text{..........................(6).}$$

This is the equation of a parabola referred to any diameter PN and its oblique ordinates QN. If S be the focus, this equation must be the same as $\xi^2 = 4\,.\,SP\cdot\eta$. We deduce the following useful rule. *The velocity at any point P of the path is that due to the distance of P from either the focus or directrix.*

Since the velocity at the highest point of the path is equal to the horizontal velocity, it follows that *one quarter of the latus rectum, i.e. AS, is equal to* $V^2\cos^2\alpha/2g$. See Art. 155.

We have also another formula to find the time of transit along any arc PQ. *Let the vertical at either end, say Q, intersect the tangent at the other end in T, then the time of describing the arc PQ is the same as that of describing QT from rest under the action of gravity. It is also the same as that of describing PT with a uniform velocity equal to that at P.*

158. *Ex.* 1. If three heavy particles be projected simultaneously from the same point in any directions with any velocities, prove that the plane passing through them will always remain parallel to itself. [Math. T. 1847.]

If gravity did not act, the plane of the particles would be always parallel to a fixed plane. When gravity acts each particle is pulled through the same vertical space in the same time, hence the theorem remains true.

Ex. 2. Two tangents PR, QR are drawn to a parabolic trajectory, prove (1) that the velocities at P and Q are proportional to the lengths of those tangents, and (2) that the vertical through R divides the arc PQ into two parts which are described in equal times.

Draw QT vertically to intersect the tangent PR in T. Then by the triangle of velocities, the sides RT, RQ, TQ represent in direction and magnitude the velocity at P, that at Q, and that added on by gravity during the time of transit. Since the diameter through R bisects the chord PQ, the results given above follow easily.

Ex. 3. Two balls A, B equal in all respects are on the same horizontal line. The ball A is projected towards B with velocity v, while at the same instant B is let fall. Prove that the balls will impinge and that after impact, the coefficient of restitution being unity, A will fall vertically and B will describe a parabola of latus rectum $2v^2/g$. [Coll. Ex. 1895.]

The balls will impinge because the straight line joining their centres moves parallel to itself. At impact they exchange their horizontal velocities.

Ex. 4. If v, v', v'' are the velocities at three points P, Q, R of the path of a projectile, where the inclinations to the horizon are α, $\alpha-\beta$, $\alpha-2\beta$, and if t, t' be the times of describing PQ, QR respectively, prove that

$$v''t=vt', \quad \frac{1}{v}+\frac{1}{v''}=\frac{2\cos\beta}{v'}. \qquad \text{[Math. T. 1847.]}$$

Resolve along and perpendicular to the middle tangent.

Ex. 5. Three heavy particles P, Q, R are projected at equal intervals of time from the same point O to describe the same parabola. Prove that the locus of the intersection of the tangents at P, R is a parabola. Prove also (1) that at any time t after the projection of Q, the tangent at Q is parallel to PR, (2) that each of these lines is parallel to the straight line joining O to the position of Q at the time $2t$.

159. *To project a particle from a given point P with a given velocity V so that it shall pass through another given point Q.*

The velocity at P being known the common directrix HK of all parabolic paths from P to Q is constructed by drawing a horizontal at an altitude $V^2/2g$ above P. With centres P, Q and radii PH, QK we describe two circles intersecting in S and S'. Then S, S' are the foci of the parabolic trajectories which could be described from P to Q. There are therefore two parabolic paths.

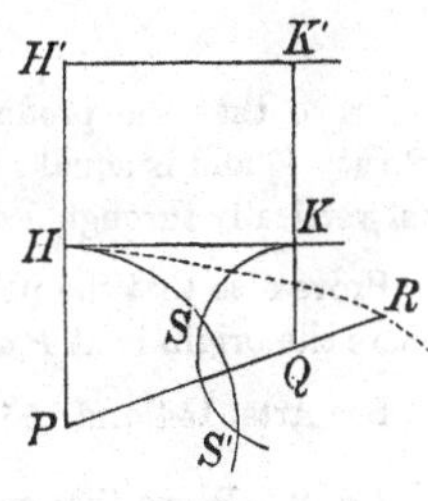

The two foci are at equal distances from the chord PQ, one lying on each side. The two directions of projection may be found by bisecting the angles HPS and HPS'. *If γ_1, γ_2 are the angles these directions of projection make with the chord PQ, and*

β the angle PQ makes with the horizon, it easily follows that $\gamma_1 + \gamma_2 = \frac{1}{2}\pi - \beta$.

We notice that the three sides of the triangle PSQ are known, viz. $PS = V^2/2g$, if y be the altitude of Q above P, $QS = PS - y$, and PQ is the known distance of Q from P.

It is clear that when PQ is greater than the sum of the radii PH, QK, the two trajectories are imaginary. The greatest possible distance of Q from P in any given direction PQ is found by making the foci S, S' coincide and lie on PQ. In this case $PH + QK = PQ$. Drawing a horizontal line $H'K'$ above HK so that $HH' = PH$, it immediately follows that $QK' = QP$. The locus of Q is therefore a parabola whose focus is P and directrix $H'K'$. This new parabola therefore touches HK at its vertex H. It is represented in the figure by the dotted line. *Unless the point Q lie within the space enclosed by this parabola, it is impossible to project a particle from P with the given velocity V, so that it shall pass through Q.*

If the particle is to be projected from P with the least velocity which will enable it to reach Q, the direction of projection must bisect the angle HPQ and $V^2 = g(r + y)$, where r is the distance PQ.

160. *Ex.* 1. A particle is projected from a point P with velocity V, so as to pass through a point Q whose coordinates referred to P as origin are x, y, the axis of y being vertical. Prove that the directions of projection are given by the quadratic

$$\tan^2 \alpha - \frac{2V^2}{gx} \tan \alpha + 1 + \frac{2V^2 y}{gx^2} = 0,$$

and that the two times of transit are the positive roots of

$$g^2 t^4 - 4(V^2 - gy)\, t^2 + 4r^2 = 0.$$

Prove that the product of the times of transit is independent of the initial velocity V and is equal to the square of the time occupied by a particle falling from rest vertically through a distance equal to PQ.

Prove also that the polar equation of the bounding parabola is $V^2/gr = 1 + \cos\theta$, where the origin is at P and θ is the angle r makes with the vertical.

See Arts. 154 and 155.

Ex. 2. Prove that every parabolic trajectory meets the bounding parabola in a point whose abscissa is $x = 2h \cot \alpha$, and whose depth below the directrix of the trajectory is $h \cot^2 \alpha$, where h is the height of that directrix above the point of projection.

If they meet, the curves must touch for otherwise it would be possible to find a trajectory which would pass through a point beyond the boundary.

Ex. 3. The point P being fixed and Q having any position, the tangents at P, Q to one parabolic path from P to Q meet in T, those to the other in T', the velocity at P being given. Prove that the locus of the middle point of TT' is the directrix of either parabola.

Prove also that for either parabolic path, the velocities at P, Q are as PT to TQ, and for the two paths the times of transit from P to Q are as PT to PT'.

Ex. 4. A fort of vertical height k stands on a plane hill-side which makes an angle α with the horizon. Prove that a gun which can fire with muzzle velocity V from the top of the fort commands a district whose shape is an ellipse of eccentricity $\sin\alpha$, and whose area is $\pi\sec\alpha\, V^2(V^2\sec^2\alpha+2kg)/g^2$. [Coll. Ex. 1896.]

The paraboloid whose focus is the top of the fort and whose directrix plane is at an altitude V^2/g is the boundary of all places which the shot can reach, Art. 159. The paraboloid cuts the plane hill-side in an ellipse whose projection on a horizontal plane is a circle. The rest follows easily.

Ex. 5. At a horizontal distance a from a gun there is a wall of height h which is greater than $a-ga^2/v^2$; prove that if the shot be fired off with a velocity v in a vertical plane at right angles to that of the wall, there will be a distance on the other side of the wall commanded by the gun equal to $\dfrac{2ha}{g\,(a^2+h^2)}(v^4-a^2g^2-2hv^2a)^{\frac{1}{2}}$, provided this expression is real. [Coll. Ex. 1893.]

Ex. 6. A particle is projected with velocity V along a straight frictionless tube of length l, inclined at an angle α to the horizontal, and after leaving the tube it describes a parabolic trajectory: prove that its range on the horizontal plane through the point of projection is $l\cos\alpha+\dfrac{V'^2}{g}\cos\alpha\sin\alpha\left\{1+\left(1+\dfrac{2gl}{V'^2\sin\alpha}\right)^{\frac{1}{2}}\right\}$, where $V'^2=V^2-2gl\sin\alpha$. [Coll. Ex. 1893.]

Ex. 7. Two smooth planes are at right angles with their edge of intersection horizontal and are equally inclined to the horizon. Prove that a perfectly elastic particle projected horizontally in a direction perpendicular to the common edge from a point vertically above it will return to its original position after two rebounds. [Coll. Ex. 1896.]

Ex. 8. Two parabolas have their axes vertical and vertices downwards and the focus of each curve is on the other. A particle, whose coefficient of restitution is unity, is projected so as to rebound from the curves at each focus in succession; prove that it will after the second rebound pass through its point of projection and follow its original path again. [Coll. Ex. 1897.]

Ex. 9. Two particles are projected from the same point at the same instant with velocities v, v', and in directions α, α'. Prove that the time which elapses between their transits through the other point which is common to both their paths is $\dfrac{2}{g}\,\dfrac{vv'\sin(\alpha-\alpha')}{v\cos\alpha+v'\cos\alpha'}$. [Math. T. 1841.]

Ex. 10. A man travelling round a circle of radius a at speed v throws a ball from his hand at height h above the ground with a relative velocity V so that it alights at the centre of the circle. Prove that the least possible value of V is given by $V^2=v^2+g\{\sqrt{(a^2+h^2)}-h\}$. [Coll. Ex. 1896.]

If the man were stationary, the least value of V^2 is given in Art. 159. To find the relative velocity we add to this $(-v)^2$.

161. *Ex.* 1. A particle is projected from a point A with a velocity V in a direction making an angle α with the horizon. After rebounding from a vertical wall, elasticity e, it hits the ground, elasticity e'. Find the condition that after the second rebound the particle may pass through A.

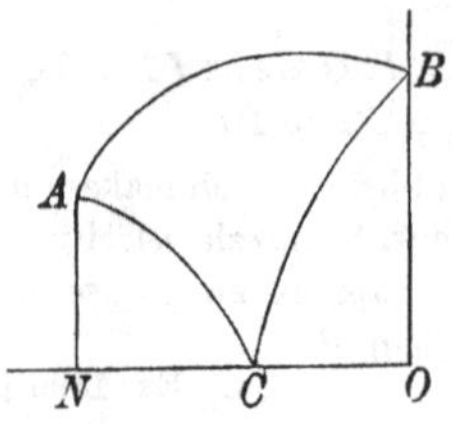

Problems of this kind are solved by considering the motion in two directions separately and equating some element (usually the time) common to both motions. Consider first the horizontal motion; the blow at C is vertical and does not affect the horizontal motion, but the blow at B must be taken account of. Let $ON=h$, and let t_1, t_2 be the times of transit along the arc AB and the broken arc BCA. Then $h=V\cos\alpha t_1$, and the horizontal velocity of the rebound at B being $eV\cos\alpha$, we have also $h=eV\cos\alpha t_2$. The whole time is $\dfrac{h}{V\cos\alpha}\left(1+\dfrac{1}{e}\right)$.

Consider next the vertical motion, the blow at B may now be neglected while that at C has to be allowed for. Let t_3, t_4 be the times of transit along ABC and CA. If $k=AN$ we have

$$-k=V\sin\alpha t_3-\tfrac{1}{2}gt_3^2.$$

One root of this quadratic is negative and the other is positive. The former indicates the time before leaving A at which the particle might have passed the level of the ground and is here inadmissible. We take the positive root. If V' be the vertical velocity of arrival at C taken positively,

$$V'^2=V^2\sin^2\alpha+2gk, \qquad k=e'V't_4-\tfrac{1}{2}gt_4^2.$$

Both the values of t_4 thus found are positive, and give the times of transit from C to A according as the particle passes through A on the up or on the down journey. Taking both these values we see that the required condition is found by equating t_1+t_2 to either of the values of t_3+t_4.

Ex. 2. A ball is projected from a point A on the floor of a room, so as to rebound from the wall (elasticity e) and hit a given point B on the floor. Let the intersection of the floor and wall be the axis of y and let A be on the axis of x. If u, v, w be the components of velocity of projection and x, y the given coordinates of B, prove that $euy=eva+vx$, and $2vw=gy$.

Ex. 3. A particle is projected from a given point O on an inclined plane in a direction making an angle γ with the plane, the inclination of the plane being β. Investigate the condition that the particle passes through O at the nth impact.

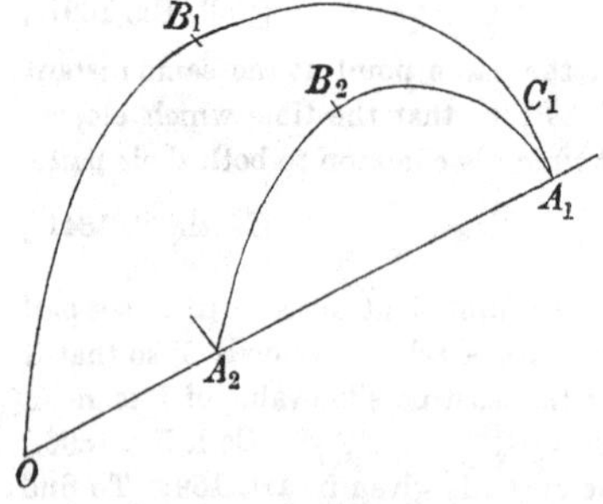

We consider the motions parallel and perpendicular to the plane separately. The motion parallel to the plane is not affected by the impacts. If T represent the whole time of transit from O to O again, we have $V\cos\gamma=\frac{1}{2}g\sin\beta T$.

The motion perpendicular to the plane is affected by each impact. The particle starts with a velocity $V\sin\gamma$, hits the plane at A_1 with the same normal velocity after a time T_1, where $V \sin\gamma = \frac{1}{2} g \cos\beta T_1$. The particle rebounds with a perpendicular velocity $eV \sin\gamma$ and the time of transit from A_1 to A_2 is found as before. The whole time of transit is therefore

$$T_1 + T_2 + \&c. = \{2V \sin\gamma / g \cos\beta\} (1 + e + \ldots + e^{n-1}).$$

Equating the two complete times, we have the condition

$$\cot\gamma \,.\, \cot\beta = (1 - e^n)/(1 - e),$$

which we notice is independent of the velocity of projection.

Let B_1, B_2, &c. be the points at which the tangents to the path are parallel to the inclined plane. The time of transit from O to B_1 is obviously equal to $\frac{1}{2} T_1$, while that from B_1 to B_2 is $\frac{1}{2}(T_1 + T_2)$, and so on. If C_1 be the point at which the tangent is perpendicular to the plane, the time from O to C_1 is clearly equal to $\frac{1}{2} T$.

Ex. 4. A ball whose elasticity is e is projected with a velocity V and rebounds from an inclined plane which passes through the point of projection. If R_1, R_2, R_3 be any three consecutive ranges on the inclined plane, prove that

$$R_3 - (e + e^2) R_2 + e^3 R_1 = 0.$$ [Math. T. 1842.]

Ex. 5. At two points A, B of a parabolic path the directions of motion are at right angles. If D be the distance AB, θ the inclination to the horizon, V the velocity at A or B, prove that $V^2 = gD(1 \pm \sin\theta)$.

Ex. 6. A particle is projected from a point on a rough horizontal plane with a velocity equal to that which would be acquired in falling freely through a height h, and in a direction making an angle a with the plane. The particle is inelastic and the coefficients of both the frictions are taken equal to unity, prove that the range from the point of projection to where the particle comes to rest is equal to

$$h(1 + \sin 2a).$$ [Coll. Ex. 1897.]

The particle describes a parabola with a range $2h \sin 2a$. On arriving at the plane, there is an impulsive friction which reduces the horizontal velocity from $v \cos a$ to $v' = v \cos a - v \sin a$. After describing a space s', when $v'^2 = 2gs'$, the particle is reduced to rest by the finite friction. The whole range is $2h \sin 2a + s'$.

Ex. 7. A perfectly elastic particle slides down a length l of a smooth fixed inclined plane, and strikes a smooth rigid horizontal plane passing through the foot of the inclined plane. Prove that the maximum range of the ensuing parabolic path, as the inclination of the inclined plane is varied, is $8l/3\sqrt{3}$. [Coll. Ex. 1896.]

Ex. 8. A smooth inclined plane of mass M, inclined to the horizon at an angle a, is free to move parallel to a vertical plane through the line of greatest slope. A particle, mass m, is projected from a point in the lowest edge, up the face of the plane with a velocity V making an angle β with the line of greatest slope. Prove that the range of the particle on the plane is $\dfrac{V^2 \sin 2\beta}{g \sin a} \cdot \dfrac{M + m \sin^2 a}{M + m}$.

[Coll. Ex. 1897.]

Ex. 9. Two inclined planes intersect in a horizontal line, their inclinations to the horizon being a and β; if a particle be projected at right angles to the former from a point in it so as to strike the other at right angles, the velocity of projection is

$$\sin\beta \,[2ga / \{\sin a - \sin\beta \cos(a + \beta)\}]^{\frac{1}{2}},$$

a being the distance of the point of projection from the intersection of the planes.

Ex. 10. A heavy particle descends the outside of a circular arc whose plane is vertical. Prove that when it leaves the circle at some point Q to describe a parabola the circle is the circle of curvature at Q of the parabola.

Thence show that the chord of intersection QR of the circle and parabola and the tangent at Q make equal angles with the vertical. Prove also that the axis of the parabola divides the chord QR in the ratio 3 : 1.

The first part follows from Art. 36. Since the pressure is zero at Q, v^2/ρ, and therefore ρ, must be the same for the circle and the parabola. The rest follows from conics.

Ex. 11. A particle projected horizontally from the lowest point A of a circle whose plane is vertical leaves the circle at C and after describing a portion of a parabola intersects the circle at D. If B is the highest point of the circle prove that the arc BD is three times the arc BC. [Despeyrous, *Cours de Méc.*]

Ex. 12. A particle is projected so as to enter in the direction of its length a smooth straight tube of small bore fixed at an angle 45° to the horizon and to pass out at the other end of the tube; prove that the latera recta of its path before entering and after leaving the tube differ by $\sqrt{2}$ times the length of the tube.

[Math. Tripos, 1887.]

Ex. 13. A man standing on the edge of a cliff throws a stone with given velocity u at a given inclination in a plane perpendicular to the edge. After an interval τ he throws from the same spot another stone, with given velocity v at an angle $\frac{1}{2}\pi+\theta$ with the line of discharge of the first stone and in the same plane. Find τ so that the stones may strike each other; and prove that the maximum value of τ for different values of θ is $2v^2/gw$, and occurs when $\sin\theta=v/u$, w being v's vertical component. [Math. Tripos, 1886.]

Ex. 14. A particle is projected from the highest point of a sphere of radius c so as to clear the sphere. Prove that the velocity of projection cannot be less than $\sqrt{(\frac{1}{2}gc)}$. [Math. Tripos, 1893.]

Resistance varies as the velocity.

162. *To determine the motion of a heavy particle when the resistance of the medium varies as the velocity.*

Let the particle be projected from any point O with a velocity V in a direction inclined at an angle α to the horizon. The equations of motion are

$$\frac{d^2x}{dt^2}=-\kappa\frac{dx}{dt},\qquad \frac{d^2y}{dt^2}=-g-\kappa\frac{dy}{dt},$$

$$\therefore\ dx/dt+\kappa x=V\cos\alpha,\quad dy/dt+\kappa y=-gt+V\sin\alpha.$$

Both these equations are of the linear form, multiplying by $e^{\kappa t}$ and integrating, we find

$$\left.\begin{aligned}\kappa x&=V\cos\alpha\,(1-e^{-\kappa t})\\ \kappa y&=-gt+(V\sin\alpha+L)(1-e^{-\kappa t})\end{aligned}\right\}\ \ldots\ldots\ldots\ldots(1),$$

where $\kappa L = g$, so that L is the limiting velocity, Art. 111. The horizontal and vertical velocities at any time t are

$$dx/dt = V\cos\alpha e^{-\kappa t}, \quad dy/dt = -L + (V\sin\alpha + L)e^{-\kappa t} \ldots (2).$$

163. From these equations we deduce the general characteristics of the motion. We notice that when t is infinite $\kappa x = V\cos\alpha$. *There is therefore a vertical asymptote at a horizontal distance* $OH = V\cos\alpha/\kappa$ *from the origin.* Let the tangent at O intersect the asymptote in T_0, then $OT_0 = V/\kappa$ and $V = \kappa . OT_0$. Since any point P may be taken as the origin, it follows that *the velocity at any point P is proportional to the length PT of the tangent at P cut off by the vertical asymptote.*

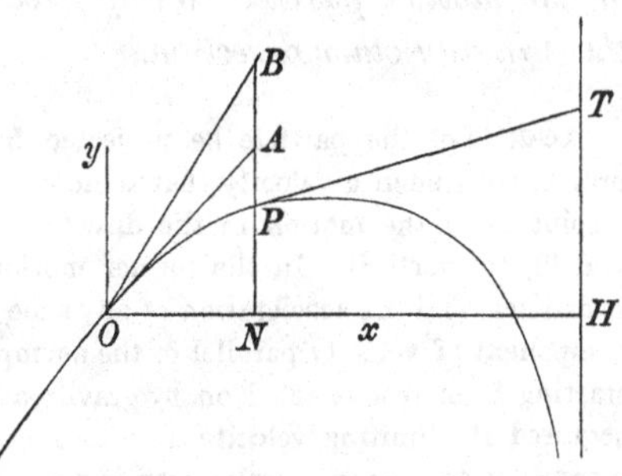

Tracing the curve backwards we make $t = -\infty$; we then find that both x and y are infinitely great. Since the exponential is infinitely greater than t, both y/x and dy/dx have ultimately the same ratio. Representing this ratio by $\tan\beta$, we have

$$\tan\beta = \tan\alpha + L/V\cos\alpha \ldots\ldots\ldots\ldots\ldots\ldots (3).$$

The curve has therefore an infinite branch, the tangent or asymptote to which makes an angle β with the horizon, determined from the initial conditions by this equation. This asymptote is at an infinite distance from the origin.

164. Eliminating the exponentials from the values of x and y, Art. 162, we find

$$y = x\tan\beta - Lt \ldots\ldots\ldots\ldots\ldots\ldots\ldots\ldots (4),$$

a linear equation which must hold throughout the motion.

Drawing a straight line OB parallel to the oblique asymptote, this equation shows that *the vertical distance of P from OB is $PB = Lt$, where L is the limiting velocity.*

The perpendicular distance of P from OB being $Lt\cos\beta$, *the resolved velocity at P perpendicular to the oblique asymptote is constant. The resultant acceleration at P is therefore parallel to BO.*

165. *General principle.* Since the resistance varies as the velocity, the resolved resistance in any direction is proportional to the resolved velocity in the same direction. The general principle proved in Art. 154 for motion in a vacuum will therefore apply to the motion with this law of resistance. *The circumstances of the motion parallel to any fixed straight line are independent of those in any other direction.*

166. Let the particle be projected from a distant point E on the oblique branch with such a velocity that it describes the trajectory. Consider the oblique resolution of the motion in the direction of (1) the tangent or asymptote at E and (2) the vertical. In the former motion the particle is acted on only by the resistance, and the acceleration at any time is therefore $-\kappa u$, where u is the oblique component of velocity parallel to the asymptote. In the latter motion the particle starting from rest is acted on by gravity as well as by the resistance and has thus acquired its limiting velocity L. This component is constant in direction and magnitude so that the acceleration is zero.

Combining these two motions, we see that in any position P of the particle, the velocity v along the tangent PT is the resultant of the vertical limiting velocity L and a velocity u parallel to the oblique asymptote. If U and u be corresponding velocities at any two points O and P of the trajectory, $u = Ue^{-\kappa t}$, where t is the time of transit from O to P; Art. 102. We also notice that the resultant acceleration at P is equal to $-\kappa u$.

Taking a parallel OB to the oblique asymptote and the vertical as axes of reference we have

$$\xi = U(1-e^{-\kappa t})/\kappa, \quad \eta = Lt \quad \text{.............................. (1)},$$

where $\xi = OB$, $\eta = BP$. If we refer the motion to the tangent at O and the vertical as axes, we have $\xi' = OA$, $\eta' = AP$. We find by considering the motions in these directions separately

$$\kappa\xi' = V(1-e^{-\kappa t}), \quad \kappa\eta' = gt - L(1-e^{-\kappa t}) \quad \text{.................... (2)}.$$

167. *Ex.* 1. Particles are projected from a given point O at the same instant with equal velocities in different directions; prove that the locus at any time is a sphere.

Refer the motion of any particle to the tangent OA and the vertical as axes of ξ, η. Both ξ, η are evidently functions of t which are independent of a. The locus is therefore a sphere whose radius is ξ and whose centre is at a depth η below O. Art. 166.

Ex. 2. Particles are projected from a given point O at the same instant with different velocities in the same direction OA, prove that at any subsequent time their locus is a straight line parallel to OA. Art. 165.

Ex. 3. If the axis of x is inclined at an angle i to the horizon and the direction of projection make an angle γ with x, prove that

$$\left.\begin{aligned} \kappa x &= -g\sin it + (V\cos\gamma + L\sin i)(1-e^{-\kappa t}) \\ \kappa y &= -g\cos it + (V\sin\gamma + L\cos i)(1-e^{-\kappa t}) \end{aligned}\right\}.$$

If M be the point at which the tangent is parallel to x, prove that the time t_1 of reaching M and the coordinates of M are

$$e^{\kappa t_1}=1+\frac{V\sin\gamma}{L\cos i},\quad \kappa y_1=-g\cos it_1+V\sin\gamma,$$

$$x_1(V\sin\gamma+L\cos i)-y_1(V\cos\gamma+L\sin i)=LVt\cos(i+\gamma);$$

the latter equation being also true for all points on the trajectory.

Ex. 4. A projectile moves under gravity in a uniform medium whose resistance varies as the velocity. Prove that the hodograph of the trajectory is a straight line and that the velocity of the point on the hodograph is proportional to the horizontal velocity of the projectile. [Coll. Ex.]

Resistance varies as the nth *power of the velocity.*

168. *To find the motion of a heavy particle, when the resistance varies as the* n^{th} *power of the velocity.*

Let ψ be the angle the tangent at any point P of the path makes with the horizontal, ρ the radius of curvature measured positively downwards so that $\rho=-ds/d\psi$. Let v be the velocity, u the horizontal component. Following John Bernoulli, 1721, we resolve the motion normally and horizontally, we thus have

$$\frac{v^2}{\rho}=g\cos\psi,\quad \frac{du}{dt}=-\kappa v^n\cos\psi.$$

Since $v\cos\psi=u$ and $\rho=-v\,dt/d\psi$, these become

$$\frac{dt}{d\psi}=-\frac{u}{g\cos^2\psi},\quad \frac{du}{dt}=-\frac{\kappa u^n}{(\cos\psi)^{n-1}}\text{ ....(A).}$$

We obtain one integral by eliminating dt,

$$\frac{du}{d\psi}=\frac{\kappa}{g}\cdot\left(\frac{u}{\cos\psi}\right)^{n+1},\quad \therefore\ \frac{1}{u^n}-\frac{1}{u_0{}^n}=-\frac{\kappa n}{g}\int_\alpha^\psi\frac{d\psi}{(\cos\psi)^{n+1}},$$

where α is the angle the initial direction of motion makes with the horizon, and u_0 the initial horizontal velocity.

To effect this integration we put $p=\tan\psi$, we then have, except $n=0$,

$$\frac{1}{u^n}-\frac{1}{u_0{}^n}=-\frac{\kappa n}{g}\int(1+p^2)^{\frac{1}{2}(n-1)}\,dp\text{ ....(B),}$$

the sign of the radical when n is even being such that the subject of integration is positive between the limits $\psi=\alpha$ and $\psi=-\frac{1}{2}\pi$, i.e. $p=p_0$ and $p=-\infty$.

We can conveniently take either u or p as the independent variable, and thus we obtain the two sets of relations,

$$\left.\begin{aligned} t &= -\frac{1}{g}\int u\,dp = -\frac{1}{\kappa}\int\frac{du}{u^n(1+p^2)^{\frac{1}{2}(n-1)}} \\ x &= -\frac{1}{g}\int u^2dp = -\frac{1}{\kappa}\int\frac{du}{u^{n-1}(1+p^2)^{\frac{1}{2}(n-1)}} \\ y &= -\frac{1}{g}\int u^2p\,dp = -\frac{1}{\kappa}\int\frac{p\,du}{u^{n-1}(1+p^2)^{\frac{1}{2}(n-1)}} \end{aligned}\right\}\text{......(C).}$$

The first follows from equation (A), the second and third from the obvious relations $dx = udt$, $dy = updt$. The limits in all the integrals being $p = p_0$ to p or $u = u_0$ to u.

In this manner all the circumstances of the motion can be expressed in terms of one independent variable which may be either p or u.

It is evident that the integral (B) has considerable importance in this theory. Putting

$$W_n = \int(1+p^2)^{\frac{1}{2}(n-1)}\,dp,$$

we see that when $n = 2$ or $n = 3$,

$$W_2 = \tfrac{1}{2}\{p(1+p^2)^{\frac{1}{2}} + \log(p + (1+p^2)^{\frac{1}{2}})\}, \quad W_3 = p + \tfrac{1}{3}p^3.$$

We may also find a general formula of reduction, viz.

$$(n+2)\,W_{n+2} = (n+1)\,W_n + p(1+p^2)^{\frac{1}{2}(n+1)} \text{.........(D).}$$

When the resistance is a constant force, say κg, $n = 0$, and the integral (B) takes the form

$$\left(\frac{u}{a}\right)^2 = \left(\frac{1+\sin\psi}{1-\sin\psi}\right)^\kappa,$$

where a is the velocity when the particle is moving horizontally.

169. The equations (C) have been applied to the calculation of the trajectories of shot in various ways*. When the angle of elevation is not more than 10° to 15°, as in the case of direct fire,

* Bashforth, Phil. Trans. 1868, *Treatise on the motion of projectiles*, 1873; supplement, 1881. Proceedings of R.A. Institution, 1871 and 1885. W. D. Niven, *On the calculation of the trajectories of shot*, Proceedings of the Royal Society, 1877. Ingall, *Exterior Ballistics*, 1885. An account of Siacci's method is given by Greenhill in the Proc. of the R. A. Institution, vol. XVII. See also *Artillery, its progress and present position* by E. W. Lloyd and A. G. Hadcock, 1893. Greenhill, *On the motion of a projectile in a resisting medium*, Proceedings of the R. A. Institution, vols. XI., XII., XIV., 1880 to 1886.

we may regard the trajectory as so flat that we can reject the square of p. Taking u as the independent variable the integration can then be effected without difficulty. When the path is more inclined we can divide the whole path into subsidiary arcs for each of which p may be regarded as approximately constant though of a different value in each arc. If the arcs were small enough the initial value of p in each arc might be taken as the proper value for that arc. For longer arcs it becomes necessary to give p a mean value taken over the whole subsidiary arc.

170. In artillery practice the values of the integrals (C) are commonly inferred from tables especially constructed for that purpose, different tables being used to find t, x and y. Opinions differ as to the best methods of constructing and using these tables. Bashforth represents the law of resistance by κv^3 where κ is a function of the velocity whose values are deduced from experiment. These values for a shot of given cross section and weight and for air of given density are tabulated for every few feet of velocity. In effecting the integrations (C) the quantity κ is regarded as constant and in a long arc a value suitable to a mean velocity over the arc has to be found. This difficulty having been overcome, the integrals (C) are tabulated for different values of κ and between certain ranges of angle.

In the Italian method a quantity allied to the velocity is taken as the independent variable. To enable the integrations to be effected the quantity p is taken as constant throughout the subsidiary arc. The integrals (C) are then determined either by the use of tables or by giving the index n the value suitable to the range of velocity in the trajectory.

An account of the methods of constructing and using these various tables would take us too far from our present subject. We must refer the reader to special treatises on Artillery.

171. Law of resistance. Many attempts have been made to discover the law of resistance to the motion of projectiles. Passing over the earlier experiments of Robins and Hutton we may mention as the most important the long-continued series made by F. Bashforth with the help of his chronograph. By this instrument the times taken by the same projectile in passing over a succession of equal spaces can be measured with great accuracy. Other experiments have also been made on the continent, for example by Mayevski in 1881. It appears from all these experiments that the resistance cannot be expressed by any one power of the velocity. The general result is that for low and high velocities the resistance varies as the square of the velocity, and for intervening velocities as the cube and even a higher power of the velocity.

To be more particular, let v be the velocity measured in feet per second, d the diameter of the ogival headed shot in inches, w the weight in pounds. Then taking the resistance to be $\beta \frac{d^2}{w}\left(\frac{v}{1000}\right)^n$, Bashforth's experiments show that

$v < 850$	$n=2$	$\beta = 61{\cdot}3$
$v > 850 < 1040$	$n=3$	$\beta = 74{\cdot}4$
$v > 1040 < 1100$	$n=6$	$\beta = 79{\cdot}2$
$v > 1100 < 1300$	$n=3$	$\beta = 108{\cdot}8$
$v > 1300 < 2780$	$n=2$	$\beta = 141{\cdot}5$.

Mayevski's experiments led to similar results except that the highest power of n was $n=5$. The values of β were also different because the shots were more pointed than in those of Bashforth.

We may notice that though the resistance for low and high velocities follows the same general law, yet the value of the coefficient β is much greater for the high than the low velocities. When the velocity of the shot approximates to that of the velocity of sound in air, we might expect a considerable change in the law of resistance and this is shown in the results given above.

172. *To discuss the motion when the resistance varies as the square of the velocity.*

In this case we can obtain two first integrals of the equations of motion. Resolving normally and horizontally as before, we find

$$\frac{v^2}{\rho} = g\cos\psi, \quad \frac{du}{dt} = -\kappa v^2 \cos\psi = -\kappa u \frac{ds}{dt} \quad\text{.....................(1).}$$

Dividing the latter equation by u and integrating

$$\log u = A - \kappa s, \quad \therefore\ u = u_0 e^{-\kappa s} \quad\text{.........................(2),}$$

where u_0 is the horizontal velocity at the point O of projection and s is measured from O.

Besides this we have the integral (B) already obtained in the general case by eliminating dt from the equations of motion. Writing $\rho = -ds/d\psi$ and $v\cos\psi = u$, in (1), we find as before

$$\frac{dt}{d\psi} = -\frac{u}{g\cos^2\psi}, \quad \frac{du}{dt} = -\frac{\kappa u^2}{\cos\psi} \quad\text{..........................(3),}$$

$$\therefore\ \frac{1}{u^2} = -\frac{2\kappa}{g}\int\frac{d\psi}{\cos^3\psi} = -\frac{2\kappa}{g}\int\sqrt{(1+p^2)}\,dp \quad\text{....................(4),}$$

where $p = \tan\psi$ as before, and the radical is to be taken positively. Integrating

$$\frac{1}{u^2} = -\frac{\kappa}{g}\{p\sqrt{(1+p^2)} + \log(p + \sqrt{(1+p^2)})\} + B \quad\text{.................(5).}$$

Eliminating u between (2) and (5) we find

$$\frac{g}{\kappa u_0^2} e^{2\kappa s} + p\sqrt{(1+p^2)} + \log\{p+\sqrt{(1+p^2)}\} = C \quad\text{..............(6)}.$$

This is the intrinsic equation of the path.

173. To discuss the form of the curve it will be convenient to place the origin at the highest point so that initially $p=0$. We then have

$$\frac{g}{\kappa u_0^2}(1-e^{2\kappa s}) = p\sqrt{(1+p^2)} + \log\{p+\sqrt{(1+p^2)}\} \quad\text{.............. (7)}.$$

When s increases to positive infinity we see from (2) and (7) that u tends to zero and p to minus infinity. Since by (3) or (C)

$$g\,dt = -u\,dp \text{ and } g\,dx = gu\,dt = -u^2dp,$$

it follows that both dt/dp and dx/dp are ultimately zero. We shall prove that while t becomes ultimately infinite, x tends to a finite limit. We therefore infer that *the curve has a vertical asymptote at a finite distance on the positive side of the highest point.*

To prove this we refer to (5) and retaining only the highest powers of p, we see that $1/u^2$ is of the order p^2. Putting $u=b/p$ when p is very great, we find

$$gt = -\int u\,dp = -b\log p, \quad gx = -\int u^2dp = b^2/p.$$

Taking these between the limits $p=p_1$ to infinity where p_1 is any large finite quantity, the first gives the time the particle takes to travel from the position defined by $p=p_1$ to that defined by $p=-\infty$, and the second gives the corresponding horizontal space. We see that the first is infinite and the second finite.

174. Consider next the other extremity of the trajectory. When the arc is negatively very great, we see by (2) that u is positive and infinite. It also follows by (7) that p tends to a limit m given by the equation

$$-g/\kappa u_0^2 + m\sqrt{(1+m^2)} + \log(m+\sqrt{(1+m^2)}) = 0 \quad\text{..................(8)}.$$

Since the left-hand side passes from a negative quantity to positive infinity as m varies from zero to infinity, it is clear that this equation has at least one positive root. If the equation could have two real roots, the differential coefficient of the left-hand side would vanish for some intervening value of m. But since the differential coefficient is $2\sqrt{(1+m^2)}$ this is impossible. It follows that *the curve on the negative side of the highest point has an asymptote inclined at a finite angle to the horizon.* We shall now prove that *this asymptote is at a finite distance from the highest point.*

To prove this we examine the limiting value of the intercept of the tangent on the axis of y, viz. $y-xp$, when $p=m$. Remembering that $gdx=-u^2dp$, $dy=pdx$, we have

$$g(y-xp) = -\int pu^2dp + p\int u^2dp,$$

the limits being $p=0$ to p. As we only wish to determine whether the limit is finite or not we shall integrate from $p=m-\xi_1$ to $p=m$, where ξ_1 is some finite quantity as small as we please. The remaining parts of the integrals will be included in two finite constants M and N. Writing $p=m-\xi$, we have

$$g(y-xp) = \int(m-\xi)\,u^2d\xi - (m-\xi)\int u^2d\xi - M + pN,$$

the limits being $\xi=\xi_1$ to 0. To find what function u is of ξ when ξ is small, we refer to equation (5). Remembering that $B=1/u_0^2$ since $u=u_0$ when $p=0$, we

7—2

write that equation in the form

$$\left(\frac{1}{u_0^2}-\frac{1}{u^2}\right)\frac{g}{\kappa}=f(p)=f(m-\xi).$$

Expanding and remembering that $df/dp=2\sqrt{(1+p^2)}$ we find after subtracting (8)

$$\frac{1}{u^2}=\frac{2\kappa}{g}\sqrt{(1+m^2)}\,\xi+A\xi^2+\ldots, \qquad \therefore\ u^2=b^2/\xi+A'+B'\xi+\ldots,$$

where b^2, A' and B' are finite constants. Substituting we find by an easy integration

$$g(y-xp)=-b^2\xi+b^2\xi\log\xi+\&c.,$$

where the &c. includes only positive powers of ξ. Taking this between the limits $\xi=\xi_1$ to 0, the result is finite.

175. *Ex.* 1. Prove that, when the resistance varies as the square of the velocity, the time of describing the infinite arc on the negative side of the highest point is finite.

Referring to equations (C) and writing $p=m-\xi$, $u^2=b^2/\xi$ we see that the time of describing the infinite arc from $p=m$ to $p=p_1$ is $M\sqrt{(m-p_1)}$, where M is a finite quantity independent of ξ or p. This result is finite; see also Art. 116.

Ex. 2. When the resistance varies as the square of the velocity, prove that the polar equation of the hodograph is $\dfrac{1}{r^2}=\cos^2\theta\left(\dfrac{1}{V^2}+\dfrac{1}{U^2}\sinh^{-1}\tan\theta\right)+\dfrac{\sin\theta}{U^2}$, where the origin is at the highest point, V is the horizontal velocity, U the terminal velocity and the initial line is horizontal and θ is measured positively downwards. [Coll. Ex. 1893.]

This is a transformation of equation (5) of Art. 172, writing r, $-\theta$, for v, ψ.

Ex. 3. When the resistance varies as the square of the velocity, prove that the radius of curvature ρ at the point where the normal makes an angle ϕ with the vertical is given by

$$2/\kappa\rho=c\sin^3\phi+2\sin^3\phi\log\cot\tfrac{1}{2}\phi+\sin 2\phi. \qquad \text{[Coll. Ex.]}$$

176. *Ex.* 1. When the resistance varies as the nth power of the velocity, prove that the curve has a vertical asymptote at a finite distance on the positive side of the highest point.

We have $v=u\sqrt{(1+p^2)}$ where u is given by equation (B). Now, by the action of gravity, p continually decreases from one end of the trajectory to the other. After the projectile has passed the summit p becomes negatively great and (B) then gives $u=L/p$, where L is the limiting velocity. We thus have $v=L$ when $p=\infty$. Substituting $u=L/p$ in (C) and integrating from $p=p_1$ to ∞, where p_1 is any large finite quantity, we find that t and y are infinite and x finite.

Ex. 2. Prove that, when the resistance varies as the nth power of the velocity, n being >2, the arc of the trajectory on the negative side of the highest point begins at a point at a finite distance from the origin. Prove also that the tangent at this point makes an angle $\tan^{-1}m$ with the horizon given by

$$\int_{p_0}^{m}(1+p^2)^{\frac{1}{2}(n-1)}\,dp=\frac{g}{\kappa n u_0^{\,n}},$$

where u_0 and p_0 are any contemporaneous values of u and p. See Art. 116. As in Art. 174, this equation has one positive root.

In the *extreme* initial position of the particle the velocity is infinite. Since $v=u\sqrt{(1+p^2)}$ we must there have either u or p infinite. If $p=\infty$, (B) gives $u=L/p$ and this makes v finite. The equation giving m is therefore obtained by putting $u=\infty$ in (B). To determine the position of the particle when this occurs we express u in terms of p and use the equations (C). Let the initial position defined by $p=p_0$ be such that $p_0=m-\xi_1$ where ξ_1 is a finite quantity as small as we please. Substituting $p=m-\xi$ in (B) and using the equation given above to find m, we have $u^n=b^n/\xi$ where b is a constant. Substituting in (C) and integrating from $\xi=\xi_1$ to ξ we find that t, x, and y are finite when $\xi=0$.

Ex. 3. When the law of resistance is the nth power of the velocity, and u, u' are the horizontal velocities at any two points of the trajectory at which the tangents make equal angles with the horizon, then $\frac{1}{u^n}+\frac{1}{u'^n}=\frac{2}{a^n}$ where a is the velocity at the highest point.

Ex. 4. When the resistance is $\kappa'+\kappa v^n$, investigate the linear equation

$$\frac{du^{-n}}{dp}+\frac{\kappa' n}{g}\frac{u^{-n}}{(1+p^2)^{\frac{1}{2}}}=-\frac{\kappa n}{g}(1+p^2)^{\frac{1}{2}(n-1)},$$

where u is the horizontal velocity and p is the tangent of the inclination to the horizon. Thence show that the determination of t, x, y may be reduced to integration. [Allegret, *Bulletin de la Société Math.* 1872.]

Ex. 5. When the resistance is constant and equal to κg, the highest point being the origin and the velocity being a, prove that the horizontal velocity u at any point of the path is $u=a(\tan\theta)^\kappa$ where $2\theta=\psi+\frac{1}{2}\pi$. Thence deduce from the integrals (C), Art. 168, the values of t, x, y in terms of $\tan\theta$.

If $\kappa<$ or $=1$, the subsequent path has a vertical asymptote which is at the finite distance $x=2\kappa a^2/g(4\kappa^2-1)$ if $\kappa>\frac{1}{2}$, but is at an infinite distance if $\kappa<\frac{1}{2}$. If $\kappa>1$ the particle arrives at a point C at which the tangent is vertical in the finite time $\kappa a/g(\kappa^2-1)$, the coordinates of C being $2\kappa a^2/g(4\kappa^2-1)$ and $-a^2/4g(\kappa^2-1)$.

On the negative side of the origin, the curve begins with a vertical asymptote which is infinitely distant and the time of describing the arc is infinite.

177. When the resistance varies as the cube of the velocity, the equation (B) of Art. 168 takes the form

$$\frac{1}{u^3}=-\frac{\kappa}{g}(p-m)(p^2+mp+m^2+3),$$

the origin being taken at the point at which the velocity is infinite and m being the corresponding value of p.

To discuss the motion we substitute this value of u in the integrals (C). For the reduction of these integrals to elliptic forms we refer the reader to a paper by Greenhill in the *Proceedings of the Royal Artillery Institution*, vol. XIV. 1886.

Ex. Show that for the cubic law of resistance the velocity is a minimum at the point given by the negative root of the quadratic $p^2-m(m^2+3)p=1$. Show also that when the direction of motion is perpendicular to the oblique asymptote, the horizontal velocity u is given by $\frac{L}{u}=m+\frac{1}{m}$ where L is the limiting velocity.

178. Some formulæ have also been given by the late Prof. Adams to determine the coordinates of a particle projected at any inclination to the horizon on the supposition that the resistance varies as the nth power of the velocity and that the path is not very curved. These were first published in the *Proceedings of the Royal Society* and proofs were given in *Nature*, vol. XLI., 1890. These appear to be long, but they admit of great abbreviation.

179. *The equation of a trajectory being given in the form* $\cos\psi = f(\rho\cos\psi)$, *it is required to find the law of resistance.*

We notice that the equation can be written in this form, except when $\rho\cos\psi$ is constant, for in that case $\rho\cos\psi$ cannot be taken as the independent variable. This excepted curve is the catenary of equal strength.

Resolving horizontally and tangentially, we have

$$\frac{d}{dt}(v\cos\psi) = -R\cos\psi, \qquad \frac{dv}{dt} = -R - g\sin\psi \;\ldots\ldots\ldots\ldots\ldots\ldots (1).$$

Eliminating dt
$$R\cos\psi = \frac{d}{dv}(v\cos\psi)(R + g\sin\psi);$$

$$\therefore\; Rv\frac{d}{dv}(\cos\psi) = -g\sin\psi\frac{d}{dv}(v\cos\psi) \;\ldots\ldots\ldots\ldots\ldots\ldots (1).$$

Remembering that the normal resolution gives $v^2/\rho = g\cos\psi$, we have $\cos\psi = f(v^2/g)$. Substituting this value of $\cos\psi$, the expression for the resistance R has been found. We may also write the expression in the form

$$Rv\frac{df}{dv} = -g(1-f^2)^{\frac{1}{2}}\frac{d}{dv}(vf) \;\ldots\ldots\ldots\ldots\ldots\ldots (2),$$

where $f = f(v^2/g)$ and the sign of the radical follows that of $\sin\psi$.

180. *Ex.* 1. Find the law of resistance when the trajectory is a cycloid with the cusps pointing downwards.

In this curve $\rho = 2a\cos\psi$, $\therefore\; f = v/\sqrt{2ag}$. We then find that the resistance $R = -2g(1 - v^2/2ag)^{\frac{1}{2}}$. Since the radical follows the sign of $\sin\psi$, R accelerates the particle on the ascending and retards it on the descending branch. Since $v = \cos\psi\sqrt{2ag}$ the particle comes to rest at the cusp. The resistance R is then acting upwards and is equal to $2g$, the particle then moves vertically. See Art. 176, Ex. 5.

Ex. 2. Find the law of resistance when the trajectory is the catenary of equal strength with the concavity downwards.

The normal and tangential resolutions show that v is constant and $R = -g\sin\psi$. R is a resistance therefore only on the descending branch.

Ex. 3. Find the law of resistance in the parabola $\rho\cos^3\psi = 2a$.

Ex. 4. Find the law of resistance in the circle $\rho = a$. The resistance is
$$-\tfrac{3}{2}g(1 - v^4/a^2g^2)^{\frac{1}{2}} \text{ and } v^2 = ag\cos\psi.$$

CHAPTER IV.

CONSTRAINED MOTION IN TWO DIMENSIONS.

Constrained Motion.

181. *A particle, constrained to describe a given smooth fixed curve, is under the action of given forces. It is required to find the velocity and the reaction between the curve and the particle.*

Let the curve be referred to fixed Cartesian coordinates and let its equation be $y=f(x)$. Let (x, y) be the position of the particle P at the time t, m its mass, X, Y the resolved forces. Let the tangent at P make an angle ψ with the axis of x, and let ρ be the radius of curvature. Let R be the pressure of the curve on the particle taken positively in the direction in which ρ is measured; this direction is generally inwards.

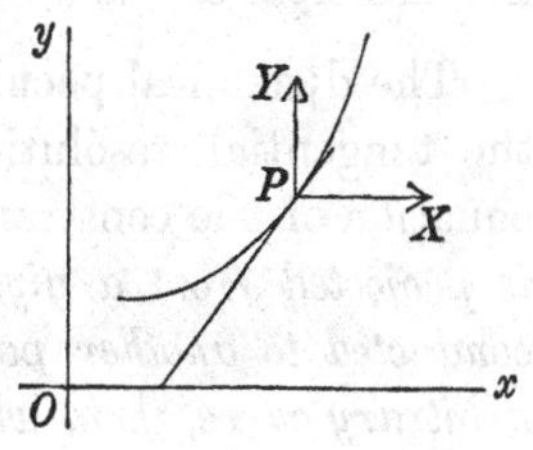

When the path of the particle is known the relations between ρ, the arc s and the other lines of the curve are also known. It is therefore generally more convenient to choose the tangent and normal as the directions in which to resolve the acceleration. Resolving in these directions, we have

$$mv\frac{dv}{ds} = X\cos\psi + Y\sin\psi \qquad \text{(1)},$$

$$\frac{mv^2}{\rho} = -X\sin\psi + Y\cos\psi + R \qquad \text{(2)}.$$

From these two equations we may deduce all the circumstances of the motion.

Considering the tangential resolution we see that since

$$\cos\psi = dx/ds, \quad \sin\psi = dy/ds,$$
$$mvdv = Xdx + Ydy \text{.......................(3).}$$

There are two cases to be considered according as the right-hand side of this equation is or is not a perfect differential of some function of x and y.

In the former case *the forces are called conservative.* Let

$$Xdx + Ydy = dU \text{.......................(4).}$$

We therefore have by integration

$$\tfrac{1}{2}mv^2 = U + C \text{.........................(5).}$$

Let (x_0, y_0) be the coordinates of the initial position A of the particle, and let U become U_0 when we write for x, y, their initial values. We therefore have

$$\tfrac{1}{2}mv^2 - \tfrac{1}{2}mv_0^2 = U - U_0 \text{....................(6).}$$

This equation is one case of a general principle usually called the Principle of Vis Viva.

The dynamical peculiarity of this case is that the equation of the tangential resolution can be integrated without using the equation of the constraining curve. It follows that *if the particle is projected from a given point A with a given velocity and if it is conducted to another point P by constraining it to move along an arbitrary curve, then, whatever the path may be, the velocity of the particle on arrival at P is always the same.*

182. When the forces are such that $Xdx + Ydy$ is not a perfect differential of any function of x and y the velocity cannot be found without using the equation of the constraining curve. Putting $y = f(x)$, we find

$$\tfrac{1}{2}mv^2 = \int\{X + Yf'(x)\}\, dx + C.$$

Since X and Y can be expressed as functions of x by the help of the equation of the curve, the integration can be effected. Let the integral be $F(x)$. We then have

$$\tfrac{1}{2}mv^2 - \tfrac{1}{2}mv_0^2 = F(x) - F(x_0).$$

In this case the change of vis viva does not conserve the same value for all paths.

183. Let us next take into consideration the equation of the normal resolution, viz.

$$\frac{mv^2}{\rho} = -X \sin\psi + Y \cos\psi + R.$$

The term mv^2/ρ is called *the centrifugal force* of the particle*. This is another name for the normal component of the effective force, Arts. 36, 68.

The force R is called *the dynamical pressure* of the curve on the particle, and $-R$ is the dynamical pressure of the particle on the curve. The two terms $-X \sin\psi + Y \cos\psi$ make up the resolved part of the acting forces along the normal to the curve and are together called *the statical pressure* of the forces on the particle. Taken with the opposite signs they are the statical pressure on the curve.

184. We are now in a position to apply the two fundamental theorems to determine the motion of a particle on any given fixed curve.

First, we use the equation of vis viva, viz.

change of kinetic energy = work of the forces.

In this way we find the velocity.

Secondly, the dynamical pressure on the particle in any position is given by the equation

$$\frac{mv^2}{\rho} = \begin{pmatrix}\text{normal}\\ \text{force inwards}\end{pmatrix} + \begin{pmatrix}\text{dynamical}\\ \text{pressure}\end{pmatrix}.$$

185. Work Function. The usual methods of finding the work of a system of forces are explained in books on Statics. As however the solution of our dynamical problems depends so much on our knowledge of these rules, it has been thought not improper to recall to mind those few which we shall here use. A more complete list applicable to a system of rigid bodies is to be found in the author's *Rigid Dynamics*.

* It is perhaps unnecessary to observe that the centrifugal force is not an actual force acting on the particle in addition to the impressed forces. It is merely a name for the quantity mv^2/ρ, and measures the amount of force which must act towards the concave side of the path to produce the curvature $1/\rho$; the mass of the particle being m and the velocity v. By the first law of motion the particle tends to move in a straight line and the force necessary to curve the path is sometimes *said* to be spent in overcoming the centrifugal force.

If X, Y are the components of a force F the work done when the particle receives a slight displacement ds from the position x, y to $x + dx$, $y + dy$ may be written in either of the equivalent forms

$$Xdx + Ydy = F\cos\phi ds \quad \text{.................(1)},$$

where ϕ is the angle the direction of F makes with the tangent to the path, see Art. 70. That the work of the two forces X, Y is equal to that of their resultant is proved in Statics. It is also seen to be true by resolving the forces along the tangent; we then have

$$X\frac{dx}{ds} + Y\frac{dy}{ds} = F\cos\phi,$$

which is equivalent to the equation (1). Either side of (1) is also called in Statics *the virtual moment of the force F*.

The integral U when used in the indefinite form

$$U = \int F\cos\phi ds + C$$

is called sometimes *the force function* and sometimes *the work function*. The definite integral $U - U_0$ is *the work done* by the force F as the particle moves from the position (x_0, y_0) to the position (x, y). Here U_0 represents the same function of x_0, y_0 that U is of x, y.

186. Work of a central force. Let the central force F be regarded as repulsive in the standard case. Let it tend from the centre S and be equal to $f(r)$ where r is the distance of the particle from S. Then since dr/ds is the cosine of the angle the distance r makes with the displacement ds of the particle, the part of the work function due to F is $\int Fdr$. The integration is to be taken from the initial position A to the final position B of the particle.

When the force under consideration is gravity the centre S is regarded as being infinitely distant. We then replace dr by $\pm dy$, the upper or lower sign being taken according as y is measured downwards or upwards. Supposing the weight of the particle to be mg and that y is measured downwards, the work of the weight is

$$\int mg\,dy = mg\,(y - y_0).$$

This rule is usually read thus, *the work done by gravity is the weight multiplied by the vertical space descended.* It should be noticed that the work is independent of the horizontal displacement. See Art. 70.

187. Work of an elastic string. The case in which the particle is attached to a fixed point S by an elastic string differs from that of a central force tending to the same point in a certain discontinuity. If l be the unstretched length, r the actual length and E Young's modulus, the tension T is given by Hooke's law $T = E\dfrac{r-l}{l}$ when the string is tight, i.e. when $r > l$, but the tension is zero when the string is slack, i.e. $r < l$.

Let the work be required when the string is stretched from a length l_1 to l_2, and let T_1, T_2 be the tensions at these lengths. If both l_1 and l_2 are greater than l, the work is

$$\int_{l_1}^{l_2} (-T)\,dr = -\frac{E}{2l}\{(l_2 - l)^2 - (l_1 - l)^2\}$$

$$= -\frac{1}{2}(T_1 + T_2)(l_2 - l_1).$$

The work done by the tension is therefore equal to *minus the arithmetic mean of the tensions multiplied by the extension.* The work done by the force which stretches the string in opposition to the tension is the same taken with the positive sign.

This rule is of considerable use when the length of the string undergoes many changes during the motion, being sometimes greater than the unstretched length and sometimes less. It is important to notice that the rule, as given above, holds in all these cases provided the string is tight in the initial and final states. If the string is slack in either terminal state, we may still use the same rule provided we suppose the string to have its natural or unstretched length in that terminal state.

188. *The equation of vis viva holds also when the particle is free from constraint and is acted on by any conservative system of forces.* For, whatever curve the particle may describe, we may suppose it to be constrained, like a bead on an imaginary wire, to describe that path. The pressure is then zero throughout the motion, but, what more immediately concerns us here, is that the equation (6) of vis viva continues to hold under these circumstances.

189. The whole area or space taken into consideration when the forces are expressed in terms of the coordinates is called *the field of force.* Such a field is usually defined by expressing the force function (when there is one) as a function of the coordinates.

It follows from the principle of vis viva that when a single particle moves in a field defined by a force function the kinetic energy of the particle in any and every position differs from the value of the force function at that point by a constant. The constant is independent of the direction of motion, so that *two particles of equal mass projected from the same point with equal velocities but in different directions will always have equal velocities whenever they pass over a given point of the field.*

190. Examples. *Ex.* 1. A particle is projected from a given point on a smooth curve and is acted on by no forces. Prove (1) that the velocity is constant and (2) that the pressure varies as the curvature.

Ex. 2. A heavy particle P describes a curve and in any position a normal PQ is drawn *outwards*, so that PQ is equal to half the radius of curvature at P. Prove that the velocity v and the pressure R on the particle measured inwards are given by

$$v^2 = 2gz, \quad R\rho = 2mgz',$$

where z, z' are the depths of P and Q below a certain horizontal straight line, which may be called *the level of no velocity.* Prove also that the particle leaves the curve when Q crosses the level of no velocity.

Supposing that the axis of y in the standard figure of Art. 181 is drawn upwards, the two fundamental equations for a heavy particle are

$$\tfrac{1}{2}mv^2 - \tfrac{1}{2}mv_0^2 = -mg(y - y_0),$$

$$mv^2/\rho = -mg\cos\psi + R.$$

If we draw a horizontal straight line at an altitude y_1, such that $gy_1 = gy_0 + \tfrac{1}{2}v_0^2$, we see that

$$z = y_1 - y, \quad z' = y_1 - y + \tfrac{1}{2}\rho\cos\psi.$$

The results to be proved follow immediately. If the particle is constrained to remain on the curve merely by the pressure R it will leave the curve when R changes sign. But this is what happens when Q crosses the level of no velocity.

Ex. 3. A particle is swung round a fixed point at the end of a string in a vertical plane. Prove that the sum of the tensions of the string when the particle is at opposite ends of a diameter is the same for all diameters.

[Coll. Exam. 1896.]

Ex. 4. A heavy particle, constrained to describe an ellipse whose plane is vertical and major axis inclined at an angle α to the horizon, is projected from the upper extremity A of the major axis with a velocity v_0. Find the velocity v_1 with which it passes the upper extremity B of the minor axis and the pressure at that point.

Since the altitude of B above A is the difference between the projections of CA and CB on the vertical, the equation of vis viva gives

$$\tfrac{1}{2}m(v_1^2-v_0^2)=-mg(b\cos\alpha-a\sin\alpha).$$

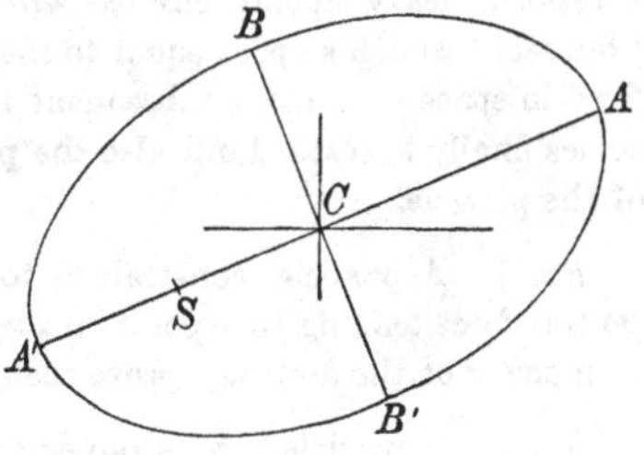

This gives two equal values of v_1 with opposite signs. One or the other is to be taken according as the particle is projected from A upwards or downwards. If the values of v_1 are imaginary the particle will not reach B.

The pressure R_1 at B is found by resolving the forces along BC inwards. We have

$$\frac{mv_1^2}{\rho_1}=mg\cos\alpha+R_1,$$

where $\rho_1=a^2/b$.

Let us suppose that in addition to its weight the particle is acted on by a centre of force at the focus S such that the attraction at a distance r is μr^n. The equation of vis viva would then have on the right-hand side the additional term $-\int\mu r^n\,dr$, the limits being the initial and the final values of r, i.e. $r=a(1+e)$ and $r=a$, Art. 186. The velocity v_1 is then given by

$$\tfrac{1}{2}m(v_1^2-v_0^2)=-mg(b\cos\alpha-a\sin\alpha)-\mu\frac{a^{n+1}}{n+1}\{1-(1+e)^{n+1}\}$$

and the pressure is determined by

$$\frac{mv_1^2}{\rho_1}=mg\cos\alpha+\mu a^n\cdot\frac{b}{a}+R_1.$$

Let us next attach the particle to the centre C by an elastic string whose natural length is l. The effect of this is to add another term to each equation. If $l<b$ and $<a$ the string is stretched throughout and the term to be added to the equation of vis viva is $-\frac{1}{2}(T_0+T_1)(b-a)$ where T_0 and T_1 are the tensions at A and B, see Art. 187. In our case $T_0=E(a-l)/l$ and $T_1=E(b-l)/l$. If however $l>b$ and $<a$ the string becomes slack at some position of the particle between A and B; the term to be added is now $-\frac{1}{2}(T_0+T_1)(l-a)$ where $T_1=0$ and T_0 has the same value as before. Lastly if $l>b$ and $>a$ the string is slack throughout and no term is to be added.

The equation of pressure will also have an additional term on the right-hand side. This term is T_1, where T_1 has the same value as in the equation of vis viva.

In this way the velocity of the particle and the pressure at any point may be found with ease no matter how complicated the forces may be.

Ex. 5. A small ring without weight can slide freely on a smooth wire bent into the form of an ellipse. An elastic string whose natural length is l also passes through the ring and has one end attached to the focus S and the other to the centre C. The ring being projected from the extremity A of the major axis, prove that the velocity v_1, and the pressure R_1 at the extremity B of the minor axis are given by

$$m(v_1^2-v_0^2)=(T_1+T_0)(a+ae-b),$$

$$\frac{mv_1^2}{\rho_1}=T_1+T_1\frac{b}{a}+R_1,$$

where $T_0=E(2a+ae-l)/l$, $T_1=E(a+b-l)/l$ provided the string is stretched at the beginning and end of the transit.

Ex. 6. A heavy bead is initially at the extremity of the horizontal diameter of a uniform heavy smooth circular wire whose plane is vertical. The system falls from rest through a space equal to the radius. The circular wire is then suddenly fixed in space. Find the subsequent motion of the bead, and determine if it ever comes finally to rest. Find also the pressure on the wire for any possible position of the particle.

Ex. 7. A particle, constrained to describe a circular wire, is acted on by a central force tending to a point on the circumference and varying inversely as the fifth power of the distance, prove that the pressure is constant.

Ex. 8. A particle is constrained to describe an equiangular spiral and is acted on by a central force tending to the pole whose acceleration is μr^n. The particle being projected with a velocity v_0 at a distance a_0 from the pole, prove that the velocity and pressure are given by

$$v^2 - v_0^2 = -\frac{2\mu}{n+1}(r^{n+1} - a^{n+1}),$$

$$\frac{R}{m} = \left(v_0^2 + \frac{2\mu}{n+1}a^{n+1}\right)\frac{\sin\alpha}{r} - \frac{n+3}{n+1}\mu r^n \sin\alpha.$$

If $n = -3$ and $v_0 = \sqrt{\mu}/a$, the pressure $R = 0$. *The spiral is therefore a free path when the force varies as the inverse cube of the distance*, and since any point may be regarded as the point of projection, *the velocity at every point is given by*

$$v = \sqrt{\mu}/r.$$

Ex. 9. A particle is constrained to move in an ellipse along which it is projected, and the straight line joining the foci attracts according to the Newtonian law. Prove that the resultant attraction varies inversely as the normal and that the velocity is constant.

Ex. 10. A particle of unit mass moves in a smooth circular tube of radius a, under the action of a centre of force which repels as the inverse square of the distance. If the centre of force be midway between the centre of the circle and the circumference, and the particle be projected from the end of the diameter through the centre of force remote from that point, with a velocity whose square is $4\mu(\sqrt{3}-1)/3a$, the particle will oscillate through an arc $2\pi a/3$ on either side of the point of projection. [Coll. Ex. 1897.]

Ex. 11. A particle is constrained to describe a lemniscate and is under the action of two central forces tending to the foci and varying inversely as the cube of the distance. Supposing the forces to be equal at equal distances from the foci, prove that the pressure at any point P varies as the distance of P from the centre of the curve.

Ex. 12. A particle slides down a smooth curve in a vertical plane. If the pressure on the curve is always λ times the weight of the particle, prove that the differential equation to the curve is $y + c = a(dx/ds - \lambda)^{\pm 2}$. [Math. Tripos, 1863.]

191. Rough Curve. When the particle slides on a rough curve the friction *acts opposite to the direction of motion and its magnitude is μ times the normal pressure taken positively.* The equations of motion are by Art. 181

$$mv\frac{dv}{ds}=X\cos\psi+Y\sin\psi\pm\mu R,$$

$$\frac{mv^2}{\rho}=-X\sin\psi+Y\cos\psi\pm R.$$

It is important to determine the signs of the terms containing R before proceeding with the solution. The *initial* value of the velocity being known the second equation determines the *initial* direction of R. *Taking R to act positively in the direction thus found*, it will continue to be positive during the subsequent motion until it vanishes. The initial direction of the velocity being known, the friction μR must be made to act in the first equation opposite to that direction. *If the particle start from rest* the friction μR must be made to act opposite to the direction of the tangential force. The sign of μ will then continue unchanged until either the pressure R or the velocity v vanishes and becomes reversed in direction.

To solve the equations of motion we in general eliminate R. Remembering that when s and ψ increase together $\rho=ds/d\psi$, we obtain an equation of the form

$$\frac{dv^2}{d\psi}\pm 2\mu v^2=P.$$

By using the geometrical properties of the curve we express P in terms of ψ. The equation being linear, we then have

$$v^2e^{\pm 2\mu\psi}=C+\int Pe^{\pm 2\mu\psi}\,d\psi.$$

The value of v being found, the value of R follows from either of the equations of motion.

192. Examples. *Ex.* 1. A particle is projected with a velocity V along a rough horizontal circle in a medium whose resistance varies as the square of the velocity. Prove that

$$\frac{1}{v}-\frac{1}{V}=\beta t,\quad v=Ve^{-\beta s},$$

where v is the velocity after a time t, s the arc described, and β is a constant.

Ex. 2. A small bead of unit mass is constrained to move along a rough wire, bent into the form of an equiangular spiral of angle a, in a medium whose resistance is $v^2\cos a/c$ and is under the action of no other forces. If the coefficient of friction is $\cot a$, prove that the time of travelling from a distance c to a distance b from the pole is $e^i(b-c)/V\cos a$ where $ci=b-c$, and V is the velocity at the first of these points and is directed from the pole.

Ex. 3. A heavy particle moves on a rough cycloid placed with its convexity upwards and vertex uppermost. The particle is started with an indefinitely small velocity at the point at which the tangent makes with the horizon an angle ϵ equal to the angle of limiting friction. Prove that the velocity at a point at which the tangent makes an angle ϕ with the horizon is $2\sqrt{ag}\sin(\phi-\epsilon)$ and that the particle will leave the curve at the point at which the velocity is $\sqrt{2ag}\,(\cos\frac{1}{2}\epsilon-\sin\frac{1}{2}\epsilon)$.
[Coll. Ex. 1889.]

Ex. 4. A particle is projected horizontally with velocity V along the inside of a rough vertical circle from the lowest point, prove that if it complete the circuit it will return to the lowest point with a velocity v given by

$$v^2=V^2e^{-4\mu\pi}-2ag\,(2\mu^2-1)\,(1-e^{-4\mu\pi})/(4\mu^2+1).\quad \text{[Coll. Ex. 1887.]}$$

193. Condition that a constrained motion is also free. It has already been pointed out that the required condition is that the pressure R must be zero throughout the motion, see Art. 190, Ex. 8. In this way we easily obtain several useful cases of free motion.

If T and N be the tangential and normal components of the accelerating force estimated positively in the directions in which the arc s and the radius of curvature ρ are measured, we may prove that the condition $R=0$ leads to the result $2T=\dfrac{d}{ds}(\rho N)$. This is obtained by eliminating v^2 between the normal and tangential resolutions in Art. 181 and differentiating the result. This form of the criterion though necessarily true is not sufficient to make $R=0$. As no notice is taken in it of the initial velocity, it is generally less convenient than the simple rule that $R=0$.

194. Examples. *Ex.* 1. A particle is constrained to describe a smooth circle under the action of two centres of force tending to fixed points S, S' on the same diameter, the accelerating forces being μ/r^5 and μ'/r'^5 where r, r' are the distances of the particle from the centres of force. If S and S' are inverse points, prove that the pressure can be made zero by giving μ'/μ and the velocity of projection suitable values.

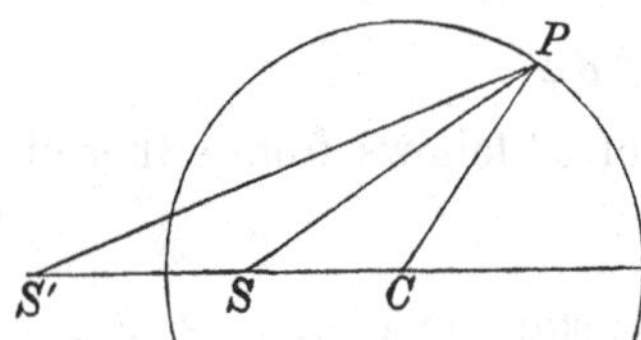

Let a be the radius; b, b' the distances of S, S' from the centre C. Since the points are inverse $bb'=a^2$. If P be the particle the triangles SPC, $S'PC$ are similar and $r'/r=a/b$. The fundamental resolutions give

$$v^2-v_0{}^2=\frac{\mu}{2}\left(\frac{1}{r^4}-\frac{1}{r_0{}^4}\right)+\frac{\mu'}{2}\left(\frac{1}{r'^4}-\frac{1}{r_0{}'^4}\right),$$

$$\frac{v^2}{a}=\frac{\mu}{r^5}\cos SPC+\frac{\mu'}{r'^5}\cos S'PC+\frac{R}{m}.$$

From these we easily obtain

$$\frac{R}{m}=\frac{1}{a}\left(v_0{}^2-\frac{\mu}{2r_0{}^4}-\frac{\mu'}{2r_0{}'^4}\right)-\frac{a^2-b^2}{2a}\left\{\mu-\mu'\left(\frac{b}{a}\right)^4\right\}\frac{1}{r^6}.$$

In order that $R=0$ we have two conditions

$$(1)\quad \mu=\mu'\left(\frac{b}{a}\right)^4, \qquad (2)\quad v_0^2=\frac{\mu}{2r_0^4}+\frac{\mu'}{2r_0'^4}.$$

Since $r'/r=a/b$, the first condition shows that the tangential accelerations due to the two forces are equal at all points of the circle. Since any point may be regarded as the point of projection the second condition gives the velocity at all points of the orbit. Since v_0 is zero at an infinite distance, this formula shows that the velocity at any point of the orbit is the same as if the particle were conducted from rest at an infinite distance to that point; Art. 181.

If the two centres of force are indefinitely near to each other the resultant attraction at any point P at a finite distance from them is the same as that of a single centre of force of double the intensity of either. Hence we arrive at Newton's theorem that *a circle can be described freely under a single centre of force whose acceleration varies as the inverse fifth power, the centre of force being on the circumference.*

When the particle comes indefinitely close to the two centres of force, they cannot be considered as one centre. The particle passes between the two centres with an infinite velocity. The two centres of force attract the particle in opposite directions with forces $\mu/(a-b)^5$ and $\mu'/(b'-a)^5$, both being infinite. The resultant force tending to the centre of the circle is therefore $\mu/a\,(a-b)^4$ which is also infinite. This last force gives the initial curvature to the subsequent path.

Ex. 2. A particle describes a catenary under the action of a force parallel to the ordinate. Show that if the pressure is zero, both the force and the velocity vary as the ordinate.

Ex. 3. Show that a particle can describe a parabola under a repulsive force in the focus varying as the distance and another force parallel to the axis always three times the magnitude of the former. Prove also that if two equal particles describe the same parabola under the action of these forces, their directions of motion will always intersect on a fixed confocal parabola. [Coll. Ex.]

Ex. 4. If a curve be described under the action of a force P tending to the pole and a normal force N, prove that

$$p^2\frac{d}{dr}\left(Nr\frac{dr}{dp}\right)+\frac{d}{dr}\left(Pp^3\frac{dr}{dp}\right)=0. \qquad \text{[Math. Tripos.]}$$

195. Does the particle leave the curve? If the particle is a small ring which slides on the curve it is obvious that it cannot separate from the curve. In this case the pressure R may have any sign.

If the particle slide on one side of the curve the pressure on the particle must tend towards that side on which the particle moves. The pressure R must therefore have the sign which suits this direction and must keep that sign throughout the motion. When therefore the analytical expression for R given by the normal resolution (Art. 184) changes sign the particle separates from the curve.

Since the forces in nature cannot be infinite the points at which R can change sign are found by putting $R=0$ in the normal resolution. Let mf be the resultant force, and let its direction make an angle ϕ with the normal. Then

$$\frac{mv^2}{\rho} = mf\cos\phi + R.$$

The possible points of separation are therefore given by

$$v^2 = f\rho\cos\phi.$$

Now $2\rho\cos\phi$ is the chord of curvature in the direction of the force mf. Representing one-fourth of this chord by c, the equation becomes $v^2 = 2fc$. Hence *the particle can leave the curve only at a point such that the velocity is that due to one-fourth the chord of curvature in the direction of the resultant force.* Art. 25.

196. Examples. *Ex.* 1. A heavy particle is suspended from a fixed point C by a string of length a. A horizontal velocity v_0 is suddenly communicated to the particle so that it begins to describe a vertical circle. It is required to determine whether the particle will oscillate or the string become slack.

The equation of vis viva shows that the velocity v at an altitude y above the lowest point of the circle is given by

$$v^2 = v_0^2 - 2gy \quad\text{.................................... (1).}$$

The tension R is given by

$$\frac{v^2}{a} = g\frac{y-a}{a} + \frac{R}{m};$$

$$\therefore\ \frac{aR}{m} = v_0^2 + ag - 3gy \quad\text{.............................. (2).}$$

If the particle oscillate the velocity is zero at the extremities of the arc of oscillation. It follows from (1) that the altitude of this point above the lowest point is $v_0^2/2g$. If the string becomes slack the tension vanishes at the point of separation. It follows from (2) that this occurs at an altitude $(v_0^2+ag)/3g$ above the lowest point. These points cannot be real points unless their altitudes are less than the diameter.

We also notice that the altitude of the first of these points is greater or less than that of the second according as v_0^2 is greater or less than $2ag$.

If $v_0^2 > 5ag$ neither point is real. The particle must describe the whole circle and the string does not become slack.

If $v_0^2 < 2ag$ the velocity vanishes at an altitude less than that at which the tension vanishes. The particle therefore oscillates and the string does not become slack.

If $v_0^2 < 5ag$ but $> 2ag$ the string becomes slack before the velocity vanishes. The particle therefore leaves the circle and describes a parabola freely in space.

If the particle, instead of being suspended by a string, were constrained to move like a bead on a vertical smooth circle of radius a the particle could not separate from the circle. It therefore oscillates or describes the whole circle according as $v_0^2 <$ or $> 4ag$.

Ex. 2. A bead can slide on a horizontal circle of radius a and is acted on only by the tension of an elastic string, the natural length of which is a, fixed to a point in the plane of the circle at a distance $2a$ from its centre; find the condition that the bead may just go round. Prove that in this case the pressures at the extremities of the diameter through the fixed point will be twice and four times the weight of the bead if that weight be such as to stretch the string to double its natural length. [Math. Tripos, 1860.]

Ex. 3. A heavy particle is allowed to slide down a smooth vertical circle of radius $27a$ from rest at the highest point. Show that on leaving the circle it moves in a parabola whose latus rectum is $16a$. [Coll. Ex. 1895.]

Ex. 4. A particle moves on the outside of a smooth elliptic cylinder whose axis is horizontal. The major axis of the principal elliptic section is vertical and the eccentricity of the section is e. If the particle start from rest on the highest generator, and move in a vertical plane, it will leave the cylinder at a point whose eccentric angle is ϕ, where $e^2 \cos^3 \phi = 3 \cos \phi - 2$. [Coll. Ex. 1892.]

Ex. 5. A particle is projected horizontally from the lowest point of a smooth elliptic arc, whose major axis $2a$ is vertical and moves under gravity along the concave side. Prove that it will quit the curve at some point if the velocity of projection V is such that V^2 lies between $2ga$ and $ga(5-e^2)$, where e is the eccentricity; and if the velocity have the latter value, prove that the particle will continue to move round the ellipse in the periodic time

$$2\left(\frac{a}{g}\right)^{\frac{1}{2}} \int_0^{\pi} \left\{\frac{1-e^2\cos^2\phi}{3-e^2+2\cos\phi}\right\}^{\frac{1}{2}} d\phi. \qquad \text{[Coll. Ex. 1892.]}$$

Ex. 6. A particle, projected inside a smooth circular tube, moves under an attractive force varying inversely as the square of the distance from a point within the rim of the tube and in its plane. Prove that the pressure cannot vanish at any point if the particle is performing complete revolutions. [Coll. Ex. 1897.]

197. Moving curves of constraint. To find the equations of motion of a particle constrained to slide on a curve moving in its own plane.

Let O be any point of the plane of the curve which it will be convenient to take as origin. Let f be the acceleration of this point, then the motion relative to O will be unchanged if we apply to every point of the curve and to the particle an acceleration equal and opposite to that of O. If we also apply to every point an initial velocity equal and opposite to that of O, we may regard O as a fixed point. *The point O is then said to have been reduced to rest.*

We shall now take O as the origin of the polar coordinates r, θ, where θ is measured from a straight line $O\xi$ fixed relatively to the curve. Let ω be the angular velocity of $O\xi$ referred to a straight line Ox fixed in space. Let ϕ be the angle the radius vector r makes with the tangent. The equations of motion are

$$\left.\begin{aligned}\frac{d^2r}{dt^2} - r\left(\frac{d\theta}{dt} + \omega\right)^2 &= \frac{P}{m} - f_1 - \frac{R}{m}\sin\phi\\ \frac{1}{r}\frac{d}{dt}r^2\left(\frac{d\theta}{dt} + \omega\right) &= \frac{Q}{m} - f_2 + \frac{R}{m}\cos\phi\end{aligned}\right\},$$

where P, Q are the components of the impressed forces, and f_1, f_2 those of f.

These equations may be written in the forms

$$\frac{d^2r}{dt^2} - r\left(\frac{d\theta}{dt}\right)^2 = \frac{P}{m} - f_1 + \omega^2 r - \left(\frac{R}{m} - 2\omega v\right)\sin\phi,$$

$$\frac{1}{r}\frac{d}{dt}\left(r^2\frac{d\theta}{dt}\right) = \frac{Q}{m} - f_2 - \frac{d\omega}{dt}r + \left(\frac{R}{m} - 2\omega v\right)\cos\phi,$$

since $$rd\theta/dt = v\sin\phi,\quad dr/dt = v\cos\phi.$$

These are the equations of motion we would have obtained if we had supposed the curve to be fixed in space and the particle to be acted on (in addition to the impressed forces) by three fictitious forces. *The introduction of these forces is said to reduce the curve to rest.*

These forces are, (1) the force $F_1 = -mf$ by which the origin is reduced to rest; (2) the force $F_2 = m\omega^2 r$ acting on the particle along the radius vector from the origin; (3) $F_3 = -mr\frac{d\omega}{dt}$ acting perpendicularly to the radius vector in the direction tending to increase θ. We also observe that the expression $R - 2m\omega v$ takes the place of the pressure of the curve on the particle.

Here v represents the velocity relatively to the curve. The velocity in space is the resultant of v and the velocity of the point of the curve occupied by the particle.

By resolving the impressed and the fictitious forces along the tangent we obtain an equation free from the reaction, and from this the velocity v of the particle relatively to the curve may be found. This equation is

$$mv\frac{dv}{ds} = (P + m\omega^2 r - mf_1)\frac{dr}{ds} + \left(Q - mr\frac{d\omega}{dt} - mf_2\right)\frac{rd\theta}{ds}.$$

By resolving the forces along the normal inwards we have

$$\frac{mv^2}{\rho} = N + R - 2m\omega v,$$

where N is the normal component of the impressed and fictitious forces. This equation gives R.

If the curve turn with a uniform angular velocity about an origin fixed in space, these equations become

$$\tfrac{1}{2}m(v^2 - v_0^2) = \int m\omega^2 r dr + \int (X d\xi + Y d\eta)$$
$$= \tfrac{1}{2}m\omega^2 r^2 + U + C,$$

$$\frac{mv^2}{\rho} = -m\omega^2 r \sin\phi + (Y\cos\psi - X\sin\psi) + R - 2m\omega v.$$

198. Examples. *Ex.* 1. A bead can slide freely on a smooth circular wire. Initially the bead is at rest at a point A. The circle then begins to turn with uniform angular velocity about a point O in the rim, where OA is a diameter. Prove that when the bead is at a distance r from O, the pressure on the curve

$$= m\omega^2(3r^2 - 4ar)/2a,$$

where a is the radius of the circle and m the mass of the bead.

To reduce the circle to rest we apply the fictitious accelerating force $F_2 = \omega^2 r$. Hence $\frac{1}{2}v^2 = \frac{1}{2}\omega^2 r^2 + C$. Since the bead is initially at rest in space, it has a velocity relatively to the curve $v = -\omega \,.\, 2a$ when $r = 2a$. Hence $C = 0$ and $v = -\omega r$ throughout the motion. To find the pressure, we have

$$\frac{v^2}{a} = -\omega^2 r \,.\, \frac{r}{2a} + \frac{R}{m} - 2\omega v.$$

Substituting for v its value, this gives the result.

Ex. 2. A bead is at rest on an equiangular spiral of angle α at a distance a from the pole. The spiral begins to turn round its pole with an angular velocity ω. Prove that the bead comes to a position of relative rest when $r = a\cos\alpha$, and that the pressure is then $\frac{1}{2}m\omega^2 a \sin 2\alpha$. Prove also that when the bead is again at its original distance from the pole, the pressure is $m\omega^2 a \sin\alpha\,(3 + \sin^2\alpha)$.

199. Time of describing an arc. *A heavy particle is in stable equilibrium at the lowest point A of a smooth fixed curve. Find the time of a small oscillation.*

Let ϕ be the angle the normal at any point P near A makes with the vertical, s the arc AP, ρ the radius of curvature at A. Then ϕ is ultimately equal to s/ρ. The equation of motion is

$$\frac{d^2s}{dt^2} = -g\sin\phi = -g\left(\frac{s}{\rho} + Bs^2 + \dots\right),$$

when $\sin\phi$ is expanded in powers of s. If the arc of oscillation is sufficiently small we may reject all the terms after the first powers of s. The time of a complete oscillation is therefore $2\pi\sqrt{\rho/g}$. *The time of oscillation is therefore the same as if the constraining curve were replaced by the circle of curvature at A.*

When it is necessary to take account of the small quantities of the order s^2, it is more convenient to replace the equation of motion by its first integral, as in Art. 200.

Ex. 1. A particle P makes small oscillations about a position of stable equilibrium at the point A of a smooth curve under the attraction of a centre of force situated at a point C on the normal OAC to the curve, the magnitude of the force being $f(r)$ where $r = CP$. Prove that the time of oscillation is $2\pi\left\{\dfrac{a\rho}{(a+\rho)F}\right\}^{\frac{1}{2}}$ where $F = f(a)$, $a = AC$ taken positively when C is on the convex side of the curve and $\rho = OA$ is the radius of curvature. Notice that the time is independent of the law of force but depends on its magnitude F at A.

Ex. 2. A smooth wire revolves with constant angular velocity ω about a fixed point in its plane and a bead is in relative equilibrium on the wire at an apse at distance a from the fixed point; prove that, if slightly disturbed, the period of a small oscillation is $\dfrac{2\pi}{\omega}\sqrt{\dfrac{\rho}{a-\rho}}$, where ρ is the radius of curvature of the wire at the apse and is less than a. [Coll. Ex. 1887.]

Reduce the curve to rest, and use Art. 199.

200. Time of describing a finite arc. By using the equation of vis viva the determination of the time can be reduced to integration. The equation of vis viva is

$$\frac{1}{2}\left(\frac{ds}{dt}\right)^2 = U + C,$$

where $U = \phi(x, y)$ is a known function of the coordinates (x, y). The constant C is known when the velocity is given at some point B whose coordinates are (h, k). We use the known equations of the curve to express any two of the variables x, y, s in terms of the third. Choosing s as this variable we have $U = \psi(s)$. Hence

$$\sqrt{2}\,.\,t = \pm\int\frac{ds}{\{\psi(s)+C\}^{\frac{1}{2}}},$$

the integration being taken from one extremity of the arc described to the other.

201. *Ex.* 1. A heavy particle is projected from a point A of a vertical circle, centre O, with such a velocity that it would come to rest at the highest point B. Prove that the time of transit from A to P is $\sqrt{\dfrac{a}{g}}\log\dfrac{\cot\frac{1}{4}\theta}{\cot\frac{1}{4}\alpha}$ where $BOA = \alpha$, $BOP = \theta$ and a is the radius. We notice that the time of arriving at the highest point is infinite.

Ex. 2. Prove that the curve such that the time of descent of a heavy particle from rest at a given point A down any arc AP is equal to the time down the chord is a lemniscate.

Taking A for origin and using polar coordinates, θ being measured from the downward vertical, the condition gives $\displaystyle\int_0^\theta\frac{ds}{\sqrt{(r\cos\theta)}} = 2\sqrt{\frac{r}{\cos\theta}}$. Differentiating both sides and solving the differential equation we find that $r^2 = A\sin 2\theta$. The condition that the lower limit on the left-hand side is zero is found on trial to be satisfied by this value of r. The required curve is therefore a lemniscate with the axis inclined at an angle of 45° to the vertical.

J. A. Serret remarks that if the ratio of the times were $k:1$, the differential equation would be

$$(k^2-1)\left(\frac{dr}{d\theta}\right)^2+2k^2\tan\theta r\frac{dr}{d\theta}+(k^2\tan^2\theta-1)r^2=0.$$

This quadratic gives $dr/rd\theta=f(\theta)$, and the solution is reduced to integration.

The history of this problem is given in the *Bulletin de la Société Mathématiques*, vol. XX. 1892. It was first solved by Euler in his *Mécanique* 1736 and afterwards by Fuss in the *Mémoires &c. de Saint Pétersbourg*, 1824. Rispal gives a geometrical proof in *Liouville*, XII. 1847.

Ex. 3. A particle is acted on by a centre of force varying as the distance. If the time of describing from rest an arc from a given point A is equal to the time of describing the chord, prove that the curve is a lemniscate. Ossian Bonnet, *Liouville*, vol. IX.

Ex. 4. If the time of descent of a heavy particle from rest at a given point A down any arc AP bears to the time of descent down the chord a ratio equal to the ratio that the length of the arc bears to k times the length of the chord, prove $s^{2-k}=Cy$, where y is the vertical ordinate of P and C is a constant.

202. Subject of integration infinite. A difficulty sometimes arises in finding the time of describing a finite arc AB if the velocity is zero at either limit. *Let a particle be projected from a point A in such a manner that the velocity of arrival at B is zero. It is required to find the time of describing the arc AB.*

Let the points A, B be determined by $s=a$, $s=b$. Since the velocity at B is zero, we have $C=-\psi(b)$. The time of describing the arc AB or BA is therefore given by

$$\sqrt{2}\,.\,t=\pm\int\frac{ds}{\{\psi(s)-\psi(b)\}^{\frac{1}{2}}},$$

the limits of integration being a, b.

The subject of integration is infinite at the limit $s=b$, but the integral itself may be finite. If we write $s=b+\sigma$, we can express the work U in a series; let

$$U-U_0=\psi(s)-\psi(b)=M\sigma^n+\dots,$$

where n is the lowest power of σ in the expansion. The part of the integral from $s=b-\sigma$ to b (σ being small) is $\pm\int\frac{d\sigma}{M^{\frac{1}{2}}\,\sigma^{\frac{1}{2}n}}$. This vanishes with σ if $n<2$ but is infinite if $n=2$ or >2.

If, as usually happens, Taylor's expansion holds true, we have $n=1$. *The time to or from a position of rest is then finite.*

If the point B is a position of equilibrium as well as of rest, we have $dU/ds = 0$ when $\sigma = 0$. It follows from Taylor's theorem that $n = 2$. *The time to a position of rest at equilibrium is therefore infinite.* If Taylor's theorem does not hold, n may lie between 1 and 2 and *the time is then finite.*

Another rule, given by Despeyrous in his *Cours de Mécanique*, is useful *when gravity is the acting force.* If B is a position of equilibrium the tangent at B is horizontal. Let ρ be the radius of curvature at B, θ the angle the normal at any point P near B makes with the vertical. The equation of vis viva is then

$$\left(\rho \frac{d\theta}{dt}\right)^2 = 2g\rho (1 - \cos\theta).$$

The time t of describing a small angle α is therefore given by

$$\left(\frac{2g}{\rho}\right)^{\frac{1}{2}} t = \int_0^\alpha \frac{d\theta}{(1-\cos\theta)^{\frac{1}{2}}} = \frac{1}{\sqrt{2}} \int_0^\alpha \frac{d\theta}{\theta}.$$

The time of transit from A to B is therefore infinite unless the radius of curvature ρ at B is zero.

203. Examples. *Ex.* 1. A heavy particle is constrained to describe the curve $x^{\frac{2}{3}} + y^{\frac{2}{3}} = a^{\frac{2}{3}}$, the axis of y being vertical. Show that the radius of curvature at every cusp is zero. Show also that a particle projected from the lowest cusp with a velocity $(2ga)^{\frac{1}{2}}$ will arrive at the next cusp in a time which is three times that of falling freely from rest at the origin to the lowest cusp.

[Despeyrous' problem.]

Ex. 2. A small ring can slide freely on a smooth wire bent into the form of a cycloid. The axes of x and y being the tangent and normal at the vertex B, the force function is given by $U = My^m$ where m is positive and <1. Prove that if the particle is projected from a point P whose ordinate is h with a velocity $(2Mh^m)^{\frac{1}{2}}$ the time of arrival at B is t where $M^{\frac{1}{2}}(1-m)\,t = 2a^{\frac{1}{2}} h^{\frac{1-n}{2}}$.

Ex. 3. If the only force acting on the particle is gravity $U = gy$. If $y = Ms^n + \ldots$ prove that $\rho = Ns^{2-n} + \ldots$ where $N^{-1} = Mn(n-1)$, provided $n > 1$. Hence $n < 2$ when $\rho = 0$ and $n = 2$ or is > 2 when ρ is finite or infinite at the position of equilibrium.

Use the theorem $\quad \rho \dfrac{d^2y}{ds^2} = \dfrac{dx}{ds} = \left\{1 - \left(\dfrac{dy}{ds}\right)^2\right\}^{\frac{1}{2}}.$

Motion in a cycloid.

204. *A heavy particle is constrained to move in a smooth fixed cycloid whose plane is vertical and vertex downwards. It is required to find the motion.*

Let A, A' be the cusps, O the vertex, OQD a circle equal to the generating circle placed with its diameter on the axis OD,

C its centre. Let PQN be a perpendicular on the axis drawn from any point P on the cycloid. The following geometrical

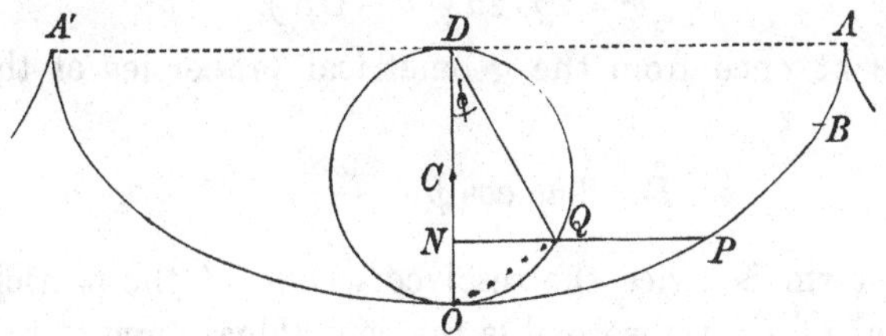

properties of the cycloid are given in treatises on the differential calculus.

(1) The tangent at P is parallel to the chord OQ and the arc OP is twice the chord OQ.

(2) The radius of curvature at P is parallel to the chord QD and is equal to twice that chord.

(3) The distance PQ is equal to the circular arc OQ.

Let the angle $QDO = \phi$, and let a be the radius of the generating circle. The tangential and normal resolutions at P give (Art. 181)

$$\left.\begin{aligned} \frac{d^2s}{dt^2} &= -g\sin\phi = -g\frac{s}{4a} \\ \frac{v^2}{\rho} &= -g\cos\phi + \frac{R}{m} \end{aligned}\right\} \quad \text{.....................(1).}$$

The first equation shows at once that the motion is oscillatory, Art. 118. The time of a complete oscillation is $4\pi\sqrt{\frac{a}{g}}$ and is independent of the arc described. Let t be measured from the instant at which the particle P passes the vertex, let c be the semi-arc OB of oscillation. The first equation gives

$$s = c\sin\left(\sqrt{\frac{g}{4a}}\,t\right).$$

It follows that if two particles oscillate in the same or in equal cycloids both starting from the vertex, the two arcs described in equal times are in a constant ratio, viz. that of the complete arcs. If therefore the circumstances of the motion of a particle oscillating from cusp to cusp are known, those of a particle oscillating in any smaller arc can be immediately deduced.

205. If b is the depth below the cusp of the extremity B of the arc of oscillation, we have by the principle of vis viva

$$v^2 = 2g(2a - b - ON).$$

It follows at once from the geometrical properties of the curve that

$$R = 2mg\cos\phi - \frac{2mgb}{\rho}.$$

The first term is twice the resolved weight of the particle along the normal at P; the second is the centrifugal force of a particle moving uniformly with the velocity due to the depth below the cusp of the extremity B of the arc of oscillation.

206. Examples. *Ex.* 1. A particle oscillates in a complete cycloid from cusp to cusp. Prove the following properties.

(1) The velocity v at any point P is equal to the resolved part of the velocity V at the vertex along the tangent at P, i.e. $v = V\cos\phi$.

(2) The time of describing an arc OP is proportional to the angle ODQ, i.e.

$$\phi = \sqrt{\frac{g}{4a}}\,.\,t.$$

(3) The particle moves as if it were rigidly attached to the generating circle, that circle being supposed to roll with a uniform angular velocity on the base AA'. This follows from the last result because $d\phi/dt$ is constant.

(4) The centrifugal force at any point P is equal to the resolved part of the weight along the normal at P, and the pressure is twice either of these.

Ex. 2. A heavy particle starts from rest at a point A of a cycloid, prove that the time T of transit from any point P to any point Q is given by

$$\cos\left(\tfrac{1}{2}T\sqrt{\frac{g}{a}}\right) = \frac{\sqrt{(pq)} + \sqrt{(l-p)(l-q)}}{l},$$

where p, q, l are the depths of P, Q and the vertex below the level of A, and a is the radius of the generating circle.

Ex. 3. A particle slides down a smooth cycloid starting from rest at the cusp. Prove that the whole acceleration at any instant is in magnitude equal to g and that its direction is towards the centre of the generating circle. [Coll. Ex.]

The required acceleration is equivalent to the resultant of g and R/m; the result follows at once from the triangle of accelerations.

Ex. 4. A smooth cycloid is placed with its axis AB inclined to the vertical, and its convexity upwards; a particle begins to slide down the arc from A, and leaves the curve at P; the perpendicular from P on AB cuts at Q the circle on AB as diameter, and QR is a diameter of this circle; prove that PR is horizontal. [Math. T. 1888.]

207. When a pendulum is removed from one place to another the number, n, of oscillations in any given time (such as a day) is altered by the change in the force of gravity and the alteration of the length l of the pendulum due to a change of temperature. Since the number of oscillations in a given time varies inversely as the time of a single oscillation, we have $n^2 = Cg/l$ where C is some constant. Taking the logarithmic differential, we find

$$2\frac{\delta n}{n} = \frac{\delta g}{g} - \frac{\delta l}{l}.$$

This formula is a very convenient first approximation to the value of δn.

208. *Ex.* 1. Prove that a seconds pendulum brought to the summit of a mountain x miles high loses about $22x$ seconds per day if the attraction of the mountain can be neglected. If the mountain is of the form of table-land, the loss is only five-eighths of the above amount. The length of the pendulum is supposed to be unaltered.

By Dr Young's rule the attraction at the top of table-land is $g\left(1 - \frac{5}{4}\frac{x}{a}\right)$ nearly where a is the radius of the earth.

Ex. 2. A railway train is running smoothly along a curve at the rate of 60 miles per hour, and a pendulum which would ordinarily oscillate seconds is observed to oscillate 121 times in two minutes. Show that the radius of the curve is approximately a quarter of a mile. [Coll. Ex. 1895.]

Ex. 3. If the moon be in the zenith, prove that a seconds pendulum would be losing at the rate of $\frac{1}{200}$th of a second per day.

The moon attracts the earth as well as the pendulum and its disturbing effect is measured by the difference of its attractions at the centre of the earth and at the pendulum. This is $\frac{2M}{E}\left(\frac{a}{r}\right)^3 g$ where $M = \frac{1}{80}E$ is the mass, and $r = 60a$ is the distance of the moon.

209. *Ex.* 1. A heavy particle oscillates on a smooth fixed curve, and the periods of oscillation in all arcs are the same. Prove that the curve is a cycloid.

Let the axis of y be measured vertically upwards from the lowest point of the curve and let $y = h$ be the initial value of y. Let the equation of the curve be $s = f(y)$, where s is the arc measured from the lowest point. Since $v^2 = 2g(h - y)$ the time t of reaching the lowest point is given by

$$\sqrt{2g}\,t = \int_0^h \frac{f'(y)\,dy}{\sqrt{(h-y)}}.$$

Put $y = hz$, then $$\sqrt{2g}t = h^{\frac{1}{2}}\int_0^1 \frac{f'(hz)\,dz}{\sqrt{(1-z)}}.$$

Since the time t is to be the same for all values of h, we have $dt/dh = 0$. Hence

$$\int_0^1 \frac{dz}{\sqrt{(1-z)}}\frac{d}{dh}\{h^{\frac{1}{2}}f'(hz)\} = 0.$$

This equation requires that the second factor under the integral sign should be zero. If this were not true we could, by taking h small enough, make that factor keep the same sign, while hz varies from $hz = 0$ to $hz = h$. Every term of the integral would then have the same sign and the sum could not be zero. Hence $h^{\frac{1}{2}}f'(hz)$ is independent of h, and therefore $f'(hz) = M(hz)^{-\frac{1}{2}}$ where M is a constant independent of h and z. We thus find by an easy integration that the arc $f(y) = 2My^{\frac{1}{2}}$. This is the equation of a cycloid having the line joining the cusps horizontal.

Ex. 2. A body of mass M can slide on a perfectly smooth horizontal plane and has attached to it a thin tube in the vertical plane containing the centre of gravity. The form of the tube is such that the periods of the oscillations of a particle of mass m placed in it are the same for all arcs. Prove that the form of the tube may be derived from a cycloid by elongating the ordinates perpendicular to the axis in the ratio $\sqrt{(M+m)}/\sqrt{M}$. This problem is due to Clairaut; *Mém. de l'Acad.*, Paris, 1742.

210. Resisting medium. If the particle *oscillate on a smooth cycloid in a medium resisting as the velocity,* the tangential equation of motion becomes

$$\frac{d^2s}{dt^2} = -n^2s - 2\kappa\frac{ds}{dt},$$

where $n^2 = g/4a$. This problem has been discussed in Arts. 121 and 126. The interval between two successive passages through the lowest point is always the same and the successive arcs of descent and ascent are in geometrical progression.

If the resistance vary as the square of the velocity, the motion is discussed in Art. 129.

211. Tautochronous curves. When a particle oscillates on a given smooth curve either in a vacuum or in a medium whose resistance varies as the velocity, we know that the oscillation is tautochronous about the position of equilibrium if the tangential force $F = m^2s$ where s is the length of the arc measured from the position of equilibrium and m is a constant, Art. 118. If therefore any rectifiable curve is given a proper force to produce a tautochronous motion can at once be assigned.

A catenary is a tautochronous curve for a force acting along the ordinate equal to m^2y because the resolved part along the tangent is obviously m^2s.

The equiangular spiral is tautochronous for a central force μr tending to the pole, because the resolved part along the tangent being m^2s where $m^2 = \mu\cos^2\alpha$, the time of arrival at the pole is the same for all arcs.

In the same way the epicycloid and hypocycloid are tautochronous curves for a central force tending from or to the centre of the fixed circle and varying as the distance, because since

$$r^2 = As^2 + B,$$

the resolved part along the tangent, viz. $\mu r dr/ds$, varies as s. In all these cases the time of arrival at the position of equilibrium is the least positive root of $\tan nt = -n/\kappa$ (Art. 121), where $2\kappa v$ is the resistance and $n^2 + \kappa^2 = m^2$. The whole time from one position of momentary rest to the next is π/n.

The properties of tautochronous curves are more fully discussed in the author's *Rigid Dynamics*. A historical summary is also there given.

212. Rough cycloid. *A particle slides from rest on a rough cycloid placed with its axis vertical in a medium whose resistance varies as the velocity. Prove that the motion is tautochronous.*

The descending motion is given by

$$\frac{dv}{dt} = \mu R - g \sin \phi - 2\kappa v, \qquad \frac{v^2}{\rho} = R - g \cos \phi \ldots\ldots\ldots(1),$$

where v is really negative. Eliminating R

$$\frac{dv}{dt} - \frac{\mu}{\rho} v^2 + 2\kappa v + \frac{g}{\cos \epsilon} \sin (\phi - \epsilon) = 0,$$

where $\tan \epsilon = \mu$. This may be written

$$\frac{d}{dt}(e^u v) + 2\kappa (e^u v) + \frac{g}{\cos \epsilon} e^u \sin (\phi - \epsilon) = 0,$$

provided $\dfrac{du}{dt} = -\mu \dfrac{v}{\rho}$, that is $u = -\mu\phi$. Put $e^u ds = dw$;

$$\therefore \quad \frac{d^2 w}{dt^2} + 2\kappa \frac{dw}{dt} + \frac{g}{\cos \epsilon} e^{-\mu\phi} \sin (\phi - \epsilon) = 0.$$

Now $\quad w = \int e^{-\mu\phi}\, 4a \cos \phi \, d\phi = 4a \cos \epsilon \, e^{-\mu\phi} \sin (\phi - \epsilon)$.
The equation therefore reduces to

$$\frac{d^2 w}{dt^2} + 2\kappa \frac{dw}{dt} + \frac{gw}{4a \cos^2 \epsilon} = 0.$$

This is the linear equation, Art. 121. We infer that at whatever point of the cycloid the particle is placed at rest, it arrives at the point E determined by $w = 0$, that is $\phi = \epsilon$, in the same time. Such a motion is called tautochronous. The point E is clearly an extreme position of equilibrium in which the limiting friction just balances gravity.

The time of arrival at E is given by the least positive root of the equation $\tan nt = -n/\kappa$ where $n^2 + \kappa^2 = g/4a\cos^2\epsilon$. The whole time from one position of momentary rest to the next is π/n.

So long as the particle is moving in the same direction the constant μ retains the same sign. The motion is therefore given by

$$e^{-\mu\phi}\sin(\phi-\epsilon) = Ae^{-\kappa t}\sin(nt+B).$$

When the particle arrives at the next position of rest, it will begin to return or will remain there at rest according as the value of ϕ at that point is greater or less than the angle of friction.

Motion in a circle.

213. *A heavy particle is constrained to move in a fixed circle whose plane is vertical. It is required to find the time of describing an arc.*

Let C be the centre, A and B the lowest and highest points of the circle, a its radius. Let P be the position of the particle at any time t, ϕ the angle CBP.

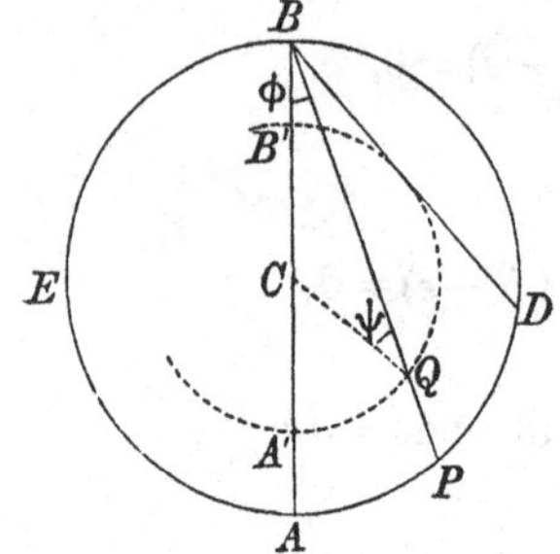

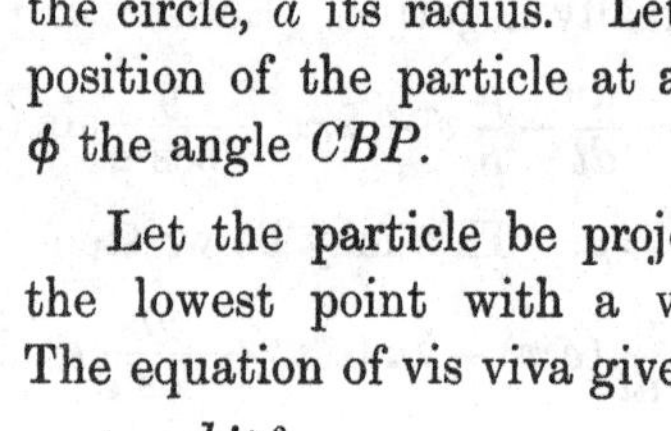

Let the particle be projected from the lowest point with a velocity V. The equation of vis viva gives

$$\left(2a\frac{d\phi}{dt}\right)^2 - V^2 = -2ga(1-\cos 2\phi).$$

Let us put $V^2 = 2gh$, so that the velocity of projection is that due to a height h; we also put $h = 2a \,.\, \kappa^2$. If $\kappa > 1$, the velocity at the lowest point is more than sufficient to carry the particle to the highest point of the circle, the particle therefore goes continually round the circle in the same direction. If $\kappa < 1$ the velocity at the lowest point is insufficient to carry the particle round the circle, the particle therefore oscillates. If $\kappa = 1$ the particle arrives at the highest point with a velocity zero, but only after an infinite time has elapsed, Art. 201.

Substituting for V^2 in the equation of vis viva, we have

$$\frac{a}{g}\left(\frac{d\phi}{dt}\right)^2 = \kappa^2 - \sin^2\phi \quad\text{.....................}\quad (2).$$

If t be the time of describing the arc AP which subtends an angle 2ϕ at the centre, we have

$$\sqrt{\frac{g}{a}}\,.\,t=\int_0^{\phi}\frac{d\phi}{\sqrt{(\kappa^2-\sin^2\phi)}}\text{ ....(3)},$$

where one radical is positive and the other has the same sign as $d\phi/dt$.

If $\kappa=1$, the integral is a known form. We have

$$\sqrt{\frac{g}{a}}\,.\,t=\int\frac{d\phi}{\cos\phi}=\log\tan\left(\frac{\pi}{4}+\frac{\phi}{2}\right)\text{ ....(4)},$$

when $\phi=\frac{1}{2}\pi$, t is infinite so that the particle takes an infinite time to reach the highest point.

If $\kappa>1$, we write the integral in the form

$$\sqrt{\frac{g}{a}}\,.\,t=\frac{1}{\kappa}\int_0^{\phi}\frac{d\phi}{\sqrt{\left(1-\dfrac{1}{\kappa^2}\sin^2\phi\right)}}\text{ ....(5)}.$$

This elliptic integral* gives the time of describing the arc which subtends an angle ϕ at the highest point of the circle. The time of arriving at the highest point is found by writing $\frac{1}{2}\pi$ for the upper limit.

214. When $\kappa<1$, we put $\kappa=\sin\alpha$. We see from (2) that $\sin\phi$ cannot exceed κ and that the velocity is zero when $\sin\phi=\kappa$; the particle therefore oscillates on each side of the lowest point through an arc AD or AE which subtends an angle α at the highest point. Let $\sin\phi=\kappa\sin\psi$, so that ψ varies from zero to $\frac{1}{2}\pi$. We then find after an easy substitution in (3)

$$\sqrt{\frac{g}{a}}\,.\,t=\int_0^{\psi}\frac{d\psi}{\cos\phi}=\int_0^{\psi}\frac{d\psi}{\sqrt{(1-\kappa^2\sin^2\psi)}}\text{ ....(6)}.$$

This elliptic integral determines the time of describing an angle ϕ where ϕ and ψ are related by the equation $\sin\phi=\kappa\sin\psi$.

We can construct the angle ψ geometrically. Describe a circle with centre C to touch BD, and let BP intersect this circle in Q; then the angle $BQC=\psi$. For another construction we draw a chord $A'P'$ equal to the chord AP, then the angle $CB'P'=\psi$.

* The reader is referred to Prof. Greenhill's *Treatise on the applications of elliptic functions.* He begins with the problem of the simple circular pendulum as being the best introduction to the theory of these functions.

In obtaining (6) we supposed the sign of $\cos\psi$ to be the same as that of the radical in (3) and therefore the same as that of $d\phi/dt$. Since $\cos\phi$ is positive, it then follows from (6) that $d\psi/dt$ is positive. The point Q therefore travels round the circle, being the lower or upper intersection of BP with the circle according as P is moving from A to D or from D to A.

215. **Series for the time of oscillation.** We may approximate very closely to the time of a complete vibration by using a series. If T be this time, the formula (6) gives $\frac{1}{4}T$ when the upper limit is $\frac{1}{2}\pi$. We have by the binomial theorem

$$(1-\kappa^2\sin^2\psi)^{-\frac{1}{2}}$$
$$=1+\tfrac{1}{2}(\kappa\sin\psi)^2+\ldots+\frac{1.3.5\ldots(2n-1)}{2.4.6\ldots 2n}(\kappa\sin\psi)^{2n}+\ldots.$$

By a theorem in the integral calculus

$$\int_0^{\frac{1}{2}\pi}(\sin\psi)^{2n}\,d\psi=\frac{1.3.5\ldots(2n-1)}{2.4.6\ldots 2n}\cdot\frac{\pi}{2}.$$

It immediately follows that

$$T=2\pi\sqrt{\frac{a}{g}}\left\{1+\left(\frac{1}{2}\right)^2\kappa^2+\ldots+\left(\frac{1.3.5\ldots(2n-1)}{2.4.6\ldots 2n}\right)^2\kappa^{2n}+\ldots\right\},$$

where $\kappa=\sin\alpha$ and α is the angle subtended at the highest point of the circle by the half-arc of oscillation. It is also useful to notice that κ is the ratio of the chord of the half-arc to the diameter of the circle.

The first term of this series represents the time of an infinitely small oscillation. The other terms are regarded as small corrections to this time, and are sometimes called the "reduction to infinitely small arcs." The second term is usually a sufficient correction. Thus suppose the arc of oscillation on each side of the vertical to subtend an angle of 36° at the point of suspension, then $\alpha=18°$ and $\kappa=\frac{3}{10}$. The second term is only about $\frac{1}{50}$th and the third $\frac{1}{900}$th of the first.

216. **Relation between continuous and oscillatory motions.** Comparing the formulæ (5) and (6) we see that the integrals are the same except that the moduli κ and $1/\kappa$ are reciprocals. This leads to a theorem by which we connect a motion all round the circle with an oscillatory motion.

Let two particles P, P' be projected from the lowest points A, A', of two circles of radii a, a', and let these be acted on by unequal gravitational forces g and g'. Let the velocities of projection V, V' be such that the moduli are reciprocals. Then κ being less than unity, we have $V^2=4ag\kappa^2$, $V'^2=4a'g'/\kappa^2$. It then follows from what precedes that the particle P' travels round the circle and P oscillates in a semi-arc equal to AD, where the angle $DBA=\alpha$ and $\kappa=\sin\alpha$.

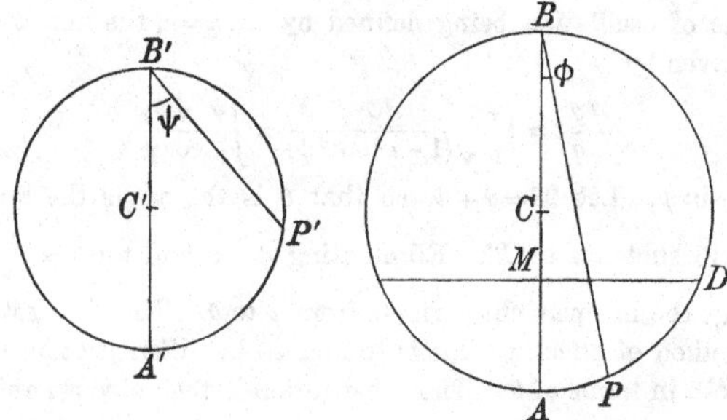

Let P, P' be the positions of the particles when the angles $ABP=\phi$, $A'B'P'=\psi$, where $\sin\phi=\kappa\sin\psi$. If t, t' be the times of describing the arcs AP, $A'P'$ we have

$$\sqrt{\frac{g'}{a'}}t'=\kappa\int_0^\psi\frac{d\xi}{\sqrt{(1-\kappa^2\sin^2\xi)}},\quad \sqrt{\frac{g}{a}}t=\int_0^\psi\frac{d\xi}{\sqrt{(1-\kappa^2\sin^2\xi)}}.$$

It follows therefore that $\sqrt{\frac{g'}{a'}}t'=\kappa\sqrt{\frac{g}{a}}t$. *The points P, P' therefore correspond to each other in the two motions, and it is easy to see that they are geometrically connected by the relation*

$$\frac{chord\ AP}{chord\ A'P'}=\frac{\kappa a}{a'}=\frac{chord\ AD}{diam.\ A'B'}.$$

It is obviously convenient that the particles should occupy corresponding points at the same instant of time. We therefore choose the constants a', g', so that $t=t'$. We then have $g'/a'=\kappa^2 g/a$. The equations of motion take the forms

$$\sqrt{\frac{a}{g}}\frac{d\phi}{dt}=\kappa\cos\psi,\quad \sqrt{\frac{a'\kappa^2}{g'}}\frac{d\psi}{dt}=\cos\phi,$$

where the coefficients on the left hand are equal.

If we make the radii equal we can suppose *both particles to describe the same circle.* We then have

$$a'=a,\quad g'=\kappa^2 g,\quad V'=\frac{1}{\kappa}V,\quad A'P'=\frac{1}{\kappa}.AP.$$

217. *Ex.* 1. If the circle described by P' has AM for its diameter, prove that P, P' move so as to be always on the same horizontal line, the gravitational forces being g and $g\kappa^4$ respectively.

Ex. 2. If the circles are equal and the arc PP' is bisected by a point Q, prove that Q moves on the circle as if it were a third heavy particle acted on by a gravitational force $g''=g\kappa$. The velocity of Q at A (and at all points) is equal to the mean of the velocities of P and P'. Prove also that Q goes half round while P' goes all round. Sang, *Edinburgh Trans.* 1865, vol. 24.

These results follow at once from Art. 216.

218. Relations between two oscillatory motions. The investigation of these relations is properly a part of the theory of elliptic integrals, but the following theorem will serve as an example.

If T, T' be the periods of oscillation corresponding to two semi-arcs which subtend angles α, α' at the highest point of the circle and so related that $\sin\alpha=(\tan\frac{1}{2}\alpha')^2$, then will $T=T'(\cos\frac{1}{2}\alpha')^2$.

The half arc of oscillation being defined by $\sin\alpha=\kappa$, the time t of describing the angle ϕ is given by

$$\sqrt{\frac{g}{a}}\,t=\int_0^\psi \frac{d\psi}{\sqrt{(1-\kappa^2\sin^2\psi)}}=\int_0^\psi \frac{d\psi}{\cos\phi};$$

where $\sin\phi=\kappa\sin\psi$. Let $2\theta=\phi+\psi$, so that θ is the angle the arc $A'Q$ in the figure of Art. 213 subtends at B'. Eliminating ϕ we find $\tan\psi=\dfrac{\sin 2\theta}{\kappa+\cos 2\theta}$. We shall now change the independent variable from ψ to θ. The simplest (though not the shortest) method of effecting this is to find $d\psi$ by differentiation and $\sin^2\psi$ by trigonometry both in terms of θ. The substitution is then obvious and we have*

$$\int_0^\psi \frac{d\psi}{\sqrt{(1-\kappa^2\sin^2\psi)}}=\frac{2}{1+\kappa}\int_0^\theta \frac{d\theta}{\sqrt{(1-\lambda^2\sin^2\theta)}},$$

where $\lambda=2\sqrt{\kappa}/(1+\kappa)$. Remembering that $\kappa=\sin\alpha$ and $\sin\phi=\kappa\sin\psi$, we now write $\lambda=\sin\alpha'$ and $\sin\phi'=\lambda\sin\theta$.

Let two particles P, P' oscillate in the circle APB through arcs AD, AD' which subtend angles α, α' at the highest point B, then the last equation shows that the times t, t', of describing corresponding angles ϕ, ϕ', are connected by the relation

$$t=2t'/(1+\kappa).$$

To compare the changes of the values of these corresponding angles we refer to the figure of Art. 213. As P moves from A to D and back to A, Q travels round the semicircle $A'QB'$, 2θ increases from 0 to π, and ϕ' increases from 0 to α'. Thus the oscillation from A to D and back to A corresponds to the oscillation A to D' only, i.e. a complete oscillation of P corresponds to half a complete oscillation of P'. If T, T' be the times of a complete oscillation of P, P', we have therefore

$$T=T'/(1+\kappa).$$

The two angles α, α' are connected by the relation

$$\sin\alpha'=\lambda=\frac{2\sqrt{\kappa}}{1+\kappa};\quad \therefore \sqrt{\kappa}=\frac{1\pm\cos\alpha'}{\sin\alpha'}.$$

Since $\kappa<1$ and $\alpha'<\frac{1}{2}\pi$ we take the lower sign in the value of $\sqrt{\kappa}$. Hence $\sin\alpha=(\tan\frac{1}{2}\alpha')^2$. It follows also that $t=2t'(\cos\frac{1}{2}\alpha')^2$.

Ex. If α_1, α_2, ... be a series of angles connected by the relation

$$\sin\alpha_{n+1}=(\tan\tfrac{1}{2}\alpha_n)^2,$$

and if T_1 be the time of a complete revolution in an arc subtending $4\alpha_1$ at the point of suspension, prove that

$$T_1=(\sec\tfrac{1}{2}\alpha_1\,.\,\sec\tfrac{1}{2}\alpha_2\,...\text{ to }\infty)^2\,.\,2\pi\sqrt{(a/g)}.$$ [Sang.]

219. Co-axial Circles. Two heavy particles, constrained to describe the same vertical circle, are projected from any two points with velocities due to their depths below the same horizontal line. It is required to prove that the straight line joining the particles always touches a co-axial circle.

---

* Cayley's *Elliptic Functions*, Art. 243.

Let Oy be the radical axis of two co-axial circles whose centres are C, C'. Let a tangent at any point T of one circle intersect the other in two points P, Q.

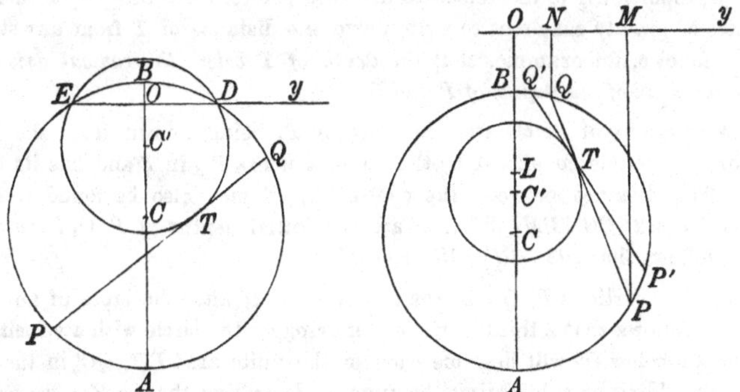

Let PM, QN be perpendiculars on the radical axis. By a known property of co-axial circles the tangents PT, QT drawn from points on the outer circle satisfy the relations*

$$PT^2 = 2 \,.\, CC' \,.\, PM, \quad QT^2 = 2 \,.\, CC' \,.\, QN.$$

In the time dt let the tangent move into the position $P'TQ'$. Then since the elementary arcs QQ', PP', make equal angles with the chord $P'Q'$, the triangles QTQ', PTP' are similar: hence

$$\text{arc } QQ'/\text{arc } PP' = QT/PT.$$

It follows from these two geometrical theorems that

$$(\text{vel. of } Q)^2/(\text{vel. of } P)^2 = QN/PM.$$

If then the point P move with a velocity equal to $(2g \,.\, PM)^{\frac{1}{2}}$, the point Q must move with a velocity equal to $(2g \,.\, QN)^{\frac{1}{2}}$. It follows that the points P, Q are the positions of two particles moving with velocities due to their depths below Oy.

If the radical axis is external to the circle described by the particles, the particles go round the circle. If the radical axis intersects the circle in the two points D and E, the particles oscillate in the same arc DAE.

In the figures the particles have been supposed to move the same way round the circle. If their directions are opposite the chord PQ envelopes a circle or a part of a co-axial circle situated above Oy.

---

* The properties of co-axial circles are fully discussed by geometrical methods in Lachlan's *Modern Pure Geometry*. The following is an analytical proof of the property $PT^2 = 2 \,.\, CC' \,.\, PM$.

Let c, c' be the distances of the centres C, C' from Oy, δ the length of a tangent drawn from O to any co-axial circle. The equations of the circles are therefore

$$x^2 - 2c'x + y^2 + \delta^2 = 0 \quad\text{.................................}\quad (1),$$

$$x^2 - 2c\,x + y^2 + \delta^2 = 0 \quad\text{.................................}\quad (2).$$

If x, y be any point P external to the first circle, PT a tangent

$$PT^2 = x^2 - 2c'x + y^2 + \delta^2.$$

If P lie on the second circle this becomes $2(c - c')\,x$ by subtracting the second equation. This is the result to be proved.

The point of contact T divides the chord PQ in the ratio of the velocities at P, Q. That point is therefore the centre of gravity of two masses placed at P, Q inversely proportional to the velocities at those points. The ordinary formulæ for the centre of gravity enable us to write down the distance of T from any straight line. It follows, for example, that *the depth of T below the radical axis is the geometrical mean of the depths of P and Q.*

Some positions of P, Q, and therefore of T, being known from the initial conditions, the circle enveloped by the chord touches PQ in T and has its centre in OA. The distance between the centres C, C' may also be found from the equation $PT^2 = 2 . CC' . PM$. If x, x' are the initial depths of P, Q, l the initial chord, it follows that $\sqrt{(2 . CC')} = l/(\sqrt{x} + \sqrt{x'})$.

Let the two particles P, Q take the positions P', Q' after the lapse of any finite time t. It follows that a third particle R moving on the circle with a velocity due to its depth below Oy will describe each of the finite arcs PP', QQ' in the same time t. By adding or subtracting the time of describing the arc $P'Q$, we see that the *times of describing PQ, $P'Q'$*, i.e. *the arcs cut off by any two tangents to a co-axial circle, are equal.*

When the radical axis is external to the system of circles there are two points L, L' one on each side of Oy which are the positions of the two co-axial circles whose radii are zero. Since L is an evanescent circle the distance OL is equal to the tangent drawn from O to any co-axial circle. Also, for the same reason, *any straight line drawn through L divides the circle APB into two parts which are described in equal times.*

220. Examples. *Ex.* 1. A circle is drawn to touch at their middle points the chord and arc of oscillation of a particle which is moving on a vertical circle under the action of gravity. Prove that a point on the first circle in the same horizontal line with the particle moves with a velocity equal to $2\sqrt{(gr)} \sin^2 \frac{1}{2}\alpha \cos \frac{1}{2}\theta$ where r is the radius of the circle on which the particle moves and α, θ are the angles which the radius drawn to the particle makes with the vertical at the instant when it is stationary and at the instant considered. [Math. Tripos.]

Ex. 2. A particle describes a vertical circle of radius a with a velocity due to its depth below the highest point B. Prove that the radius of the circle enveloped by the chord joining any two positions of the particle at a constant time interval T is $a/\cosh^2(T\sqrt{g/a})$. Prove also that the depth of the point of contact of the chord and its envelope below B is $2a/\cosh \xi_1 \cosh \xi_2$ where $\xi_1\sqrt{a/g}$ and $\xi_2\sqrt{a/g}$ are the times from the lowest point of the extremities of the chord. [Coll. Ex. 1897.]

Ex. 3. Prove that if a particle move round a circle so that its velocity is proportional to the product of its distances from two fixed points in the plane, one inside and one outside, any circle drawn through them divides the orbit into two parts which are described in equal times. State the corresponding result when the points are both inside, or both outside. [Math. Tripos, 1888.]

Describe two consecutive circles through the fixed points A, B to cut the given circle in the points P, P' and Q, Q'; we shall prove that the times of describing the elementary arcs PP', QQ' are equal.

The distance between any two parallel tangents to these co-axial circles is easily seen to be proportional to the product $AP . BP$ where P is the point of contact of either. If then PR, QS are any two normals to the circle $APQB$ intersecting the consecutive circle in R and S, the time of moving from P to R is equal to the time from Q to S.

Because the given circle and the circle $APBQ$ are symmetrical about the straight line joining their centres, the tangents PP', QQ' make equal angles with the normals PR, QS; the lengths PP', QQ' are therefore proportional to PR, QS. The arcs PP', QQ', therefore, are also described in equal times.

Let $ABCD$ be any one co-axial circle cutting the given circle in C, D. Then describing all the co-axial circles, each elementary arc PP' in the larger arc CD has a corresponding elementary arc QQ' in the smaller arc CD, and these are described in equal times. The times therefore of describing the smaller and larger arcs CD are equal.

Wherever A, B may be, let two of the co-axial circles cut the given circle in C, D and C', D'. It follows from what precedes that the times of describing the arcs CC', DD' are equal.

Ex. 4. A particle oscillates in a circular arc EAD, see fig. of Art. 219. A tangent is drawn from A to the co-axial circle to cut the arc of oscillation in X. A horizontal tangent to the same co-axial cuts the same arc in Y. It follows from the theorem of Art. 219, that the time of moving from A to X is twice that from A to Y. Prove that this is equivalent to the theorem

$$\int_0^{\psi'} \frac{d\psi}{\sqrt{(1-\kappa^2\sin^2\psi)}} = 2\int_0^{\psi} \frac{d\psi}{\sqrt{(1-\kappa^2\sin^2\psi)}}\,.$$

where $$\sin\psi' = 2\sin\psi\cos\psi\,(1-\kappa^2\sin^2\psi)^{\frac{1}{2}}\,(1-\kappa^2\sin^4\psi)^{-1}.$$

[Cayley's *Elliptic Functions*, Art. 249.]

CHAPTER V.

MOTION IN ONE PLANE.

Moving Axes.

221. THE components of velocity and acceleration along the axes of coordinates, the tangent and normal to the path and in some other directions have been already considered in Chapter I. The solution of the more difficult problems in dynamics requires however that we should have at our command a greater power of resolution than is given by these. We shall now investigate the general components for any moving axes in one plane.

222. To avoid the continual repetition of the same argument, we shall use the term *vector* to represent the subject under consideration, whether it be a velocity or an acceleration.

Let us understand by a vector *any quantity which has direction as well as magnitude, and which obeys the parallelogram law.* Thus the radius vector of a point P is a vector and its resolved parts along the axes are the coordinates x and y. Again the velocity of P is a vector, and its resolved parts along the axes are dx/dt and dy/dt. The acceleration of P is also a vector and the resolved parts are d^2x/dt^2 and d^2y/dt^2. Lastly if R be any vector whose direction makes an angle ψ with the axis of x, its components along the axes, supposed to be rectangular, are $R\cos\psi$ and $R\sin\psi$.

223. Fundamental theorem. *A vector R having been resolved in the directions of two rectangular axes $O\xi$, $O\eta$ which turn round a fixed origin in a given manner, it is required to find the rates at which these components are increasing with the time.*

Let P be the position of the moving point at any time t. Draw a straight line PQ to represent the instantaneous direction and magnitude of the vector R. Let u, v be the resolved parts of the vector in the directions of the axes $O\xi$, $O\eta$.

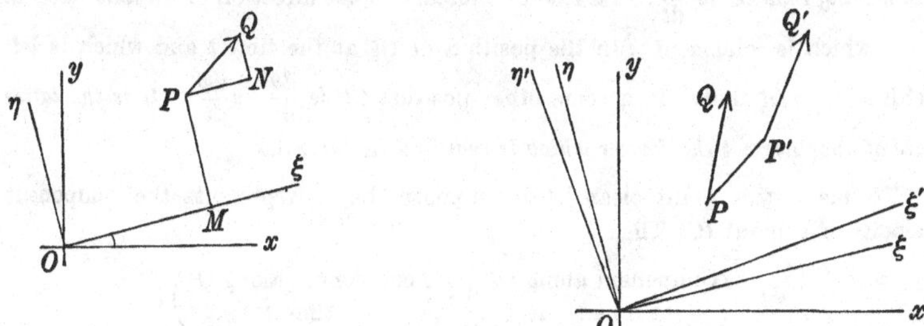

After a time dt, the point P will occupy a position P', the vector R will become $R+dR$ and may be represented by the straight line $P'Q'$. The axes $O\xi$, $O\eta$ will turn round O through a small angle $d\phi$ and will take the positions $O\xi'$, $O\eta'$. The resolved parts of $R+dR$ along these new axes will be $u+du$, and $v+dv$.

At the time t the component of the vector in the direction $O\xi$ is u. At the time $t+dt$ the component in the same direction (i.e. in the direction $O\xi$ not $O\xi'$) is

$$(u+du)\cos d\phi-(v+dv)\sin d\phi.$$

The rate of increase of u in the direction $O\xi$ is found by subtracting the component at the time t from that at the time $t+dt$ and dividing by dt.

If we represent the rate of increase in the direction $O\xi$ by u_1, we have

$$u_1=\frac{[(u+du)\cos d\phi-(v+dv)\sin d\phi]-u}{dt}.$$

When we reject the squares of small quantities according to the rules of the differential calculus, we write unity for $\cos d\phi$ and $d\phi$ for $\sin d\phi$. We therefore have

$$u_1=\frac{du}{dt}-v\frac{d\phi}{dt}.$$

In the same way if the rate of increase in the direction $O\eta$ be v_1, we have

$$v_1=\frac{(u+du)\sin d\phi+(v+dv)\cos d\phi-v}{dt}$$

$$=u\frac{d\phi}{dt}+\frac{dv}{dt}.$$

224. This theorem is of great importance and particular attention should be given to the meaning of the letters. The rate of increase of u in the direction of the *moving* axis $O\xi$ is $\frac{du}{dt}$. Its rate of increase in the direction of an axis *fixed in space* which is coincident with the position of $O\xi$ at the time t and which is left behind when $O\xi$ moves into some other position $O\xi'$ is $\frac{du}{dt} - v\frac{d\phi}{dt}$. *It is the latter rate of increase not the former which is required in dynamics.*

To make this point clear let us suppose that u represents the component velocity of a point P. Then

$$du = \begin{pmatrix}\text{component along } O\xi' \\ \text{at time } t+dt\end{pmatrix} - \begin{pmatrix}\text{component along } O\xi \\ \text{time } t\end{pmatrix},$$

$$du - v\,d\phi = \begin{pmatrix}\text{comp. along } O\xi \\ \text{time } t+dt\end{pmatrix} - \begin{pmatrix}\text{comp. along } O\xi \\ \text{time } t\end{pmatrix}.$$

When it is necessary to distinguish between these two we may call the first the *relative rate* and the second the *space rate* of increase of the vector.

225. There is *another method of establishing the fundamental theorem* which is very generally used and which puts the argument into a more algebraic form.

Let the moving axis $O\xi$ make an angle ϕ with an arbitrary direction Ox fixed in space. Then if U be the component of the vector along Ox,

$$U = u\cos\phi - v\sin\phi;$$

$$\therefore \quad \frac{dU}{dt} = \left(\frac{du}{dt} - v\frac{d\phi}{dt}\right)\cos\phi - \left(u\frac{d\phi}{dt} + \frac{dv}{dt}\right)\sin\phi.$$

This gives the rate of increase in the direction of the fixed axis Ox. Let Ox coincide with $O\xi$ and be left behind when $O\xi$ moves into the position $O\xi'$, then $\phi = 0$ though $d\phi/dt$ is not zero. By definition $dU/dt = u_1$, and therefore

$$u_1 = \frac{du}{dt} - v\frac{d\phi}{dt}.$$

Again let Ox coincide with $O\eta$ and let it be left behind when $O\eta$ moves to $O\eta'$. Since ϕ is the angle $O\xi$ makes with Ox measured from Ox round O positively in the direction $\xi\eta$, the instantaneous value of ϕ is $-\frac{1}{2}\pi$ though as before it is increasing at the rate $d\phi/dt$. By definition dU/dt is now v_1, and hence

$$v_1 = \frac{dv}{dt} + u\frac{d\phi}{dt}.$$

226. *Ex.* 1. *To deduce the components of velocity and acceleration along and perpendicular to the radius vector*, Art. 35.

We take the arbitrary axis of ξ to coincide with the radius vector, then $\phi=\theta$. Regarding $\xi=r$, $\eta=0$ as the components of the vector r, the space components of the velocity are

$$u=\frac{d\xi}{dt}-\eta\frac{d\theta}{dt}=\frac{dr}{dt},\qquad v=\frac{d\eta}{dt}+\xi\frac{d\theta}{dt}=r\frac{d\theta}{dt}.$$

Taking the velocity as a second vector, the components are $u=dr/dt$, $v=rd\theta/dt$, and the space components of the acceleration are

$$u_1=\frac{du}{dt}-v\frac{d\theta}{dt}=\frac{d^2r}{dt^2}-r\left(\frac{d\theta}{dt}\right)^2,$$

$$v_1=\frac{dv}{dt}+u\frac{d\theta}{dt}=\frac{1}{r}\frac{d}{dt}\left(r^2\frac{d\theta}{dt}\right).$$

Ex. 2. *To deduce the components of acceleration along the tangent and normal*, Art. 36.

Taking the axis of ξ parallel to the tangent, we have $\phi=\psi$. Let the velocity be the vector, then u represents the velocity and $v=0$. The components of acceleration are therefore

$$u_1=\frac{du}{dt}-v\frac{d\psi}{dt}=\frac{du}{dt},\qquad v_1=\frac{dv}{dt}+u\frac{d\psi}{dt}=u\frac{d\psi}{dt}.$$

227. *To find the components of velocity and acceleration with regard to moving axes.*

Let the position of the moving point P be given by its coordinates (ξ, η) with regard to two rectangular axes $O\xi$, $O\eta$ which turn round a fixed origin O with an angular velocity $d\phi/dt$. Let (u, v) be the components of the velocity of P parallel to the instantaneous positions of $O\xi$, $O\eta$. Let (X, Y) be the components of the acceleration of P. The relations between (ξ, η), (u, v), (X, Y) follow at once from the general theorem. We have

$$u=\frac{d\xi}{dt}-\eta\frac{d\phi}{dt},\qquad v=\frac{d\eta}{dt}+\xi\frac{d\phi}{dt}\ \ldots\ldots\ldots\ldots(A);$$

$$X=\frac{du}{dt}-v\frac{d\phi}{dt},\qquad Y=\frac{dv}{dt}+u\frac{d\phi}{dt}\ \ldots\ldots\ldots\ldots(B).$$

Substituting for u, v in the latter expressions their values given by the former, we have

$$\left.\begin{aligned} X&=\frac{d^2\xi}{dt^2}-\xi\left(\frac{d\phi}{dt}\right)^2-\frac{1}{\eta}\frac{d}{dt}\left(\eta^2\frac{d\phi}{dt}\right)\\ Y&=\frac{d^2\eta}{dt^2}-\eta\left(\frac{d\phi}{dt}\right)^2+\frac{1}{\xi}\frac{d}{dt}\left(\xi^2\frac{d\phi}{dt}\right)\end{aligned}\right\}\ \ldots\ldots\ldots\ldots(C).$$

If the origin O is also in motion, these equations require some modification. Let p, q be the components of the space velocity of the origin in the directions of the axes. Let u, v continue to represent the components of the space velocity of the point P.

To find u, v we add to the expressions (A) for the relative component velocities the component velocities of O, Art. 10. We thus have

$$u = p + \frac{d\xi}{dt} - \eta\frac{d\phi}{dt}, \qquad v = q + \frac{d\eta}{dt} + \xi\frac{d\phi}{dt} \text{......(A').}$$

These equations give the motion of P referred to a system of moving axes having any fixed origin but always remaining parallel to the original moving axes. With these values of u, v, the accelerations X, Y will continue to be expressed by the formulæ (B).

228. We may deduce the expressions (C) for the accelerations X, Y in terms of the coordinates ξ, η from *the theory of relative motion*, explained in Art. 10.

The motion of P in space is made up of the velocity relative to M together with that of M in space; see fig. of Art. 223. Now OM is the radius vector of M, and the component velocities in the directions OM, MP are ξ' and $\xi\phi'$, while the accelerations in the same directions are

$$\xi'' - \xi\phi'^2 \text{ and } \frac{1}{\xi}\frac{d}{dt}(\xi^2\phi')$$

where accents represent differentiations with regard to the time. Again regarding M as fixed, MP is the radius vector of P, hence the component velocities of P along MP and parallel to MO (not OM) are η' and $\eta\phi'$, while the accelerations in the same directions are $\eta'' - \eta\phi'^2$ and $\dfrac{1}{\eta}\dfrac{d}{dt}(\eta^2\phi')$. Adding together these components, we obviously obtain the values of u, v; X, Y already given in Art. 227.

229. Relative and actual path. When the motion of a point is referred to moving axes $O\xi$, $O\eta$ it is necessary to distinguish between *the path in space* and *the path relative to the moving axes.* Suppose a sheet of paper to be attached to the moving axes and to turn round the fixed point O with them. The point P traces out on this sheet the relative path which is not the same as that traced out on a sheet fixed in space.

The coordinates of P in the relative motion are (ξ, η) and the displacements parallel to these axes are $d\xi$ and $d\eta$. The direction of the tangent of the relative path and the radius of curvature of that path are therefore found by the ordinary rules of the differential calculus. The coordinates of P in the path in space are also (ξ, η), but the displacements have just been proved to be $d\xi - \eta\, d\phi$ and $d\eta + \xi\, d\phi$. *These must be used instead of dx and dy in the formulæ of the differential calculus.*

Let us represent by accents the differential coefficients with regard to any independent variable t. *The formulæ of the differential calculus giving the space motion of P referred to fixed axes may be adapted to moving axes by writing u, v for x', y' respectively, where* $u = \xi' - \eta\phi'$, $v = \eta' + \xi\phi'$, *and* u_1, v_1 *for* x'', y'' *where*

$$u_1 = u' - v\phi', \quad v_1 = v' + u\phi'.$$

Thus, if ψ, χ be the angles the tangents to the relative and actual paths make with $O\xi$, and ρ, R be the radii of curvature of these paths, we have

$$\tan\psi = \frac{\eta'}{\xi'}, \qquad \tan\chi = \frac{\eta' + \xi\phi'}{\xi' - \eta\phi'},$$

$$\frac{(\xi'^2 + \eta'^2)^{\frac{3}{2}}}{\rho} = \xi'\eta'' - \eta'\xi'', \qquad \frac{(u^2+v^2)^{\frac{3}{2}}}{R} = uv' - vu' + (u^2+v^2)\,\phi'.$$

When we apply kinematical theorems to purely geometrical properties in which the idea of time is absent, we regard t as an auxiliary arbitrary quantity introduced to represent the independent variable. If we wish the arc s to be the independent variable, we write $t = s$.

The effect of these changes may be exhibited in a figure. Let P, P' be the positions in space of the moving point at the times t, $t + dt$, and $O\xi$, $O\xi'$ the positions of the axis of reference at the same times. If PM, $P'N$ be perpendiculars on $O\xi$, $O\xi'$, we have

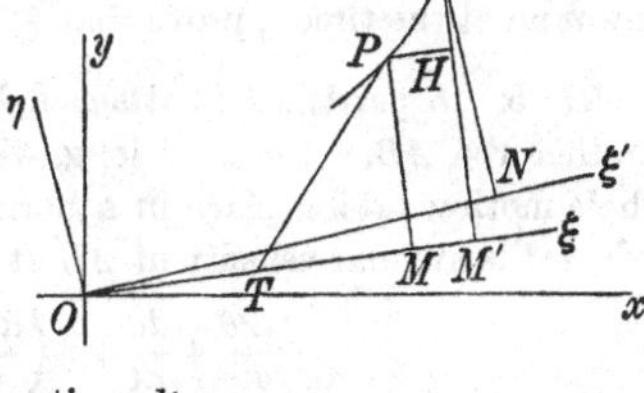

$$OM = \xi, \quad ON = \xi + d\xi, \quad MP = \eta,$$

$$NP' = \eta + d\eta \;\ldots\ldots\; \text{(A)}.$$

Let $P'M'$, PH be perpendiculars on $O\xi$ and $P'M'$ respectively. The coordinates of P, P' referred to axes $O\xi$, $O\eta$ fixed in space for a time dt are

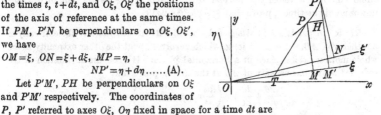

$$OM = \xi, \quad OM' = \xi + d\xi - \eta\, d\phi, \quad MP = \eta, \quad M'P' = \eta + d\eta + \xi\, d\phi \;\ldots\ldots\ldots\; \text{(B)}.$$

These values of MM', $P'H$ follow at once from Art. 223, but they may be obtained by projecting the broken line ON, NP' on $O\xi$, $O\eta$. If χ be the angle the tangent PP' makes with $O\xi$ and $d\sigma$ the arc PP', we have $\tan\chi = P'H/PH$ and $(d\sigma)^2 = (P'H)^2 + (PH)^2$, and these by substitution from (B) lead to the same results as before.

230. Many of the formulæ used in the differential calculus may be inferred by resolving the accelerations in different directions. For example, the formulæ for the radius of curvature in polar coordinates may be written down by simply resolving the polar accelerations of Art. 35 along the tangent and equating the result to V^2/R. The expressions for R in Cartesian moving and fixed axes may be obtained in the same way.

231. Examples. *Ex.* 1. The position of a point P is referred to rectangular axes $Q\xi$, $Q\eta$ which move so that Q describes a given curve AQ while $Q\xi$ is always a tangent to the curve. Prove that the component velocities and accelerations of P are

$$u = s' + \xi' - \eta\phi', \qquad v = \eta' + \xi\phi',$$

$$X = u' - v\phi', \qquad Y = v' + u\phi',$$

where ϕ' is the angular velocity of ξ, and $\phi' = s'/\rho$.

Deduce an expression for the radius of curvature of the space locus of P.

Ex. 2. A particle P is attached to the extremity of a string of length l which is being wound on to a fixed curve after the manner of an involute. Prove that the component accelerations of P along and perpendicular to the straight portion ξ of the string are respectively

$$X = -\xi\phi'^2, \quad Y = \xi'\phi' + \xi\phi'',$$

where ϕ' is the angular velocity of ξ. Also $\phi' = -\xi'/\rho$.

Ex. 3. Assuming the earth to be uniformly describing a circle of radius a about the sun with velocity U, and the sun to be moving in a straight line in the plane of the earth's orbit with a uniform velocity V, prove that the radius of curvature at any point of the earth's orbit in space is $\dfrac{(V^2+2VU\sin\theta+U^2)^{\frac{3}{2}}a}{U^2(U+V\sin\theta)}$, where θ is the angle the line joining the earth and sun makes with the direction of the sun's motion. [Coll. Ex. 1892.]

Ex. 4. A fine string wound round a circle has a particle P attached to its extremity and the circle is constrained to *turn round its centre* in its own plane with a uniform angular velocity ω. The particle is initially in contact with the circle and has a velocity V normal to the circle. If ξ be the length of string unwound at the time t, prove that $\xi^2 = a^2\omega^2t^2 + 2aVt$.

Ex. 5. A particle P is attached by a rod PA without mass to the extremity of another rod AB, n times as long, which revolves about the other extremity B, the whole motion taking place in a horizontal plane. If θ be the inclination of the rods, ω the angular velocity of AB at the time t, prove that

$$\frac{d^2\theta}{dt^2} + \frac{d\omega}{dt} + n\left(\frac{d\omega}{dt}\cos\theta + \omega^2\sin\theta\right) = 0. \quad \text{[Math. Tripos, 1860.]}$$

232. Oblique axes. The general method of finding the resolved velocities and accelerations of a point referred to moving axes may be extended to oblique axes. These extensions however are not of any great importance because oblique axes are seldom used in mechanics.

Let $O\xi$, $O\eta$ be any two axes which make angles θ, ϕ with an axis Ox fixed in space. These angles we shall suppose to be perfectly arbitrary so that the angle $\xi O\eta$ between the axes is not necessarily constant. See figure of Art. 223.

Let PQ represent any vector; u, v its components obtained by oblique resolution according to the parallelogram law. Let u_1, v_1, represent as before the rates of increase of the components of the vector in directions fixed in space but coincident with the positions of $O\xi$, $O\eta$ at the time t.

Let us resolve the vector in a direction perpendicular to $O\xi$. The resolved parts of u_1 and v_1 are clearly zero and $v_1\sin(\phi-\theta)$. Since $\xi O\xi' = d\theta$, $\eta O\eta' = d\phi$, the resolution gives

$$v_1\sin(\phi-\theta) = \frac{[(u+du)\sin d\theta + (v+dv)\sin(\phi-\theta+d\phi)] - [v\sin(\phi-\theta)]}{dt}$$

$$= u\frac{d\theta}{dt} + \frac{dv}{dt}\sin(\phi-\theta) + v\cos(\phi-\theta)\frac{d\phi}{dt}.$$

By resolving in a direction perpendicular to $O\eta$ we obtain in the same way

$$u_1\sin(\phi-\theta) = -v\frac{d\phi}{dt} + \frac{du}{dt}\sin(\phi-\theta) - u\cos(\phi-\theta)\frac{d\theta}{dt}.$$

If ξ, η are the oblique coordinates of P, the space velocities u, v of P are similarly

$$v \sin(\phi - \theta) = \xi \frac{d\theta}{dt} + \frac{d\eta}{dt} \sin(\phi - \theta) + \eta \cos(\phi - \theta) \frac{d\phi}{dt},$$

$$u \sin(\phi - \theta) = -\eta \frac{d\phi}{dt} + \frac{d\xi}{dt} \sin(\phi - \theta) - \xi \cos(\phi - \theta) \frac{d\theta}{dt}.$$

The advantage of resolving perpendicularly to $O\xi$ and $O\eta$ is that only one of the components u_1, v_1, enters into the resolution. We thus obtain each independently of the other. If we resolve in the directions $O\xi$, $O\eta$ we obtain the values of $u_1 + v_1 \cos(\phi - \theta)$ and $v_1 + u_1 \cos(\phi - \theta)$ and, from these, u_1 and v_1 can be obtained by solving the equations.

These values of u_1, v_1 were first given by H. W. Watson in the Math. Tripos of 1861.

233. Hyper-accelerations. It is seldom that we use higher differential coefficients with regard to the time than the second, Art. 21. When these are required the general theorem on vectors (Art. 223) gives the components for differential coefficients of any order.

Let x, y be the coordinates of a moving point referred to fixed axes, then $X_n = d^n x/dt^n$, $Y_n = d^n y/dt^n$ are the components of the space hyper-acceleration of the nth order, Art. 21. Let $O\xi$, $O\eta$ be any set of moving axes, the relations between the space components of two successive orders of acceleration are

$$X_{n+1} = \frac{dX_n}{dt} - Y_n \frac{d\phi}{dt}, \qquad Y_{n+1} = \frac{dY_n}{dt} + X_n \frac{d\phi}{dt}.$$

The reader may consult a Note *Sur les Principes de la Mécanique* by Abel Transon, *Liouville's Journal*, vol. x. 1845, for another mode of treatment.

Ex. 1. A point moves along a curve with velocity u, prove that the components along the tangent and normal of the acceleration of the third order are respectively

$$\frac{d^2u}{dt^2} - \frac{u^3}{\rho^2}; \quad \frac{d}{ds}\left(\frac{u^3}{\rho}\right).$$

Ex. 2. A point P moves along a curve with uniform velocity. Prove that $\tan\delta = \frac{1}{3}\cot\delta'$ where δ, δ' are the angles the diameter of the parabola of closest contact and the direction of the hyper-acceleration make with the normal at P. Show also that the semi-latus rectum of this parabola is $\rho\cos^3\delta$.

D'Alembert's Principle.

234. When a single particle moves under the action of given forces the equations of motion may in general be found by resolving the forces in some convenient directions. In the case of a system of particles the mutual reactions must also be taken into the account; these are in general unknown and will have to be eliminated from the equations. It is important to be able to write down some of the results of this elimination without first forming the equations of motion of every particle. Various methods have been given to effect this either completely or partially.

When in a statical problem, we wish to avoid introducing into our equations the mutual reactions of two bodies, we treat the two as one system. *We resolve and take moments for the two bodies as if they were one.* We may adopt the same method in dynamics.

235. In applying this principle to dynamics, it will be found convenient to use the term *effective force.* This may be defined as follows. When a particle is moving as part of a system, it is acted on by the external forces and the reactions of the other particles. If we consider this particle to be separated from the system and all these forces removed there is some one force which, with the same initial conditions, would make it move in the same way as before. This force is called the effective force on the particle.

It follows that the effective force is statically equivalent to the impressed forces which act on the particle and the reactions of the rest of the system, but *is differently expressed.* Let m be the mass of the particle, (x, y) the Cartesian coordinates; the components of the force which must act to produce any given motion have been proved to be md^2x/dt^2 and md^2y/dt^2, these then are the components of the effective force. In the same way if v be the velocity and $1/\rho$ the curvature of the path, the tangential and normal components of the effective force are mdv/dt and mv^2/ρ. See Art. 68.

236. Considering any one particle of the system, we know that the resolved parts of the effective forces in any directions are equal to the corresponding resolved parts of the impressed forces and the reactions. It immediately follows that the effective forces on each particle, *if reversed*, are in equilibrium with the impressed forces and the reactions. But, by Newton's third law, the mutual reactions of any two particles are in equilibrium. *Making then any selection of the particles of a system, the reversed effective forces of those particles are in equilibrium with the external forces which act on them, excluding their mutual reactions, but including the pressures (if any) of the remainder of the system.*

Some of the equations of motion may therefore be found (1) by equating the sum of *the resolved parts* of the effective forces in any convenient directions to the sum of the resolved parts of the external forces, (2) by equating the sum of *the moments* of the effective forces about any point to the sum of the moments of the external forces.

The resolved parts and moments of the external forces may be written down by the rules of statics. The components of the effective forces in various directions have been found in the preceding articles. The moment about any point O then follows by multiplying that component by the length of the perpendicular from O, Art. 6.

If (x_1, y_1), (x_2, y_2) &c. are the Cartesian coordinates of a system of mutually attracting particles whose masses are m_1, m_2 &c., and if these are acted on by the external accelerating forces (X_1, Y_1), (X_2, Y_2) &c., the equations of resolution and moments are

$$\Sigma m\left(\frac{d^2x}{dt^2}\right)=\Sigma mX,\quad \Sigma m\left(\frac{d^2y}{dt^2}\right)=\Sigma mY,$$

$$\Sigma m\left(x\frac{d^2y}{dt^2}-y\frac{d^2x}{dt^2}\right)=\Sigma m\,(xY-yX),$$

where the Σ implies summation for all the particles.

237. So long as we confine our attention to resolutions and moments it is unnecessary to include the mutual actions of the particles under consideration. If however we use the principle of virtual velocities to express the conditions of equilibrium we must remember that the particles may not be rigidly connected together. Now the work of two equal and opposite forces F, $-F$, acting on two particles distant r from each other is proved in statics to be Fdr. It is obvious that this does not vanish unless the distance r is invariable. This point is important in using the principle of vis viva.

The most convenient way of applying the principle of Virtual Velocities to Dynamical problems is to use Lagrange's equations.

238. When the selected system of particles is a rigid body, the mutual distances of the particles composing it are invariable.

It is proved in statics that the position of such a body in space of two dimensions can be defined by three quantities usually called *coordinates*. For example, these might be the Cartesian coordinates of some point and the angle which some straight line fixed in the body makes with some straight line fixed in space. Three independent equations of motion, free from mutual reactions, are therefore necessary and sufficient to determine the position of the system at any time t. These three are supplied by the two resolutions and the equation of moments above described.

It is proved in statics that a system of forces can be reduced to a single force R acting at some convenient point O and a couple G. The components of the force R are equal to the sums of the components of all the forces of the system, and the couple G is equal to the sum of their moments about O. This is usually called Poinsot's method of compounding forces. We shall now apply this method to find the resultants of a system of effective forces.

239. A system of particles, rigidly connected, moves in space of two dimensions. The coordinates of the centre of gravity are $(\bar{x}, \bar{y})$, the angle which a straight line fixed in the body makes with a straight line fixed in space is ϕ and the whole mass is M. It is required to prove that *the effective forces of the whole system are equivalent to two effective forces* $M\dfrac{d^2\bar{x}}{dt^2}$, $M\dfrac{d^2\bar{y}}{dt^2}$ *acting at the centre of gravity, and an effective couple* $Mk^2\dfrac{d^2\phi}{dt^2}$, *where* Mk^2 *is a constant which depends on the form and structure of the body or system.*

Let m be the mass of any particle of the body, $x=\bar{x}+\xi$, $y=\bar{y}+\eta$ be its coordinates. Then since $\Sigma m\xi/\Sigma m$, $\Sigma m\eta/\Sigma m$ are the coordinates of the centre of gravity referred to the centre of gravity as origin, it is clear that $\Sigma m\xi=0$, $\Sigma m\eta=0$.

The sum of the resolved parts of the effective forces parallel to the axis of x is

$$\Sigma m\frac{d^2x}{dt^2}=\frac{d^2\bar{x}}{dt^2}\Sigma m+\frac{d^2}{dt^2}(\Sigma m\xi)=M\frac{d^2\bar{x}}{dt^2}.$$

The resolved part parallel to the axis of y may be found in the same way. *These two effective forces are the same as the effective forces of a particle whose mass is M placed at the centre of gravity and moving with that point in space.*

240. To find the effective couple we take moments about the centre of gravity. Remembering that ξ, η are the coordinates of the particle m when referred to the centre of gravity, the couple is

$$\Sigma m\left(\xi\frac{d^2y}{dt^2}-\eta\frac{d^2x}{dt^2}\right)=\Sigma m\left(\xi\frac{d^2\eta}{dt^2}-\eta\frac{d^2\xi}{dt^2}\right)+\frac{d^2\bar{y}}{dt^2}\Sigma m\xi-\frac{d^2\bar{x}}{dt^2}\Sigma m\eta.$$

Since $\Sigma m\xi=0$, $\Sigma m\eta=0$, the right-hand side reduces to the first term. Let ρ, θ be the polar coordinates of the particle m, referred to the centre of gravity as origin, then $\xi d\eta-\eta d\xi=\rho d\theta$. The couple is therefore

$$\frac{d}{dt}\Sigma m\left(\xi\frac{d\eta}{dt}-\eta\frac{d\xi}{dt}\right)=\frac{d}{dt}\left\{\Sigma m\rho^2\frac{d\theta}{dt}\right\}.$$

We shall now introduce the condition that the particles are rigidly connected together. When this is the case the $d\theta/dt$ of every particle is equal to $d\phi/dt$, and the length of every ρ is constant during the motion. For, let α be the angle the radius vector ρ of any particle m makes with the straight line fixed in the body, then $\theta=\phi+\alpha$. Though α may be different for every particle, yet its value does not change during the motion, hence $d\alpha/dt=0$, and $d\theta/dt=d\phi/dt$. The effective couple is $(\Sigma m\rho^2)\dfrac{d^2\phi}{dt^2}$.

241. The constant $\Sigma m\rho^2$ is called the moment of inertia of the system about an axis drawn through the centre of gravity perpendicularly to the plane containing the particles.

To find the moment of inertia of any system about any axis, we multiply the mass of every particle by the square of its distance from the axis and add the results together.

When the particles are so close together that they form a continuous body, the sum is an integral. Thus for a circular area of radius a and density D, the area of any element is $\rho\, d\theta\, d\rho$; hence the moment of inertia about an axis drawn through the centre perpendicular to its plane is

$$\Sigma m\rho^2=\iint D\rho\, d\theta\, d\rho\,.\,\rho^2=D\left[\tfrac{1}{4}\rho^4\right].\left[\theta\right],$$

where the square brackets imply that the quantity is to be taken between the limits of integration. These limits being $\rho=0$ to a, and $\theta=0$ to 2π, the moment of inertia about the centre is $\frac{1}{2}Ma^2$.

In the same way the moment of inertia of a rectangle whose sides are $2a$ and $2b$ about an axis drawn through the centre of gravity perpendicular to its plane is $\frac{1}{3}M(a^2+b^2)$.

The moment of inertia of a sphere of radius a about a diameter is $\frac{2}{5}Ma^2$.

The moment of inertia of a triangular area about any axis is the same as that of three particles each one-third of its mass placed at the middle points of the sides.

242. The moment of inertia is of special importance in rotational motions, for, in a certain sense, it measures the dynamical significance of the form and structure of the moving body. Thus all free bodies having equal moments of inertia *rotate* with equal angular accelerations when acted on by equal couples. The translational motion depends on the mass and the position of the centre of gravity, Arts. 92, 239.

243. Sufficiency of the equations. The equations of motion of a particle moving freely are

$$\frac{d^2x}{dt^2} = X, \quad \frac{d^2y}{dt^2} = Y,$$

where X, Y are the accelerating components of the forces, Arts. 68, 73. We shall now prove that when the initial values of x, y, dx/dt, dy/dt are also given, these equations are sufficient to find x, y as functions of t.

To prove this we replace the proposition by a more general theorem, the limiting case of which is the proposition to be established. Let τ be any very small time which we shall afterwards replace by dt. Let $x = \phi(t)$, $y = \psi(t)$; the equations may be written in the functional forms

$$\left.\begin{aligned} \phi(t+2\tau) - 2\phi(t+\tau) + \phi(t) &= X\tau^2 \\ \psi(t+2\tau) - 2\psi(t+\tau) + \psi(t) &= Y\tau^2 \end{aligned}\right\} \quad \ldots\ldots\ldots\ldots\ldots\ldots (1),$$

where X, Y are known functions of $\phi(t)$ and $\psi(t)$.

Representing the initial time by $t=0$, we suppose that the four initial values

$$\phi(0), \ \psi(0), \ \phi(\tau) - \phi(0), \ \psi(\tau) - \psi(0) \ldots\ldots\ldots\ldots\ldots\ldots (2)$$

are given. Putting $t=0$ in (1) we deduce the values of $\phi(2\tau)$, $\psi(2\tau)$; again putting $t=\tau$ we obtain $\phi(3\tau)$, $\psi(3\tau)$, and so on. Thus by a continual repetition of the process the values of $\phi(n\tau)$, $\psi(n\tau)$ and therefore of $\phi(t)$, $\psi(t)$ can be found.

That the solution of the two equations of motion of the second order leads to results which contain four arbitrary constants (to be determined by the initial conditions) is also proved in treatises on differential equations; see Forsyth's *Differential Equations*, Art. 173.

244. On general and particular integrals. The Cartesian equations of motion of a free particle are

$$x'' = X, \quad y'' = Y \ \ldots\ldots\ldots\ldots\ldots\ldots\ldots\ldots (1),$$

where accents denote differential coefficients with regard to the time. These are usually solved by combining them together so as to obtain a perfect differential. We then have by integration

$$F(x, y, x', y', t) = C \ \ldots\ldots\ldots\ldots\ldots\ldots\ldots\ldots (2),$$

where C is a constant. When an integral is obtained in this manner there is nothing to limit the initial conditions. However the particle may be projected the equation (2), after determining the proper value of C, must be true throughout the whole motion. Such an integral is called *a general integral.* An integral which is true only for special initial conditions is called *a particular integral.*

245. If any equation such as (2) be arbitrarily written down containing one arbitrary constant we may enquire *what the dynamical problem is of which that equation is a general integral.*

To answer this we differentiate (2) and substitute from (1). We then have

$$\frac{dF}{dx}x' + \frac{dF}{dy}y' + \frac{dF}{dx'}X + \frac{dF}{dy'}Y + \frac{dF}{dt} = 0 \quad\text{.......................} \quad (3).$$

Since the state of motion at any time t may be taken as the arbitrary initial motion the quantities x, y, x', y' are really arbitrary. The forces X, Y must therefore be such as to make (3) an identity.

To determine X, Y we differentiate (3) partially with regard to any of the four letters x, y, x', y', treating the others as constants. Supposing that X, Y are intended to be functions of x, y only, they are constants when we differentiate partially with regard to x', y'. In this way we may obtain, by successive differentiations, several equations each containing X, Y in the first degree.

If these equations lead to inconsistent values of X, Y we infer that the given equation cannot be a general integral.

It may also happen that all the equations to find X, Y are identical, and in this case the forces X, Y are to a certain extent arbitrary. Bertrand has shown that this can happen only when the integral (2) has the form

$$(xy' - x'y)^2 + f\left(\frac{y}{x}\right) = C \quad\text{................................} \quad (4).$$

This therefore, when X, Y are functions of x, y only, is the only general integral which can be common to several dynamical problems. *Liouville's Journal*, 1852.

Ex. 1. If $x'^2 + y'^2 - 2f(x, y) = C$ be taken as the general integral, prove that $X = df/dx$, $Y = df/dy$. This is the equation of vis viva.

Ex. 2. Prove that $xy' \pm x'y = C$ with the upper sign cannot be a general integral; but, with the lower sign, is a general integral when the resultant force tends to the origin.

The Principle of Vis Viva.

246. *To investigate the principle of vis viva for a system of particles.*

Besides the external forces which act on the several particles we must here take into account their mutual actions and reactions.

Let m be the mass of any one particle; x, y its coordinates; let X, Y be the components of *all* the forces which act on that particle. The equations of motion of that particle are

$$m\frac{d^2x}{dt^2}=X,\quad m\frac{d^2y}{dt^2}=Y\quad\text{.....................(1).}$$

Multiplying these by dx/dt and dy/dt respectively and adding the results, we have

$$m\left(\frac{dx}{dt}\frac{d^2x}{dt^2}+\frac{dy}{dt}\frac{d^2y}{dt^2}\right)=\left(X\frac{dx}{dt}+Y\frac{dy}{dt}\right)\quad\text{...........(2).}$$

Summing this for all the particles of the system, we have

$$\Sigma m\left(\frac{dx}{dt}\frac{d^2x}{dt^2}+\frac{dy}{dt}\frac{d^2y}{dt^2}\right)=\Sigma\left(X\frac{dx}{dt}+Y\frac{dy}{dt}\right)\quad\text{......(3).}$$

The right-hand side of this equation, after multiplication by dt, is the work done by the forces as the system makes a small displacement, Art. 185.

Amongst the forces X, Y are included the unknown reactions on the several particles, but it is clear that *we may omit from the right-hand side all the reactions which would disappear in the principle of work in statics.*

When the *remaining forces* are such that the work integral

$$\int\Sigma\,(Xdx+Ydy)=U+C\quad\text{.....................(4),}$$

where U is a known function of the coordinates of the particles, *these forces are said to form a conservative system.* Art. 181.

Representing by v the velocity of the particle m, the integral of (3) becomes

$$\tfrac{1}{2}\Sigma mv^2=U+C\quad\text{.........................(5).}$$

Let U_0 be the same function of the initial coordinates that U is of the coordinates at the time t, and let v_0 be the initial value of v. The equation of vis viva may also be written in the form

$$\tfrac{1}{2}\Sigma mv^2-\tfrac{1}{2}\Sigma mv_0^2=U-U_0\quad\text{....................(6).}$$

247. The principle of vis viva is important for several reasons.

(1) The principle is of general application. The forces in nature are such that there is a work function, and the unknown reactions, in general, disappear from the equation.

(2) When there is only one way in which the system can move, that motion is determined by the principle.

(3) The principle gives a relation between the circumstances of the motion in any stated position of the system and those at the initial stage. When the intermediate motion is not required this is particularly important.

248. The force function. The equation of vis viva can be usefully employed only when the integrations necessary to obtain the force function U can be effected. It is also important to notice beforehand what forces and reactions may be omitted in forming that equation.

The acting forces may be classified thus,

(1) the external forces which act on the particles,

(2) the mutual actions of such of the particles as are rigidly connected together,

(3) the mutual attractions of independent particles,

(4) the pressures due to any fixed curve or surface on which some of the particles are constrained to move.

The external forces are in general central forces tending to or from fixed points. It follows from Art. 186 that, when each force is some function of the distance from the fixed point, the contribution of each to the work function can be integrated.

Let R be the mutual action between two particles whose instantaneous distance apart is r, and let R be measured positively when the action tends to increase r. It is proved in statics that the work of both the action and reaction is Rdr.

It follows from this that the reaction between any two particles which keep an invariable distance from each other throughout the motion disappears from the equation of vis viva, for in such a case $dr=0$.

If any two independent particles repel each other with a force R which is a known function of their distance r, the contribution of this force to the work function can be integrated.

If two particles are connected together by a tight string, even if bent by passing over smooth pulleys, fixed or moveable, the work of the tension is $-Tdl$, where l is the whole length of the string. If the length of the string is invariable the work is zero. The action of an inextensible string may therefore be omitted in the equation of vis viva. If the string is extensible and the tension obeys Hooke's law, the corresponding work can be found by integrating $-Tdl$, see Art. 187.

249. If one of the particles is *constrained to move on a smooth fixed curve whose equation is* $f(x, y)=0$, let R be the normal pressure. The work of R is $R\cos\phi\, ds$; this is zero because ϕ, being the angle between the direction of R and the arc of the path, is $\frac{1}{2}\pi$. If however *the curve is itself constrained to move*, the angle ϕ is not necessarily a right angle and the work may not be zero. Since the equation of the moving curve will contain t, this is usually expressed by saying that *the geometrical relations must not contain the time explicitly, if the reactions are to disappear.*

If *the curve or surface is rough*, the friction acts along the tangent to the path, and the work is zero only when the particle in contact is not in motion.

250. Energy. Selecting some geometrically possible arrangement of the particles as a standard position, the work done by the forces as the particles move or are moved from any other given arrangement to the standard position is called the *potential energy* in the given position.

Let the standard position be called S; let the system move from some given initial position A and at the time t let its position be P. It has already been proved (Arts. 69, 246) that

$$\text{Kin. En. at } P - \text{Kin. En. at } A = \text{work } A \text{ to } P.$$

But
$$\text{Pot. En. at } P = \text{work } P \text{ to } S,$$
$$\text{Pot. En. at } A = \text{work } A \text{ to } S.$$

$$\therefore \text{Kin. En. at } P + \text{Pot. En. at } P = \text{Kin. En. at } A + \text{Pot. En. at } A.$$

It follows therefore that *the sum of the kinetic and potential energies is constant throughout the motion.* This sum is called the energy of the system, and it has just been proved that *the energy of the system is constant* and equal to its initial value.

This theorem is true whatever standard position may be chosen, but it will be found convenient to so choose this position that the system may finally arrive there. When this choice is made the potential energy represents the whole work which can be obtained from the forces as the system moves to its final position.

251. As a simple example, let a heavy particle fall from rest at the ceiling of a room to the floor; the kinetic energy after falling a distance z is $\frac{1}{2}mv^2 = mgz$. Let us take the floor (i.e. $z = h$) as the standard position, because the particle cannot descend any lower; the potential energy at the depth z is $mg(h-z)$. The whole energy is therefore mgh, which is constant throughout the motion. At the ceiling the energy is wholly potential because the particle starts from rest; on arriving at the floor the energy is wholly kinetic, all the available potential energy having been changed into kinetic energy.

252. Degrees of freedom. If a system contain n particles free to move in space of two dimensions, its position can only be defined by the use of the $2n$ coordinates of the particles. There are evidently just $2n$ different ways in which the particles can be moved, all other displacements being compounded of these.

The system is then said to have $2n$ degrees of freedom. If some of the particles are constrained to move on κ given curves, or more generally if there are κ given relations between the $2n$ coordinates, only $2n-\kappa$ coordinates are necessary to fix the position of the system and there are then $2n-\kappa$ degrees of freedom. *The degrees of freedom of a system may be defined to be the number of coordinates required to fix its position.*

253. Vis viva of a rigid body. When some or all of the particles of a system are rigidly connected together a simple and useful expression for the vis viva can be found. Let $(\bar{x}, \bar{y})$ be the coordinates of the centre of gravity, ϕ the angle which a straight line fixed in the body makes with a straight line fixed in space, and M the mass. The vis viva is then

$$\Sigma mv^2 = M\left\{\left(\frac{d\bar{x}}{dt}\right)^2 + \left(\frac{d\bar{y}}{dt}\right)^2\right\} + Mk^2\left(\frac{d\phi}{dt}\right)^2,$$

where Mk^2 is the constant called the moment of inertia of the body about the centre of gravity, see Art. 241.

To prove this, let $x = \bar{x} + \xi$, $y = \bar{y} + \eta$ be the coordinates of any particle m, then

$$\Sigma m\left(\frac{dx}{dt}\right)^2 = (\Sigma m)\left(\frac{d\bar{x}}{dt}\right)^2 + 2\left(\Sigma m\frac{d\xi}{dt}\right)\frac{d\bar{x}}{dt} + \Sigma\left\{m\left(\frac{d\xi}{dt}\right)^2\right\}.$$

Since $\Sigma m\xi = 0$ as in Art. 240 the middle term is zero. Hence

$$\Sigma mv^2 = M\left\{\left(\frac{d\bar{x}}{dt}\right)^2 + \left(\frac{d\bar{y}}{dt}\right)^2\right\} + \Sigma m\left\{\left(\frac{d\xi}{dt}\right)^2 + \left(\frac{d\eta}{dt}\right)^2\right\}.$$

This equation expresses the proposition that *the whole vis viva of a moving system, whether rigid or not, is equal to that of a particle of mass M moving with the centre of gravity together with the vis viva of the motion relative to the centre of gravity.*

To introduce the condition that the system is rigid we change to polar coordinates by writing

$$(d\xi)^2 + (d\eta)^2 = (ds)^2 = (d\rho)^2 + (\rho d\theta)^2.$$

Remembering that $d\theta/dt$ is now the same for all the particles and equal to $d\phi/dt$ (Art. 240) and that $d\rho/dt$ is zero, we find

$$\Sigma m\left\{\left(\frac{d\xi}{dt}\right)^2 + \left(\frac{d\eta}{dt}\right)^2\right\} = (\Sigma m\rho^2)\left(\frac{d\theta}{dt}\right)^2 = Mk^2\left(\frac{d\phi}{dt}\right)^2.$$

254. Examples. *Ex.* 1. An endless light string of length $2l$, on which are threaded beads of masses M and m, passes over two small smooth pegs A and B in the same horizontal line and at a distance apart a, one bead lying in each of the festoons into which the string is divided by the pegs. The lighter bead m is raised to the mid-point of AB and then let go. Show that the beads will just meet if $\frac{M+m}{M}=2\left(\frac{l}{l+a}\right)^{\frac{1}{2}}$. [Math. Tripos, 1897.]

We notice that only two positions of the system are contemplated in the problem, viz. (1) the initial position in which the bead m lies in AB, and (2) the position in which the beads are in contact. In both these cases the kinetic energy is zero. The principle of vis viva asserts that *the change of kinetic energy is equal to the work*. It immediately follows that the work done when the system passes from the first to the second position is zero. Let x be the depth below AB at which the beads meet. Then omitting the tension, Art. 248, we have

$$mgx+M\{x-\sqrt{(l^2-al)}\}=0.$$

We also have by geometry $4x^2+a^2=l^2$. Eliminating x we obtain the result.

The circumstances of the motion when the beads m, M are at any depths y, η below AB may also be deduced from the principle. We have

$$\tfrac{1}{2}(mv^2+Mv'^2)=mgy+Mg\{\eta-\sqrt{(l^2-al)}\} \quad\text{.....................(1).}$$

Since the sum of lengths joining m and M to A is l, we have the geometrical equation

$$\sqrt{(\tfrac{1}{4}a^2+y^2)}+\sqrt{(\tfrac{1}{4}a^2+\eta^2)}=l \quad\text{.............................(2).}$$

Differentiating the second equation, we have

$$\frac{yv}{\sqrt{(a^2+4y^2)}}+\frac{\eta v'}{\sqrt{(a^2+4\eta^2)}}=0 \quad\text{.............................(3).}$$

Joining this to (1) we have the values of v, v' when y and η have any values not inconsistent with (2).

Ex. 2. A particle of mass m has attached to it two equal weights by means of strings passing over pulleys in the same horizontal line and is initially at rest half way between them. Prove that if the distance between the pulleys be $2a$, the velocity of m will be zero when it has fallen through a space $\frac{4mm'a}{4m'^2-m^2}$.

[Coll. Exam.]

Ex. 3. Two pails of weights W, w, are suspended at the ends of a rope which is coiled round the perfectly rough rim of a uniform circular disc of radius a supported in a vertical plane on a smooth horizontal axis, and the pails can descend into a well so that when one comes up the other goes down. If the pails be allowed to move freely under gravity, and, when the heavier has descended a distance b from rest, a drop of water be thrown off from the highest point of the rim of the disc, prove that this drop will strike the ground at a horizontal distance x from the axis of the disc given by

$$x^2(\tfrac{1}{2}W'+W+w)=4hb(W-w),$$

where W' is the weight of the disc, and h is the vertical distance above the ground of the highest point of the rim of the disc. [Math. Tripos, 1897.]

The equation of vis viva gives

$$M'k^2\omega^2+(M+m)v^2=2(M-m)gb.$$

The theory of parabolic motion gives $x=vt$, and $h=\frac{1}{2}gt^2$. Putting $\omega=v/a$ and $k^2=\frac{1}{2}a^2$, we obtain the required value of x.

Ex. 4. Two small holes A, B are made in a smooth horizontal table, the distance apart being $2a$. A particle of mass M rests on the table midway between A and B; and a particle of mass m hangs beneath the table, suspended from M by two equal weightless and inextensible strings, passing through the two holes. The length of each string is $a(1+\sec\alpha)$. A blow J is applied to M in a direction perpendicular to AB; show that if $J^2>2Mmag\tan\alpha$, M will oscillate to and fro through a distance $2a\tan\alpha$. But if J^2 is less than this quantity and equal to $2Mmag(\tan\alpha-\tan\beta)$, the distance through which M oscillates will be

$$2a\{p(p+2)\}^{\frac{1}{2}}, \text{ where } p=\sec\alpha-\sec\beta.$$ [Coll. Ex. 1895.]

The effect of the blow J is to communicate an initial velocity $V=J/M$ to the mass M, leaving m initially at rest.

Ex. 5. Two particles M, m are connected by a string passing over a smooth pulley, the lesser mass m hangs vertically, and M rests on a plane inclined at an angle α to the vertical. M starts without initial velocity from the point of the inclined plane vertically under the pulley. Prove that M will oscillate through a distance $\dfrac{2m(M-m)h\cos\alpha}{m^2-M^2\cos^2\alpha}$ where h is the height of the pulley above the initial position of M, m is greater than $M\cos\alpha$ but less than M. [Coll. Ex. 1897.]

Ex. 6. Two equal particles connected by a string are placed in a circular tube. In the circumference is a centre of force varying as the inverse distance. One particle is initially at rest at its greatest distance from the centre of force, prove that if v, v' be the velocities with which they pass through a point 90° from the centre of force, $e^{-v^2/\mu}+e^{-v'^2/\mu}=1$. [Coll. Exam.]

Ex. 7. A thin spherical shell of mass M is driven out symmetrically by an internal explosion. Prove that if when the shell has a radius a the outward velocity of each particle be V, the fragments can never be collected by their mutual attraction unless $V^2<M/a$. [Coll. Exam.]

The attraction of a thin spherical shell on an element of itself is the same as if *half* the mass of the shell were collected at the centre.

Ex. 8. Three equal and similar particles repelling each other with forces varying as the distance are connected by equal inextensible strings and are at rest; if one string be cut, the subsequent angular velocity of either of the other strings will vary as $\sqrt{\dfrac{1-2\cos\theta}{2+\cos\theta}}$, θ being the angle between them. [Christ's Coll.]

Ex. 9. An elastic string of mass m and modulus E rests unstretched in the form of a circle of radius a. It is now acted on by a repulsive force situated in its centre whose magnitude is $\mu\,(\text{distance})^{-2}$. Prove that the radius of the circle when it next comes to rest is a root of the quadratic $r^2-ar=m\mu/E\pi$. [Coll. Exam.]

Ex. 10. A circular hoop of radius b, without mass, has a heavy particle rigidly attached to it at a point distant c from its centre, and its inner surface is constrained to roll on the outer surface of a fixed circle of radius a (b being greater than a), under the action of a repelling force from the centre of the fixed circle equal to μ times the distance. Prove that the period of small oscillations of the hoop will be $2\pi\dfrac{b+c}{a}\left(\dfrac{b-a}{c\mu}\right)^{\frac{1}{2}}$.

Prove that when $c=b$, all oscillations large or small have the same period; and prove further that in the general case the hoop may be started so that it will continue to roll with uniform angular velocity equal to $\{\mu b/(b-a)\}^{\frac{1}{2}}$.

[Math. Tripos, 1886.]

The following is a simple (but not necessarily the shortest) method of writing down the equation of vis viva in problems of this kind. Having selected some independent variable to fix the position of the system, say, the inclination θ of the straight line joining the centres C, O of the two circles to the vertical, we find the coordinates x, y of the particle in terms of θ by projecting OC, CP on the vertical and horizontal. The vis viva, being the sum of $m(dx/dt)^2$ and $m(dy/dt)^2$, follows immediately. Equating the half of this sum to the force function $\frac{1}{2}m\mu \, . \, CO^2+C$ we have an equation giving $d\theta/dt$ in terms of θ.

It is then easily seen that, if the constant C be properly chosen, the value of $d\theta/dt$ reduces to the constant given in the question. To find the small oscillations, we differentiate the equation of vis viva and reject the squares of θ.

When $c=a$, the path of the particle is an epicycloid and the oscillations large or small are, by Art. 211, tautochronous.

255. Rotating field of force. When a particle moves in a field of force which rotates round the origin O with a uniform angular velocity n, an integral of the equations of motion can be found which reduces to that of vis viva when $n=0$.

Let $O\xi$, $O\eta$ be two rectangular axes which rotate with the field of force, and let X, Y be the component accelerating forces. We then have by Art. 227

$$\left.\begin{aligned}\frac{d^2\xi}{dt^2}-2n\frac{d\eta}{dt}-n^2\xi=X\\ \frac{d^2\eta}{dt^2}+2n\frac{d\xi}{dt}-n^2\eta=Y\end{aligned}\right\}\text{.....................(1).}$$

Multiplying these by $d\xi/dt$ and $d\eta/dt$ and adding, we find

$$\frac{d\xi}{dt}\frac{d^2\xi}{dt^2}+\frac{d\eta}{dt}\frac{d^2\eta}{dt^2}-n^2\left(\xi\frac{d\xi}{dt}+\eta\frac{d\eta}{dt}\right)=X\frac{d\xi}{dt}+Y\frac{d\eta}{dt},$$

$$\therefore\ \frac{1}{2}\left\{\left(\frac{d\xi}{dt}\right)^2+\left(\frac{d\eta}{dt}\right)^2\right\}-\frac{n^2}{2}(\xi^2+\eta^2)=\int(Xd\xi+Yd\eta)\text{......(2).}$$

We introduce the condition that the field of force rotates by making X, Y such functions of ξ, η only that $X=dU/d\xi$ and $Y=dU/d\eta$. Then U is a function of ξ, η only and not of t. The equation then becomes

$$\tfrac{1}{2}(v^2-n^2r^2)=U+C\text{.....................(3),}$$

where v is the velocity of the particle relatively to the moving axes and r is the radius vector.

We may notice that if U be expressed in terms of the coordinates x, y referred to fixed axes, the expression will contain t also, except when the force is central and tends to O.

The equation, when written in the form (2), is a slight extension of that given by Jacobi in the *Comptes Rendus*, Tome III. p. 59, 1836.

If V be the space velocity of the particle, A the angular momentum about O referred to a unit of mass, then

$$V^2 - 2nA = v^2 - r^2n^2 \quad \text{.......................(4).}$$

The equation of Jacobi then becomes

$$\tfrac{1}{2}V^2 - nA = U + C \quad \text{.......................(5).}$$

To prove the relation (4), let p be the perpendicular from O on the tangent to the relative path. Since V is the resultant of v and nr, (the latter being perpendicular to r), we have

$$V^2 = v^2 + n^2r^2 - 2v \, . \, np, \quad A = vp + nr^2,$$

the second equation being obtained by taking moments about O. The equation (4) follows at once.

An example of a rotating field of force is met with in astronomy. If the components of a binary star describe circles about their common centre of gravity, the force is always the same at the same point of the rotating plane. Jacobi's integral will therefore apply to the motion of a satellite moving in that plane, provided it is of such insignificant mass that the motions of the primaries are undisturbed by its attraction.

256. When the particle moves in space of two dimensions and the field of force rotates about a perpendicular axis with a variable angular velocity ϕ' we may obtain an extension of the equations.

We know that $\frac{1}{2}dV^2/dt$ is equal to the sum of the virtual moments of the forces divided by dt, (Art. 246), hence

$$\begin{aligned}\tfrac{1}{2}dV^2/dt &= Xu + Yv \\ &= X\xi' + Y\eta' + \phi'(\xi Y - \eta X).\end{aligned}$$

But $dA/dt = \xi Y - \eta X$ by taking moments about the origin, hence

$$\frac{1}{2}\frac{dV^2}{dt} - \phi' \, . \, \frac{dA}{dt} = \frac{dU}{dt} \quad \text{..............................(6),}$$

where U is a function of the moving coordinates ξ, η, z. When ϕ' is constant, this can be integrated and we obtain the equation (5).

When *a system of particles* moving in a given rotating field of force is under consideration, we have for each an equation similar to (6). Multiplying these by the masses of the particles and adding the products, we have an extended equation of vis viva. If $2T$ be the vis viva, A the angular momentum of the system, U the force function, this equation is

$$T - \phi' A = U + C \quad\text{..............................}\quad (7),$$

where ϕ' is the angular velocity of the field supposed to be constant. In this form we may omit from U all the actions and reactions which disappear in the principle of virtual work.

257. Coriolis' theorem on relative vis viva. A system of particles is referred to moving axes $O\xi$, $O\eta$. Supposing the system at any instant to become fixed to the moving axes, let us calculate what would *then* be the effective forces on the system. If we apply these as additional impressed forces on the system, but reversed in direction, we may use the equation of vis viva to determine the relative motion as if the axes were fixed in space.

Let m_1, m_2, &c. be the masses of the particles; (X_1, Y_1), (X_2, Y_2), &c. the components of the impressed forces. Let also p, q be the resolved velocities of the origin, then, including these as explained in Art. 227, the equations of motion of any representative particle m are

$$\left.\begin{aligned} m\left\{\frac{d^2\xi}{dt^2} - \omega^2\xi - \frac{1}{\eta}\frac{d}{dt}(\eta^2\omega) + \frac{dp}{dt} - q\omega\right\} &= X \\ m\left\{\frac{d^2\eta}{dt^2} - \omega^2\eta + \frac{1}{\xi}\frac{d}{dt}(\xi^2\omega) + \frac{dq}{dt} + p\omega\right\} &= Y \end{aligned}\right\}\quad\text{....................}\quad (1),$$

where $\omega = d\phi/dt$.

The left-hand sides of these equations measure the components of *the effective forces* on the particle m, Art. 227. The corresponding components on an imaginary particle of the same mass m attached to the moving axes and momentarily coinciding with the real particle are found by treating ξ, η as constants. These are

$$\left.\begin{aligned} \left\{- \omega^2\xi - \eta\frac{d\omega}{dt} + \frac{dp}{dt} - q\omega\right\} &= X_0 \\ \left\{- \omega^2\eta + \xi\frac{d\omega}{dt} + \frac{dq}{dt} + p\omega\right\} &= Y_0 \end{aligned}\right\}\quad\text{........................}\quad (2).$$

These we represent by X_0, Y_0 for the sake of brevity.

Transposing these terms to the other sides of the equations of motion, we have

$$\left.\begin{aligned} m\left(\frac{d^2\xi}{dt^2} - \omega\frac{d\eta}{dt}\right) &= X - X_0 \\ m\left(\frac{d^2\eta}{dt^2} + \omega\frac{d\xi}{dt}\right) &= Y - Y_0 \end{aligned}\right\}\quad\text{...........................}\quad (3).$$

These equations may also be used to supply another proof of the theorem in Art. 197.

Multiplying these respectively by $d\xi/dt$, $d\eta/dt$ and adding, we have, as in Art. 255,

$$m\left\{\frac{d\xi}{dt}\frac{d^2\xi}{dt^2} + \frac{d\eta}{dt}\frac{d^2\eta}{dt^2}\right\} = (X - X_0)\frac{d\xi}{dt} + (Y - Y_0)\frac{d\eta}{dt}.$$

Summing this representative equation for all the particles and integrating

$$\tfrac{1}{2}\Sigma m\left\{\left(\frac{d\xi}{dt}\right)^2 + \left(\frac{d\eta}{dt}\right)^2\right\} = \Sigma\int\{(X - X_0)\,d\xi + (Y - Y_0)\,d\eta\}\quad\text{...........}\quad (4).$$

If the axes rotate round a fixed origin with a uniform angular velocity, ω is constant and p, q are zero. The equation of Coriolis then takes the simpler form

$$\tfrac{1}{2}\Sigma mv^2 = U + \tfrac{1}{2}\omega^2\Sigma mr^2 + C\quad\text{............................}\quad (5),$$

where r is the distance of the particle m from the origin and v is its velocity *relatively to the axes*. For a single particle this is the same as Jacobi's integral.

If the angular velocity ω is not uniform and p, q not zero, the system of additional forces (X_0, Y_0) is not conservative and the integration in (4) cannot be effected except in special cases. The equation is however still important, for the first step in the integration of the equations (1) must be to eliminate the unknown reactions, if any such exist. Now the equation (4) is free from all the reactions which would disappear in the principle of vertical work, and that equation therefore supplies us at once with one result at least of the elimination.

For the purposes of this proposition the forces measured by X_0, Y_0 are called *the forces of moving space*. When the origin of coordinates is fixed, these take the simple form

$$X_0 = -\omega^2\xi - \eta\frac{d\omega}{dt}, \qquad Y_0 = -\omega^2\eta + \xi\frac{d\omega}{dt} \quad\text{.....................} \quad (6).$$

This theorem is due to Coriolis; see the *Journal Polytechnique*, 1831.

258. Laisant's theorem. *Ex.* A particle moves under the action of a force whose Cartesian components are $X = v^n\frac{dU}{dx}$, $Y = v^n\frac{dU}{dy}$, where v is the velocity. Prove that the equation of vis viva is $v^{2-n} = (2-n)U + C$.

See the *Bulletin de la Société Mathématique*, 1893, vol. XXI.

Moments and Resolutions.

259. The equation of Moments. If P, Q are the components of the force on a single particle resolved along and transverse to the radius vector, it is clear that Qr is equal to the moment of the forces about the origin. Representing this moment by M, the transverse polar equation of motion becomes

$$m\frac{d}{dt}\left(r^2\frac{d\theta}{dt}\right) = M \quad\text{.....................}(1).$$

260. When a system of *mutually attracting particles* moves under the action of external forces we have by adding together the transverse polar equations of each particle

$$\Sigma m\frac{d}{dt}\left(r^2\frac{d\theta}{dt}\right) = \Sigma M \quad\text{.....................}(2).$$

If R be the attraction of m_1 on m_2, the reaction of m_2 on m_1 is $-R$, and the sum of the moments of these two must disappear from the right-hand side. If then the external forces are such that their resultant passes through the origin, we have $\Sigma M = 0$, and therefore by integration

$$\Sigma mr^2\frac{d\theta}{dt} = H \quad\text{.....................}(3),$$

where H is a constant. This equation expresses the proposition that *when a system of mutually attracting particles moves under the action of external forces such that the sum of the moments about a fixed point is zero, the sum of the angular momenta of all the particles about that point is constant.* For example, if any number of mutually attracting planets move under the influence of a fixed sun, the sum of their angular momenta is constant. See also Art. 93.

Since $x\,dy - y\,dx = r^2 d\theta$ (Art. 7), the equation (3) of moments when written in Cartesian coordinates takes the form

$$\Sigma m\left(x\frac{dy}{dt} - y\frac{dx}{dt}\right) = H \qquad \text{.................. (4).}$$

261. Rigid system. When a system of particles is rigid it is useful to have an expression for the resultant angular momentum about the origin. Let $(\bar{x}, \bar{y})$ be the coordinates of the centre of gravity, ϕ the angle a straight line fixed in the body makes with a straight line fixed in space, and M the mass. *The angular momentum of the whole mass is then*

$$H = M\left(\bar{x}\frac{d\bar{y}}{dt} - \bar{y}\frac{d\bar{x}}{dt}\right) + Mk^2\frac{d\phi}{dt},$$

where Mk^2 is the moment of inertia about the centre of gravity. See Art. 241.

To prove this, let (x, y) be the coordinates of the particle m, then $x = \bar{x} + \xi$, $y = \bar{y} + \eta$. Remembering that $\Sigma m\xi = 0$, $\Sigma m\eta = 0$ as in Art. 239, we find by substitution that

$$\Sigma m\left(x\frac{dy}{dt} - y\frac{dx}{dt}\right) = (\Sigma m)\left(\bar{x}\frac{d\bar{y}}{dt} - \bar{y}\frac{d\bar{x}}{dt}\right) + \Sigma m\left(\xi\frac{d\eta}{dt} - \eta\frac{d\xi}{dt}\right).$$

Since $d\bar{x}/dt$, $d\bar{y}/dt$ are the components of the velocity of the centre of gravity, the first term is the moment of the velocity of a particle of mass M placed at the centre of gravity and moving with it. The equation therefore asserts that *the angular momentum about any point is equal to that of the whole mass collected at the centre of gravity together with the angular momentum round the centre of gravity of the relative motion.*

To introduce the condition that the system is rigid we change to polar coordinates by writing $\xi d\eta - \eta d\xi = \rho^2 d\theta$. The second term then becomes $\Sigma m\rho^2 \dfrac{d\theta}{dt}$. Remembering that $d\theta/dt$ is the same for every particle and equal to $d\phi/dt$ (Art. 240), this term becomes $Mk^2 \dfrac{d\phi}{dt}$.

It follows that, when a rigid body is acted on by any forces whose moment about the origin is G, the equation of moments is

$$\frac{d}{dt}\left[M\left(\bar{x}\frac{d\bar{y}}{dt}-\bar{y}\frac{d\bar{x}}{dt}\right)+Mk^2\frac{d\phi}{dt}\right]=G.$$

262. *Ex.* 1. A particle moves in a field of force defined by the force function

$$U=mf(r)+\frac{mF(\theta)}{r^2}.$$

Show how to find the coordinates r, θ in terms of the time.

The force transverse to the radius vector is $Q=dU/rd\theta$. The equation of moments therefore becomes $\dfrac{d}{dt}\left(r^2\dfrac{d\theta}{dt}\right)=\dfrac{1}{r^2}\dfrac{dF}{d\theta}$. Multiplying by $r^2d\theta/dt$, the integration can be effected and we find

$$\left(r^2\frac{d\theta}{dt}\right)^2=2F(\theta)+A \;\;\ldots\ldots\ldots\ldots\ldots\ldots\ldots\ldots (1),$$

where A is an arbitrary constant. This integral is equivalent to a result given by both Jacobi and Bertrand.

The equation of vis viva is

$$\left(\frac{dr}{dt}\right)^2+r^2\left(\frac{d\theta}{dt}\right)^2=2f(r)+\frac{2F(\theta)}{r^2}+C \;\;\ldots\ldots\ldots\ldots\ldots (2).$$

Eliminating $d\theta/dt$ by the help of (1) we arrive at an equation giving dt/dr as a function of r. The determination of t in terms of r has thus been reduced to an integration. The relation between θ and t may then be found from (1) by another integration.

Ex. 2. A particle is placed at rest at the point $x=0$, $r=a$ in a field defined by $U=m\dfrac{a^4x}{r^3}$. Show by writing down the equations of vis viva and moments that the path is a circle.

263. The equation of resolution. If a system of particles moves under the action of external forces, we have by resolving parallel to the axis of x, (Art. 236),

$$\Sigma m\frac{d^2x}{dt^2}=\Sigma mX,$$

where X is the typical accelerating force on the particle m. In this equation we may omit the mutual attractions of the particles, for the action and reaction being equal and opposite, these disappear in the resolution.

If any direction fixed in space exist such that the sum of the components of the impressed forces in that direction is zero, we can take the axis of x parallel to that direction. We then have

$$\Sigma mX = 0, \quad \therefore \ \Sigma m \frac{dx}{dt} = A,$$

where A is a constant. This result is the same as that already arrived at, and more fully stated, in Art. 92.

264. Summary of methods of integration. When the system of particles moves in a given field of force the equation of vis viva in general supplies one integral of the equations of motion. If the system has only one degree of freedom, this integral is sufficient to determine the motion.

When another integral is required, there is no general method of proceeding. We usually search if there is any direction fixed in space in which the sum of the resolved parts of the forces is zero, or any fixed point about which the sum of the moments is zero. In either of these cases an additional integral is supplied by the methods of Arts. 263 and 260. The first case usually occurs when the acting force is gravity, the second when the force is central.

When these methods fail we have recourse to some artifice suited to the problem. Suppose that we have some reason for believing that a particle describes a certain path, we constrain the particle by a smooth curve. If the pressure can be made zero by the proper initial conditions, the constraint may be removed and the particle will describe the path freely, Art. 193.

265. Examples. *Ex.* 1. Two particles, of masses m, M, placed on a smooth table, are connected by a string of length $a+b$, which passes through a fine ring fixed at a point O on the table. The particles are projected with velocities U and V perpendicularly to the portions of the string attached to them, and the initial lengths are respectively a and b. Find the motion.

Let (r, θ), (ρ, ϕ) be the polar coordinates of m and M at the time t. By the principles of angular momentum and vis viva, we have

$$r^2 \frac{d\theta}{dt} = Ua, \qquad \rho^2 \frac{d\phi}{dt} = Vb \quad \text{.............................. (1)},$$

$$m \left\{ \left(\frac{dr}{dt}\right)^2 + r^2 \left(\frac{d\theta}{dt}\right)^2 \right\} + M \left\{ \left(\frac{d\rho}{dt}\right)^2 + \rho^2 \left(\frac{d\phi}{dt}\right)^2 \right\} = mU^2 + MV^2 \quad \text{......... (2)}.$$

We have also the geometrical equation

$$r + \rho = a + b \quad \text{... (3)}.$$

Eliminating ρ, θ, ϕ, we find

$$(M+m)\left(\frac{dr}{dt}\right)^2 + \frac{mU^2a^2}{r^2} + \frac{MV^2b^2}{(a+b-r)^2} = mU^2 + MV^2 \quad \text{.............. (4)}.$$

In this differential equation, the variables can be separated and thus t can be expressed in terms of r by an integral. The integration cannot be generally effected.

If the system oscillate, the extreme positions are determined by putting $dr/dt=0$. We thus have

$$\frac{mU^2a^2}{r^2}+\frac{MV^2b^2}{(a+b-r)^2}-(mU^2+MV^2)=0 \quad\text{.................... (5).}$$

Since the left-hand side is positive when $r=0$ and $r=a+b$ and vanishes when $r=a$ there is a second positive root less than $a+b$. This second root may be proved to be greater or less than a according as mU^2/a is greater or less than MV^2/b. These values of r determine the extreme positions of the system. We notice that if V be very small, the second root is very nearly equal to $a+b$.

If $V=0$ the particle M arrives at the origin, but the appearance when $r=a+b$ of the singular form 0/0 in the equation (5) is a warning that the motion changes its character in this case. In fact if the third term on the left-hand side of (4) is removed, the velocity of arrival at O is finite instead of being infinitely great.

To find the tension T of the string, we use the radial equation of motion for one of the particles. This gives

$$\frac{d^2r}{dt^2}-r\left(\frac{d\theta}{dt}\right)^2=-\frac{T}{m}.$$

Differentiating (4) we find dr/dt in terms of r and after some slight reductions

$$T=\frac{Mm}{M+m}\left(\frac{U^2a^2}{r^3}+\frac{V^2b^2}{(a+b-r)^3}\right).$$

The string therefore does not become slack.

Ex. 2. Two particles whose masses are in the ratio 1 : 2 lie on a smooth horizontal table, and are connected by a string that passes through a small ring in the table: the string is stretched and the particles are equidistant from the ring: the lighter particle is then projected at right angles to its portion of the string. Prove that the other particle will strike the ring with half the initial velocity of the first particle. [Coll. Ex. 1896.]

Ex. 3. One A of two particles of equal mass, without weight, and connected by an inelastic string moves in a straight groove. The other B is projected parallel to the groove, the string being stretched. Prove that the greatest tension is four times the least. [Coll. Ex.]

Reduce A to rest, then B is acted on by T and $T\cos\theta$, the latter being parallel to the groove, where θ is the angle AB makes with the groove. The particle B now describes a circle, and the normal and tangential resolutions give the angular velocity and the tension.

Ex. 4. Two particles m, M, are connected by a string, of length $a+b$, which passes through a hole in a smooth table; M hangs vertically at a depth b below the hole, m is projected horizontally and perpendicularly to the string with velocity V from a point on the table distant a from the hole. Prove that if M just rise to the table, $mV^2(2ab+b^2)=2Mgb(a+b)^2$. Prove also that if M oscillates,

$$mV^2+2Mga>3(M^2mV^2g^2a^2)^{\frac{1}{3}}.$$

What is the motion if $mV^2=Mga$?

Ex. 5. Two *small* spheres of masses m and $2m$ are fixed at the ends of a weightless rigid rod AB which is free to turn about its middle point O; the heavier sphere rests on a horizontal table, the rod making an angle $30°$ with it. If a sphere of mass m falling vertically with velocity u strike the lighter sphere directly, prove that the impulse which the heavier sphere ultimately gives to the table is $\frac{4}{5} mu (1+e)$, where e is the coefficient of restitution between the two spheres, the table being perfectly inelastic. [Coll. Ex. 1893.]

At the first impact we take moments for the two particles m, $2m$ about O to avoid the reaction at O. We therefore have $3mv'a = Ra \cos \alpha$, $m(u'-u) = -R$ where $\alpha = 30°$. At the moment of greatest compression the velocity of approach of the centres is zero, $\therefore u' = v' \cos \alpha$, and $R = \frac{4}{5} mu$. Since the complete value of R is found by multiplying this by $1+e$, the velocity of either end of the rod after impact is $\frac{4}{15} u \cos \alpha (1+e)$. The balls m and $2m$ rotate with the rod round O through some angle, and $2m$ finally hits the table with a velocity v'. Taking the same equation of moments as before $R'a \cos \alpha = 3mv'a$, $\therefore R' = \frac{4}{5} mu (1+e)$.

Ex. 6. One end of a string of length l is attached to a small ring of mass m which can slide freely on a smooth horizontal wire, and the other end supports a heavy particle of mass m'. If this particle be held displaced in the vertical plane containing the groove, the string being straight and then let go, prove that the path of m' is part of an ellipse whose semi-axes are l, $lm/(m+m')$, the major axis being vertical. [Coll. Ex. 1896.]

Only the horizontal resolution and the geometrical equation are required.

Ex. 7. A rectangular block of wood of mass M is free to slide between two smooth horizontal planes, and in it is inserted a smooth tube in the shape of a quadrant of a circle of radius a, one of the bounding radii lying along the lower plane, and the other being vertical. A particle of mass m is shot into the tube horizontally with velocity V, rebounds from the lower plane, and leaves the tube again with a relative velocity V', prove that

$$V'^2 = e^2 V^2 - 2ga(1-e^2)(M+m)/M,$$

where e is the coefficient of restitution for the lower plane. [Coll. Ex. 1895.]

Ex. 8. If in the case of three equal particles the units are so chosen that the energy integral is $\frac{1}{2}(v_1^2 + v_2^2 + v_3^2) = \frac{1}{r_{23}} + \frac{1}{r_{31}} + \frac{1}{r_{12}} - \frac{1}{r}$, where r_{12} is the distance between the particles whose velocities are v_1 and v_2, and if r is a positive constant, the greatest possible value of the angular momentum of the system about its centre of inertia is $\frac{2}{3}\sqrt{(2r)}$. [Math. Tripos, 1893.]

Ex. 9. Two equal particles are initially at rest in two smooth tubes at right angles to each other. Prove that whatever be their positions and whatever their law of attraction, they will reach the intersection of the tubes together.

[Coll. Ex.]

Ex. 10. Three mutually attracting particles, of masses m_1, m_2, m_3, are placed at rest within three fixed smooth tubes Ox, Oy, Oz at right angles to each other. The attraction between any two, say m_1, m_2, is $\mu m_1 m_2 r_3^{\kappa}$ where r_3 is the distance. If the triangle joining the particles always remains similar to its initial form, prove that the initial distances satisfy the equations

$$\frac{m_2 m_3 r_1^{\kappa-1}}{m_2+m_3-m_1} = \frac{m_3 m_1 r_2^{\kappa-1}}{m_3+m_1-m_2} = \frac{m_1 m_2 r_3^{\kappa-1}}{m_1+m_2-m_3}.$$

266. Double answers. *Ex.* A cube, of mass M, constrained to slide on a smooth horizontal table, has a fine tube ACB cut through it in the vertical plane through its centre of gravity, the extremities A, B being on the same horizontal line and the tangents at A, B horizontal. A particle, of mass m, is projected into the tube at A with velocity V, deduce analytically from the equations of linear momentum and vis viva that the velocity of emergence at B is also V.

Let u, v be the velocities of the cube and particle at emergence. The principles referred to give

$$Mu+mv=mV, \qquad Mu^2+mv^2=mV^2.$$

These give two solutions, viz. (1) $u=0$, $v=V$, and (2) $u=2mV/S$, $v=(m-M)V/S$, where $S=m+M$. To interpret these we notice that there are two sets of initial conditions which give the same linear momentum and vis viva. These are determined by the values of u, v just written down. We have therefore really solved two problems and have thus obtained two results.

To distinguish the solutions, we investigate the intermediate motion. Let P be any point in the tube and let p be the tangent of the angle the tangent makes with the horizon. If u, v now represent the horizontal velocities at P, the same two principles give

$$Mu+mv=mV, \qquad Mu^2+m(v^2+p^2x'^2)=mV^2,$$

where $x'=v-u$ is the relative velocity. These give

$$v=\frac{V}{M+m}\left\{m \pm M\left(1+p^2\frac{M+m}{M}\right)^{-\frac{1}{2}}\right\}.$$

Now $v=V$ initially when $p=0$, hence the radical must have the positive sign and must keep that sign until it vanishes. On emergence therefore, when p is again zero, $v=V$. The negative sign of the radical evidently gives the initial conditions of the other problem.

267. Bodies without mass. *Ex.* 1. A heavy bead is free to slide along a rod whose ends move without friction on a horizontal circle; prove that when the mass of the rod is negligible compared with that of the bead, the bead will, when started, continue to slide along the rod with an acceleration varying inversely as the cube of its distance from the middle point. [Math. Tripos, 1887.]

The reaction between the rod and the particle is zero because the rod has no mass. To prove this, let R be the reaction, M the mass of the rod, then, taking moments about the centre O of the circle, we have $M\kappa^2 d\omega/dt=Rp$, where ω is the angular velocity of the rod. Hence $R=0$ when $M=0$.

The particle P, being not acted on by any horizontal force, describes a straight line in space with uniform velocity b. If x be the distance of P from the middle point C of the rod; a, c, the perpendiculars from O on the path and on the rod, we have $x^2+c^3=OP^2=a^2+b^2t^2$.

This gives $$d^2x/dt^2=b^2(a^2-c^2)/x^3.$$

Ex. 2. A rigid wire without mass is formed into an arc of an equiangular spiral and carries a heavy particle fixed in the pole. If the convexity of the wire be placed in contact with a perfectly rough horizontal plane prove that the point of contact will move with a uniform acceleration equal to $g\cot\alpha$, where α is the angle of the spiral. [Math. Tripos, 1860.]

268. Equation of the path. Let P, Q be the resolved accelerating forces acting on the particle respectively along and perpendicular to the radius vector. Let P be regarded as *positive when acting towards the origin.* The equations of motion are

$$\frac{d^2r}{dt^2} - r\left(\frac{d\theta}{dt}\right)^2 = -P, \qquad \frac{1}{r}\frac{d}{dt}\left(r^2\frac{d\theta}{dt}\right) = Q \quad\text{.....................(1).}$$

To find the path we eliminate t. The second equation, after multiplication by $r^3 d\theta/dt$ and integration, as in Art. 262, becomes

$$\left(r^2\frac{d\theta}{dt}\right)^2 = h^2 + 2\int Qr^3 d\theta \quad\text{..............................(2).}$$

For the sake of brevity we represent the right-hand side by H^2. Putting also $u=1/r$, we find $d\theta/dt = Hu^2$. We then have

$$\frac{dr}{dt} = -\frac{1}{u^2}\frac{du}{d\theta}\frac{d\theta}{dt} = -H\frac{du}{d\theta},$$

$$\therefore \frac{d^2r}{dt^2} = -\frac{d}{d\theta}\left(H\frac{du}{d\theta}\right) . Hu^2.$$

Substituting in the first equation of motion

$$Hu^2\frac{d}{d\theta}\left(H\frac{du}{d\theta}\right) + H^2u^3 = P,$$

$$\therefore H^2u^2\left(\frac{d^2u}{d\theta^2} + u\right) + \tfrac{1}{2}u^2\frac{du}{d\theta}\frac{dH^2}{d\theta} = P$$

Replacing H^2 by its value given in (2),

$$\left(\frac{d^2u}{d\theta^2} + u\right)\left(h^2 + 2\int\frac{Q}{u^3}d\theta\right) + \frac{Q}{u^3}\frac{du}{d\theta} = \frac{P}{u^2} \quad\text{.....................(3).}$$

This is *Laplace's differential equation of the path of the particle.* The forces P, Q being given in terms of the coordinates u, θ, of the moving particle, this equation, when solved, will determine u as a function of θ, and thus lead to the equation of the path. To find the motion along the path we use equation (2). Substituting in that equation the value of u in terms of θ we find by integration the time t at which the particle occupies any given position.

The polar differential equation of the path cannot be integrated except for special forms of the forces P, Q. If $Q=0$, the equation takes the form

$$\frac{d^2u}{d\theta^2} + u = \frac{P}{h^2u^2} \quad\text{.......................................(4).}$$

This can be integrated when P is a function of u alone, a case which is considered in the chapter on central forces. It can also be integrated when $P=u^2F(\theta)$, the method of solution being that shown in Art. 122.

When $P=u^3F(\theta)$ the equation is linear. If one solution of the differential equation is known, say $u=\phi(\theta)$, the general integral may be determined by substituting $u=z\phi(\theta)$. After integration we find $z=A+B\int[\phi(\theta)]^{-2}d\theta$.

269. When $P=u^3F(\theta)$, $Q=u^3f'(\theta)$, *the differential equation of the path takes the linear form*

$$\left(\frac{d^2u}{d\theta^2} + u\right)\{h^2 + 2f(\theta)\} + f'(\theta)\frac{du}{d\theta} - F(\theta)u = 0 \quad\text{................ (5).}$$

The various cases in which this equation can be integrated are enumerated in treatises on Differential Equations.

By multiplying the equation by the proper factor we can make the left-hand side a perfect differential. Conversely choosing any factor, we can find the relation between P and Q that this may be the proper integrating factor. If we wish *the relation between P, Q to be independent of the initial conditions*, the terms containing h^2 as a factor must be made a perfect differential independently of the remaining terms. The coefficient of h^2 is $\frac{d^2u}{d\theta^2}+u$ and this is made a perfect differential by either of the factors $\sin\theta$ or $\cos\theta$. The remaining terms must therefore also become a perfect differential by the same factor. The condition that $L\frac{d^2u}{d\theta^2}+M\frac{du}{d\theta}+Nu$ is a perfect differential is $N-\frac{dM}{d\theta}+\frac{d^2L}{d\theta^2}=0$, and the integral is known to be $L\frac{du}{d\theta}+\left(M-\frac{dL}{d\theta}\right)u$.

Multiplying equation (5) by $\sin\theta$, the product is a perfect differential if

$$\{2f(\theta)-F(\theta)\}\sin\theta-\frac{d}{d\theta}\{\sin\theta f'(\theta)\}+2\frac{d^2}{d\theta^2}\{\sin\theta f(\theta)\}=0,$$

which reduces at once to
$$\frac{P}{u^3}=\frac{d}{d\theta}\frac{Q}{u^3}+3\cot\theta\frac{Q}{u^3}\;\ldots\ldots\ldots\ldots(6).$$

The integral, since $f'(\theta)=Q/u^3$, becomes

$$\left(h^2+2\int\frac{Q}{u^3}d\theta\right)\left(\sin\theta\frac{du}{d\theta}-\cos\theta u\right)-\frac{Q}{u^3}\sin\theta u=C\;\ldots\ldots\ldots\ldots(7),$$

where C is a constant. This is a linear equation of the first order and can be integrated a second time when Q/u^3 is given as a function of θ. *The determination of the path can therefore be reduced to integration when the relation* (6) *is satisfied.*

In the same way, if we multiply (5) by $\cos\theta$, we find that the product is a perfect differential if
$$\frac{P}{u^3}=\frac{d}{d\theta}\frac{Q}{u^3}-3\tan\theta\frac{Q}{u^3}\;\ldots\ldots\ldots\ldots(8),$$

and the integral is
$$\left(h^2+2\int\frac{Q}{u^3}d\theta\right)\left(\cos\theta\frac{du}{d\theta}+\sin\theta u\right)-\frac{Q}{u^3}\cos\theta u=C'\;\ldots\ldots(9),$$
which is linear and can be integrated a second time.

Another case in which the integration of (3) can be effected may be deduced from Art. 262. The equation (3) is

$$\frac{d}{d\theta}\left[H^2\left\{\left(\frac{du}{d\theta}\right)^2+u^2\right\}\right]=\frac{P}{u^3}\frac{2u\,du}{d\theta}+\frac{2Q}{u^3}u^2.$$

If then $\frac{P}{u^3}=f(u)+2\int\frac{Q}{u^3}d\theta$, the integral is

$$\left\{h^2+2\int\frac{Q}{u^3}d\theta\right\}\left\{\left(\frac{du}{d\theta}\right)^2+u^2\right\}=2\int f(u)\,u\,du+2u^2\int\frac{Q}{u^3}d\theta+C\;\ldots\ldots(10).$$

270. *Ex.* 1. If $P=u^3F(\theta)$ and $Q=P\tan\theta$, prove that $u=A\sin\theta$ is a particular solution of the linear equation (5). Thence obtain the general integral by putting $u=z\sin\theta$, where z is a function of θ which is determined by solving a linear equation of the first order.

Ex. 2. A particle moves under the forces

$$P=\mu u^3(3+5\cos 2\theta),\qquad Q=\mu u^3\sin 2\theta;$$

prove that an integral of its motion is

$$h^2\left\{\frac{du}{d\theta}\sin\theta-u\cos\theta\right\}+\mu\left\{\tfrac{1}{2}(\sin\theta-\sin 3\theta)\frac{du}{d\theta}+\cos 3\theta u\right\}=C.$$

Obtain also a similar integral if

$$P=\mu u^3 \cos n\theta \left\{n+\frac{3\tan n\theta}{\tan\theta}\right\}, \qquad Q=\mu u^3 \sin n\theta.$$

[Coll. Exam. 1892.]

Ex. **3. If the Cartesian accelerating forces X, Y are unrestricted, prove that the differential equation of the path is**

$$(A+2\int X dx)\frac{d^2y}{dx^2}+X\frac{dy}{dx}-Y=0,$$

where A is a constant depending on the initial conditions.

Prove also that the determination of y as a function of x can be reduced to integration when both X, Y are functions of x only.

Ex. **4. If X and Y/y are functions of x only, the differential equation of the path is linear. Prove that it can be integrated when $Y=y\frac{dX}{dx}$, and that the first integral is**

$$(A+2\int X dx)\frac{dy}{dx}-Xy=C.$$

Prove also that when $\frac{Y}{y}=\frac{dX}{dx}+\frac{3X}{x}$, the differential equation can be integrated and that the first integral is

$$(A+2\int X dx)\,x\frac{dy}{dx}-(A+2\int X dx+xX)\,y=C.$$

Ex. **5. Prove that the Cartesian equations of motion can be completely integrated when the force function satisfies**

$$\frac{d^2U}{dx^2}-\frac{d^2U}{dy^2}=\kappa\frac{d^2U}{dxdy}.$$

To prove this we notice that $\qquad U=\phi(y+ax)+\psi(y+a'x),$

where a, a' are the roots of $a^2-\kappa a=1$. We then change the variables to $\xi=y+ax$ and $\eta=y+a'x$. The new coordinates ξ, η are also rectangular. The equations of motion become $d^2\xi/dt^2=\phi'(\xi)$, $d^2\eta/dt^2=\psi'(\eta)$, which may be solved as in Art. 122.

Ex. **6. If the direction of the acting force is always a tangent to the direction of motion, as in the case of a resisting medium, prove that the path is a straight line. Consider the resolution along the normal.**

Ex. **7. If the direction of the force is always perpendicular to the path, prove that the velocity is constant.**

Superposition of Motions.

271. A particle is constrained to describe a fixed curve. When projected from a point A with a velocity u_1 under the action of any forces the velocity and pressure at any point P are v_1 and R_1. When projected with a velocity u_2 from the same point A under a second system of forces the velocity and pressure at P are v_2

and R_2. When the particle is projected from A with a velocity u such that $u^2 = u_1^2 + u_2^2$, and moves under the action of both systems of forces, the velocity and pressure at P are v and R. It is required to prove that

$$v^2 = v_1^2 + v_2^2, \quad R = R_1 + R_2.$$

To prove this we write down the two equations for each of the three types of motion. Representing for the sake of brevity the normal components of accelerating force by N_1, N_2, $N_1 + N_2$, we have

$$v_1^2 - u_1^2 = 2\int(X_1 dx + Y_1 dy), \qquad v_1^2/\rho = N_1 + R_1/m,$$
$$v_2^2 - u_2^2 = 2\int(X_2 dx + Y_2 dy), \qquad v_2^2/\rho = N_2 + R_2/m,$$
$$v^2 - u^2 = 2\int\{(X_1 + X_2)\, dx + (Y_1 + Y_2)\, dy\}, \quad v^2/\rho = N_1 + N_2 + R/m,$$

the limits of integration being always from the point A to P.

The results follow at once by subtracting from the third equation the sum of the other two.

272. The following corollary will be found useful.

A particle can describe a curve freely under the action of certain forces, the velocity at some point A being u_1. If the particle is now constrained to describe the same curve the velocity at A being changed to u_2, then the pressure at any point P is C/ρ, where ρ is the radius of curvature at P, and C is the constant $m\,(u_2^2 - u_1^2)$.

To prove this we notice that when the velocity at A is u_1 and the forces act on the particle, the pressure is $R_1 = 0$. If the velocity at A were u' and no forces acted on the particle, the pressure at P would be mu'^2/ρ. Superimposing these two states and putting $u'^2 = u_2^2 - u_1^2$, the theorem follows at once.

273. We may also deduce the following theorem due to Ossian Bonnet. If a particle can freely describe the same curve under two different systems of forces, the velocities at some point A being respectively u_1 and u_2, then the particle can describe the same path under both systems of forces provided the velocity at A is u, where $u^2 = u_1^2 + u_2^2$. Since any point may be taken as the point of projection this relation between the velocities holds at all points of the curve. *Liouville's Journal*, Tome IX. page 113.

274. The following example of Ossian Bonnet's theorem is important. It will be shown in the chapter on central forces that a particle P will describe an ellipse freely about a centre of force in one focus H_1, whose law of attraction is

μ_1/r_1^2, provided the velocity of projection at any point A is given by

$$v_1^2 = \mu_1\left(\frac{2}{r_1} - \frac{1}{a}\right).$$

The same ellipse can also be described about a centre of force in the other focus H_2 whose law of attraction is μ_2/r_2^2 provided the velocity v_2 has the corresponding value. It immediately follows that *the particle can describe the ellipse freely about both centres of force acting simultaneously*, provided (1) the velocity v at any point A is given by

$$v^2 = \mu_1\left(\frac{2}{r_1} - \frac{1}{a}\right) + \mu_2\left(\frac{2}{r_2} - \frac{1}{a}\right),$$

and (2) the direction of projection at A bisects externally the angle between the focal distances.

According to this mode of proof both the centres of force should be attractive, for it is evident that an ellipse could not be freely described about a single centre of repulsive force situated in either focus. But the law of continuity shows that this limitation is unnecessary. Supposing μ_1 and μ_2 to have arbitrary positive values, it has been proved that the equations of motion of a particle moving freely under both centres of force become satisfied when this value of v^2 is substituted in them. The equations contain only the first powers of μ_1 and μ_2 (see Art. 271) and can be satisfied only by the vanishing of the coefficients of these quantities. They will therefore still be satisfied if we change the signs of either μ_1 or μ_2.

In the same way we may introduce other changes into the theorem, provided always we can obtain a dynamical interpretation of the result.

275. *Ex.* 1. Prove that a particle can describe an ellipse freely under the action of three centres of force; one in each focus attracting as the inverse square and the third in the centre attracting as the direct distance. Find also the velocity of projection.

Ex. 2. Particles of masses m_1, m_2, &c. projected from the same point in the same direction with velocities u_1, u_2, &c. under the action of given forces F_1, F_2, &c. describe the same curve. Show that a particle of mass M projected in the same direction with a velocity V under the simultaneous action of all the forces F_1, F_2, &c. will also describe the same curve, provided

$$MV^2 = m_1u_1^2 + m_2u_2^2 + \dots.$$

Ossian Bonnet, Note IV. to Lagrange's *Mécanique.*

Ex. 3. A bead is projected along a smooth elliptical wire under the action of two centres of force, one in each focus, and attracting inversely as the square of the distance. If TP, TQ be any two tangents to the ellipse, prove that the pressure when the bead is at P : pressure when the bead is at Q :: $TQ^3 : TP^3$.

Initial Tensions and radii of Curvature.

276. *Particles, of given masses, are connected together by inelastic rods or strings of given lengths and are projected in any given manner consistent with these constraints. It is required to find the initial values of the tensions and the radii of curvatures of the paths.*

The peculiarity of the problems on initial motion is that the velocities and directions of motion of all the particles are known. It will thus not be necessary to integrate the differential equations of motion, for the results of these integrations are given.

Supposing that there are n particles, we shall require besides the $2n$ equations of motion a geometrical equation corresponding to each reaction.

To show how the geometrical equations may be formed, let us suppose that two particles m_1, m_2 are connected by a rod or straight string of length l. The component velocities of the two particles in the direction of the string being necessarily equal, their *relative velocity* is the difference of their component velocities perpendicular to the rod; let these be V_1, V_2. If ϕ be the angle the rod makes with some fixed straight line, the geometrical equation is $l\dfrac{d\phi}{dt} = V_2 - V_1$.

The simplest method of obtaining the relative equations of motion is perhaps to reduce m_1 to rest. To effect this we apply to both particles (1) an acceleration equal and opposite to that of m_1, and (2) an initial velocity equal and opposite to that of m_1. The path of m_2 being now a circle whose centre is at m_1 and whose radius is l, the relative accelerations are those for a circular motion. (Art. 39.)

Let X_1, X_2 be the components along the rod of junction of all the forces and tensions which act on m_1, m_2 respectively. We then have (Art. 35)

$$-l\left(\frac{d\phi}{dt}\right)^2 = -\frac{(V_2 - V_1)^2}{l} = \frac{X_2}{m_2} - \frac{X_1}{m_1} \quad\text{...............(1).}$$

In this way we may form as many equations as there are reactions. By solving these the initial values of the reactions become known.

If the angular accelerations of the rods are also required, let Y_1, Y_2 be the component forces perpendicular to the rod which act on m_1, m_2. Then

$$l\frac{d^2\phi}{dt^2} = \frac{Y_2}{m_2} - \frac{Y_1}{m_1} \quad\text{.......................(2).}$$

277. *To find the curvatures of the paths, we refer to the equations of motion in space.* The velocity and direction of motion of each particle being known, we may conveniently use the tangential and normal resolutions. We thus have $2n$ equations of the form

$$m\frac{v^2}{\rho}=N,\quad m\frac{dv}{dt}=T \dots\dots\dots\dots\dots\dots (3),$$

where N, T are linear functions of the forces and tensions which act on the particle m.

These reactions having been found by considering the relative motion, we substitute in (3). The first of these determines the radius of curvature ρ of the path of m, and the second the tangential acceleration, if that be required.

When any *one of the particles is constrained to describe a given curve*, the initial pressure of that curve is one of the unknown reactions. This pressure will be determined by the normal resolution of (3) since the radius of curvature of the path is the same as that of the constraining curve.

278. *If some or all the particles start from rest*, the equations of relative motion are simplified, for we then have $\phi'=0$ where the accent denotes d/dt. Since however *the direction of motion of a free particle at rest is not given, the tangential and normal resolutions are then inappropriate.* We can however use the Cartesian or polar resolutions in space. Since $\theta'=0$, the polar resolutions reduce to r'' and $r\theta''$ which are very simple forms. We must however bear in mind that if we require to differentiate the equations of motion this simplification must not be introduced until all the differentiations have been effected, Art. 281. We may also use Lagrange's equations, when the curvatures and not the tensions are required. These modifications of the general method are more especially useful in Rigid Dynamics and are discussed in the first volume of the author's treatise on that subject.

279. Examples. *Ex.* 1. Particles are attached to a string at unequal distances, and placed in the form of an unclosed polygon on a smooth table. The particles are then set in motion without impacts and are acted on by any forces. It is required to find the initial tensions and curvatures.

Let $ABCD$ &c. be any consecutive particles, and let the tensions of AB, BC, &c. be T_1, T_2, &c. Let the given forces be F_1, F_2, &c. and let them act in directions making angles α, β, &c. with AB, BC, &c. Let $l_1d\phi_1/dt$, $l_2d\phi_2/dt$, &c. stand for the known difference of the velocities of the consecutive particles resolved perpendicular to the rod or string joining them.

The particle B being reduced to rest, C is acted on by T_3/m_3 along CD, T_2/m_3 along CB, T_2/m_2 along CB, T_1/m_2 parallel to AB. Besides these there are the

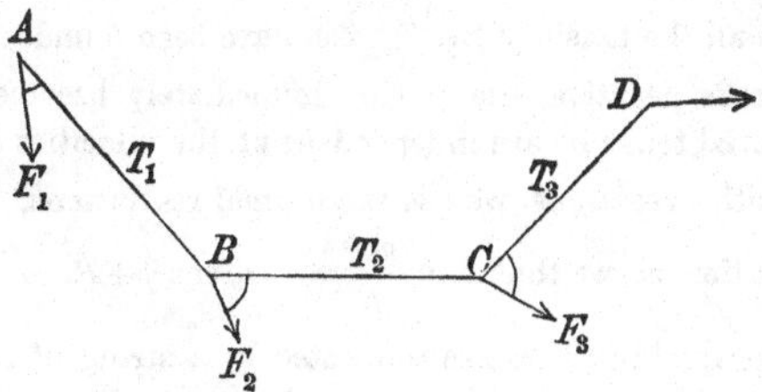

impressed accelerating forces F_3/m_3 and $-F_2/m_2$. Since C describes a circle relatively to B, we have for the particle C

$$l_2\left(\frac{d\phi_2}{dt}\right)^2=\frac{T_3}{m_3}\cos C+T_2\left(\frac{1}{m_2}+\frac{1}{m_3}\right)+\frac{T_1}{m_2}\cos B+\frac{F_3}{m_3}\cos(C+\gamma)+\frac{F_2}{m_2}\cos\beta,$$

$$l_2\frac{d^2\phi}{dt^2}=\frac{T_3}{m_3}\sin C-\frac{T_1}{m_2}\sin B+\frac{F_3}{m_3}\sin(C+\gamma)+\frac{F_2}{m_2}\sin\beta,$$

where A, B, C, &c. are the internal angles of the polygon. The second resolution may be omitted if the angular accelerations of the several portions of string are not required.

An equation, corresponding to the first of these, can be written down for each of the n particles, beginning at either end, except the last. We thus form $(n-1)$ equations to find the $(n-1)$ tensions.

To find the initial radius of curvature of the path in space of any particle C we resolve along the normal to the path. Let the directions of motion of the particles be AA', BB', &c. and let v_1, v_2, &c. be the velocities of the particles. Then

$$\frac{m_3v_3^2}{\rho_3}=T_3\sin DCC'+T_2\sin BCC'-F_3\sin(DCC'-\gamma).$$

If the particle m_3 is initially at rest, $v_3=0$ and the last equation fails to determine ρ_3. The initial tensions may still be deduced from the first equation. The initial direction of motion of the particle coincides with the direction of the resultant force and is therefore known when the initial tensions have been found. The tangential acceleration is also known for the same reason. The determination of the radius of curvature requires further consideration.

Ex. 2. Heavy particles, whose masses beginning at the lowest are m_1, m_2, &c., are placed with their connecting strings on a smooth curve in a vertical plane. Find the initial tensions.

In this problem the arc between any two particles remains constant, so that the tangential accelerations of all the strings are equal. Let this common acceleration be f. Taking all the particles as one system, the tensions do not appear in the resulting equation, we have therefore

$$(m_1+m_2+\&c.)f=-m_1g\sin\psi_1-m_2g\sin\psi_2-\&c.,$$

where ψ_1, ψ_2, &c. are the angles the tangents at the particles make with the horizon.

Considering the lowest particle, we have

$$m_1f=-m_1g\sin\psi_1+T_1.$$

Considering the two lowest,

$$(m_1+m_2)f = -m_1 g \sin\psi_1 - m_2 g \sin\psi_2 + T_2,$$

and so on. Thus all the tensions T_1, T_2, &c. have been found.

If any tension is negative, that string immediately becomes slack. We also notice that the initial tensions are independent of the velocities of the particles.

To find the initial reactions, we use the normal resolutions. If v be the initial velocity of the particle m, we thus find $\dfrac{mv^2}{\rho} = -mg\cos\psi + R$.

Ex. 3. Three equal particles are connected by a string of length $a+b$ so that one of them is at distances a, b from the other two. This one is held fixed and the others are describing circles about it with the same angular velocity so that the string is straight. Prove that if the particle that was held fixed is set free the tensions in the two parts of the string are altered in the ratios $2a+b : 3a$ and $2b+a : 3b$. [Coll. Ex. 1897.]

Ex. 4. Three equal particles tied together by three equal threads are rotating about their centre of gravity. Prove that if one of the threads break, the curvatures of the paths instantaneously become 3/5, 6/5, 3/5ths respectively, of their former common value. [Coll. Ex. 1892.]

Ex. 5. Two particles are fastened at two adjacent points of a closed loop of string without weight which hangs in equilibrium over two smooth horizontal parallel rails. Prove that when the short piece of string between the particles is cut the product of the tensions before and after the cutting is equal to the product of the weights of the particles. [Coll. Ex. 1896.]

Ex. 6. Two particles of equal weight are connected by a string of length l which becomes straight just when it is vertical. Immediately before this instant the upper particle is moving horizontally with velocity $\sqrt{gl}$, and the lower is moving vertically downwards with the same velocity. Prove that the radius of curvature of the curve which the upper particle begins to describe is $\frac{5}{12}\sqrt{5}l$. [Coll. Ex. 1897.]

Just after the impulse the upper particle begins to move in a direction inclined $\tan^{-1} 1/2$ to the horizon.

Ex. 7. Two equal particles A, B, are connected by a string of length l, the middle point C of which is held at rest on a smooth horizontal table. The particles describe the same circle on the table with the same velocity in the same direction, and the angle ACB is right. The point C being released, prove that the radii of curvature of their paths just after the string becomes tight are $5\sqrt{5}l/4$ and infinity.

Ex. 8. Four small smooth rings of equal mass are attached at equal intervals to a string, and rest on a smooth circular wire whose plane is vertical and whose radius is equal to one-third of the length of the string, so that the string joining the two uppermost is horizontal, and the line joining the other two is the horizontal diameter. If the string is cut between one of the extreme particles and the nearer of the middle ones, prove that the tension in the horizontal part of the string is immediately diminished in the ratio 9 : 5. [Coll. Ex. 1895.]

Ex. 9. Six equal rings are attached at equal intervals to points of a uniform weightless string, and the extreme rings are free to slide on a smooth horizontal rod. If the extreme rings are initially held so that the parts of the string attached to them make angles α with the vertical, and then let go, the tension in the horizontal part of the string will be instantaneously diminished in the ratio of $\cos^2\alpha$ to $1+\sin^2\alpha$. [Coll. Ex. 1889.]

Ex. 10. Three particles A, B, C are in a straight line attached to points on a string and are moving in a plane with equal velocities at right angles to this line, their masses being m, m', m respectively. If B come in contact with a perfectly elastic fixed obstacle, prove that the initial radius of curvature of the paths which A and C begin to describe is $\frac{1}{4}a$, where $AB=BC=a$. [Coll. Ex. 1892.]

The particle B rebounds with velocity v. By considering the relative motion of A and B we have $4v^2/a=T/m$. By considering the space motion of A, $v^2/\rho=T/m$.

Ex. 11. A tight string without mass passes through two smooth rings A, B, on a horizontal table. Particles of masses p, q respectively are attached to the ends and a particle of mass m to a point O between A and B. If m be projected horizontally perpendicularly to the string, the initial radius of curvature ρ of its path is given by $(m+p+q)/\rho=p/a-q/b$, where $OA=a$, $OB=b$. [Coll. Ex. 1893.]

Ex. 12. A circular wire of mass M is held at rest in a vertical plane, on a smooth horizontal table, a smooth ring of mass m being supported on it by a string which passes round the wire to its highest point and from there horizontally to a fixed point to which it is attached. If the wire be set free, show that the pressure of the ring on it is immediately diminished by amount $\dfrac{2m^2g\sin^2\theta}{M+4m\sin^2\frac{1}{2}\theta}$, where θ is the angular distance of the ring from the highest point of the wire.

[Coll. Ex. 1897.]

Ex. 13. Two particles P, P' of masses m, m' respectively are attached to the ends of a string passing over a pulley A and are held respectively on two inclined planes each of angle α placed back to back with their highest edge vertically under the pulley. If each string makes an angle β with the plane, prove that the heavier particle will at once pull the other off the plane if

$$m'/m<2\tan\alpha\tan\beta-1. \qquad \text{[Coll. Ex. 1896.]}$$

Ex. 14. Two particles of masses m, M are attached at the points B, C of a string ABC, the end A being fixed. The two portions AB, BC rest on a smooth horizontal table, the angle at B being α. The particle M has a velocity communicated to it in a direction perpendicular to BC. Prove that if the strings remain tight, the initial radius of curvature of the locus of M is $a(1+n\sin^2\alpha)$, where $n=M/m$ and $BC=a$. [Coll. Ex. 1895.]

280. *To find the initial radius of curvature when the particle starts from rest.* In this problem it may be necessary to use differential coefficients of a higher order than the second. Let x, y be the Cartesian coordinates of a particle, then representing differential coefficients with regard to the time by accents

$$\rho=\frac{(x'^2+y'^2)^{\frac{3}{2}}}{x'y''-y'x''},$$

which takes a singular form when the component velocities x', y'

are zero. Putting $u = x'y'' - y'x''$, we have after differentiation

$$\begin{aligned} u' &= x'y''' - y'x''', \\ u'' &= x'y^{\text{iv}} - y'x^{\text{iv}} + x''y''' - y''x''', \\ u''' &= x'y^{\text{v}} - y'x^{\text{v}} + 2(x''y^{\text{iv}} - y''x^{\text{iv}}). \end{aligned}$$

For the sake of brevity let the initial value of any quantity be denoted by the suffix zero, thus x_0'' represents the initial value of x''. Using Taylor's theorem and remembering that $x_0' = 0$, $y_0' = 0$, we have

$$x'y'' - y'x'' = \tfrac{1}{2}(x_0''y_0''' - x_0'''y_0'')\,t^2 + \tfrac{1}{3}(x_0''y_0^{\text{iv}} - x_0^{\text{iv}}y_0'')\,t^3 + \&c.$$

Similarly $\qquad (x'^2 + y'^2)^{\frac{3}{2}} = (x_0''^2 + y_0''^2)^{\frac{3}{2}}\,t^3 + \&c.$

If the particle start from rest the initial radius of curvature is therefore zero. But if the circumstances of the problem are such that $x_0''y_0''' - x_0'''y_0'' = 0$, the radius of curvature is given by

$$\rho = \frac{3(x_0''^2 + y_0''^2)^{\frac{3}{2}}}{x_0''y_0^{\text{iv}} - x_0^{\text{iv}}y_0''}.$$

This is the general formula when the axes of x, y have any positions.

If the axis of y be taken in the direction of the resultant force $x_0'' = 0$, and if we then also have $x_0''' = 0$, the expression for the radius of curvature takes the simple form

$$\rho = 3\,\frac{y_0''^2}{x_0^{\text{iv}}}.$$

If Y_0 be the initial resultant force on the particle, X the transverse force, the formula when $X_0 = 0$, $X_0' = 0$ may be written

$$\rho = 3\,\frac{Y_0^2}{X_0''}.$$

The corresponding formula for ρ in polar coordinates may be obtained in the same way. We have when $r(r''\theta''' - r'''\theta'') = 0$ initially,

$$\frac{3(r^2\theta''^2 + r''^2)^{\frac{3}{2}}}{\rho} = 3r^2\theta''^3 + 6r''^2\theta'' + rr''\theta^{\text{iv}} - r\theta''r^{\text{iv}},$$

where the letters are supposed to have their initial values. If the initial value of $r'' = 0$, this takes the simpler form

$$3\left(\frac{1}{\rho} - \frac{1}{r}\right) = -\frac{r^{\text{iv}}}{r^2\theta''^2}.$$

281. Let n particles P_1, P_2, &c. at rest, be acted on by given forces and be connected by κ geometrical relations. To find the initial radius of curvature of the path of any one particle P we proceed in the following manner, though *in special cases a simpler process may be used.* We differentiate the dynamical equations twice and reduce each to its initial form by writing for all the coordinates (x_1, y_1), (x_2, y_2), &c. their initial values, and for (x_1', y_1'), &c. zero. We differentiate the geometrical equations four times and reduce each to its initial form. We then have sufficient equations to find the initial values of x'', x''', x^{iv}, &c., R, R', R'', &c. where R is any reaction. Lastly solving these for the coordinates of the particular particle under consideration we substitute in the standard formula for ρ.

This process may sometimes be shortened by eliminating the tensions (if these are not required) before differentiation. We thus avoid introducing their differential coefficients into the work.

282. Shorter Methods. We can sometimes simplify the geometrical relations by introducing subsidiary quantities, say θ, ϕ, &c. In this way we can express all the coordinates (x_1, y_1), &c. in terms of θ, ϕ, &c. by equations of the form

$$x = f(\theta, \phi, \text{\&c.}), \qquad y = F(\theta, \phi, \text{\&c.}) \dots\dots\dots\dots(1),$$

where θ, ϕ, &c. are independent variables. Substituting in the dynamical equations and eliminating the reactions, we have $2n-\kappa$ equations of the second order to determine θ, ϕ, &c. in terms of t. *These eliminations may be avoided and the results shortly written down by using Lagrange's equations.* Lagrange's method is described in chap. VII.

These equations, however obtained, contain θ, θ', θ''; ϕ, ϕ', ϕ'', &c. and by differentiation we can find as many higher differential equations as are required.

Since θ', ϕ', &c. are zero, we find by differentiation

$$x'' = f_\theta \theta'' + f_\phi \phi'' + \dots,$$
$$x''' = f_\theta \theta''' + f_\phi \phi''' + \dots,$$
$$x^{iv} = 3\left(f_{\theta\theta}\theta''^2 + 2f_{\theta\phi}\theta''\phi'' + \dots\right) + f_\theta \theta^{iv} + f_\phi \phi^{iv} + \dots,$$

where suffixes as usual indicate partial differential coefficients, thus $f_\theta = df/d\theta$. There are similar expressions for the differential coefficients of y. Substituting in the standard form for ρ, we obtain the required radius of curvature.

283. We notice that if the partial differential coefficients f_θ, f_ϕ, &c. are zero the initial value of x^{iv} does not depend on any higher differential coefficients of θ, ϕ, &c., than the second, and these are given at once by the equations of motion. Since $\rho = 3y''^2/x^{iv}$, when the axis of y is taken parallel to the resultant force on the particle, *the radius of curvature can then be found without differentiating the equations of motion.*

Since $$\frac{dx}{dt} = f_\theta \frac{d\theta}{dt} + f_\phi \frac{d\phi}{dt} + \dots,$$

the geometrical meaning of the equations $f_\theta = 0$, $f_\phi = 0$, &c. clearly is that $dx/dt = 0$ for every geometrically possible displacement of the system. The point, whose initial radius of curvature is required, must begin to move parallel to the axis of y however the system is displaced.

284. Examples. *Ex.* 1. A particle is placed at rest at the origin and is acted on by forces X, Y parallel to the axes. If X, Y are expanded in powers of t and the lowest powers are $X = ft$, $Y = g$, show that the path near the origin is $y^3 = mx^2$ and that the radius of curvature is zero. If $X = \frac{1}{2}ft^2$, $Y = g$, the path is a parabola whose radius of curvature is $3g^2/f$. We notice that in the first of these cases X' is finite, in the second zero.

Ex. 2. A particle is at rest on a plane, and forces X, Y in the plane begin to act on it. If these forces are functions of the coordinates x, y only, prove that the initial radius of curvature of the path is

$$3(X^2+Y^2)^{\frac{3}{2}} \Big/ \left\{ X\left(X\frac{dY}{dx} + Y\frac{dY}{dy}\right) - Y\left(X\frac{dX}{dx} + Y\frac{dX}{dy}\right) \right\}.$$

[Coll. Ex. 1895.]

This result follows from Art. 280.

Ex. 3. Two heavy particles are attached to two points B, C of a string, one end A being fixed. Prove that if the string ABC is initially horizontal, the initial radii of curvature of the paths of B and C are equal.

Prove also that if there are n particles on the horizontal string, all the initial radii of curvature are equal. If AB, BC were two equal heavy rods, hinged at B, and having A fixed, prove that the initial radii of curvature at B and C are unequal.

In this problem we see beforehand that it will be unnecessary to differentiate the equations of motion. Take the angles θ, ϕ, which the strings make with the initial position ABC as the independent variables, Art. 283.

Ex. 4. Two heavy particles P, Q, are connected by a string which passes through a smooth fixed ring O, the portions OP, OQ of the string making angles θ, ϕ, with the vertical. If the masses m, M of P, Q, satisfy the condition $m\cos\theta = M\cos\phi$, the initial radius of curvature of the path of P is given by

$$\frac{M+m}{M}\frac{\sin^2\theta}{\rho} = \frac{\sin^2\theta}{r} + \frac{\sin^2\phi}{l-r},$$

where $r = OP$ and l is the length of the string.

Take the polar equations of motion, eliminate the tension and differentiate twice. We thus find the initial values of θ'', r'', r^{iv}; since $r''=0$ the polar formula for ρ is much simplified.

Ex. 5. A uniform rod, moveable about one end O which is fixed, is held in a horizontal position by being passed through a small ring of equal weight; show that if the ring is initially at the middle point of the rod, when it is released the initial radius of curvature of its path is 9 times the length of the rod.

[Coll. Ex. 1887.]

Taking O as origin, the polar equation of motion of the particle shows that the initial values of r'', r''' are zero, while that of $r^{\text{iv}} = g\theta'' + 2r\theta''^2$. Taking moments about O, Art. 261, we have $\frac{d}{dt}[(Mk^2+mr^2)\,\theta'] = (Ma+mr)\,g\cos\theta$. This gives the initial value of $\theta''=6g/7a$. The length of the radius of curvature follows by the differential calculus, Art. 280.

Ex. 6. Three particles whose masses are m_1, m_2, m_3 are placed at rest at the corners of a triangle ABC, and mutually attract each other with forces which vary according to some power of the distance. If $m_1m_2cF_3$, $m_2m_3aF_1$, $m_3m_1bF_2$ are the forces, prove that the initial radius of curvature ρ of the path of C is given by

$$\frac{3R^2}{\rho} = -m_2a\sin\phi\,\{-m_3F_1{}^2+m_1F_3(F_2-F_1)-PF_1'\}$$
$$+m_1b\sin\theta\,\{-m_3F_2{}^2+m_2F_3(F_1-F_2)-QF_2'\},$$

where θ, ϕ are the angles CA, CB make with the resultant force on C,

$$F_1'=dF_1/da, \quad F_2'=dF_2/db,$$
$$P=(m_2+m_3)\,aF_1+m_1(F_3c\cos B+F_2b\cos C),$$
$$Q=(m_1+m_3)\,bF_2+m_2(F_3c\cos A+F_1a\cos C),$$

and R is the resultant force on C.

Deduce that the initial radii of curvature of the three paths are infinite when the triangle is equilateral.

Small oscillations with one degree of freedom.

285. The theory of small oscillations has already been discussed in the chapter on Rectilinear Motion so far as systems with one degree of freedom are concerned. In this section a series of examples will be found showing the method of proceeding in cases somewhat more extended.

The particle, or system of particles, is supposed to be either in equilibrium or in some given state of motion. A slight disturbance being given, we express the displacements of the several particles at any subsequent time t from their positions in the state of equilibrium or motion by quantities x, y, &c. These are supposed to be so small that their squares can be neglected. If required, corrections are afterwards introduced for the errors thus caused.

We form the equations of motion either by resolving and taking moments or by Lagrange's method. By neglecting the squares of the displacements these equations are made linear in x, y, z, &c. They are also linear in regard to the reactions between the several particles. Eliminating the latter we obtain linear equations which can in general be completely solved. The solution when obtained will enable us to determine whether the system oscillates about its undisturbed state or departs widely from it on the slightest disturbance.

The principle of vis viva supplies an equation which has the advantage of being free from the unknown reactions, but it has the disadvantage that its terms contain the *squares* of the velocities, that is, the terms may be of the order we neglect. Being an accurate equation, it may sometimes be restored to the first order by differentiating it with regard to t and dividing by some small quantity. Generally the solution is more easily arrived at by using the equations of motion which contain the second differential coefficients with regard to t.

286. Examples. *Ex.* 1. Two particles whose masses are m, m' are connected by a string which passes through a small hole in a smooth horizontal table. The particle m' hangs vertically, while m is projected on the table perpendicularly to the string with such a velocity that m' is stationary. If a small disturbance is given to the system so that m' makes vertical oscillations, prove that the period is $2\pi\sqrt{\dfrac{(m+m')\,c}{mg}}$ where c is the mean radius vector of the path of m.

Let r, θ be the polar coordinates of m, z the depth of m', l the length of the string and T the tension. The equations of motion after the disturbance are

$$\frac{d^2r}{dt^2}-r\left(\frac{d\theta}{dt}\right)^2=-\frac{T}{m}, \qquad \frac{1}{r}\frac{d}{dt}\left(r^2\frac{d\theta}{dt}\right)=0,$$

$$\frac{d^2z}{dt^2}=g-\frac{T}{m'}, \qquad r+z=l.$$

The second equation gives $r^2d\theta/dt=h$, where h is a constant whose magnitude depends on the disturbance. Eliminating T, z and $d\theta/dt$ we find

$$(m+m')\frac{d^2r}{dt^2}-\frac{mh^2}{r^3}=-m'g.$$

Let $r=c+\xi$ where c is a constant which is as yet arbitrary except that the variable ξ is so small that its square can be neglected.

$$\therefore\ (m+m')\frac{d^2\xi}{dt^2}+m\frac{3h^2}{c^4}\xi=\frac{mh^2}{c^3}-m'g.$$

Let us now choose c to be such that the right-hand side of the equation is zero; then $mh^2=m'c^3g$. Substituting for h we find

$$\xi=A\sin(nt+\alpha), \quad n^2=\frac{3m'}{m+m'}\frac{g}{c}.$$

Since ξ is wholly periodic and has no constant term, its mean value is zero, when taken either for any long time or for the period of oscillation. It follows that $r=c$ is the mean radius vector of the path of m after the disturbance. This is not necessarily the same as the radius of the circle described before disturbance; whether it is so or not depends on the nature of the disturbance given to the system.

Let the particle m before disturbance be describing a circle of radius a with velocity V, then $mV^2/a=m'g$, each being the tension of the string; and the angular momentum of m is mVa. If the disturbance be given by a vertical blow B applied to the particle m', this reacts on m by an impulsive tension, and, the moment of this about O being zero, the angular momentum of m is unaltered. In this case we have $h = Va$ and we find $c = a$. If the disturbance be given by a transverse blow B applied at m, the velocity of m is changed to V' where $V' - V = B/m$. In this case $h = V'a$ and c is not equal to a.

Ex. 2. A particle of mass m is attached to two points A, B by two elastic strings each having the same modulus E and natural length l. If the particle be displaced parallel to this line, prove that the time of oscillation is $2\pi\sqrt{(ml/2E)}$. [Coll. Ex. 1895.]

Ex. 3. A heavy particle hangs in equilibrium suspended by an elastic string whose modulus is three times the weight of the particle. The particle is slightly displaced in a direction making an angle $\cot^{-1} 4$ with the horizontal and is then released. Prove that the particle will oscillate in an arc of a small parabola terminated by the ends of the latus rectum. [Math. Tripos, 1897.]

Ex. 4. A straight rod AB without weight is in a vertical position, with its lower end A hinged to a fixed point, and a weight attached to the upper end B. To B are attached three similar elastic strings equally stretched to a length k times their natural length and equally inclined to one another, their other ends being attached to three fixed points in the horizontal plane through B. Show that, when the strings obey Hooke's law, the condition for stability of equilibrium is that the weight must not exceed that which, when suspended by one of the strings, would cause an increase of length equal to $\frac{3}{2}(2 - 1/k)AB$. Show that, when this condition is fulfilled, the system can perform small vibrations parallel to any vertical plane. [Math. Tripos, 1888.]

Ex. 5. A smooth ring P can slide freely on a string which is suspended from two fixed points A and B not in the same horizontal line. If P be disturbed, find the time of a small oscillation in the vertical plane passing through A and B. If T be the time, $(T/2\pi)^2 g = 4(rr')^{\frac{3}{2}}/(r+r')\{(r+r')^2 - 4c^2\}^{\frac{1}{2}}$, where r, r' are the distances AP, BP in equilibrium and $AB = 2c$.

Ex. 6. A rod of mass M hangs in a horizontal position supported by two equal vertical elastic strings, modulus λ and natural length a. Prove that if the rod receive a small displacement parallel to itself, the period of a horizontal oscillation is $2\pi\left(\frac{a}{g} + \frac{Ma}{2\lambda}\right)^{\frac{1}{2}}$. [Coll. Ex. 1897.]

Ex. 7. A particle of mass m is attached to an elastic string stretched between two points fixed in a smooth board of mass M, and the board is free to slide on a smooth table. Prove that the period in which the particle oscillates is less than it would be if the board were fixed in the ratio $1 : \sqrt{(1 + m/M)}$. [Coll. Ex. 1895.]

Reduce the board to rest.

Ex. 8. A ring of mass nm is free to slide on a smooth horizontal wire, and a string tied to it passes through a small ring vertically below the wire at a depth h, and supports a particle of mass m. Prove that if the first mass be released when the upper part of the string makes an angle α with the vertical, and if θ be the inclination after a time t, the equation of motion is

$$h(n + \sin^2\theta)(d\theta/dt)^2 = 2g\cos^4\theta(\sec\alpha - \sec\theta).$$

Prove hence that the small oscillations about the position of equilibrium will be synchronous with a simple pendulum of length nh. [Coll. Ex. 1896.]

Ex. 9. A crane is lowering a heavy body and the chain is paid out with a uniform velocity V. Prove that the small lateral oscillations of the body are determined by

$$r\frac{d^2\theta}{dr^2}+2\frac{d\theta}{dr}+\frac{g\theta}{V^2}=0,$$

where r is the length of the chain at any time and θ its inclination to the vertical, the weight of the chain being neglected.

Also if $\theta\sqrt{r}=y$, $2\sqrt{gr}=xV$, prove that

$$x^2\frac{d^2y}{dx^2}+x\frac{dy}{dx}+(x^2-1)\,y=0.$$

This equation can be solved by the use of Bessel's functions. See Gray and Mathews' *Treatise on Bessel's Functions*. [Coll. Ex. 1895.]

Ex. 10. A gravitating solid of revolution is cut by a plane perpendicular to the axis. A particle is fastened by a fine string of length l to a point in the prolongation of the axis, so that when the string is perpendicular to the plane section the particle just does not touch the plane at its centre O. Assuming the conditions such that when the particle is slightly disturbed the motion is that of a simple pendulum, prove that the time T of a small oscillation is given by $l(2\pi/T)^2=R+\frac{1}{2}lR'$ where R is the force exerted by the solid on a unit mass at O and R' is the space variation of the force at O, taken outside the solid, along the axis. [Coll. Ex. 1892.]

Small oscillations with two or more degrees of freedom.

287. Oscillations about equilibrium. *A particle is in equilibrium under the action of forces X, Y which are given functions of the coordinates. A slight disturbance being given, it is required to determine whether the particle oscillates and the nature of the motion.*

Let a, b be the coordinates of the position of equilibrium, $a+x$, $b+y$, the coordinates at any time t. We shall assume as the standard case that x and y are small throughout the motion. Solving the equations of motion we shall express x, y in terms of t. By examining the results we shall determine whether and how nearly the subsequent motion follows the standard form.

We shall suppose that the forces X, Y can be expanded in integer powers of x, y, viz.

$$X=Ax+By,\quad Y=B'x+Cy\text{..............(1)},$$

where we have rejected the higher powers in our first approximation. There are no constant terms because X, Y vanish in the position of equilibrium. Taking the mass of the particle as unity, the equations of motion are

$$\frac{d^2x}{dt^2}=Ax+By,\quad \frac{d^2y}{dt^2}=B'x+Cy\text{..............(2)}.$$

To solve these we let δ represent d/dt,

$$\therefore\ (\delta^2 - A)x - By = 0, \quad -B'x + (\delta^2 - C)y = 0 \;\ldots(3).$$

Eliminating y, we have the two forms

$$\begin{vmatrix} \delta^2 - A, & -B \\ -B', & \delta^2 - C \end{vmatrix} x = 0, \quad By = (\delta^2 - A)x \;\ldots\ldots\ldots(4).$$

The first of these is a differential equation with constant coefficients. Its solution can be written down by the usual rules given in treatises on differential equations. The solution contains four arbitrary constants, and the value of y follows from that of x, without the introduction of any new constants.

The usual method is to assume as a trial solution $x = Le^{mt}$. Substituting we arrive at the biquadratic

$$m^4 - (A + C)m^2 + AC - BB' = 0 \;\ldots\ldots\ldots\ldots(5);$$

$$\therefore\ m^2 = \tfrac{1}{2}\left[A + C \pm \sqrt{\{(A - C)^2 + 4BB'\}}\right].$$

Assuming that no two roots are equal, let the four values of m be $\pm m$, $\pm n$; then

$$x = L_1 e^{mt} + L_2 e^{-mt} + M_1 e^{nt} + M_2 e^{-nt} \;\ldots\ldots\ldots\ldots(6),$$

where L_1, L_2 &c. are four arbitrary constants and the values of m may be real or imaginary.

It is at once obvious, if m be positive or of the form $r \pm p\sqrt{-1}$, where r is positive, that the value of x will become large by efflux of time. *It is therefore necessary for an oscillatory motion that all the real roots and the real parts of the imaginary roots of the determinantal equation* (5) *should be negative.*

Since the sum of the four roots of (5) is zero, some of the real parts must be positive unless the four roots are of the form $\pm p\sqrt{-1}$. *It is therefore necessary for an oscillatory motion that both the roots of the quadratic* (5) *should be real and negative.* The algebraical conditions for this are, that both $(A - C)^2 + 4BB'$ and $AC - BB'$ should be positive and $A + C$ negative.

As our solution represents the motion only when x and y remain small, it is unnecessary for us here to consider any case except that in which the roots of (5) take the forms $m^2 = -p^2$, $n^2 = -q^2$. The motion is then given by

$$\left.\begin{aligned} x &= L \sin(pt + \alpha) + M \sin(qt + \beta) \\ y &= L' \sin(pt + \alpha) + M' \sin(qt + \beta) \end{aligned}\right\} \ldots\ldots\ldots\ldots(7),$$

where $BL' = -(p^2 + A)L$ and $BM' = -(q^2 + A)M$. The quantities p^2, q^2 are the roots of

$$(p^2 + A)(p^2 + C) - BB' = 0 \quad\ldots\ldots\ldots\ldots(8).$$

288. *If B, B' have the same sign, the roots of the quadratic* (8) *are separated by each of the values $p^2 = -A$, $p^2 = -C$.* To prove this, it is sufficient to notice that the left-hand side of that equation is positive when $p^2 = \pm\infty$ and is negative when p^2 has either of the separating values.

It is also sometimes useful to notice that the roots cannot be equal unless the two separating values A and C are equal and that the equal roots are then $p^2 = -A = -C$. If $AC - BB' = 0$ the biquadratic (5) has two equal zero roots, though the roots of the same equation regarded as a quadratic are unequal.

289. *To find the four arbitrary constants L, M, α, β, we solve the equations* (7) *with regard to the trigonometrical terms.* We thus find

$$\left.\begin{aligned} By + (q^2 + A)x &= -(p^2 - q^2)\, L \sin(pt + \alpha) \\ By + (p^2 + A)x &= (p^2 - q^2)\, M \sin(qt + \beta) \end{aligned}\right\} \quad\ldots\ldots\ldots(9).$$

Putting $t = 0$, we at once have the values of $L \sin\alpha$, $M \sin\beta$ in terms of the initial values of the coordinates. Differentiating with regard to t and again putting $t = 0$, we find $L \cos\alpha$, $M \cos\beta$ in terms of the initial velocities.

290. Equal roots. The case in which the equation (5) has equal roots has been excepted. This occurs when either $(A - C)^2 + 4BB' = 0$ or $AC - BB' = 0$. When B, B' have the same sign the first alternative requires $A = C$ and either B or B' equal to zero. In the second alternative the equation has two zero roots.

Excepting when both B and B' are zero, the solution of the dynamical equations (2) is known to contain terms of the form $(Lt + L')\, e^{mt}$. If m is positive or zero (or has its real part positive or zero), this term will increase indefinitely with t. If however the real part of m is negative and not zero, say equal to $-r$, the maximum value of Lte^{-rt} is L/re. Since L is so small that its square can be neglected, this term in the solution will always remain small except when r also is small. *The existence of equal roots in the determinantal equation* (5) *does not therefore necessarily imply that the oscillation becomes large.*

291. Before disturbance the particle P was in equilibrium at the origin under the influence of the forces X, Y given by (1) Art. 287. When $AC = BB'$, the equations $X = 0$, $Y = 0$ are satisfied by values of x, y other than zero. These lie on the straight line $Ax + By = 0$. *The dynamical significance of the condition $AC = BB'$ is therefore that there are other positions of equilibrium in the immediate neighbourhood of the origin.* The roots of equation (8) being $p^2 = 0$, $q^2 = -A - C$, the values of x, y take the form

$$x = L_1 t + L_2 + M \sin(qt + \beta),$$
$$By = -A(L_1 t + L_2) - CM \sin(qt + \beta).$$

The first terms represent a uniform motion along the line of equilibrium, while the trigonometrical terms represent an oscillation in the direction $By = -Cx$. Whether the particle will travel far or not along the line of equilibrium will depend on the nature of the forces when x, y become large.

292. Principal oscillations. Let the type of motion be that represented by such equations as (7). By giving the particle the proper initial conditions it may be made to move in either of the ways defined by the following partial solutions

$$x = L \sin(pt + \alpha), \quad y = L' \sin(pt + \alpha) \ldots\ldots\ldots(10),$$

$$x = M \sin(qt + \beta), \quad y = M' \sin(qt + \beta) \ldots\ldots\ldots(11).$$

Each of these is called a principal oscillation and all the modes of oscillation included in (7) are compounded of these two. *The dynamical peculiarity of a principal oscillation is the singleness of the period.*

The solution (10) is sometimes taken as the trial solution instead of the exponential used in obtaining (5). Practically we then begin the solution by finding the principal oscillations and finally combine these into the general solution (7).

The paths of the particle when describing the principal oscillations are the two straight lines

$$Ly = L'x, \quad My = M'x \ldots\ldots\ldots\ldots\ldots\ldots(12).$$

In each oscillation the ratio of the coordinates, being equal to L'/L or M'/M, is constant throughout the motion. We have by (7), using the values of $p^2 + q^2$, p^2q^2, given by the coefficients of the quadratic (8),

$$\frac{L'M'}{LM} = \frac{(p^2 + A)(q^2 + A)}{B^2} = -\frac{B'}{B} \ldots\ldots\ldots\ldots(13).$$

It follows that when B, B' have the same sign, the ratios L'/L, M'/M have opposite signs. In one principal oscillation, the coordinates x, y increase together; in the other, when one increases the other decreases.

We also notice that when $B' = B$, the two straight lines (12) are at right angles.

The directions of these rectilinear oscillations may be obtained without investigating the motion. The lines must be so placed that if the particle be displaced along either, the perpendicular force must be zero. The lines are therefore given by

$$Xy - Yx = 0; \quad \therefore By^2 + (A - C)xy - B'x^2 = 0.$$

These lines are real when $(A - C)^2 + 4BB'$ is positive. This condition is satisfied when the roots of the determinantal equation (5) are real or of the form $p\sqrt{-1}$.

293. When the coordinates are such that only one varies along each principal oscillation, they are called *principal coordinates.*

Referring to the equations (9), we see that if we put

$$By + (q^2 + A)x = \xi, \quad By + (p^2 + A)x = \eta,$$

ξ, η will be the principal coordinates. This transformation of coordinates is always possible, so long as p^2 and q^2 are real and unequal.

We may also discover the principal coordinates without previously finding the values of p^2, q^2. We deduce from the equations (2)

$$\frac{d^2}{dt^2}(x+\lambda y) = (A+\lambda B')\left(x + \frac{B+\lambda C}{A+\lambda B'}y\right),$$

by using an indeterminate multiplier λ. If now we write $(B+\lambda C)/(A+\lambda B')=\lambda$, we see that $x+\lambda y$ will be a trigonometrical function with one period. We have a quadratic to find λ; representing the roots by λ_1, λ_2, the principal coordinates are $\xi = x+\lambda_1 y$, $\eta = x+\lambda_2 y$, or any multiples of these.

294. Conservative forces. *When the forces which act on the particle are conservative*, the solution admits of some simplifications. Let U be the force function, then, since dU/dx and dU/dy vanish in the position of equilibrium, we have by Taylor's theorem,

$$U = U_0 + \tfrac{1}{2}(Ax^2 + 2Bxy + Cy^2) + \ldots \quad \ldots\ldots\ldots\ldots(1).$$

It follows that the equations of motion are

$$\frac{d^2x}{dt^2} = X = Ax + By, \quad \frac{d^2y}{dt^2} = Y = Bx + Cy \quad \ldots\ldots\ldots(2).$$

Comparing these with the former values of X, Y, we see that $B' = B$.

If we turn the axes round the origin we know by conics that the equation (1) can be always cleared of the term containing the product xy. Representing the new coordinates by ξ, η, let the expression for U become

$$U = U_0 + \tfrac{1}{2}(A'\xi^2 + C'\eta^2) + \ldots \quad \ldots\ldots\ldots\ldots\ldots(3),$$

where $A' + C' = A + C$, $A'C' = AC - B^2$. The equations of motion are then

$$\frac{d^2\xi}{dt^2} = A'\xi, \quad \frac{d^2\eta}{dt^2} = C'\eta \quad \ldots\ldots\ldots\ldots\ldots\ldots\ldots(4).$$

The motion is oscillatory for all displacements or for none according as A', C' are both negative or both positive. If A' is negative and C' positive, the motion is oscillatory for a displacement along the axis of ξ and not wholly oscillatory for other displacements.

The level curves of the field of force are obtained by equating U to a constant; in the neighbourhood of the position of equilibrium, these become the conics

$$Ax^2 + 2Bxy + Cy^2 = N, \text{ or } A'\xi^2 + C'\eta^2 = N.$$

The lines of the principal oscillations are the directions of the principal diameters of the limiting level conic, and the periods of the principal oscillations are proportional to the lengths of the diameters along which the particle moves.

295. The representative particle. The investigation of the small oscillations of a particle in a given field of force has a more extended application to dynamical problems than appears at first sight. Suppose, for example, that a system, consisting of several particles connected together by geometrical relations, has two degrees of freedom. Let the position of this system be defined by the two coordinates x, y. The equations giving the small oscillations, after the elimination of the reactions, take the form

$$\frac{d^2x}{dt^2} = Ax + By, \qquad \frac{d^2y}{dt^2} = B'x + Cy,$$

because the squares of x and y are neglected. If $B = B'$ these are the equations of motion of a single particle moving in the field of force defined by

$$U - U_0 = \tfrac{1}{2}(Ax^2 + 2Bxy + Cy^2).$$

The investigations given in Art. 292 and Art. 294 apply therefore to both problems.

To exhibit the motion of an oscillating system to the eye, we take its coordinates x, y to be also the Cartesian coordinates of an imaginary particle which moves freely in the field of force U. We represent by a figure the level conics, the path of this representative particle, and sketch the positions of the principal oscillations. The special peculiarities of the motion will then become apparent in the figure.

296. Test of stability*. Let the field of force in which the particle moves be given by the function U. Since dU/dx and dU/dy vanish in the position of equilibrium, U must be at that point a maximum or a minimum. In the neighbourhood we have

$$U - U_0 = \tfrac{1}{2}(Ax^2 + 2Bxy + Cy^2) + \dots$$

If $AC - B^2$ is positive, U is a maximum or a minimum for all displacements according as the common sign of A and C is negative or positive, and if $AC - B^2$ is negative, U is a maximum for

* The energy test of the stability of a position of equilibrium is given by Lagrange in the *Mécanique Analytique*. He gives both this proof and that in Art. 297. The demonstration for the general case of a system of bodies has been much simplified by Lejeune-Dirichlet in *Crelle's Journal*, 1846, and *Liouville's Journal*, 1847. See the author's *Rigid Dynamics*, vol. I.; the corresponding test for the stability of a state of motion is in vol. II.

some and a minimum for other displacements. It follows from Art. 294 that *the motion of the particle, when disturbed from its position of equilibrium, will be wholly oscillatory if U is a real maximum at that point. The particle will oscillate for some displacements and not for others if U has a stationary value, and will not oscillate for any displacement if U is a real minimum.*

We have here assumed that *all* the coefficients A, B, C are not zero. When this happens the cubic terms in the expression for U govern the series. The equations of motion (2) of Art. 295 will then have terms of the second order of small quantities on their right-hand sides.

Besides this if $AC - B^2 = 0$, the quadratic terms of the expression for U take the form of a perfect square, viz. $(Ax + By)^2/A$. In this case the forces $X = dU/dx$ and $Y = dU/dy$ contain the common factor $Ax + By$ so that there are other positions of equilibrium in the neighbourhood of the origin, see Art. 291. To determine the motion, even approximately, it is necessary to take account of the powers of x, y of the higher orders.

The geometrical theory of maxima and minima has a corresponding peculiarity, for it is shown in the Differential Calculus that further conditions, involving the higher powers, are necessary for a maximum or minimum.

The following investigation shows how far this correspondence extends.

297. Let a particle be in equilibrium at a point P_0 whose coordinates are x_0, y_0, and let $U = f(x, y)$ be the work function. Let the particle be projected with a small velocity v_1 from a point P_1, whose coordinates are x_1, y_1, very near to P_0. The equation of vis viva gives (Art. 246)

$$v^2 = v_1^2 + 2(U - U_1) \quad\quad (1),$$

$$= v_0^2 + 2(U - U_0) \quad\quad (2),$$

where

$$v_0^2 = v_1^2 + 2(U_0 - U_1) \quad\quad (3).$$

Let U be a maximum at the point P_0 for all directions of displacement, then $U_1 < U_0$ and v_0^2 is a small positive quantity. As the particle recedes from P_0, $U_0 - U$ increases, but the equation (2) shows that the particle cannot go so far that $U_0 - U$ becomes greater than the small quantity $\frac{1}{2}v_0^2$. The equilibrium is therefore stable for displacements in all directions.

Let U be a minimum at P_0 for all directions of displacement, then as the particle moves from P_0 the difference $U - U_0$ increases. So far as the principle of vis viva is concerned, there is nothing to prevent the particle from receding indefinitely from P_0.

Let U be a maximum for some directions of displacement and a minimum for others. The particle cannot recede far from P_0 in the directions for which U is a maximum, but there is nothing to restrict the motion in the other directions.

298. *Ex.* A particle P is in equilibrium under the action of a system of fixed attracting bodies situated in one plane, the law of attraction being the inverse κth power of the distance. Prove that, if $\kappa > 1$, the equilibrium of P cannot be stable for all displacements in that plane, though it may be stable for some and unstable for other displacements. If $\kappa < 1$, the equilibrium cannot be unstable for all displacements in that plane.

To prove this let m_1 be any particle of the attracting mass, coordinates f, g; let x, y be the coordinates of P. The potential of m_1 at P is by definition $U_1 = \dfrac{m_1}{(\kappa - 1) r_1^{\kappa-1}}$, where r_1 is the distance of m_1 from P. We then find by a partial differentiation

$$\frac{d^2U_1}{dx^2} + \frac{d^2U_1}{dy^2} = \frac{(\kappa - 1) m_1}{r_1^{\kappa+1}}.$$

Summing this for all the particles of the attracting mass and writing $U = \Sigma U_1$, we find

$$\frac{d^2U}{dx^2} + \frac{d^2U}{dy^2} = (\kappa - 1) \Sigma \frac{m}{r^{\kappa+1}}.$$

The right-hand side is positive or negative according as $\kappa > 1$ or $\kappa < 1$.

Taking the equilibrium position of P for the origin and the principal directions of motions for the axes, Art. 294, we see by Taylor's Theorem

$$U = U_0 + \tfrac{1}{2}(A'\xi^2 + C'\eta^2) + \dots,$$

where $A' = d^2U/dx^2$, $C' = d^2U/dy^2$. It is evident that U cannot be a maximum for all displacements in the plane of xy if $A' + C'$ is positive and cannot be a minimum for all displacements in the plane if this sum is negative. The result also follows from Art. 296.

299. Barrier curves. It is clear that this line of argument may be extended to apply to cases in which there is no given position of equilibrium in the neighbourhood of the point of projection. Let the particle be projected from any point P_1 with any velocity v_1 in any direction. Throughout the subsequent motion we have

$$v^2 = v_1^2 + 2(U - U_1),$$

where U is a given function of x, y and U_1 is its value at the point of projection.

If we equate the right-hand side of this equation to zero, we obtain the equation of a curve traced on the field of force at which the velocity of the particle, if it arrive there, is zero. *This curve is therefore a barrier to the motion, which the particle cannot pass.*

If the barrier curve be closed as in Art. 297, the particle is, as it were, imprisoned, and cannot recede from its initial position beyond the limits of the curve. Some applications of this theorem will be given in the chapter on central forces.

The right-hand side of the equation will in general have opposite signs on the two sides of the barrier. When this is the case the particle, if it reach the barrier in any finite time, must necessarily return, because the left-hand side of the equation cannot be negative.

If the right-hand side of the equation have the same sign on both sides of the barrier, that sign must be positive, and U must be a minimum at all points of the barrier. The particle is therefore approaching a position of equilibrium and arrives there with velocity equal to zero. The particle therefore will remain on the barrier, see Art. 99.

The barrier is evidently a level curve of the field of force and, as the particle approaches it, the resultant force must be normal to the barrier. Just before the particle arrives at its position of zero velocity, the tangential component of the velocity must be zero, for this component cannot be destroyed by the force. The path cannot therefore touch the barrier, but must meet it perpendicularly or at a cusp.

300. Examples. *Ex.* 1. Two heavy particles of masses m, m', are attached to the points A, B of a light elastic string. The upper extremity O is fixed and the string is in equilibrium in a vertical position. A small vertical disturbance being given, find the oscillations.

Let x, y be the depths of m, m' below O; a, b the unstretched lengths of OA, AB, E the coefficient of elasticity. The equations of motion reduce to

$$\left.\begin{aligned} m\frac{d^2x}{dt^2} + \left(\frac{E}{a}+\frac{E}{b}\right)x - \frac{E}{b}y = mg \\ -\frac{E}{b}x + m'\frac{d^2y}{dt^2} + \frac{E}{b}y = m'g + E \end{aligned}\right\} \quad \text{......................} (1).$$

To solve these we put

$$x-h=L\sin(pt+\alpha),\qquad y-k=M\sin(pt+\alpha)\qquad\text{.....................(2)},$$

the constants h, k being introduced to cancel the right-hand sides of the equations of motion. Since $x=h$, $y=k$ make $d^2x/dt^2=0$, $d^2y/dt^2=0$, these constants are the equilibrium values of x, y. We then find

$$\left(mp^2-\frac{E}{a}-\frac{E}{b}\right)\left(m'p^2-\frac{E}{b}\right)=\frac{E^2}{b^2},\qquad \frac{L}{M}=1-\frac{m'b}{E}p^2\qquad\text{..............(3)}.$$

One principal oscillation is given by (2) and the other by using instead of p^2, the other root of the quadratic. It follows that in one oscillation the two particles are always moving in the same directions, that is both are moving upwards or both downwards. In the other when one moves upwards the other moves downwards.

Ex. 2. Two heavy particles, of masses m, M, are attached to the points A, B of a light inextensible string, the upper extremity O being fixed. Prove that the periods of the small lateral oscillations are $2\pi/p$ and $2\pi/q$ where p and q are the roots of

$$\frac{1}{p^4}-\frac{a+b}{g}\frac{1}{p^2}+\frac{m}{M+m}\frac{ab}{g^2}=0,$$

and $OA=a$, $AB=b$. Prove also that the magnitudes of the principal oscillations in the inclinations of the upper and lower strings to the vertical are in the ratio $(g-bp^2)/ap^2$. Show that in one principal oscillation the two particles are on the same side of the vertical through O and in the other on opposite sides.

Ex. 3. Two particles M, m, are connected by a fine string, a second string connects the particle m to a fixed point, and the strings hang vertically; (1) m is held slightly pulled aside a distance h from the position of equilibrium, and, being let go, the system performs small oscillations; (2) M is held slightly pulled aside a distance k, without disturbance of m, and being let go the system performs small oscillations. Prove that the angular motion of the lower string in the first case will be the same as that of the upper string in the second if $Mk=(M+m)h$.

[Math. Tripos, 1888.]

Ex. 4. Three beads, the masses of which are m, m', m'', can slide along the sides of a smooth triangle ABC and attract each other with forces which vary as the distance. Find the positions of equilibrium and prove that if slightly disturbed the periods $2\pi/p$ of oscillation are given by

$$(p^2-\alpha)(p^2-\beta)(p^2-\gamma)-m'm''(p^2-\alpha)\cos^2A-m''m(p^2-\beta)\cos^2B$$
$$-mm'(p^2-\gamma)\cos^2C-2mm'm''\cos A\cos B\cos C=0,$$

where α, β, γ represent $m''+m'$, $m+m''$, $m'+m$ respectively.

Ex. 5. A particle P of unit mass is placed at the centre of a smooth circular horizontal table of radius a. Three strings, attached to the particle, pass over smooth pulleys A, B, C at the edge of the table and support three particles of masses m_1, m_2, m_3; the pulleys being so placed that the particle P is in equilibrium. A small disturbance being given, prove that the periods of the oscillations are $2\pi/p$, where

$$\left\{\frac{p^2(1+\sigma)}{p^2+g/a}\right\}^2-\frac{p^2(1+\sigma)\sigma}{p^2+g/a}+\frac{\sigma^2H}{4m_1m_2m_3}=0,$$

$$H=(m_1+m_2-m_3)(m_2+m_3-m_1)(m_1+m_3-m_2),$$
$$\sigma=m_1+m_2+m_3.$$

Ex. 6. A heavy particle P is suspended by a string of length l to a point A which describes a horizontal circle of radius a with a slow angular velocity n. Prove that the two periods of the oscillatory motion are $2\pi/n$ and $2\pi\sqrt{l/g}$.

301. Particle on a surface. *Ex.* 1. A heavy particle rests in equilibrium on the inside of a fixed smooth surface at a point O, at which the surface has only one tangent plane. The particle being slightly disturbed, it is required to find the oscillations.

Taking the point O as origin and the tangent plane as the plane of xy, the equation of the surface may be written

$$z = \tfrac{1}{2}(ax^2 + by^2) + \ldots,$$

where the axes of x, y are the tangents to the principal sections and $1/a$, $1/b$ are the radii of curvature of those sections. By the principles of solid geometry the direction cosines of the normal at any point P become $(ax, by, 1)$ when the squares of x, y are neglected. The equations of motion are therefore

$$m\frac{d^2x}{dt^2} = -Rax, \qquad m\frac{d^2y}{dt^2} = -Rby, \qquad m\frac{d^2z}{dt^2} = -mg + R.$$

Since z is of the second order of small quantities the third equation shows that $R = mg$, and the other two become

$$\frac{d^2x}{dt^2} = -agx, \qquad \frac{d^2y}{dt^2} = -bgy.$$

If a and b are positive, that is if both the principal sections are concave upwards, the motion is oscillatory and the two periods of oscillations are $2\pi/\sqrt{ag}$ and $2\pi/\sqrt{bg}$. The particle, by definition, performs a principal oscillation when its motion has but one period. This occurs when

(1) $x=0$, $y = B\sin(\sqrt{bg}t + \beta)$, (2) $y = 0$, $x = A\sin(\sqrt{ag}t + \alpha)$.

The directions of these oscillations are the tangents to the principal sections.

Ex. 2. A particle rests on a smooth surface which is made to revolve with uniform angular velocity ω about the vertical normal which passes through the particle. Show that the equilibrium is stable (1) if the curvature is synclastic upwards, and ω does not lie between certain limits, or (2) if the curvature is anticlastic and the downward principal radius is greater than the upward principal radius, and ω exceeds a certain limit. Find the limits of ω in each case.

[Math. Tripos, 1888.]

Taking as axes the tangents to the principal sections, the equations of motion (Art. 227) reduce to

$$\frac{d^2x}{dt^2} - \omega^2 x - 2\omega\frac{dy}{dt} = -gax, \qquad \frac{d^2y}{dt^2} - \omega^2 y + 2\omega\frac{dx}{dt} = -gby.$$

To solve these we put $x = L\sin(pt + \alpha)$, $y = L'\cos(pt+\alpha)$. We then obtain a quadratic for p^2 and the ratio L'/L.

The path of the particle relatively to the moving surface when performing the principal oscillation defined by either value of p^2 is the ellipse $\left(\frac{x}{L}\right)^2 + \left(\frac{y}{L'}\right)^2 = 1$. The two ellipses are coaxial.

302. The insufficiency of the first approximation. In forming the equations of motion in Arts. 287, 294, we have rejected the squares of x and y.

But unless the extent of the oscillation is indefinitely small, the rejected terms have some values, and it may be, that they sensibly affect the results of the first approximation. See Art. 141.

303. To find a second approximation we include in the equations (2) of Art. 287 the terms of the second order. We write these in the form

$$\left.\begin{aligned}(\delta^2 - A)\,x - By &= E_1x^2 + 2E_2xy + E_3y^2\\ -B'x + (\delta^2 - C)\,y &= F_1x^2 + 2F_2xy + F_3y^2\end{aligned}\right\} \quad\quad (1).$$

Taking as our first approximation

$$\left.\begin{aligned}x &= L\sin(pt+\alpha) + M\sin(qt+\beta)\\ y &= L'\sin(pt+\alpha) + M'\sin(qt+\beta)\end{aligned}\right\} \quad\quad (2),$$

we substitute these in the right-hand sides of (1). The equations take the form

$$\left.\begin{aligned}(\delta^2 - A)\,x - By &= \Sigma P\sin(\lambda t+\mu)\\ -B'x + (\delta^2 - C)\,y &= \Sigma Q\sin(\lambda t+\mu)\end{aligned}\right\} \quad\quad (3),$$

where λ may have any one of the values 0, $2p$, $2q$, $p \pm q$ and P, Q contain the squares of the small quantities L, M, L', M'. To solve these equations, we consider only the specimen term of (3) and assume

$$\left.\begin{aligned}x &= L\sin(pt+\alpha) + M\sin(qt+\beta) + R\sin(\lambda t+\mu)\\ y &= L'\sin(pt+\alpha) + M'\sin(qt+\beta) + R'\sin(\lambda t+\mu)\end{aligned}\right\} \quad\quad (4).$$

We find by an easy substitution

$$R(\lambda^2 + A) + BR' = -P, \qquad B'R + R'(\lambda^2 + C) = -Q;$$

$$\therefore\ R = \frac{-P(\lambda^2 + C) + QB}{(\lambda^2 + A)(\lambda^2 + C) - BB'}, \qquad R' = \frac{PB' - Q(\lambda^2 + A)}{(\lambda^2 + A)(\lambda^2 + C) - BB'}.$$

It appears that R, R' are very small quantities of the second order, except when λ is such that the common denominator is small, and in this case R, R' may become very great. The roots of the denominator are $\lambda^2 = p^2$, $\lambda^2 = q^2$, and the denominator is small when λ is nearly equal to either p or q. This requires either that one of the two frequencies p, q should be small or that one should be nearly double the other.

If for example p is nearly equal to $2q$ and the numerators of R, R', are not thereby made small, the terms defined by $\lambda = p - q$ and $\lambda = 2q$ will considerably influence the motion, the other terms producing no perceptible effect. If $p = 2q$ exactly the denominator is zero and both R, R' take infinite values. The dynamical meaning of the infinite term is that the expressions (2) do not represent the motion with sufficient accuracy (except initially) to be a first approximation. The corrections to these expressions are found to become infinite and if we desire a solution we must seek some other first approximation.

304. Oscillation about steady motion. *Ex.* 1. The constituents of a multiple star describe circles about their centre of gravity O with a uniform angular velocity n, the several bodies always keeping at the same distances from each other. A planet P, *of insignificant mass*, freely describes a circle of radius a, centre O, with the same angular velocity, under the attraction of the other bodies. It is required to find the oscillations of P when disturbed from this state of motion.

Let $r = a(1 + x)$, $\theta = nt + y$ be the polar coordinates of the planet P at any time t. Let the work function in the revolving field of force be

$$U - U_0 = A_0x + B_0y + \tfrac{1}{2}(Ax^2 + 2Bxy + Cy^2) + \&c. \quad\quad (1),$$

at all points in the neighbourhood of the circular motion. Since that motion is possible only in that part of the field in which the force tends to O and is equal to n^2a, it is clear that $A_0 = -a^2n^2$ and $B_0 = 0$.

Substituting the values of r, θ in the polar equations

$$\frac{d^2r}{dt^2} - r\left(\frac{d\theta}{dt}\right)^2 = \frac{dU}{a\,dx}, \qquad \frac{1}{r}\frac{d}{dt}\left(r^2\frac{d\theta}{dt}\right) = \frac{dU}{r\,dy} \quad \text{..................(2),}$$

we find the linear equations

$$\left.\begin{aligned}(a^2\delta^2 - a^2n^2 - A)\,x - (2a^2n\delta + B)\,y = 0\\ (2a^2n\delta - B)\,x + (a^2\delta^2 - C)\,y = 0\end{aligned}\right\} \quad \text{..................(3).}$$

A principal oscillation is therefore given by

$$x = L\cos pt + L'\sin pt, \qquad y = M\cos pt + M'\sin pt \quad \text{..................(4),}$$

$$M = \frac{2a^2npL' - BL}{a^2p^2 + C}, \qquad M' = \frac{-2a^2npL - BL'}{a^2p^2 + C} \quad \text{..................(5),}$$

$$(a^2p^2 + A + a^2n^2)(a^2p^2 + C) - B^2 - 4a^4n^2p^2 = 0 \quad \text{..................(6).}$$

The path of the particle when describing a principal oscillation relatively to its undisturbed path is the conic

$$(a^2p^2 + A + a^2n^2)\,x^2 + 2Bxy + (a^2p^2 + C)\,y^2 = \frac{4a^2n^2p^2}{a^2p^2 + C}(L^2 + L'^2) \quad \text{........(7),}$$

the ratio and directions of the axes being independent of the disturbance. In the limiting case in which $n = 0$ the conic reduces to two straight lines.

When the multiple star has *two constituents* A, B, whose masses are M, M', the planet P can describe a circular orbit only when $M\rho^{-\kappa}\sin APO = M'\rho'^{-\kappa}\sin BPO$, where $\rho = AP$, $\rho' = BP$ and the law of force is the inverse κth power of the distance. Since O is the centre of gravity of M, M' this proves that either the angle APO is zero or $\rho = \rho'$, except when $\kappa = -1$. The planet P must therefore be either in the straight line AB or at the corner C of the equilateral triangle ABC.

When the planet P is in the straight line AB at a point C such that the sum of the attractions of A and B on it is equal to $n^2 \,.\, OC$, the planet can describe a circle about O with the same periodic time as A and B. This motion is unstable.

When the planet P is at the third corner C of the equilateral triangle ABC, the circular motion is stable when $\dfrac{(M+M')^2}{MM'} > 3\left(\dfrac{1+\kappa}{3-\kappa}\right)^2$.

These two results may be obtained in several ways. Putting ρ, ρ' for the distances of P from the two primaries the work function is

$$U = \frac{1}{\kappa - 1}\left(\frac{M}{\rho^{\kappa-1}} + \frac{M'}{\rho'^{\kappa-1}}\right).$$

Expressing this in terms of r, θ, and expanding in powers of x, y, including the terms of the second order, the values of A, B, C in equation (1) become known. The periods are then given by (6).

Instead of using the work function, we may determine the forces dU/adx and dU/rdy by resolving the attractions of the primaries along and perpendicular to the radius vector of P. *This method has the advantage that the task of calculating the terms of the second order becomes unnecessary.*

Lastly, we may use the Cartesian equations referred to moving axes which rotate round O with a uniform angular velocity n, OC being the axis of ξ; Art. 227.

In all these methods, the assumption that the mass of the planet P is insignificant compared with that of either of the attracting bodies greatly simplifies the analysis. It does not seem necessary to examine these cases more fully here, as the results and the method of proceeding when this assumption is not made will be considered further on.

Ex. 2. If in the last example the attracting primaries either coincide or are so arranged that the field of force is represented by $U - U_0 = A_0 x + \frac{1}{2}Ax^2$; prove that other circular orbits in the immediate neighbourhood of the given one are possible paths for the particle P, Art. 291. Prove also that after disturbance the oscillation of P about the *mean circular path* is given by

$$x = L\cos(pt + \alpha), \qquad py = -2nL\sin(pt+\alpha),$$

where $p^2 = 3n^2 - A/a^2$, the oscillation having only one period.

Ex. 3. Two equal centres of force S, S', whose attraction is $\mu\rho^\kappa$, rotate round the middle point O of the line of junction with a uniform angular velocity n. A particle in equilibrium at O is slightly disturbed, prove that the periods of the small oscillation are given by $(p^2 + n^2 - \beta)(p^2 + n^2 - \kappa\beta) = 4n^2p^2$ where $\beta = 2\mu b^{\kappa-1}$ and $SS' = 2b$. Thence deduce the conditions that the equilibrium should be stable.

Problems requiring Finite Differences.

305. *Ex.* 1. A light elastic string of length nl and coefficient of elasticity E is loaded with n particles each of mass m, ranged at intervals l along it beginning at one extremity. If it be hung up by the other extremity, prove that the periods of its vertical oscillations will be given by the formula

$$\pi\sqrt{\frac{lm}{E}}\cdot \operatorname{cosec}\frac{2i+1}{2n+1}\frac{\pi}{2}, \text{ where } i = 0, 1, 2 \ldots n-1^*. \qquad \text{[Math. Tripos, 1871.]}$$

Let x_κ be the distance of the κth particle from the fixed end O; T_κ the tension above, $T_{\kappa+1}$ that below, the particle. We then have

$$mx_\kappa'' = mg + T_{\kappa+1} - T_\kappa \qquad (1),$$

and by Hooke's law for elastic strings

$$T_\kappa = E\left(\frac{x_\kappa - x_{\kappa-1}}{l} - 1\right) \qquad (2).$$

The equation of motion is therefore

$$x_\kappa'' - g = c^2(x_{\kappa+1} - 2x_\kappa + x_{\kappa-1}) \qquad (3),$$

where $c^2 = E/lm$. We assume as the trial solution

$$x_\kappa = h_\kappa + X_\kappa \sin(pt + \epsilon) \qquad (4),$$

where h_κ and X_κ are two functions of κ which are independent of t, and p, ϵ are independent of both κ and t. Substituting we find

$$\left.\begin{aligned} X_{\kappa+1} - 2X_\kappa + X_{\kappa-1} &= -\frac{p^2}{c^2}X_\kappa \\ h_{\kappa+1} - 2h_\kappa + h_{\kappa-1} &= -\frac{1}{c^2}g \end{aligned}\right\} \qquad (5).$$

* The solution is given at greater length than is necessary for this example, in order to illustrate the various cases which may arise.

To solve the first of these linear equations of differences we follow the usual rules. Taking $X_\kappa = Aa^\kappa$ as a trial solution, where A and a are two constants, we get after substitution and reduction

$$a - 2 + \frac{1}{a} = -\frac{p^2}{c^2} \quad\text{.......(6)},$$

$$\therefore \sqrt{a} = \pm\sqrt{\left(1 - \frac{p^2}{4c^2}\right)} + \frac{p}{2c}\sqrt{-1} \quad\text{.......(7)}.$$

Let these values of a be called α and β. Then

$$X_\kappa = A\alpha^\kappa + B\beta^\kappa \quad\text{.......(8)}.$$

We notice that when either $p=0$ or $2c$ the equation (6) has *equal roots*, viz. $a=1$ or -1. The theory of linear equations shows that the terms depending on these values of p take a different form, viz.

$$X_\kappa = (A + B\kappa)(\pm 1)^\kappa \quad\text{.......(9)}.$$

The complete value of x_κ may be written in the form

$$x_\kappa = h_\kappa + A_0 + B_0\kappa + (A_{2c} + B_{2c}\kappa)(-1)^\kappa \sin(2ct + \epsilon_{2c})$$
$$+ \Sigma (A_p\alpha^\kappa + B_p\beta^\kappa) \sin(pt + \epsilon_p) \quad\text{......(10)},$$

where Σ implies summation for all existing values of p.

We have yet to examine the conditions at the extremities of the string. The formula (2) does not express the tension of the highest string unless we suppose that $x_0 = 0$. Again the tension below the lowest particle must be zero and this requires that $T_{n+1} = 0$. The equation (3) will therefore express the motion of every particle from $\kappa=1$ to $\kappa=n$ only if we make

$$x_0 = 0, \quad x_{n+1} - x_n = l \quad\text{.......(11)}.$$

Since $x_0 = 0$ *for all values of* t, it follows from (10) that

$$h_0 + A_0 = 0, \quad A_{2c} = 0, \quad A_p + B_p = 0 \quad\text{.......(12)}.$$

Since $x_{n+1} - x_n = l$, we see in the same way that

$$h_{n+1} - h_n + B_0 = l, \quad B_{2c} = 0, \quad A_p\alpha^{n+1} + B_p\beta^{n+1} = A_p\alpha^n + B_p\beta^n \quad\text{.........(13)}.$$

Eliminating the ratio A_p/B_p we have

$$\alpha^{n+1} - \beta^{n+1} = \alpha^n - \beta^n \quad\text{.......(14)}.$$

If $p > 2c$ we see by (7) that both α and β are real negative quantities. The equation (14) has then one side positive and the other negative, since the integers n, $n+1$ cannot be both even or both odd. Hence p must be less than $2c$, let $p = 2c\sin\theta$, hence

$$\alpha = \cos 2\theta + \sin 2\theta\sqrt{-1}, \quad \beta = \cos 2\theta - \sin 2\theta\sqrt{-1} \quad\text{.......(15)}.$$

The equation (14) now gives $\sin(2n+2)\theta = \sin 2n\theta$, excluding $p=0$ we have

$$\theta = \frac{2i+1}{2n+1}\frac{\pi}{2}, \qquad \frac{p}{2c} = \sin\frac{2i+1}{2n+1}\frac{\pi}{2} \quad\text{.......(16)},$$

where i has any integer value. It is however only necessary to include the values $i=0$ to $i=n-1$. The values of θ indicated by $i=i'$ and $2n-i'$ are supplementary, while the values of $\sin\theta$ indicated by $i=i'$ and $i'+2n+1$ are equal with opposite signs. The value $i=n$ is excluded because the value $p=2c$ has been already taken account of.

The oscillations of the κth particle are therefore given by

$$x_\kappa = H_\kappa + \Sigma C_p \sin 2\kappa\theta \sin(pt + \epsilon_p) \quad\text{.......(17)},$$

where

$$H_\kappa = h_\kappa + A_0 + B_0\kappa, \quad\text{and}\quad C_p = 2A_p\sqrt{-1}.$$

The value of h_κ might be determined by solving the second equation of differences (5), using the rules of linear equations adapted to that equation. But it is evident that in the position of equilibrium of the system, when there is no oscillation, every $C_p=0$, and therefore that position is determined by $x_\kappa=H_\kappa$. This enables us to deduce H_κ from the elementary rules of Statics.

We notice that in equilibrium, $T_n=mg$, $T_{n-1}=2mg$, &c., $T_\kappa=(n+1-\kappa)\,mg$. Hence by Hooke's law

$$H_\kappa-H_{\kappa-1}=l+(n+1-\kappa)\,g/c^2.$$

Adding these for all values of κ from $\kappa=1$ to $\kappa=\kappa$, and remembering that $H_0=0$ by (12), we find

$$H_\kappa=\left\{l+\frac{2n+1}{2}\frac{g}{c^2}\right\}\kappa-\frac{g}{2c^2}\kappa^2 \quad\ldots\ldots\ldots\ldots(18).$$

The equation (17) shows that the motion of every particle is compounded of n principal or simple harmonic oscillations. The periods of these are unequal and are represented by $2\pi/p$ where p has the values given in (16).

Suppose the system to be performing the principal oscillation defined by the value of $\theta=\pi/2\gamma$. By considering the signs of $\sin 2\kappa\theta$ in (17) we see that all the particles determined by $\kappa<\gamma$ are moving in the same direction as the highest particle, those determined by $\kappa>\gamma$ but $<2\gamma$ are moving in the opposite direction, those given by $\kappa>2\gamma$ but $<3\gamma$ are moving at any time in the same direction, and so on.

Ex. 2. A smooth circular cylinder is fixed with its axis horizontal at a height h above the edge of a table. A light string has a series of particles attached to it over a part of its length, the particles being each of mass m and distant a apart. The portion of the string to which the particles are attached is coiled up on the table, and the rest is carried over the cylinder, and a mass M attached to the further end of it. The system is held so that the first particle is just in contact with the table, the free portions of the string being vertical, and is then allowed to move from rest; prove that if v be the velocity of the system immediately after the nth particle is dragged into motion ($na<h$), then

$$v^2=\frac{(n-1)\,ga}{3}\cdot\frac{6M^2-n\,(2n-1)\,m^2}{(M+nm)^2}.$$

Supposing the string of particles to be replaced by a uniform chain deduce from the above result the velocity of the system after a length x of the chain has been dragged into motion. If l be the length of the chain and μ the mass, then, if l be less than h, the amount of energy that will have been dissipated by the time the chain leaves the table will be $\dfrac{\mu gl}{6}\dfrac{3M-\mu}{M+\mu}$. [Coll. Ex. 1887.]

If v_n represent the velocity required, we deduce from vis viva and linear momentum at the next impact the equation

$$\{M+(n+1)\,m\}^2\,v^2_{n+1}-\{M+nm\}^2\,v^2_n=2ga\,(M^2-n^2m^2).$$

Writing the left-hand side $\phi\,(n+1)-\phi\,(n)$, we find $\phi\,(n+1)-\phi\,(1)$ by summing from $n=1$ to n. Remembering that $v_1=0$, this gives v_n. The energy dissipated is found by subtracting the semi vis viva, viz. $\frac{1}{2}(M+\mu)\,v^2$, from the work done by gravity, viz. $(M-\frac{1}{2}\mu)\,lg$.

Ex. 3. A train of an engine and n carriages running with a velocity u, is brought to rest by applying the brakes to the engine alone, the steam being cut off. There is a succession of impacts between the buffers of each carriage and the next following. Prove that the velocity v of the engine immediately after the rth impact is given by

$$(M+rm)^2(v-u)^2=Mafr\{2M+m(r-1)\},$$

where m is the mass of any carriage, M that of the engine, a the distance between the successive buffers when the coupling chains are tight, f the retardation the brake would produce in the engine alone. [Coll. Ex.]

Ex. 4. A heavy particle falls from rest at a given altitude h in a medium whose resistance varies as the square of the velocity. On arriving at the ground it is immediately reflected upwards with a coefficient of elasticity β. Show that the whole space described from the initial position to the ground at the nth impact is $\frac{L^2}{g}\log\left\{1+\frac{1-\beta^{2n}}{1-\beta^2}\left(e^{\frac{2gh}{L^2}}-1\right)\right\}-h.$

If u_n be the height described just after the nth rebound, we show

$$u_n(u_{n+1}-1)=\beta^2(u_n-1).$$

To solve this equation of differences we put $u_n=1+1/w_n$. The equation then takes a standard form with constant coefficients. The whole space described is found by taking the logarithm of the product $u_0u_1u_2\ldots u_{n-1}$.

This problem was first solved by Euler in his *Mechanica*, vol. I. prop. 58, for the case in which $\beta=1$. An extension by Dordoni, *Memorie della Societa Italiana*, 1816, page 162, is mentioned in Walton's *Mechanical Problems*, chap. II. page 247.

CHAPTER VI.

CENTRAL FORCES.

Elementary Theorems.

306. *To find the polar equations of motion of a particle describing an orbit about a centre of force.*

Let the plane of the motion be the plane of reference and let the origin be at the centre of force. Let F be the accelerating force at any point measured positively towards the origin. Then by Art. 35,

$$\frac{d^2r}{dt^2} - r\left(\frac{d\theta}{dt}\right)^2 = -F, \quad \frac{1}{r}\frac{d}{dt}\left(r^2\frac{d\theta}{dt}\right) = 0 \text{............(1).}$$

The latter equation gives by integration

$$r^2d\theta/dt = h \text{ ....(2),}$$

where h is an arbitrary constant whose value depends on the initial conditions.

This important equation can be put into other forms of which much use is made. Let v be the velocity of the particle, p the perpendicular drawn from the origin on the tangent. Let A be the area described by the polar radius as it moves from some initial position to that which it has at the time t. Then (Art. 7)

$$r^2d\theta = 2dA = pds.$$

Remembering that $v = ds/dt$, we see that the equation (2) may be written in either of the forms

$$v = \frac{h}{p}, \qquad \frac{dA}{dt} = \tfrac{1}{2}h \text{ ....(3).}$$

The first of these shows that *the velocity at any point of the orbit is inversely proportional to the perpendicular drawn from the centre on the tangent.* The second, by integration between the limits $t = t_0$ to t, shows that *the polar area traced out by the radius vector is proportional to the time of describing it.* We also see that the constant h represents twice the polar area described in a unit of time. Both these are Newtonian theorems.

We also infer that in a central orbit, the angular velocity $d\theta/dt$ always keeps one sign and never vanishes at a finite distance from the origin. The radius vector therefore continually turns round the origin in the same direction.

307. Conversely, we may show that if a particle so move that the radius vector drawn from the origin describes areas proportional to the time the resultant force always tends to the origin and is therefore a central force. To prove this let F and G be the components of the accelerating force along and perpendicular to the radius vector. Taking the transversal resolution, we have

$$\frac{1}{r}\frac{d}{dt}\left(r^2\frac{d\theta}{dt}\right) = G.$$

As already explained $r^2 d\theta = 2dA$, and if the area A bear a constant ratio to the time, say $A = at$, we have at once $r^2 d\theta/dt = 2a$ and therefore $G = 0$.

308. If m is the mass of the particle, its linear momentum is mv and this being directed along the tangent to the path, the moment of the momentum about the centre of force is $mv \cdot p$. The moment of the momentum is called *the angular momentum* (Art. 79) and we see that *in a central orbit the angular momentum about the centre of force is constant and equal to mh.* When we are concerned only with a single particle its mass is usually taken to be unity, and h then represents the angular momentum.

309. *To find the polar equation of the orbit* we must eliminate t from the equations (1). Let $r = 1/u$, then, as in Art. 268,

$$\frac{dr}{dt} = -\frac{1}{u^2}\frac{du}{d\theta}\frac{d\theta}{dt} = -h\frac{du}{d\theta},$$

$$\frac{d^2r}{dt^2} = -h\frac{d^2u}{d\theta^2}\frac{d\theta}{dt} = -h^2u^2\frac{d^2u}{d\theta^2}.$$

Substituting this value of d^2r/dt^2 and the value of $d\theta/dt = hu^2$ given by (2) in $\dfrac{d^2r}{dt^2} - r\left(\dfrac{d\theta}{dt}\right)^2 = -F$, we have

$$\frac{d^2u}{d\theta^2} + u = \frac{F}{h^2u^2} \quad\text{......................}(4).$$

When the polar equation of the path is given in the form $u=f(\theta)$ the equation (4) determines F in terms of u and θ. Since the attractive forces of the bodies which form the solar system are in general functions of the distance only we should eliminate θ by using the known polar equation of the path. We thus find F as a function of u only.

Strictly this expression for F only holds for points situated on the given path, but *if the initial conditions are arbitrary*, the path may be varied and the law of force may be extended to hold for other parts of space.

When the force F is given as a function of r or $1/u$, the equation (4) is a differential equation of the form $\dfrac{d^2u}{d\theta^2}=f(u)$. This differential equation has been already solved in Art. 97.

It is evident from dynamical considerations that when the central force is attractive, i.e. when F is positive, the orbit must be concave to the centre of force, and when F is negative the orbit must be convex. By looking at equation (4) we immediately verify the theorem in the differential calculus that a curve is concave or convex to the origin according as $\dfrac{d^2u}{d\theta^2}+u$ is positive or negative.

310. *To apply the tangential and normal resolutions to a central orbit.*

Referring to Art. 36 we have the two equations

$$v\frac{dv}{ds}=-F\cos\phi,\quad \frac{v^2}{\rho}=F\sin\phi \quad\text{..............(5)},$$

where ϕ is the angle behind the radius vector when the particle moves in the direction in which s is measured. Writing dr/ds for $\cos\phi$ and integrating we have

$$v^2=C-2\int F\,dr \quad\text{......................(6)},$$

where C is a constant whose value depends on the initial conditions. This equation is obviously the equation of vis viva, Art. 246. The integral has a minus sign because the central force is, as usual, measured positively towards the origin, while the radius vector is measured positively from the origin.

If we substitute for v its value h/p given by (3) and differentiate we deduce

$$F = -\tfrac{1}{2} h^2 \frac{d}{dr}\left(\frac{1}{p^2}\right) \text{ ....(7).}$$

This expression for the central force F is very useful when the orbit is given in the form $p = f(r)$.

311. Considering the normal resolution (5), we have an expression for v which is useful when both the law of force and the path are known. It has the advantage of giving the velocity without requiring the previous determination of either of the constants C or h. If χ is one-quarter of the chord of curvature of the path drawn in the direction of the centre of force we may write the equation in either of the forms

$$v^2 = F\rho \sin\phi = 2F\chi \text{ ....(8).}$$

This is usually read; *the velocity at any point is that due to one-quarter of the chord of curvature.*

When the particle describes a circle about a centre of force in the centre $\sin\phi = 1$ and ρ is the radius r. The velocity given by the normal resolution, viz. $v^2/r = F$, is often called *the velocity in a circle at a distance r from the centre of force.*

312. The velocity acquired by a particle which travels from rest at an infinite distance from the centre of force to any given position P is called *the velocity from infinity.* Referring to the equation of vis viva (6), let

$$F = \frac{\mu}{r^n}; \quad \therefore\ v^2 = C + \frac{2\mu}{n-1}\frac{1}{r^{n-1}}.$$

Now $v = 0$ when $r = \infty$; hence, if n is greater than unity, we have $C = 0$. The velocity from infinity to the distance $r = R$ is therefore given by $v^2 = \dfrac{2\mu}{n-1}\dfrac{1}{R^{n-1}}$. See Art. 181.

If n is less than unity the value of C is infinite. Instead of the velocity from infinity we use *the velocity acquired by the particle in travelling from rest at the given point P to the origin* under the attraction of the central force. In this case $v = 0$ when $r = R$; hence (since $n < 1$) $C = \dfrac{2\mu}{1-n} R^{1-n}$. The velocity to the origin (where $r = 0$) is then given by $v^2 = \dfrac{2\mu}{1-n} R^{1-n}$.

When the force varies as the inverse cube of the distance, i.e. $F=\mu/r^3$, we notice that the velocity in a circle and the velocity from infinity are equal. When the force varies as the distance, i.e. $F=\mu r$, the velocity in a circle is equal to that to the origin. When the force varies inversely as the distance, i.e. $F=\mu/r$, both the velocity from infinity and the velocity to the origin are infinite.

313. The constants. The two constants h and C may be determined from the initial conditions when these are known. Let the particle be projected from a point P at an initial distance R from the origin with a velocity V, let β be the angle the direction of projection makes with the initial radius vector. The tangent at P makes two angles with the radius vector OP, respectively equal to β and $\pi-\beta$. When a distinction has to be made it is usual to take β equal to the angle *behind the radius vector* when P travels along the curve in the positive direction (i.e. the direction which makes the independent variable increase). The angle β is called *the angle of projection*. We evidently have $h=vp=VR\sin\beta$. If $F=\mu/r^n$, we have $v^2=C+\dfrac{2\mu}{n-1}\dfrac{1}{r^{n-1}}$.

It follows that, if $n>1$ and the velocity from infinity is V_1, $C=V^2-V_1^2$; if $n<1$, $C=V^2+V_0^2$ where V_0 is the velocity to the origin.

We may obtain another interpretation for the constant C. Selecting any standard distance $r=a$, the potential energy at a distance r is

$$\int_r^a(-F)\,dr=\frac{\mu}{n-1}\left(\frac{1}{a^{n-1}}-\frac{1}{r^{n-1}}\right)=\frac{\mu}{(n-1)\,a^{n-1}}+\frac{C}{2}-\frac{v^2}{2}.$$

See Art. 250. It follows that $\frac{1}{2}C$ plus $\dfrac{\mu}{n-1}\dfrac{1}{a^{n-1}}$ is equal to the whole energy of the motion. *Hence by taking the standard position at infinity or the origin according as n is greater or less than unity, we may make $\frac{1}{2}C$ equal to the whole energy.*

314. When a point P on the orbit is such that the radius vector OP is perpendicular to the tangent, the point P is called *an apse*.

When OP is a maximum the apse is sometimes called *an apocentre*, and when a minimum *a pericentre*.

315. Summary. As the formulæ we have arrived at are the fundamental ones in the theory of central forces, it is useful to make a short summary before proceeding further. There are three elements to be considered: (1) the law of force, (2) the equations of the path, (3) the velocity and time of describing an arc. Any one of these elements being given, the other two can be deduced by dynamical considerations. There are therefore three sets of equations; firstly, equations (4) and (7) connect the force and path, so that either being known the other can be deduced; secondly, equation (6) connects the force and velocity; thirdly, equations (2) and (3) connect the path with the motion in that path.

The equations of one of these sets are mere algebraic transformations of each other, any one being given the others can be found from it by reasoning which is purely mathematical. But an equation of one set cannot be deduced from an equation of another set in this manner, because each set depends on different dynamical facts.

316. Dimensions. It is important to notice the dimensions of the various symbols used. The accelerating force F, like that of gravity, i.e. g, is one dimension in space and -2 in time. We see this by examining any formula which contains F or g, say $s=\frac{1}{2}gt^2$ or $-F\cos\phi=d^2s/dt^2$. The force F will in general vary as some power of the distance from the centre of force, say $F=\mu/r^n$ where μ is a constant which measures the strength of the central force. The quantity $\mu=Fr^n$ is therefore $n+1$ dimensions in space and -2 in time. The velocity $v=ds/dt$ is one dimension in space and -1 in time. The constant $h=vp$ is 2 dimensions in space and -1 in time. See Art. 151.

317. Force given, find the orbit. *Ex.* 1. The force being

$$F=\mu u^3(2a^2u^2+1),$$

a particle is projected from an initial distance a, with a velocity which is to the velocity in a circle at the same distance as $\sqrt{2}$ to $\sqrt{3}$, the angle of projection being 45°. Find the path described.

Putting $a=1/c$ the differential equation of motion is, by Art. 309,

$$h^2\left(\frac{d^2u}{d\theta^2}+u\right)=\frac{2\mu}{c^2}u^3+\mu u;$$

$$\therefore\ v^2=h^2\left\{\left(\frac{du}{d\theta}\right)^2+u^2\right\}=\frac{\mu}{c^2}u^4+\mu u^2+C.$$

When $u=c$, the conditions of the question give $v^2=\frac{2}{3}F/c$ and $h=v\sin\beta/c$ where $\sin^2\beta=\frac{1}{2}$, see Arts. 311, 313. We therefore have $C=0$, $h^2=\mu$. The equation now reduces to

$$\left(\frac{du}{d\theta}\right)^2=\frac{u^4}{c^2}; \quad \therefore \int\frac{du}{u^2}=\pm\frac{\theta}{c}+A.$$

Replacing u by $1/r$ and measuring θ from the initial radius OA in such a direction that r and θ increase together, this leads to $r=a(1+\theta)$.

From the equation $r^2d\theta/dt=h$, we infer that the time from a distance a to r is $(r^3-a^3)/3a\sqrt{\mu}$.

Ex. 2. A particle moves under the action of a central force $\mu(u^5-\frac{1}{8}a^2u^7)$, the velocity of projection being $(25\mu/8a^4)^{\frac{1}{2}}$, and the angle of projection $\sin^{-1}\frac{4}{5}$. Prove that the polar equation of the path is $3a^2=(4r^2-a^2)(\theta+C)^2$. [Coll. Ex. 1892.]

Ex. 3. When the central acceleration is $\mu(u^3+a^2u^5)$ and the velocity at the apsidal distance a is equal to $\sqrt{\mu/a}$, prove that the orbit is $r=a\,\mathrm{cn}\,\theta\;(\mathrm{mod}\;\sqrt{\frac{1}{2}})$. [Coll. Ex. 1897.]

Ex. 4. The central force being $F=2\mu u^3(1-a^2u^2)$, the particle is projected from an apse at a distance a with a velocity $\sqrt{\mu/a}$. Prove that it will be at a distance r after a time $\frac{1}{2\sqrt{\mu}}\left\{a^2\log\frac{r+\sqrt{(r^2-a^2)}}{a}+r\sqrt{(r^2-a^2)}\right\}$. [Math. Tripos.]

Ex. 5. A particle, acted on by two centres of force both situated at the origin respectively $F=\mu u^3$ and $F'=\mu' u^5$, is projected from an initial distance a with a velocity equal to that from infinity, the angle of projection being $\tan^{-1}\sqrt{2}$. If the forces are equal at the point of projection, the path is $a\theta=(r-a)\sqrt{2}$.

Ex. 6. A particle, acted on by the central force $F=u^2f(\theta)$, is initially projected in any manner. Prove that the radius vector can be expressed as a function of θ if the integrals of $\cos\theta f(\theta)$ and $\sin\theta f(\theta)$ can be found. [Use the method of Art. 122.]

318. Orbit given, find the force. *Ex.* 1. *A particle describes a given circle about a centre of force on the circumference. It is required to find the law of force and the motion. Newton's problem.*

Let O be the centre of force, C the centre of the circle, P the particle at the time t. Let a be the radius of the circle, $OP=r$. If $p=OY$ be the perpendicular on the tangent, we have (since the angles OPY, OAP are equal) $p=r^2/2a$. Hence using (7) of Art. 310, we have

$$F=-\tfrac{1}{2}h^2\frac{d}{dr}\frac{1}{p^2}=\frac{8h^2a^2}{r^5}\quad\text{(1)}.$$

If we suppose the magnitude of the force to be given at a unit of distance from the centre of force we write this in the form $F=\frac{\mu}{r^5}$, where μ is a known constant, sometimes called the magnitude or strength of the force. The constant h is then determined by the equation

$$8h^2a^2=\mu\quad\text{(2)}.$$

The velocity at any point P is found by the normal resolution, Art. 310,

$$\frac{v^2}{a}=F\sin OPY=\frac{\mu}{r^5}\frac{r}{2a};\qquad\therefore v=\sqrt{\frac{\mu}{2}}\cdot\frac{1}{r^2}\quad\text{(3)}.$$

By Art. 312 this velocity is equal to that from infinity.

To find the time of describing any arc AP, where A is the extremity of the diameter opposite to the centre of force, we use the equation $A=\frac{1}{2}ht$, Art. 306. Since the area AOP is made up of the triangle OCP and the sector ACP, we have

$$\tfrac{1}{2}ht=A=\tfrac{1}{2}a^2(2\theta+\sin 2\theta),$$

where $\theta=$ the angle AOP. Substituting for h

$$t=2a^3\sqrt{\frac{2}{\mu}}(2\theta+\sin 2\theta) \quad\dots\dots\dots\dots (4).$$

It appears from this that the particle will arrive at the centre of force after a finite time obtained by writing $\theta=\frac{1}{2}\pi$. The particle arrives with an infinite velocity due to the infinite force at that point.

Let the force at all points of space act towards the point O and vary as the inverse fifth power of the distance from O. It is required to find *the necessary and sufficient condition that a particle projected from a given point P in a given direction PT with a given velocity V may describe a circle passing through O*. It is obvious from (3) that it is necessary that $V^2=\frac{1}{2}\mu/r^4$ where $r=OP$; we shall now prove that this is also sufficient.

Describe the circle which passes through O and touches PT at P. The particle which describes this circle freely satisfies the given conditions at P. If then the given particle does not also describe the circle we should have two particles projected from P in the same direction, with equal velocities, acted on by the same forces, describing different paths; which is impossible; Art. 243.

We notice that a change in the direction of projection PT affects the size of the circle described, but not the fact that the path is a circle.

Ex. 2. A particle moves in a circle about a centre of force in the circumference, the force being attractive and equal to μr^n. Prove that the resistance of the medium in which the particle moves is $\frac{1}{4}\mu(n+5)r^n\sin\theta$, where $\cos\theta=r/2a$.

Use the normal and tangential resolutions. [Coll. Ex.]

Ex. 3. A particle of unit mass describes a circle about a given centre of force O situated on the circumference. If the particle at any point P is acted on by an impulse $2v\cos\phi$ in a direction making an angle $\pi-\phi$ with the direction of motion PT, show that the new orbit is also a circle and prove that the ratio of the radii is $\cos 2\phi+\sin 2\phi\cot\theta$, where θ is the angle OPT.

Ex. 4. The force being $F=\mu u^5$, a particle when projected from a point P with an initial velocity V, equal to that from infinity, describes the circle $r=2a\cos\theta$; investigate the path when the initial velocity is $V(1+\gamma)$, where γ is so small that its square can be neglected.

Proceeding as in Art. 317, we find

$$v^2=h^2\left\{\left(\frac{du}{d\theta}\right)^2+u^2\right\}=\frac{\mu}{2}u^4+C.$$

The conditions of the question give

$$C=\frac{\mu}{2}\frac{c^4}{\cos^4\alpha}(2\gamma+\gamma^2), \qquad h^2=\frac{\mu}{2}c^2(1+\gamma)^2,$$

where $c=1/2a$ and $\theta=\alpha$ initially. Putting $u=c\sec\theta+c\eta$ and neglecting the squares of η and γ, we arrive at

$$\frac{\cos^2\theta}{\sin\theta}\frac{d\eta}{d\theta}+\frac{\cos^3\theta-2\cos\theta}{\sin^2\theta}\eta=\frac{-\gamma}{\sin^2\theta}+\frac{\gamma}{\cos^4\alpha}\frac{\cos^4\theta}{\sin^2\theta}.$$

Each side being a perfect differential, we find

$$\frac{\cos^2\theta}{\sin\theta}\eta=\kappa+\gamma\cot\theta-\frac{\gamma}{\cos^4\alpha}(\cot\theta+\tfrac{3}{2}\theta+\tfrac{1}{2}\sin\theta\cos\theta),$$

and κ is determined from the condition that $\eta=0$ when $\theta=\alpha$;

$$\therefore\ \kappa=-\gamma\cot\alpha+\frac{\gamma}{\cos^4\alpha}(\cot\alpha+\tfrac{3}{2}\alpha+\tfrac{1}{2}\sin\alpha\cos\alpha).$$

Putting $u=1/r$, we have $r=2a\cos\theta\,(1-\eta\cos\theta)$,

$$\therefore\ \frac{r}{2a}=\cos\theta-\kappa\sin\theta-\gamma\cos\theta+\frac{\gamma}{\cos^4\alpha}(\cos\theta+\tfrac{3}{2}\theta\sin\theta+\tfrac{1}{2}\sin^2\theta\cos\theta).$$

It has been assumed that $\cos\alpha$ is not small, the point P must therefore not be close to the centre of force. It easily follows that when

$$\theta=\tfrac{1}{2}\pi-\kappa+\tfrac{3}{4}\pi\gamma\sec^4\alpha,$$

the distance of the particle from the centre of force is of the order of small quantities neglected above.

Ex. 5. Any number of particles are projected in all directions from a given point P each with the velocity from infinity, the central force being $F=\mu u^5$. Prove that their locus at any instant is (θ being measured from OP)

$$\frac{(r^2+c^2-2cr\cos\theta)^{\frac{3}{2}}}{\sin^3\theta}\left\{\theta-\sin\theta\,\frac{(r^2+c^2)\cos\theta-2cr}{r^2+c^2-2cr\cos\theta}\right\}=A,$$

where $OP=c$ and A is a constant depending on the time elapsed.

319. *Ex.* 1. *A particle describes an equiangular spiral of angle α under the action of a centre of force in the pole, prove that*

$$F=\frac{\mu}{r^3},\qquad h=\sin\alpha\sqrt{\mu},\qquad v=\frac{\sqrt{\mu}}{r},\qquad 2\cos\alpha t\sqrt{\mu}=r_1^{\,2}-r_0^{\,2},$$

where t is the time of describing the arc bounded by the radii vectores r_0, r_1. Conversely, a particle being projected from any point in any direction will describe an equiangular spiral about a centre of force whose law is $F=\mu/r^3$, provided the velocity of projection is $\sqrt{\mu}/r$, i.e. is equal to that from infinity.

Assuming $p=r\sin\alpha$ we follow the same line of reasoning as in Ex. 1 of Art. 318.

Ex. 2. A particle acted on by a central force moves in a medium in which the resistance is κ(vel.)^2, and describes an equiangular spiral, the pole being the centre of force. Prove that the central force varies as $\dfrac{1}{r^3}e^{-2\kappa r\sec\alpha}$, where α is the angle of the spiral. [Math. Tripos, 1860.]

320. *Ex. A particle describes the curve $r^m=a\cos n\theta+b\sin n\theta$, under the action of a centre of force in the origin. Prove that*

$$F=\frac{\mu}{r^{2m+3}}+\frac{\mu'}{r^3},\qquad v^2=\frac{\mu}{m+1}\,\frac{1}{r^{2m+2}}+\frac{\mu'}{r^2}.$$

We notice (1) that the exponents of r are independent of n, (2) that, when $m+1$ is positive, the velocity at any point is that due to infinity. Art. 312.

Supposing the law of force and the velocity of projection to be given by these formulæ, let the particle be projected from any point P in any direction PT. The four constants h^2, n, a, b are determined by

$$h^2(m+1)\left(\frac{n}{m}\right)^2(a^2+b^2)=\mu, \qquad h^2\left(1-\frac{n^2}{m^2}\right)=\mu',$$

joined to the conditions that the curve must pass through P and touch PT.

We find that n^2 and $\frac{\mu}{m+1}-\mu' R^{2m}\cot^2\phi$ have the same sign, where $R=OP$ and ϕ is the angle of projection. When the sign of n^2 thus determined becomes negative or zero the curve obviously changes into

$$r^m=a'e^{n\theta}+b'e^{-n\theta}, \text{ or } r^m=a+b''\theta,$$

where $4a'b'=a^2-b^2$ and b'' is the limit of bn when b is infinite and n zero.

It is useful to notice the following geometrical properties of the curve. If p be the perpendicular on the tangent, ϕ the angle the radius vector makes with the tangent

$$\tan\phi=-\frac{m}{n}\cot n\theta, \qquad \frac{1}{p^2}=\frac{n^2}{m^2}\frac{a^2+b^2}{r^{2m+2}}+\frac{m^2-n^2}{m^2}\frac{1}{r^2}.$$

This example includes many interesting cases. Putting $m=2$, $n=2$, we see that the lemniscate of Bernoulli could be described about a centre of force in the node varying as the inverse seventh power of the distance. Putting $m=n$, we have the path when the force varies as the inverse $(2m+3)$th power and the velocity is that from infinity. Writing $m=\frac{1}{2}$, $n=\frac{1}{2}$, we find the path is a cardioid when the central force varies as the inverse fourth power and the velocity is that from infinity. Writing $m=1$, $n=1$, the path is a circle described about a centre of force on the circumference.

321. *Ex.* 1. *A particle describes a circle about a centre of force situated in its plane. It is required to find the law of force and the motion.*

Let O be the centre of force, C the centre of the circle, a its radius and $CO=c$. Taking the equations of Art. 310, we have

$$v=\frac{h}{p}, \qquad \frac{v^2}{a}=F\frac{p}{r}, \qquad \therefore F=\frac{h^2}{a}\frac{r}{p^3}.$$

Since in a circle $2ap=r^2+a^2-c^2$, we can, by substitution, express F and v in terms of r alone. We have

$$v=\left(\frac{\mu}{2}\right)^{\frac{1}{2}}\frac{1}{r^2+B}, \qquad F=\frac{\mu r}{(r^2+B)^3},$$

where $8a^2h^2=\mu$ and $B=a^2-c^2$. When $B=0$, the law of force reduces to the inverse fifth power, and the velocity becomes the same as that found in Art. 318.

If this law of force be supposed to hold throughout the plane of the circle, the values of μ and B are given. In order that the orbit may be a circle it is necessary that the velocity of projection should satisfy the above value of v, i.e. should be equal to the velocity from infinity. The direction of projection being also given, the angular momentum h (Art. 313) is also known. The values of a and c follow at once from the equations given above and must be real.

Newton, when discussing this problem, supposes that the centre of force lies inside the circle. It follows that B is positive, and at no point of space can either the force or velocity be infinite.

When the centre of force is outside the circle, one portion of the orbit is concave and the other convex to the centre of force. We must therefore suppose that the force is attractive in the first and repulsive in the other part. Writing $B=-b^2$, we have $b^2=c^2-a^2$, and therefore b is the length of either of the tangents drawn from the centre of force to the circle, and the force changes sign through infinity when the particle passes the circle whose radius is b.

Sylvester, in the *Phil. Mag.* 1865, points out that the resultant attraction of a circular plate, whose elements attract according to the law of the inverse fifth power, at an external point P situated in its plane, is $\dfrac{\mu r}{(r^2-b^2)^3}$ where μ is the mass of the plate, b its radius and r the distance of P from the centre. The circle described by P under the attraction of this plate cuts the rim orthogonally.

Let the particle P be constrained to move on a smooth plane under the action of a centre of force situated at a point C distant b' from the plane, the law of force being the inverse fifth power. The component of force in the plane is $F=\dfrac{\mu r}{(r^2+b'^2)^3}$, where r is the distance of P from the projection O of the centre of force on the plane. Putting $B=b'^2$, it appears from what precedes that, if the velocity of projection is equal to that from infinity, the path of the particle on the plane is a circle. The length of the chord bisected by the point O is constant for all the circles and equal to $2b'$.

Ex. 2. A particle moves under the action of a centre of force $F=\mu u^5$. Prove that all the circles which can be described either pass through a fixed point or have a fixed point for centre.

322. *Ex.* 1. *A particle moves under the action of a centre of force whose attraction is* $F=\dfrac{\mu r}{(r^2+B)^2}$ *and the velocity at any point is equal to that from infinity. It is required to find the path.*

The equation of vis viva (Art. 310) gives

$$v^2=C-2\int F\,dr=C+\frac{\mu}{r^2+B} \quad\quad (1).$$

Since this formula is independent of the path and it is given that v is zero when r is infinite we see that $C=0$. Substituting for v its value h/p, the equation of the path becomes

$$r^2+B=ip^2, \qquad ih^2=\mu \quad\quad (2).$$

The curve required is therefore such that a linear relation exists between p^2 and r^2. There are several species of curves which possess this property distinguished from each other by the values of B and i.

One such curve is known to be an epicycloid. Supposing the radii of the fixed and rolling circles to be a and b, we have at the cusp $r=a$, $p=0$ and at the vertex p and r are each equal to $a+2b$. We thus find

$$B=-a^2, \qquad \frac{\mu}{h^2}=i=\frac{(a+2b)^2-a^2}{(a+2b)^2} \quad\quad (3).$$

The law of force and the conditions of projection being given both B and h^2 are known. If the force is attractive, B negative, and μ/h^2 less than unity, the path is an epicycloid, the values of a and b being given by (3).

Changing the sign of b the epicycloid becomes a hypocycloid and in this case we learn from (3) that i and μ are negative. When therefore the force is repulsive, and B negative, the path is a hypocycloid.

The remaining species are more easily separated by putting the equation (2) into the form $\rho = ip$, a result which follows at once from the identity $\rho = r\,dr/dp$. Remembering that $\rho = p + d^2p/d\psi^2$ the differential equation becomes

$$\frac{d^2p}{d\psi^2} - (i-1)p = 0 \qquad (4).$$

When i is less than unity or is negative we easily deduce the cycloidal species given above. If $\beta^2 = 1 - i$, we find

$$p = L\sin\beta\psi + M\cos\beta\psi.$$

If the axis of x pass through the cusp, we have $p = 0$ when $\psi = 0$ and $p = a + 2b$ when $\beta\psi = \frac{1}{2}\pi$. Hence $L = a + 2b$ and $M = 0$.

When i is greater than unity we have the forms

$$p = Le^{\alpha\psi} + Me^{-\alpha\psi}, \qquad p = L\psi + M \qquad (5),$$

where $\alpha^2 = i - 1$ and the second form occurs when $i = 1$. Since in any curve the projection of the radius vector on the tangent is $dp/d\psi$, we find by elementary geometry

$$r^2 = p^2 + \left(\frac{dp}{d\psi}\right)^2, \qquad \tan\phi = \frac{p\,d\psi}{dp} \qquad (6),$$

where ϕ is the angle behind the radius vector. Since $\phi = \psi - \theta$, we can in this way express the polar coordinates r and θ in terms of the subsidiary angle ψ.

Substituting in (2) we find that $4\alpha^2LM = B$, so that L and M have the same or opposite signs according as the given quantity B is positive or negative. When $B = 0$, either L or M is zero, and since, by (6), $\tan\phi$ is then constant the curve is an equiangular spiral.

To trace the forms of the exponential spirals it is convenient to turn the axis

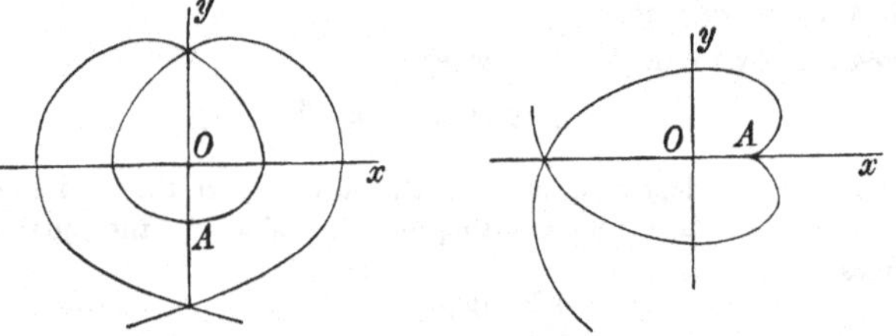

of x round the origin so that the equation (5) may assume a symmetrical form. We then have

$$p = \tfrac{1}{2}c\,(e^{\alpha\psi} \pm e^{-\alpha\psi}) \qquad (7),$$

where the upper or lower sign is to be taken according as B is positive or negative. When B is positive there is an apse whose position is found by putting $p = r$ in (2), whence $(i-1)\,r^2 = B$. When B is negative there is a cusp at the point determined by $p = 0$, i.e. at $r^2 = -B$. These spirals were first discussed by Puisseux (with a different object in view) in *Liouville's Journal*, 1844.

By using a proposition in the theory of attractions we may put some of the preceding problems in another light. It may be shown that the resultant attraction of a thin circular ring, whose elements attract according to the law of the inverse cube, at any point P in the plane of the ring is $\dfrac{\pm\mu r}{(r^2 - c^2)^2}$, where μ is the mass of the ring, c its radius and r the distance of P from the centre. The plus or minus sign is to be taken according as P is without or within the ring, (see Townsend in the *Quarterly Journal*, 1879). The path of the particle P moving under the attraction of the ring has now been found provided the velocity of projection is equal to that from infinity.

Again, when a particle P is constrained to move on a smooth plane under the action of a centre of force C situated at a distance c from the plane, the law of force being the inverse cube, the component of attraction in the plane is $\dfrac{\mu r}{(r^2+c^2)^2}$, where r is the distance of P from the projection O of the centre of force on the plane.

Ex. 2. If s be the arc AP of any path measured from a fixed point A, show that $s(i-1)/i$ differs from the projection of the radius vector OP on the tangent at P by a constant quantity which is zero when A is an apse.

Ex. 3. Show that the polar area traced out by a radius vector OP is equal to i times the corresponding polar area of the pedal. Thence show that the time of describing any arc is given by $ht=i\int p^2 d\psi$.

323. Parallel forces. *Ex.* 1. *A particle describes a central conic under the action of a force F tending always in a fixed direction. It is required to find F.*

Let the conic be referred to conjugate diameters OA, OB; the force acting

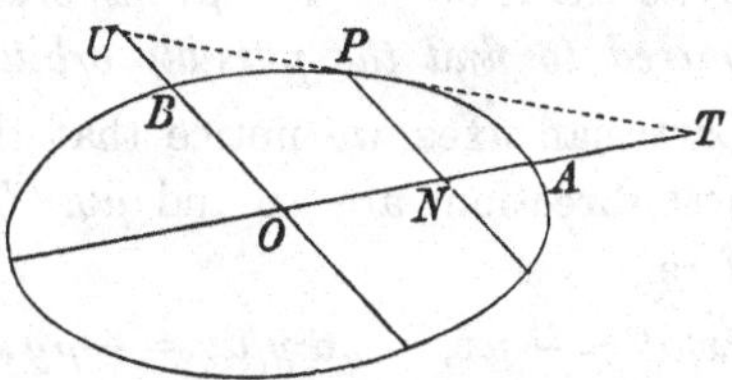

parallel to BO. Let the angle $AOB=\omega$, $OA=a'$, $OB=b'$. Let $ON=x$, $PN=y$ be the coordinates of P. Then

$$d^2x/dt^2=0, \qquad d^2y/dt^2=-F.$$

The first equation gives $x=At$, where A is the *oblique* component of velocity parallel to x. Hence A is the resultant velocity at B. We then have

$$y=\frac{b'}{a'}(a'^2-x^2)^{\frac{1}{2}}, \qquad \therefore \frac{d^2y}{dt^2}=-\frac{b'^4A^2}{a'^2}\frac{1}{y^3}.$$

The component of velocity at right angles to the force is constant. Representing this component by V, and remembering that the resultant velocity at B is A, we find $V=A\sin\omega$.

If a, b are the semi-axes of the conic the expression for the force becomes

$$F=\frac{V^2b'^4}{a'^2\sin^2\omega}\frac{1}{y^3}=\frac{V^2b'^6}{a^2b^2}\frac{1}{y^3}.$$

It follows that *the force tending in a given direction by which a conic can be described varies inversely as the cube of the chord along which the force acts.* This result may also be obtained without difficulty by taking the normal resolution of force.

Ex. 2. If the tangent to the conic at P intersect the conjugate diameters in T and U, prove that the velocity at P is $v=Ax\,.\,TU/a'^2$.

Ex. 3. A particle describes the curve $y=f(x)$ freely under the action of a force F whose direction is parallel to the axis of y; prove $F=A^2d^2y/dx^2$.

Ex. 4. Show that a particle can describe a complete cycloid freely under the action of a force tending towards the straight line joining the cusps and varying inversely as the square of the distance. Prove also that the square of the velocity varies inversely as the distance.

324. *Ex.* Two masses M, m are connected by a string which passes through a hole in a smooth horizontal plane, the mass m hanging vertically. Prove that M describes on the plane a curve whose differential equation is

$$\left(1+\frac{m}{M}\right)\frac{d^2u}{d\theta^2}+u=\frac{mg}{M}\frac{1}{h^2u^2}.$$

Prove also that the tension of the string is $\frac{Mm}{M+m}(g+h^2u^3)$. [Coll. Exam.]

Law of the direct distance.

325. *A particle is acted on by a centre of force situated in the origin whose acceleration is $F=\mu r$ where r is the radius vector. It is required to find the possible orbits.*

Taking any Cartesian axes, we notice that the resolved parts of the force in these directions are μx and μy. The equations of motion are therefore

$$d^2x/dt^2=-\mu x,\qquad d^2y/dt^2=-\mu y\ \ldots\ldots\ldots\ldots(1).$$

We observe that though the axes of coordinates are arbitrary, the equations (1) are independent; one containing only x, the other only y. We infer that the general principle enunciated for parabolic motion may also be applied here. *The circumstances of the motion parallel to any fixed direction are independent of those in other directions and may be deduced from the corresponding formulæ for rectilinear motion.*

Supposing that the force is attractive in the standard case, μ is positive and the solutions of (1) are

$$x=A\cos\sqrt{\mu}t+A'\sin\sqrt{\mu}t,\quad y=B\cos\sqrt{\mu}t+B'\sin\sqrt{\mu}t.$$

As there is nothing to prevent us from using oblique axes, let us take the initial radius vector as the axis of x and let the axis of y be parallel to the direction of initial motion. If R and V be the initial distance and velocity, we have when $t=0$,

$$x=R,\quad dx/dt=0;\quad y=0,\quad dy/dt=V.$$

These give $R=A,\quad 0=A',\quad 0=B,\quad V=B'\sqrt{\mu}.$

The motion is therefore determined by

$$x = R\cos\sqrt{\mu}t, \quad y = R'\sin\sqrt{\mu}t,$$

where $V = R'\sqrt{\mu}$. Eliminating t, we obviously arrive at the equation of a conic having its centre at the centre of force and R, R' for semi-conjugate diameters.

If μ is positive, the centre of force is attractive and the orbit must be at every point concave to the origin. *The orbit is therefore an ellipse. If μ is negative,* the central force repels, *and the orbit,* being convex to the origin, *is a hyperbola.* Since the centre of the conic is always at the centre of force the orbit can be a parabola only when the centre of force is infinitely distant. If the force at the particle is then finite, the coefficient μ must be zero. The finite changes of r as the particle moves about do not affect the value of μr. The force on the particle is then constant in magnitude and fixed in direction.

When μ is negative, we put $\mu = -\mu'$. The solution of the differential equations then becomes

$$x = \tfrac{1}{2}R\left(e^{\sqrt{\mu'}t} + e^{-\sqrt{\mu'}t}\right), \quad y = \frac{i}{2}R'\left(e^{\sqrt{\mu'}t} - e^{-\sqrt{\mu'}t}\right),$$

where $V = iR'\sqrt{\mu'}$ and $i = \sqrt{-1}$. It is evident that iR' is real.

326. Since any point of the orbit may be taken as the point of projection, we deduce from the equation $V = \sqrt{\mu}R'$, that *the velocity v at any point P of the ellipse is given by $v = \sqrt{\mu}R'$ where R' is semi-conjugate of OP.* If r be the radius vector of the moving particle this equation may also be written $v^2 = \mu(a^2 + b^2 - r^2)$ where a and b are the semi-axes.

Since $vp = h$ and $pR' = ab$, we see that *the constant h is $h = \sqrt{\mu}ab$.*

If the principal diameters are taken as the axes of coordinates, we have $x = a\cos\phi$, $y = b\sin\phi$, where ϕ is the eccentric angle of the particle. It immediately follows that *the particle so moves that $\phi = \sqrt{\mu}t$.* When ϕ has increased by 2π the particle has made a complete circuit and returned to its former position. *The periodic time is therefore $2\pi/\sqrt{\mu}$.* It appears from this that *the periodic time is independent of all the conditions of projection and is the same for all ellipses. It depends solely on the strength μ of the central force.*

In general the time of describing any arc PP' is the difference of the eccentric angles at P and P' divided by $\sqrt{\mu}$.

When the orbit is a hyperbola we have

$$x = \tfrac{1}{2} a \left(e^{\phi'} + e^{-\phi'}\right), \quad y = \tfrac{1}{2} b \left(e^{\phi'} - e^{-\phi'}\right),$$

where ϕ' is an auxiliary angle. It immediately follows that $\phi' = \sqrt{\mu'}t$ where μ' is positive and equal to $-\mu$.

327. When the velocity V and angle β of projection as well as the initial distance R are given, the semi-axes a, b of the conic described may be deduced from the equations

$$a^2b^2 = \frac{h^2}{\mu} = \frac{V^2R^2 \sin^2\beta}{\mu}, \qquad a^2 + b^2 = R^2 + \frac{V^2}{\mu}.$$

These give real values to a^2 and b^2. The angle θ which the major axis makes with the initial distance is given by

$$\frac{\cos^2\theta}{a^2} + \frac{\sin^2\theta}{b^2} = \frac{1}{R^2}; \quad \therefore \tan^2\theta = \frac{b^2}{a^2} \cdot \frac{a^2 - R^2}{R^2 - b^2}.$$

Since $V = \sqrt{\mu}R'$, it is evident that the problem of finding the particular conic described when R and V are given is the same as *the geometrical problem of constructing a conic when two semi-conjugate diameters R, R' are given in position and magnitude.* This useful construction is given in most books on geometrical conics.

328. Referring to the equations (1) of Art. 325 we see that the motion in an ellipse about a centre of force $F = \mu r$ is the resultant of two rectilinear harmonic oscillations along two arbitrary directions Ox, Oy represented by

$$X = -\mu x, \; Y = -\mu y.$$

The resultant of any number of rectilinear harmonic oscillations (performed in equal times) along arbitrary straight lines OA, OB, &c. may be found by resolving the displacements of each along two arbitrary axes and compounding the sums of the components. The resulting motion is therefore an elliptic motion with O for centre.

Ex. Investigate the conditions that the resultant of two rectilinear harmonic oscillations, of equal periods, whose directions make an angle θ, should be (1) a rectilinear, (2) a circular motion. Prove that in the first case their angles or phases must be equal; in the second their amplitudes must be equal and their phases differ by $\pi - \theta$. The radius is $a \sin\theta$.

329. *Ex.* 1. If OP, OQ are conjugate diameters of an ellipse, prove that the time from P to Q is one-quarter of the whole periodic time. This follows at once from the fact that the area POQ is one-quarter of the area of the ellipse.

Ex. 2. Prove that in a hyperbolic orbit the time from the extremity of the major axis to a point whose distance from that axis is equal to the minor axis is the same for all hyperbolas.

Ex. 3. If the circle of curvature at any point P of an ellipse cut the curve again in Q, and A is the extremity of the major axis nearest to P, prove that the time from Q to A is three times the time from A to P.

Since $\phi = \sqrt{\mu} t$, Art. 326, the theorems in conics which, like this one, are concerned with eccentric angles may at once be translated into dynamics.

Ex. 4. Two tangents TP, TQ are drawn to an ellipse, prove that the velocities at P and Q are proportional to the lengths of the tangents. [For these tangents are known to be proportional to the parallel diameters.]

330. Point to Point. *To find the directions in which a particle must be projected from a given point P with a given velocity V, so as to pass through another given point Q.*

Let r_1, r_2 be the distances of P, Q from the centre of force O. Let OP be produced to D where D is such that the velocity V of projection at P is equal to

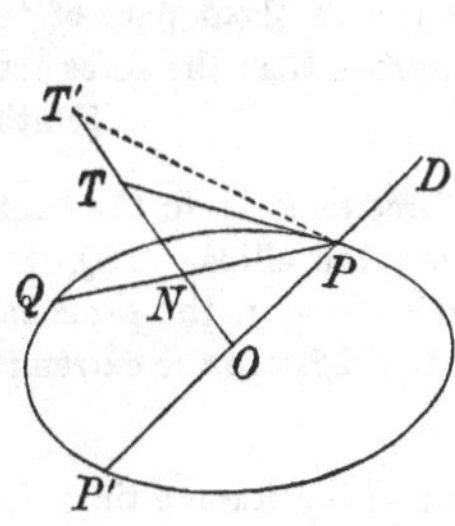

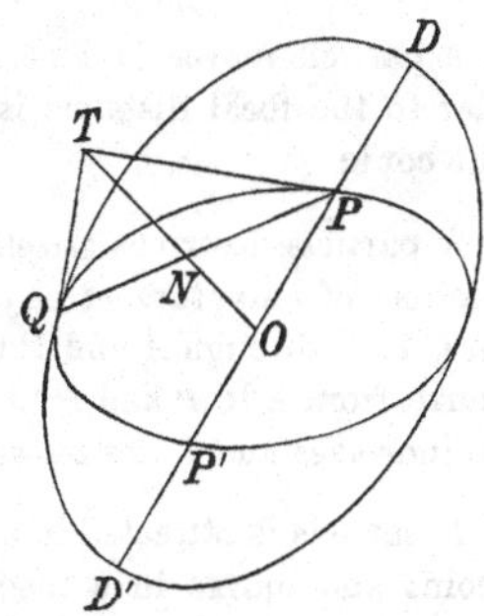

that acquired by a particle starting from rest at D and moving to P under the action of the centre of force. Let $OD = k$. Then since $V^2 = \mu(a^2 + b^2 - r_1^2)$, the sum of the squares of any two semi-conjugates of the trajectory is k^2.

Bisect PQ in N and let $ON = x$, $NP = NQ = y$. From the equation of the ellipse,

$$\frac{x^2}{a^2} + \frac{y^2}{k^2 - a^2} = 1;$$

$$\therefore\ a^4 - a^2(x^2 - y^2 + k^2) + k^2x^2 = 0 \qquad \text{(1)}.$$

Since x, y, k are given, this quadratic gives two values of a^2, showing that there are two directions of projection which satisfy the given conditions.

Let these directions of projection from P intersect ON produced in T and T', then since $a^2 = ON \,.\, OT$, the quadratic gives the positions of T and T'. We also have $OT \,.\, OT' = k^2$, and $NT \,.\, NT' = y^2$.

The roots of the quadratic (1) are imaginary if $x + y > k$. Produce PO to P' where $OP' = OP$, the roots of the quadratic are imaginary unless Q lie within the ellipse whose foci are P, P' and semi-major axis $a' = k$. *This ellipse is the boundary of all the positions of Q which can be reached by a particle projected from P with the given velocity.* It is also the envelope of all the trajectories.

Ex. 1. If two circles be described having their centres at O and N and their radii equal to k and y respectively, prove (1) that their radical axis will intersect ON produced in the middle point R of TT'; (2) that RT^2 is equal to the product of the segments of any chord drawn from R to either circle.

Ex. 2. Show that the greatest range $r=PQ$ on any straight line PQ making a given angle θ with $OP=r_1$ is determined by $(k^2-r_1^2)/r=k-r_1\cos\theta$.

Show also that in this case $OT=k$, and $NT=NP=NQ$. Thence deduce that the common tangent at Q to the trajectory and the envelope intersects the direction of projection from P at right angles in a point T which lies on the circle whose centre is O and radius k.

The first part follows from the focal polar equation of the ellipse and the second from known geometrical properties of the ellipse.

331. Examples. *Ex.* 1. If the sun were broken up into an indefinite number of fragments, uniformly filling the sphere of which the earth's orbit is a great circle, prove that each would revolve in a year. [Coll. Ex.]

The attractions of a homogeneous solid sphere on the particles composing it are proportional to their distances from the centre.

Ex. 2. A particle moves in a conic so that the resolved part of the velocity perpendicular to the focal distance is constant, prove that the force tends to the centre of the conic. [Math. Tripos.]

Ex. 3. A particle describes an ellipse, the force tending to the centre; prove that if the circle of curvature at any point P cut the ellipse in Q, the times of transit from Q to P through A and P to Q through B are in the same ratio as the times of transit from A to P and P to B, where A and B are the extremities of the major and minor axes and P lies between A and B.

Ex. 4. A particle is attracted to a fixed point with a force μ times its distance from the point and moves in a medium in which the resistance is k times the velocity; prove that, if the particle is projected with velocity v at a distance a from the fixed point, the equation of the path when referred to axes along the initial radius and parallel to the direction of projection is

$$k\tan^{-1} 2any/(2vx-aky)+n\log(x^2/a^2+\mu y^2/v^2-kxy/av)=0,$$

where $n^2=\mu-k^2/4$. [Coll. Ex. 1887.]

Ex. 5. Three centres of force of equal intensity are situated one at each corner of a triangle ABC and attract according to the direct distance. A particle moving under their combined influence describes an ellipse which touches the sides of the triangle ABC. Prove that the points of contact are the middle points of the sides, and that the velocities at these points are proportional to the sides.

[Math. Tripos, 1893.]

Ex. 6. If any number of particles be moving in an ellipse about a force in the centre, and the force suddenly cease to act, show that after the lapse of $(1/2\pi)$th part of the period of a complete revolution all the particles will be in a similar concentric and similarly situated ellipse. [Math. Tripos, 1850.]

Ex. 7. A particle moves in an ellipse under a centre of force in the centre. When the particle arrives at the extremity of the major axis the force ceases to act until the particle has moved through a distance equal to the semi-minor axis; it then acts for a quarter of the periodic time in the ellipse. Prove that if it again ceases to act for the same time as before, the particle will have arrived at the other end of the major axis. [Art. 325.] [Math. Tripos, 1860.]

Ex. 8. An elastic string passes through a smooth straight tube whose length is the natural length of the string. It is then pulled out equally at both ends until its length is increased by $\sqrt{2}$ times its original length. Two equal perfectly elastic balls are attached to the extremities and projected with equal velocities at right angles to the string, and so as to impinge on each other. Prove that the time of impact is independent of the velocity of projection, and that after impact each ball will move in a straight line, assuming that the tension of the string is proportional to the extension throughout the motion. [Math. Tripos, 1860.]

Ex. 9. A point is moving in an equiangular spiral, its acceleration always tending to the pole S; when it arrives at a point P the law of acceleration is changed to that of the direct distance, the actual acceleration being unaltered. Prove that the point P will now move in an ellipse whose axes make equal angles with SP and the tangent to the spiral at P, and that the ratio of these axes is $\tan \frac{1}{2}\alpha : 1$ where α is the angle of the spiral.

Ex. 10. A series of particles which attract one another with forces varying directly as the masses and distance are under the attraction of a fixed centre of force also varying directly as the distance; prove that if they are projected in parallel directions from points lying on a radius vector passing through the centre of force with velocities inversely proportional to their distances from the centre of force, they will at any subsequent time lie on a hyperbola. [Math. Tripos, 1888.]

Ex. 11. A particle starting from rest at a point A moves under the action of a centre of force situated at S whose magnitude is equal to μ. (distance from S). It arrives at A after an interval T and the centre of force is then suddenly transferred to some other point S' without altering its magnitude. If the particle be at a point B at the termination of a second interval T equal to the former, prove that the straight lines SS' and AB bisect each other. If at this instant the centre of force be suddenly transferred back to its original position S, prove that at the end of a third interval T the particle will be at S'. If at that instant the centre of force ceased to act, the particle will describe a path which passes through its original position A.

Ex. 12. If the central force is attractive and proportional to $u^2/(cu+\cos\theta)^3$, prove that the orbit is one of the conics given by the equation

$$(cu+\cos\theta)^2 = a + b\cos 2(\theta+\alpha). \qquad \text{[Coll. Ex. 1896.]}$$

Putting $cu+\cos\theta = U$, the differential equation of the path becomes the same as that for a central force varying as the distance $1/U$. The solution is therefore known to be the form given above.

Ex. 13. A particle moves under a central force $F = \mu u^2(1+k^2\sin^2\theta)^{-\frac{3}{2}}$. Find the orbit and interpret the result geometrically. [Math. Tripos.]

Ex. 14. A smooth horizontal plane revolves with angular velocity ω about a vertical axis to a point of which is attached the end of a weightless string, extensible according to Hooke's law and of natural length d just sufficient to reach the plane. The string is stretched and after passing through a small ring at the point where the axis meets the plane is attached to a particle of mass m which moves on the plane. Show that, if the mass be initially at rest relative to the plane, it will describe on the plane a hypocycloid generated by the rolling of a circle of radius $\frac{1}{2}a\{1-\omega(md\lambda^{-1})^{\frac{1}{2}}\}$ on a circle of radius a, where a is the initial extension and λ the coefficient of elasticity of the string.

[Math. Tripos, 1887.]

The accelerating tension is $\lambda r/md = \mu r$ (say). The path in space is therefore an ellipse having a and $b = \omega a/\sqrt{\mu}$ for semi-axes. To find the path relative to the rotating plane we apply to the particle a velocity ωr transverse to r backwards. If p' be the perpendicular from the centre on the resultant of v and ωr, we have by taking moments about the centre

$$(v^2 - 2v\omega p + \omega^2 r^2)\, p'^2 = (vp - \omega r^2)^2.$$

Substituting for v^2 and vp their values in elliptic motion we find

$$b^2 (a^2 - r^2) = p'^2 (a^2 - b^2).$$

This is a linear relation between r^2 and p'^2 and the curve will be an epicycloid if the radii of the corresponding circles are real (Art. 322). To find the radius of the fixed circle, we put $p' = 0$; this gives the radius $r = a$. To find the radius c of the rolling circle, we put $p' = r$, and $r = a + 2c$; this gives the required value of c. If c is negative the curve is a hypocycloid.

Law of the inverse square of the distance.

332. *A particle is acted on by a centre of force situated in the origin whose acceleration is $F = \mu u^2$ where u is the reciprocal of the radius vector. It is required to find the possible orbits.*

We have the differential equation (Art. 309)

$$\frac{d^2u}{d\theta^2} + u = \frac{F}{h^2u^2} = \frac{\mu}{h^2} \quad \text{.....................(1)};$$

$$\therefore\ u = \frac{\mu}{h^2} + A \cos(\theta - \alpha),$$

where A and α are the constants of integration. Comparing this with the equation of a conic

$$lu = 1 + e \cos(\theta - \alpha) \quad \text{.....................(2)},$$

where l is the semi-latus rectum, we see that *the orbit is a conic having one focus at the centre of force. We also have $h^2 = \mu l$.*

Conversely, if the orbit is a conic with the centre of force in one focus, the law of force must be the inverse square. To prove this, we let (2) be the given equation of the orbit; substituting in the left-hand side of equation (1) we find $F = \mu u^2$, where μ has been written for the constant h^2/l.

333. The velocity. The relations between the conic and the force are more easily deduced from the equation

$$F = -\tfrac{1}{2}h^2 \frac{d}{dr}\frac{1}{p^2} = \frac{\mu}{r^2},$$

the force being attractive in the standard case,

$$\therefore \frac{h^2}{p^2} = \frac{2\mu}{r} + C,$$

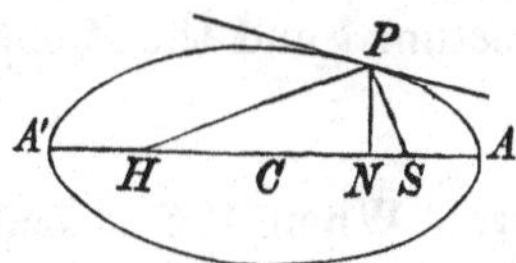

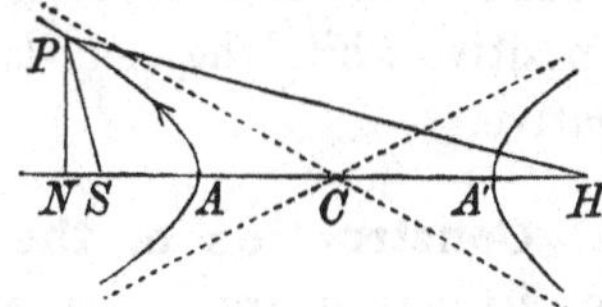

where C is the constant of integration. The p and r equation of an ellipse having a focus S at the origin is

$$\frac{l}{p^2} = \frac{2}{r} - \frac{1}{a},$$

where $l = b^2/a$ is the semi-latus rectum. Comparing these equations, we have the standard formulæ

$$h^2 = \mu l, \quad C = -\frac{\mu}{a}, \quad \therefore v^2 = \mu\left(\frac{2}{r} - \frac{1}{a}\right) \text{............(A)}.$$

We change from the ellipse to the hyperbola by making the centre C pass through infinity to the other side of the origin S, we therefore put $-a'$ for a; also b^2 becomes $-b'^2$, the semi-latus rectum remaining positive and equal to b'^2/a'. We now have

$$h^2 = \mu l, \quad C = \frac{\mu}{a'}, \quad \therefore v^2 = \mu\left(\frac{2}{r} + \frac{1}{a'}\right) \text{..............(B)}.$$

In passing from that branch of the hyperbola which is concave to the centre of force to the convex branch, the radius vector r changes sign through infinity from positive to negative. Before comparing the equation of the orbit with that of the hyperbola we should write $-r'$ for r in the latter. Also since this branch is convex to the origin the force is repulsive and μ is negative, let us put $\mu = -\mu'$. Comparing the formulæ

$$v^2 = \frac{h^2}{p^2} = -\frac{2\mu'}{r'} + C, \quad \frac{l}{p^2} = -\frac{2}{r'} + \frac{1}{a'},$$

we have

$$h^2 = \mu' l, \quad C = \frac{\mu'}{a'}, \quad \therefore v^2 = \mu'\left(-\frac{2}{r'} + \frac{1}{a'}\right) \text{............(C)}.$$

In the parabola, a is infinite, and

$$h^2 = \mu l, \quad C = 0, \quad v^2 = \mu\frac{2}{r} \text{..................(D)}.$$

All these formulæ may be included in the standard form (A) of the ellipse if we understand that on the concave branch of the hyperbola the major axis is by interpretation negative; on the convex branch, the radius vector being made positive, the major axis is positive while the semi-latus rectum l and the strength μ are negative.

334. Construction of the orbit. When the velocity V and the distance R are known at any point P of the orbit (say, the initial position), we may determine the curve in the following manner. *Let the force be attractive.* The orbit is now concave to the centre of force and μ is positive. Comparing the formulæ (A), (B) and (D) and remembering that the velocity V_1 from infinity to the initial position is given by $V_1^2 = 2\mu/R_1$ (Art. 312), we see that *the orbit is an ellipse, parabola or the concave branch of a hyperbola according as the velocity is less than, equal to, or greater than that from infinity.* We notice that this criterion is independent of the angle of projection at P. *Let the force be repulsive.* Since the path is convex to the centre of force *the orbit is the convex branch of a hyperbola.*

335. Having ascertained the nature of the orbit we have next to determine the lengths of the major axis and latus rectum. Supposing the ellipse to be the standard case, we have by (A), $\frac{1}{a} = \frac{2}{R} - \frac{V^2}{\mu}$. We notice that the length a is independent of the angle of projection. *If then particles are projected from the same point with equal velocities the major axes of the orbits described are equal.*

If β be the angle of projection (Art. 313) we have $p = R\sin\beta$ and $h = Vp$. The constant h and the semi-latus rectum l are therefore found from $h = VR\sin\beta$, $h^2 = \mu l$.

336. *The position in space of the major axis may be found in various ways.* Let S be the focus occupied by the centre of force and A the extremity of the major axis nearest to S.

We may find θ from the analytical equation of the curve

$$l/r = 1 + e\cos\theta,$$

where θ is the angle the initial radius vector SP makes with SA.

We may also use a geometrical construction. The focus S and the tangent PT at P being known, we can draw a straight line PH so that SP, PH make equal angles with PT, the direction of PH depending on whether the curve is an ellipse or hyperbola. If the point H is then determined so that $SP+PH=2a$, where a has been already found, it is clear that H is the empty focus. If the curve is a hyperbola, these lengths (as already explained) must have their proper signs. The position of the major axis is then found by joining S and H, and a being known the eccentricity e is equal to $SH/2a$.

337. *Ex.* 1. The initial distance of a particle from the centre of force being r, and the initial radial and transverse velocities being V_1 and V_2, prove that the latus rectum $2l$ and the angle θ which the radius vector r makes with the major axis are given by $\dfrac{l}{r^2}=\dfrac{V_2^2}{\mu}$, $\tan\theta=\dfrac{V_1V_2}{V_2^2-\mu/r}$.

Ex. 2. Prove that there are two directions in which a particle can be projected from a given point P with a given velocity V, so that the line of apses may have a given direction Sx in space, and find a geometrical construction for these directions.

Since V is given, a is known. With centre P and radius $2a-r$ describe a circle cutting Sx in H, H'. The required directions bisect externally the angles SPH, SPH'.

Let β be either of the angles the direction of projection at P makes with SP, Art. 313. The quadratic giving the two values of $\tan\beta$ is

$$\cot^2\beta+\left(2-\frac{r}{a}\right)\cot\theta\cot\beta+\frac{r}{a}-1=0,$$

where θ is the angle PSx. This follows from Ex. 1 by writing $V_1=V\cos\beta$, $V_2=V\sin\beta$. The quadratic may also be written in the form

$$\tan(\theta+\beta)=\left(\frac{r}{a}-1\right)\tan\beta.$$

Ex. 3. Three focal radii SP, SQ, SR of an elliptic orbit and the angles between them are given. Show that the ellipticity may be found from the equation $b\Delta=a\Delta'$, where Δ is the area PQR, Δ' the area of a triangle whose sides are $2SQ^{\frac{1}{2}}\,.\,SR^{\frac{1}{2}}\sin\frac{1}{2}QSR$ and two similar expressions. [Math. Tripos, 1893.]

Let P', Q', R' be the points on the auxiliary circle which correspond to P, Q, R. We first find by elementary conics the length of the side $Q'R'$ in terms of SQ, SR and the contained angle. The result shows that the side $Q'R'$ is equal to the corresponding side of the triangle Δ' after multiplication by a/b. Since the areas of the triangles PQR, $P'Q'R'$ are known to be in the ratio b/a, the result follows at once.

Ex. 4. Two particles P, Q describe the same orbit about a centre of force O. Prove that throughout the motion the area contained by the radii vectores OP, OQ is constant.

Thence deduce that if a ring of meteors (not attracting each other) describe a closed orbit, the angular distance between consecutive meteors varies inversely as the square of their distance from O.

Ex. 5. Two particles P, Q describe adjacent elliptic orbits of small eccentricity in equal times, the centre of force being in the focus and the major axes coincident in direction. Supposing the particles to be simultaneously at corresponding apses, prove that the angle ψ which PQ makes with the line of apses is given by $\cot\psi = -3 \operatorname{cosec} 2nt + \cot 2nt$, and find when ψ is a maximum.

338. **Elements of an orbit.** To fix the position in space of an elliptic orbit described about a focus we must know the values of *six constants*, called the elements of the orbit.

These are (1) the angle which the radius vector from the given focus to the nearer extremity of the major axis makes with some determinate line in the plane of the orbit, the angle being measured in the positive direction; (2) the length of the major axis; (3) the eccentricity; (4) a constant usually called the epoch to fix the longitude of the particle at the time $t=0$. This constant will be considered later on.

To determine the plane of the orbit we require two more constants. Taking the focus as origin, let some rectangular axes be given in position. Let the plane of the orbit intersect the plane of xy in the straight line $N'SN$. This line is called the line of nodes, and that node at which the particle passes to the positive side of the plane of xy is called the ascending node. We require (5) the angle the radius vector to the ascending node makes with the axis of x, and (6) the inclination of the plane of the orbit to the plane of xy.

339. **Point to Point.** *To project a particle with a given velocity V from a given point P so that it shall pass through another given point Q.*

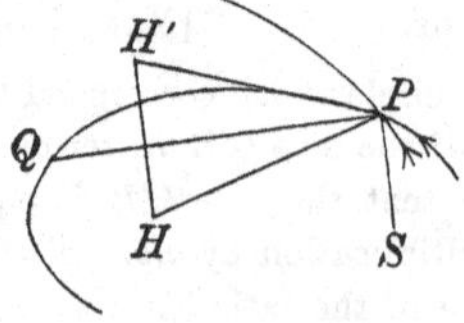

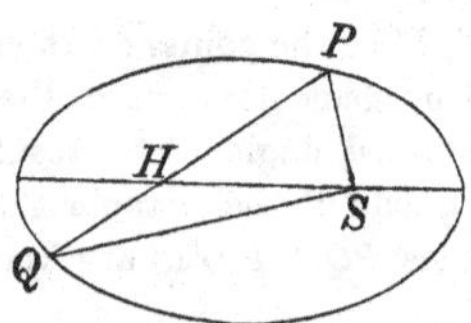

Let r_1, r_2 be the distances SP, SQ. The velocity at P being given, the major axis $2a$ is also known from the formula $V^2 = \mu\left(\frac{2}{r_1} - \frac{1}{a}\right)$.

With centres P and Q, describe two circles of radii $2a-r_1$, $2a-r_2$; these intersect in two points H, H'. Either of these may be the empty focus. *The three sides of the equal triangles PQH, PQH′ are therefore known.*

There are two directions of projection which satisfy the given conditions. These directions are the bisectors of the supplements of the angles SPH, SPH'. Let β, β' be the angles of projection at P (measured behind the radius vector SP, see Art. 313), then $\beta+\beta'$ is equal to the supplement of SPQ, and $\beta-\beta'$ is equal to the known angle HPQ.

The range PQ on a given straight line is the greatest possible when H, H' coincide and lie on the straight line PQ. We then have

$$PQ = PH + QH = 4a - r_1 - r_2.$$

This equation requires that the semi-major axis should be one-quarter of the perimeter of the triangle SPQ.

Since two consecutive trajectories whose foci are in the neighbourhood of PQ intersect in Q, the locus of Q as the range PQ turns round P is *the envelope of all trajectories from a given point P with a given velocity. Since* $PQ+QS=4a-r_1$ *this locus is another ellipse having its foci at P and S.* Each trajectory touches the enveloping ellipse in the point where the straight line joining P to the empty focus of the trajectory cuts either curve.

340. *Ex.* 1. Prove that the semi-major axis a', the eccentricity e' and the semi-latus rectum l' of the enveloping ellipse are given by

$$2a' = 4a - r_1, \qquad e' = \frac{r_1}{4a - r_1}, \qquad l'^2 = 2a(2a - r_1).$$

Ex. 2. If the variation of gravity is taken account of and the resistance of the air neglected, prove that the least velocity with which a shot could be projected from the pole so as to meet the earth's surface at the equator is about $4\frac{1}{2}$ miles per second, and that the angle of elevation is $22\frac{1}{2}°$. [Coll. Ex. 1892.]

Ex. 3. If a particle when projected from P_1 passes through two other points P_2, P_3, prove that the semi-latus rectum l is given by either of the equalities

$$l\Delta = r_1A_1 + r_2A_2 + r_3A_3 = 2r_1r_2r_3 \sin\alpha_1 \sin\alpha_2 \sin\alpha_3,$$

where r_1, r_2, r_3, are the distances SP_1, SP_2, SP_3; A_1, A_2, A_3 are the areas of the triangles P_2SP_3, P_3SP_1, P_1SP_2; α_1, α_2, α_3 the angles at the focus S and Δ is the area of the triangle $P_1P_2P_3$. Prove also that the eccentricity is given by

$$e^2(\Sigma A)^2 = \Sigma(A \,.\, \sec\alpha)^2 - 2\Sigma(A_1A_2 \sec\alpha_1 \sec\alpha_2 \cos\alpha_3).$$

341. Time of describing any arc. The time of describing the whole ellipse, usually called the periodic time, can be deduced at once from the formula $A = \frac{1}{2}ht$, (Art. 306). Putting $A = \pi ab$ and $h^2 = \mu b^2/a$, (Art. 332), we find that *the periodic time* $= \frac{2\pi}{\sqrt{\mu}} a^{\frac{3}{2}}$.

It appears from this that the period is independent of the minor axis and depends only on the strength μ of the centre of force and on the length of the major axis.

If n be the mean angular velocity in the orbit, the mean being taken with regard to time, the period is $2\pi/n$. It follows that $n^2 = \frac{\mu}{a^3}$.

342. *To find the time of describing any arc AP of an elliptic orbit.*

Let S be the focus occupied by the centre of force, AQA' the auxiliary circle and QPN an ordinate. If A is the extremity of the major axis nearest to S, the angle ASP is called *the true anomaly* and is sometimes represented by the letter v, i.e. the angle $ASP = v$. The angle ACQ is the eccentric angle of P and in astronomy is called *the eccentric anomaly*; it is usually represented by u, i.e. the angle $ACQ = u$. Thus the true anomaly v is measured at the centre of force, the eccentric anomaly u at the centre of the orbit.

When the particle is a planet the extremities A, A' of the major axis are called *the perihelion* and *aphelion*; when the particle is the moon the same points are called perigee and apogee. They are also called *the apses*, Art. 314.

Representing the time of describing the arc AP by t, and the mean angular velocity of the particle by n, the product nt is called *the mean anomaly*, and is generally represented by m, i.e. $m = nt$. To represent this angle geometrically we let a second particle describe a circle, having its centre at S, with a uniform motion in the same period as the given particle describes the ellipse. The actual angular velocity of this particle is therefore n. If A and Q' are its positions at the times $t = 0$ and $t = t$, the angle $ASQ' = nt$.

The true and mean anomalies are the important angles in the theory of elliptic motion. The eccentric anomaly is introduced as an auxiliary angle because, by its help, very simple expressions can be found for the other two anomalies and for the radius vector.

The difference between the true and the mean anomaly, or $v - m$, is called *the equation of the centre*, and is positive from the nearer apse to the farther and negative from the farther to the nearer.

Using the geometrical theorem that the ratio of the area ASP of the ellipse to the corresponding area ASQ of the circle is constant for all positions of P and equal to b/a, we have, if A = area ASP,

$$A = \frac{b}{a}(\text{area } ACQ - \text{area } SCQ)$$

$$= \frac{1}{2}\frac{b}{a}(a^2u - a^2e \sin u).$$

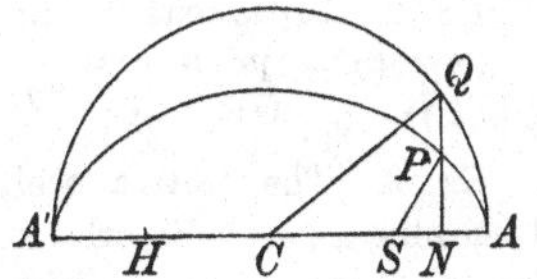

Since $A = \frac{1}{2}ht$, $h^2 = \mu b^2/a$, $n^2 = \mu/a^3$, this gives

$$nt = u - e \sin u \quad \text{......................(A)}.$$

We may obtain this relation between u and t without using any figure. Taking the focus S for origin, we have

$$x' = -ae + a \cos u, \qquad y' = b \sin u,$$

$$hdt = 2dA = x'dy' - y'dx'.$$

Substituting for x' and y' we obtain t in terms of u by an easy integration.

343. To find the relation between the true and eccentric anomalies we notice that $CS = ae$, $CN = x$, $SP = r = a - ex$.

$$\therefore 1 - \cos v = 1 + \frac{ae - x}{a - ex} = \frac{(1+e)(a-x)}{r},$$

$$1 + \cos v = 1 - \frac{ae - x}{a - ex} = \frac{(1-e)(a+x)}{r}.$$

Remembering that $x = a \cos u$, these give at once

$$\sqrt{\frac{r}{a}} \sin\frac{v}{2} = \sqrt{(1+e)} \sin\frac{u}{2}, \quad \sqrt{\frac{r}{a}} \cos\frac{v}{2} = \sqrt{(1-e)} \cos\frac{u}{2};$$

$$\therefore \tan\frac{v}{2} = \sqrt{\frac{1+e}{1-e}} \tan\frac{u}{2} \quad \text{..................(B)}.$$

Eliminating u between (A) and (B) we have

$$nt = 2\tan^{-1}\left\{\sqrt{\frac{1-e}{1+e}}\tan\frac{v}{2}\right\} - e\sqrt{(1-e^2)}\frac{\sin v}{1 + e\cos v}.$$

The expression for the time in terms of the longitude θ may also be found by integration. Since $r^2d\theta/dt = h$, we have $t = \frac{l^2}{h}\int\frac{f^2d\theta}{(f+\cos\theta)^2}$, where $f = 1/e$. But it is known that $\int \frac{d\theta}{f+\cos\theta} = \frac{2}{\sqrt{(f^2-1)}} \tan^{-1}\left(\sqrt{\frac{f-1}{f+1}} \tan\frac{\theta}{2}\right)$. By differentiating this with regard to f, the value of t follows at once.

Ex. Prove that $r\frac{dv}{du} = \sqrt{(al)}$, and $r\frac{du}{dt} = an$.

344. *Ex.* 1. Prove that the mean distance of a planet from the sun is a or $a(1+\frac{1}{2}e^2)$ according as the mean is taken with reference to the longitude or the time. [These means are respectively $\int r d\theta/2\pi$ and $\int r dt/T$, where T is the periodic time.]

Ex. 2. Prove that the mean value of r^n with regard to time for a planet is $\frac{a^n}{L(n+1)} \frac{(f^2-1)^{n+3/2}}{(-f)^{n+1}} \frac{d^{n+1}}{df^{n+1}} (f^2-1)^{-1/2}$, where $f=1/e$ and $L(n) = 1 . 2 . 3 \ldots n$.

Ex. 3. The earth's orbit being regarded as a circle, prove that a comet, describing a parabolic orbit in the same plane, cannot remain within the circumference of the earth's orbit longer than the $(2/3\pi)$th part of a year. [Coll. Ex.]

Ex. 4. A particle is projected from the earth's surface so as to describe a portion of an ellipse whose major axis is $1\frac{1}{2}$ times the earth's radius. If the direction of projection make an angle of 30° with the vertical, prove that the time of flight is $\frac{3}{4}(3a/g)^{\frac{1}{2}}\{\tan^{-1}\sqrt{6} + \sqrt{\frac{2}{3}}\}$ where a is the earth's radius. [Coll. Ex. 1895.]

345. Orbits of small eccentricity. The equations (A) and (B) of Arts. 342, 343 determine the time of describing any given angle v in an elliptic orbit of any eccentricity, the equation (B) giving u when v is known while the equation (A) then determines t. The converse problem of finding the polar coordinates r and v when t is given is usually called *Kepler's problem*. One solution by which u and v are expressed in terms of t by series arranged in ascending powers of e will be presently considered. It is enough here to notice that in a planetary orbit, where e is small, the value of u when t is given can be found by successive approximation. The value of v then follows from (B) by using the trigonometrical tables.

346. To solve $\phi(u) = u - e\sin u - m = 0$ by Newton's rule, when m, i.e. nt, is given.

Supposing u_1, u_2 to be two successive approximations to the value of u, that rule gives

$$u_2 - u_1 = -\frac{\phi(u_1)}{\phi'(u_1)} = \frac{m - m_1}{1 - e\cos u_1},$$

where $m_1 = u_1 - e\sin u_1$. To find a first approximation we notice that u lies between m and $m \pm e$, the upper or lower sign being taken according as m is $<\pi$ or $>$. We choose some value of u, lying between these limits, which is an integer number of minutes so that its trigonometrical functions can be found from the tables without interpolation. By Fourier's addition to Newton's rule this first approximation should be such that $\phi(u)$ and $\phi''(u)$ have the same sign.

Substituting this first approximation for u_1, the formula gives a second approximation. Substituting again this second approximation for u_1, we obtain a third, and so on. When e is very small the first computed value of the denominator is sometimes sufficiently accurate for all the approximations required. See Encke, *Berliner Astronomisches Jahrbuch*, 1838. Gauss, *Theoria Motus* &c., translated by C. H. Davis. Adams's *Collected Works*, vol. I. p. 289.

Ex. Prove that if we choose $u_1=m+e$ as the first approximation, the error of the value of u_2 is of the order e^3.

347. *Ex.* 1. *Leverrier's rule.* If terms of the order e^4 can be neglected, prove

$$u=m+\frac{e\sin m}{1-e\cos m}-\frac{1}{2}\left(\frac{e\sin m}{1-e\cos m}\right)^3.$$

Glaisher remarks that if we replace the third term by $-\frac{1}{2}(e\sin m)^3(1-e\cos m)^{-\frac{10}{3}}$ the formula is correct when terms of the order e^5 are neglected. He also gives a series for u correct up to e^8. *Monthly Notices of the Astronomical Society*, 1877.

Ex. 2. Prove that $\cot u=\cot m-\dfrac{e\operatorname{cosec} m}{f(u-m)}$ where

$$f(x)=\frac{x}{\sin x}=1+\frac{1}{6}\sin^2 x+\frac{3}{40}\sin^4 x+\frac{15}{336}\sin^6 x+\&c.$$

Putting $u=m+e$ on the right-hand side of the first equation we obtain an approximation for $\cot u$ whose error is of the order e^3. This is Zenger's solution of Kepler's problem. He has tabulated the values of $f(e)$ for the eight principal planets. Some improvements of the method have been suggested by J. C. Adams. Both papers are to be found in the *Monthly Notices of the Astronomical Society*, 1882, vol. XLII. p. 446, vol. XLIII. p. 47.

Ex. 3. Prove the following graphical solution of Kepler's problem. Construct the curve of sines $y=\sin x$, measure a distance $OM=m$ along the axis of x and draw MP making the angle PMx equal to $\cot^{-1} e$. If MP cut the curve in P, the abscissa of P is the value of u.

This method was described by J. C. Adams at the meeting of the B. Association in 1849. It is also given by See in the *Astronomical Notices*, 1895, who also refers to Klinkerfues and Dubois. Another graphical solution, using a trochoid, is given by Plummer, *Astronomical Notices*, 1895, 1896.

Ex. 4. The equation $u-e\sin u=m$ has only one real value of u when m is given.

This follows from the graphical construction. If the ordinate MP could cut the curve in a second point Q, move the straight line PQ parallel to itself until P and Q coincide. We should then have a tangent to the curve making an angle $\tan^{-1} 1/e$ with the axis of x. But if $e<1$ this is impossible, for in the curve of sines the greatest value of the angle is 45°.

Ex. 5. By using Lagrange's theorem we may expand $f(u)$ in a series of ascending powers of the eccentricity, the coefficients being functions of m. Prove that if the form of the function $f(u)$ be so chosen that the coefficient of e^2 is zero, we obtain the series

$$\cot u=\cot m-e\operatorname{cosec} m+\tfrac{1}{6}e^3\sin m+\&c.,$$

which takes a very simple form, when the cubes of e can be neglected. This equation is due to Rob. Bryant, *Astronomical Notices*, 1886.

Ex. 6. Prove that when e^4 can be neglected

$$\sin\tfrac{1}{2}(u-m)=\tfrac{1}{2}e\sin m+\tfrac{1}{4}e^2\sin 2m+\tfrac{5}{12}e^3\sin 3m+\&c. \quad \text{[R. Bryant.]}$$

Ex. 7. If θ' be the longitude of a planet seen from the empty focus and measured from an apse, prove that

$$\theta'=nt+\tfrac{1}{4}e^2\sin 2nt+\&c.,$$

the error being of the order e^4. It follows that the angular velocity round the empty focus is very nearly constant.

348. We may apply the method of Art. 342 to find the time of describing an arc of the concave branch of the hyperbola. Taking the focus as origin the equation of a hyperbola may be written

$$x' = ae - \frac{a}{2}(f^u + f^{-u}), \qquad y' = \frac{b}{2}(f^u - f^{-u}),$$

where u is an auxiliary quantity and f a constant which will be immediately chosen to be the base of the Napierian logarithms;

$$\therefore\ hdt = 2dA = x'dy' - y'dx' = ab\left\{\frac{e}{2}(f^u + f^{-u}) - 1\right\} du.$$

Since $h^2 = \mu b^2/a$ we have, putting $\mu/a^3 = n^2$,

$$nt = -u + e \sinh u \text{ .... (A).}$$

Again, as in Art. 343, we have $x = CN = \frac{a}{2}(f^u + f^{-u})$;

$$\therefore\ \cos v = \frac{ae - x}{ex - a}; \quad \therefore\ \tan\frac{v}{2} = \sqrt{\frac{e+1}{e-1}}\tanh\frac{u}{2} \text{ .... (B),}$$

where $v = \angle ASP$. If we eliminate u, we have

$$nt = \log\frac{\sqrt{(e+1)} - \sqrt{(e-1)}\tan\frac{1}{2}v}{\sqrt{(e+1)} + \sqrt{(e-1)}\tan\frac{1}{2}v} + e\sqrt{(e^2-1)}\frac{\sin v}{1 + e\cos v}.$$

To find a geometrical interpretation for the auxiliary quantity u, let us describe a rectangular hyperbola having the same major axis and produce the ordinate NP to cut the rectangular hyperbola in Q. Then $\tan QCN = \tanh u$.

Ex. A particle describes the convex branch of the hyperbola, and $\mu = -\mu'$ is negative. Prove

$$nt = u + e\sinh u, \qquad \tan\frac{v}{2} = \sqrt{\frac{e-1}{e+1}}\tanh\frac{u}{2},$$

where $v = ASP$, $\mu'/a^3 = n^2$.

349. The time in a parabolic orbit may be more easily found by using the equation $r^2 d\theta = hdt$.

Putting $l/r = 1 + \cos v$ where l is the semi-latus rectum, and $h^2 = \mu l$, we have

$$\sqrt{\frac{\mu}{l^3}}.t = \int\frac{dv}{(1+\cos v)^2} = \frac{1}{2}\int\left(1 + \tan^2\frac{v}{2}\right)d\tan\frac{v}{2}$$

$$= \frac{1}{2}\left(\tan\frac{v}{2} + \frac{1}{3}\tan^3\frac{v}{2}\right).$$

This formula gives the time t of describing the true anomaly $v = ASP$.

If c be the radius of the earth's orbit, and p the perihelion distance of the particle *expressed as a fraction of* c, we have $l = 2pc$. To eliminate μ, let $T = 2\pi\sqrt{c^3/\mu}$ be the length of a year. Then

$$\frac{\pi\sqrt{2}}{T}.t = p^{\frac{3}{2}}\left\{\tan\frac{v}{2} + \frac{1}{3}\tan^3\frac{v}{2}\right\}.$$

If we write $T = 365{\cdot}256$ this gives t in days.

When a formula like this has to be frequently used we construct a table to save the continual repetition of the same arithmetical work. Let the values of $\{\tan \frac{1}{2}v + \frac{1}{3}\tan^3 \frac{1}{2}v\}$ be calculated for values of v from 0 to 180°, with differences for interpolation. When p is known for any comet moving in a parabolic orbit, the table can be used with equal ease to find the time when the true anomaly is given or the true anomaly when the time is known.

350. Euler's theorem. *A particle describes a parabola under the action of a centre of force in the focus S. It is required to prove that the time of describing an arc PP′ is given by*

$$6\sqrt{\mu}t = (r + r' + k)^{\frac{3}{2}} - (r + r' - k)^{\frac{3}{2}},$$

where r, r' are the focal distances of P, P' and k is the chord joining P, P'.

Let x, y; x', y' be the coordinates of P, P', then since $y^2 = 4ax$,

$$k^2 = (x - x')^2 + (y - y')^2 = (y - y')^2 \left\{1 + \left(\frac{y + y'}{4a}\right)^2\right\}.$$

As we wish to make the right-hand side a perfect square, we put

$$y + y' = 4a \tan\theta, \quad y - y' = 4a \tan\phi \quad \text{............(1)}.$$

We shall suppose that in the standard case y is positive and y' numerically less than y; then θ and ϕ are positive,

$$\therefore k = 4a \tan\phi \sec\theta \quad \text{......................(2)}.$$

Also $\qquad r + r' = 2a + x + x' = 2a(\sec^2\theta + \tan^2\phi);$

$$\left.\begin{aligned} \therefore r + r' + k &= 2a(\sec\theta + \tan\phi)^2 \\ r + r' - k &= 2a(\sec\theta - \tan\phi)^2 \end{aligned}\right\};$$

$$\begin{aligned} \therefore (r + r' + k)^{\frac{3}{2}} - (r + r' - k)^{\frac{3}{2}} \\ &= (2a)^{\frac{3}{2}} \{(\sec\theta + \tan\phi)^3 - (\sec\theta - \tan\phi)^3\} \\ &= 2(2a)^{\frac{3}{2}} \{3 + 3\tan^2\theta + \tan^2\phi\} \tan\phi. \end{aligned}$$

Drawing the ordinates PN, $P'N'$, we see that

$$\begin{aligned} \text{area } PSP' &= APN - AP'N' + SP'N' - SPN \\ &= \tfrac{2}{3}(xy - x'y') + \tfrac{1}{2}(x' - a)y' - \tfrac{1}{2}(x - a)y \\ &= \frac{1}{24a}(y^3 - y'^3) + \frac{a}{2}(y - y') \\ &= \tfrac{2}{3}a^2 \tan\phi \{3\tan^2\theta + \tan^2\phi + 3\}. \end{aligned}$$

Since the area $PSP' = \frac{1}{2}ht = \frac{1}{2}\sqrt{(2a\mu)}\,t$ the result to be proved follows at once.

The arc PP' gradually increases as P' moves towards and past the apse. The quantity $r+r'-k$ decreases and vanishes when the chord passes through the focus. To determine whether the radical changes sign we notice that this can happen only when it vanishes. We can therefore without loss of generality so move the points P, P', that, when the chord crosses the focus, PP' is a double ordinate. We then have

$$6\sqrt{\mu}t=(2r+2y)^{\frac{3}{2}}-(2r-2y)^{\frac{3}{2}}=\{(2a+y)^3\pm(2a-y)^3\}/(2a)^{\frac{3}{2}}.$$

Comparing this with the ordinary parabolic expression for twice the area ASP it is evident that the last term should change sign where y increases past $2a$ and that the double sign should be a minus. *The second radical in Euler's equation must be taken positively when the angle PSP' is greater than* 180°.

351. *Ex.* 1. If the ordinate $P'N'$ cut the parabola again in Q'; prove that θ, ϕ are the acute angles made by the chords PP', PQ' with the axis of y.

Ex. 2. Show that there are two parabolas which can pass through the given points P, P', and have the same focus. Show also that in using Euler's theorem to find the time P to P', the second radical has opposite signs in the two paths.

To find the parabolas we describe two circles, centres P, P' and radii SP, SP'. These circles intersect in S and the two real common tangents are the directrices. These tangents intersect on PP' and make equal angles with it on opposite sides. The concavities of the parabolas are in opposite directions, and the angles described are PSP' and $360°-PSP'$. If then one angle is greater than 180°, the other must be less.

Ex. 3. A parabolic path is described about the focus. Show that the squares of the times of describing arcs cut off by focal chords are proportional to the cubes of the chords.

352. Lambert's Theorem*. *If t is the time of describing any arc $P'P$ of an ellipse, and k is the chord of the arc, then*

$$nt=(\phi-\sin\phi)-(\phi'-\sin\phi'),$$

where $$\sin\tfrac{1}{2}\phi=\tfrac{1}{2}\sqrt{\frac{r+r'+k}{a}},\quad \sin\tfrac{1}{2}\phi'=\tfrac{1}{2}\sqrt{\frac{r+r'-k}{a}}, \quad\text{.........(A).}$$

Let u, u' be the eccentric anomalies of P, P',

$$\therefore\ k^2=a^2(\cos u-\cos u')^2+a^2(1-e^2)(\sin u-\sin u')^2$$
$$=4a^2\sin^2\tfrac{1}{2}(u-u')\{1-e^2\cos^2\tfrac{1}{2}(u+u')\}\ \text{.....................}\ (1),$$

* This proof of Lambert's theorem is due to J. C. Adams, *British Association Report*, 1877, or *Collected Works*, p. 410. He also gives the corresponding theorem for the hyperbola, using hyperbolic sines. In the *Astronomical Notices*, vol. XXIX., 1869, Cayley gives a discussion of the signs of the angles ϕ, ϕ'. The theorem for the parabola was discovered by Euler (*Miscell. Berolin.* t. VII.), but the extension to the other conic sections is due to Lambert.

$$r+r'=2a-ae\cos u-ae\cos u'$$
$$=2a\{1-e\cos\tfrac{1}{2}(u+u')\cos\tfrac{1}{2}(u-u')\}\ \ldots\ldots\ldots\ldots\ldots\ldots\ldots\ldots\ (2),$$
$$nt=u-u'-e(\sin u-\sin u')$$
$$=u-u'-2e\cos\tfrac{1}{2}(u+u')\sin\tfrac{1}{2}(u-u')\ \ldots\ldots\ldots\ldots\ldots\ldots\ldots\ (3).$$

Hence we see that if a, and therefore also n, are given, then $r+r'$, k, and t are functions of the two quantities $u-u'$, and $e\cos\frac{1}{2}(u+u')$. Let

$$u-u'=2\alpha,\qquad e\cos\tfrac{1}{2}(u+u')=\cos\beta\ \ldots\ldots\ldots\ldots\ldots\ldots\ldots\ (4).$$
$$\therefore\ k=2a\sin\alpha\sin\beta\ \ldots\ldots\ldots\ldots\ldots\ldots\ldots\ldots\ldots\ldots\ldots\ (5),$$
$$r+r'+k=2a\{1-\cos(\beta+\alpha)\}\ \ldots\ldots\ldots\ldots\ldots\ldots\ldots\ldots\ (6),$$
$$r+r'-k=2a\{1-\cos(\beta-\alpha)\}\ \ldots\ldots\ldots\ldots\ldots\ldots\ldots\ldots\ (7),$$
$$nt=2\alpha-2\sin\alpha\cos\beta\ \ldots\ldots\ldots\ldots\ldots\ldots\ldots\ldots\ldots\ldots\ (8).$$

If we put $\beta+\alpha=\phi$, $\beta-\alpha=\phi'$, the equations (6) and (7) lead to the expressions for $\sin\frac{1}{2}\phi$, $\sin\frac{1}{2}\phi'$ given above, while (8) when put into the form

$$nt=\{\beta+\alpha-\sin(\beta+\alpha)\}-\{\beta-\alpha-\sin(\beta-\alpha)\}$$

gives at once the required value of nt.

353. Let us trace the values of ϕ, ϕ' as the point P travels round the ellipse in the positive direction beginning at a fixed point P'. We suppose that u increases from u' to $2\pi+u'$.

The positive sign has been given to the square root k. Since k can vanish only when P coincides with P', and α begins positively, we see that both α and β lie between 0 and π for all positions of P. The latter is also restricted to lie between $\cos^{-1}e$ and $\pi-\cos^{-1}e$.

We have by differentiating (4)

$$d\phi=d\beta+d\alpha=\ \ \tfrac{1}{2}du\{1+e\operatorname{cosec}\beta\sin\tfrac{1}{2}(u+u')\},$$
$$d\phi'=d\beta-d\alpha=-\tfrac{1}{2}du\{1-e\operatorname{cosec}\beta\sin\tfrac{1}{2}(u+u')\}.$$

Since $\sin^2\beta=e^2\sin^2\frac{1}{2}(u+u')+1-e^2$, and $e^2<1$, it follows that $d\phi$ is always positive and $d\phi'$ always negative. If β_0 be the least value of β which satisfies $\cos\beta=e\cos u'$, ϕ continually increases from β_0 to $2\pi-\beta_0$ and ϕ' decreases from β_0 to $-\beta_0$.

When $\phi=\pi$, $r+r'+k=4a$, and the chord $P'P$ passes through the empty focus H. Let it cut the ellipse in Q. It follows that ϕ is less or greater than π according as P lies in the arc $P'Q$ or QP'.

When $\phi'=0$, $r+r'-k=0$, and the chord $P'P$ passes through the centre of force S. Let it cut the ellipse in R. Then ϕ' is positive or negative according as P lies in the arc $P'R$ or RP'.

The values of ϕ, ϕ' are determined by the radicals (A). Each of these gives more than one value of the angle, thus ϕ may be greater or less than π and ϕ' may be positive or negative. This ambiguity disappears (as explained above) when the position of P' on the ellipse is known. Thus $\sin\phi$ and $\sin\phi'$ have the same sign when the two foci are on the same side of the chord PP' and opposite signs when the chord passes between the foci.

354. *Ex.* 1. Prove that the time t of describing an arc $P'P$ of a hyperbola is given by

$$t\sqrt{\frac{\mu}{a^3}}=-\phi+\phi'+\sinh\phi-\sinh\phi',$$

where $$\sinh\frac{\phi}{2}=\sqrt{\frac{r+r'+k}{4a}},\quad \sinh\frac{\phi'}{2}=\sqrt{\frac{r+r'-k}{4a}}.$$

and k is the chord of the arc. [Adams.]

Ex. 2. The length of the major axis being given, two ellipses can be drawn through the given points P, P' and having one focus at the centre of force. Prove that the times of describing these arcs, as given by Lambert's theorem, are in general unequal.

To find the ellipses we describe two circles with the centres at P, P' and the radii equal to $2a-SP$, and $2a-SP'$. These intersect in two points H, H', either of which may be the empty focus, and these lie on opposite sides of the chord PP'.

355. Two centres of force. *Ex.* 1. An ellipse is described under the action of two centres of force, one in each focus. If these forces are $F_1(r_1)$ and $F_2(r_2)$, prove that $\frac{1}{r_1^2}\frac{d}{dr_1}(r_1^2F_1)=\frac{1}{r_2^2}\frac{d}{dr_2}(r_2^2F_2)$. If one force follow the Newtonian law, prove that the other must do so also.

These results follow from the normal and tangential resolutions.

Ex. 2. A particle describes an elliptic orbit under the influence of two equal forces, one directed to each focus. Show that the force varies inversely as the product of the distances of the particle from the foci. [Coll. Ex.]

Ex. 3. A particle describes an ellipse under two forces tending to the foci, which are one to another at any point inversely as the focal distances; prove that the velocity varies as the perpendicular from the centre on the tangent, and that the periodic time is $\pi(a^2+b^2)/kab$, ka, kb being the velocities at the extremities of the axes. [Coll. Ex.]

Ex. 4. A particle describes an ellipse under the simultaneous action of two centres of force situated in the two foci and each varying as (distance)$^{-2}$. Prove that the relation between the time and the eccentric anomaly is

$$\left(\frac{du}{dt}\right)^2=\frac{\mu}{a^3}\frac{1}{(1-e\cos u)^2}+\frac{\mu'}{a^3}\frac{1}{(1+e\cos u)^2}.$$

[Cayley, *Math. Messenger*, 1871.]

The inverse cube and the inverse n^{th} powers of the distance.

356. The law of the inverse cube. *A particle projected in any given manner describes an orbit about a centre of force whose attraction varies as the inverse cube of the distance. It is required to find the motion*.*

* The orbits when the force $F=\mu u^3$ were first completely discussed by Cotes in the *Harmonia Mensurarum* (1722) and the curves have consequently been called Cotes' spirals. The motion for $F=\mu u^n$ when the velocity is equal to that from infinity is generally given in treatises on this subject. The paths for several other laws of force are considered by Legendre (*Théorie des Fonctions Elliptiques*, 1825), and by Stader (*Crelle*, 1852); see also Cayley's *Report to the British Association*, 1863. Some special paths when $F=\mu u^n$, for integer values of n from $n=4$ to $n=9$, are discussed by Greenhill (*Proceedings of the Mathematical Society*, 1888), one case when $n=5$, being given in Tait and Steele's *Dynamics*.

Let attraction be taken as the standard case and let the accelerating force be $F=\mu u^3$. We have

$$\frac{d^2u}{d\theta^2}+u=\frac{F}{h^2u^2}=\frac{\mu}{h^2}u,$$

$$\therefore \frac{d^2u}{d\theta^2}+\left(1-\frac{\mu}{h^2}\right)u=0.$$

The solution depends on the sign of the coefficient of u. Let V be the velocity of the particle at any point of its path (say the point of projection), β the angle and R the distance of projection, then $h=VR\sin\beta$; (Art. 313). Let V_1 be the velocity from infinity, then $V_1^2=\mu/R^2$. It follows that h^2 is $>$ or $<\mu$ according as $V\sin\beta$ is $>$ or $<V_1$; i.e. the coefficient of u is positive or negative according as the transverse velocity at any point is greater or less than the velocity from infinity. If the force is repulsive the coefficient is always positive.

Case 1. Let $h^2>\mu$, we put $1-\mu/h^2=n^2$, then $n<1$ or >1 according as the force is attractive or repulsive. The equation of the path is (Art. 119)

$$u=a\cos n(\theta-\alpha).$$

The curve consists of a series of branches tending to asymptotes, each of which makes an angle π/n with the next.

When the curve is given the motion may be deduced from the following relations (Art. 306),

$$h^2=\frac{\mu}{1-n^2},\qquad v^2=\mu\left(\frac{a^2n^2}{1-n^2}+u^2\right).$$

Also by integrating $d\theta/dt=hu^2$, and putting $a=1/b$, we find that the time of describing the angle $\theta=\alpha$ to θ, i.e. $r=b$ to r, is given by

$$\tan n(\theta-\alpha)=\frac{hnt}{b^2},\qquad r^2-b^2=\frac{h^2n^2t^2}{b^2}.$$

357. Case 2. Let μ be positive and $>h^2$, we put $1-\mu/h^2=-n^2$. The equation of the path is then $u=Ae^{n\theta}+Be^{-n\theta}$. The values of the constants A, B are to be deduced from the initial values of u and $du/d\theta$. Two cases therefore arise, according as A and B have the same or opposite signs. In the former case, u cannot vanish and therefore the orbit has no branches which go to infinity; in the latter case there is an asymptote. If we write $\theta=\theta_1+\alpha$ and choose α so that $Ae^{n\alpha}=\mp Be^{-n\alpha}$, we may reduce the equation to one of the three standard forms

$$u = \frac{a}{2}(e^{n\theta_1} \pm e^{-n\theta_1}), \quad u = Ae^{n\theta},$$

where $2n\alpha = \log(\pm B/A)$, $a = 2\sqrt{(\pm AB)}$, the upper or lower signs being taken according as A, B have the same or opposite signs.

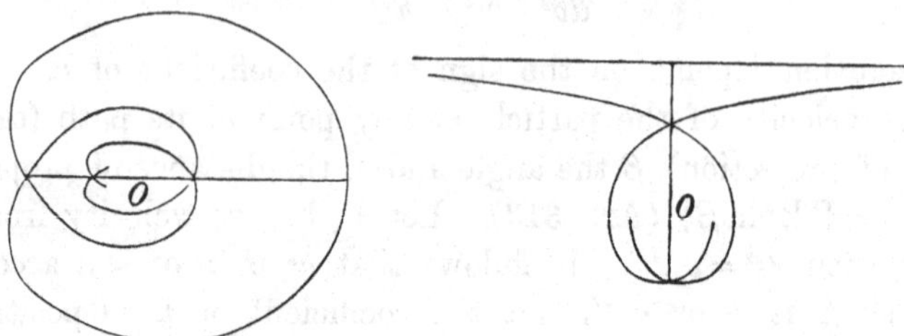

The third case occurs when $B = 0$; the orbit is then the equiangular spiral already considered in Art. 319.

When the curve is given the motion may be deduced from the following relations

$$h^2 = \frac{\mu}{1+n^2}, \quad v^2 = \mu\left(\mp\frac{a^2n^2}{1+n^2} + u^2\right), \quad b^2 \mp r^2 = \left(\frac{nht}{b} + C\right)^2,$$

where C is determined by making t vanish when r has its initial value and $b = 1/a$.

When A and B have the same sign the two branches beginning at the point $\theta_1 = 0$, i.e. $\theta = \alpha$, wind symmetrically round the origin in opposite directions. When A and B have opposite signs the two branches begin at opposite ends of an asymptote, whose distance from the origin is $y = 1/an$, and then wind round the origin. As the particle approaches the centre of force, the convolutions of either branch become more and more nearly those of an equiangular spiral whose angle is given by $\cot\phi = \pm n$, the upper or lower sign being taken according as $\theta = \pm\infty$. The particle arrives at the pole with an infinite velocity at the end of a finite time.

358. Case 3. Let μ be positive and $= h^2$. The orbit is

$$u = a(\theta - \alpha).$$

When the path is known the motion is given by

$$h^2 = \mu, \quad v^2 = \mu(u^2 + a^2), \quad t\sqrt{\mu} = br,$$

where t is the time from a distance r to the centre of force and $b = 1/a$. We notice that the radial velocity is constant.

Beginning at the opposite extremities of an asymptote the two branches wind round the origin and ultimately when $\theta = \pm\infty$ cut the radius vector at right angles. If OZ is drawn perpendicular to the radius vector OP to meet the tangent at P in Z, we may show that OZ is constant and equal to $1/a$.

359. *Ex.* The motion for a force $F=f(u)$ being known, show how to deduce that for a force $F=f(u)+\mu u^3$ and give a geometrical interpretation. [Newton.]

The differential equations are

$$\frac{d^2u}{d\theta^2}+\left(1-\frac{\mu}{h^2}\right)u=\frac{f(u)}{h^2}, \qquad \frac{d\theta}{dt}=hu^2.$$

These may be reduced to the forms used when $F=f(u)$ by writing $c\theta=\theta'$, $ch=h'$, where $c^2=1-\mu/h^2$.

To construct the path $u=\phi(c\theta)$, when $u=\phi(\theta)$ is known, we make the axis of x together with the latter curve revolve round the centre of force with an angular velocity $d\omega/dt$, where $c\theta=\theta-\omega$. The axis of x therefore advances or regredes according as c is less or greater than unity.

360. Law of the inverse nth power. *It is required to find the path of a particle when the central force $F=\mu u^n$.* See Art. 320. We have

$$\frac{d^2u}{d\theta^2}+u=\frac{F}{h^2u^2}=\frac{\mu}{h^2}u^{n-2};$$

$$\therefore v^2=h^2\left\{\left(\frac{du}{d\theta}\right)^2+u^2\right\}=\frac{2\mu}{n-1}u^{n-1}+C \quad\ldots\ldots\ldots\ldots\ldots\ldots\ (1),$$

except when $n=1$, for then the right-hand side takes a logarithmic form.

The integration of this equation can be reduced to elementary forms when $C=0$; this requires that $n>1$ for otherwise v^2 would be negative. The equation then shows that *at every point of the orbit the velocity is equal to that from infinity*, Art. 312.

If V be the velocity, R and β the distance and angle of projection, we have

$$V^2=\frac{2\mu}{n-1}\left(\frac{1}{R}\right)^{n-1}, \qquad h=VR\sin\beta\ \ldots\ldots\ldots\ldots\ldots\ldots\ (2).$$

Representing $\dfrac{2\mu}{h^2(n-1)}=\dfrac{R^{n-3}}{\sin^2\beta}$ by c^{n-3}, we have

$$\frac{du}{u\sqrt{\{(cu)^{n-3}-1\}}}=\mp d\theta\ldots\ldots\ldots\ldots\ldots\ldots\ldots\ldots(3),$$

where the upper or lower sign is to be taken according as $du/d\theta$ is initially negative or positive, i.e. according as the angle β is acute or obtuse.

To integrate this put $cu=x^\kappa$ where κ is to be chosen to suit our convenience. Taking the logarithmic differential we find $du/u=\kappa\,dx/x$, and the integral equation (3) becomes

$$\frac{\kappa\,dx}{x\sqrt{(x^{\kappa(n-3)}-1)}}=\mp d\theta.$$

We now see that if we put $\kappa(n-3)=-2$ the integration can be effected at once, but this supposition is impossible if $n=3$. We find

$$\kappa\cos^{-1}x=\pm(\theta-\alpha), \qquad \therefore \left(\frac{r}{c}\right)^{\frac{n-3}{2}}=\cos\frac{n-3}{2}(\theta-\alpha).$$

Conversely, when the path is given, we have

$$h^2=\frac{2\mu}{n-1}\frac{1}{c^{n-3}}, \qquad v^2=\frac{2\mu}{n-1}\frac{1}{r^{n-1}}.$$

It appears that the orbit takes different forms according as $n>$ or <3. In the former case the curve has a series of loops with the origin for the common node and $r=c$ for the maximum radius vector. In the latter case the curve has infinite branches, and $r=c$ for the minimum radius vector.

361. If the force is repulsive, we write $F=-\mu' u^n$. We then have

$$v^2=h^2\left\{\left(\frac{du}{d\theta}\right)^2+u^2\right\}=\frac{2\mu'}{1-n}r^{1-n}+C.$$

If $C=0$, we must have $n<1$. *The velocity at every point is equal to that from rest at the centre of force.* Proceeding as before, we have

$$\left(\frac{c}{r}\right)^{\frac{3-n}{2}}=\cos\frac{3-n}{2}(\theta-\alpha).$$

362. *Ex.* The law of attraction being $F=\mu u^n$, show that the time t of describing a loop is

$$t\sqrt{\frac{\mu}{2(n-1)c^{n+1}}}=\int\left(\cos\frac{n-3}{2}\theta\right)^{\frac{4}{n-3}}d\theta=\frac{\sqrt{\pi}}{n-3}\frac{\Gamma(p)}{\Gamma(q)},$$

where the limits are $\theta=0$ to $\pi/(n-3)$ and $2(n-3)p=n+1$, $(n-3)q=(n-1)$. The integrations can be effected when $n-3=\pm 4/i$ and $q-p=\pm i$ where i is any integer.

363. Examples. *Ex.* 1. Prove the following geometrical properties of the curve $(r/c)^m=\cos m\theta$ (Art. 320),

$$\phi=\frac{\pi}{2}+m\theta, \qquad p=\frac{r^{m+1}}{c^m}, \qquad \left(\frac{r'}{c}\right)^{\frac{m}{m+1}}=\cos\frac{m\theta'}{m+1},$$

where ϕ is the angle the radius vector makes with the tangent, and r', θ' are the coordinates of a point on the pedal curve.

Since equation (1) of Art. 360 becomes $p^2=\frac{(n-1)h^2}{2\mu}r^{n-1}$ when $C=0$, the second of these geometrical results enables us to write down the equation of the required path and thus to avoid the integration of (3).

Ex. 2. A perpendicular OY is drawn from the origin O on the tangent at P to the lemniscate $r^2=a^2\cos 2\theta$. If the locus of Y be described by a particle under the action of a central force tending to O, prove that this force varies inversely as $OY^{13/3}$. [Coll. Ex.]

Ex. 3. A particle is describing the curve $(r/c)^m=\cos m\theta$ under the action of the central force $F=\mu u^n$, where $m=\frac{1}{2}(n-3)$. Prove that, if the velocity at the point $\theta=a$ is suddenly increased in the ratio 1 to $1+\gamma$ where γ is very small, the subsequent path is

$$(r/c)^m=\cos m\theta\,\{1-m\xi\,(\cos m\theta)^{\frac{1}{m}}\},$$

$$\frac{(\cos m\theta)^{1+\frac{1}{m}}}{(\sin m\theta)}\,\xi=\gamma\,\frac{\cot m\theta}{m}+\frac{\gamma}{(\cos ma)^{\frac{2m+2}{m}}}\int\frac{(\cos m\theta)^{2+\frac{2}{m}}\,d\theta}{(\sin m\theta)^2},$$

where the limits are $\theta=0$ to a.

Substitute $r/c=(\cos m\theta)^{\frac{1}{m}}+\xi$, in the differential equation of the path, Art. 309, and neglect the squares of ξ.

364. The inverse fifth power. The equation (1), Art. 360, has the form

$$\left(\frac{du}{d\theta}\right)^2=\frac{\mu}{2h^2}u^4-u^2+\frac{C}{h^2}\ \dots\dots\dots\dots\dots\dots\dots\dots\ (1).$$

This can be reduced to elliptic integrals as explained in Cayley's *Elliptic Functions*, Art. 400, or Greenhill, *The Elliptic Functions*, Art. 70.

The integration can be effected in two cases: (1) when velocity of projection is equal to that from infinity, and (2) when the initial conditions are such that $h^4=2\mu C$. In the latter case the right-hand side of (1) is a perfect square.

Ex. 1. Prove that the integration when $h^4=2\mu C$ leads to the curves $\tanh(\theta/\sqrt{2})=r/c$ or c/r, which have a common asymptotic circle $r=c$ where $c=\sqrt{\mu}/h$. Prove also that the velocity V of projection is given (Art. 313) by

$$V^2\sin^4\beta=2V'^2\{1\pm\sqrt{(1-\sin^4\beta)}\},$$

where V' is the velocity from rest at infinity, and the upper or lower sign is to be taken according as the path is outside or inside the asymptotic circle.

Ex. 2. Prove that, if the central force $F=\mu u^5$, the inverse of any path with regard to the origin is another possible path provided the total energy of the motion exceed the potential energy at infinity by a positive constant E reckoned per unit mass and also that for the two paths $Eh'^4=E'h^4$.

Prove that when $h^4>4\mu E>0$ the path is of the form $r=a\,\mathrm{sn}\left(K-\dfrac{\theta}{\sqrt{(1+k^2)}}\right)$ modulus k or the inverse form. [Math. Tripos, 1894.]

According to the notation of Art. 313, $2E=C$.

365. The inverse fourth power. The equation (1) of Art. 360 is

$$\left(\frac{du}{d\theta}\right)^2=\frac{2\mu}{3h^2}\left(u^3-\frac{3h^2}{2\mu}u^2+\frac{3C}{2\mu}\right)\dots\dots\dots\dots\dots\dots\dots\ (1).$$

This cubic can always be written in the form

$$\left(\frac{du}{d\theta}\right)^2=\frac{2\mu}{3h^2}(u+a)(u^2+Au+B),$$

and the integration can be reduced to forms similar to those in Art. 364 by writing $u+a=\xi^2$.

The integration can be effected when the initial conditions are such that $h^6=3\mu^2C$. In this case the right-hand side has the factor $(u-h^2/\mu)^2$.

Ex. Show that the integration leads to the curves $u=\dfrac{h^2}{\mu}\dfrac{\cosh\theta\pm2}{\cosh\theta\mp1}$, the upper signs being taken together and the lower together. These curves have a common asymptotic circle $r=\mu/h^2$, one curve being within and the other outside.

366. Other powers. *Ex.* If the force $F=\mu u^7$, and the initial conditions are such that $2h^3=3C\sqrt{\mu}$, prove that the equation (1) of Art. 360 takes the form

$$\left(\frac{du}{d\theta}\right)^2=\frac{1}{3b^4}(u^2-b^2)^2(u^2+2b^2),$$

where $b^2=h/\sqrt{\mu}$. Thence deduce the integrals $\dfrac{u^2}{b^2}=\dfrac{\cosh 2\theta \mp 1}{\cosh 2\theta \pm 2}$, having a common asymptotic circle. The Lemniscate can also be described under this law of force, if the velocity is equal to that from infinity; Arts. 320, 360.

367. Nearly circular orbits. *To find the motion approximately, when the central force $F=\mu u^n$ and the orbit is nearly circular.*

Beginning as in Art. 360 with the equation

$$\frac{d^2u}{d\theta^2}+u=\frac{F}{h^2u^2}=\frac{\mu}{h^2}u^{n-2}\text{ ....(1)},$$

we put $u=c(1+x)$ where c is some constant to be presently chosen but subject to the condition that x is to be a small fraction. We thus find

$$\frac{d^2x}{d\theta^2}+x=-1+\frac{\mu c^{n-3}}{h^2}\left\{1+(n-2)x+\frac{(n-2)(n-3)}{2}x^2+\text{\&c.}\right\}\text{ (2).}$$

We see now that the right-hand side of the equation will be simplified if we choose c so that the constant term is zero, i.e. we put $h^2=\mu c^{n-3}$. The equation then becomes

$$\frac{d^2x}{d\theta^2}+x=(n-2)x+\tfrac{1}{2}(n-2)(n-3)x^2+\text{\&c.}\text{ ......(3).}$$

As a first approximation, we assume

$$x=M\cos(p\theta+\alpha)\text{........................(4)},$$

where M is a small quantity. Substituting and rejecting the squares of M we find

$$(1-p^2)M\cos(p\theta+\alpha)=(n-2)M\cos(p\theta+\alpha)\text{......(5).}$$

The differential equation is therefore satisfied to the first order, if we put $p^2=3-n$. In this case we have as the equation of the path

$$u=c\{1+M\cos(p\theta+\alpha)\}\text{ .. ..(6).}$$

If $n<3$, the equation (6) represents a real first approximate solution of the differential equation (1). We notice that the particle oscillates between the two circles $u=c(1+M)$ and $u=c(1-M)$. The meaning of the constant c is now apparent; geometrically, $1/c$ is the harmonic mean of the radii of the bounding circles; dynamically, $1/c$ is the radius of that circle which would be described about the centre of force with the given angular momentum h.

The positions of the apses are found by equating $du/d\theta$ to zero. This gives $p\theta + \alpha = i\pi$, the angle at the centre of force between two successive apses is therefore π/p.

If $n > 3$, the value of p is imaginary, and the trigonometrical expression takes a real exponential form, Art. 120. The quantity x therefore becomes large when θ increases, and the particle, instead of remaining in the immediate neighbourhood of the circumference of the circle, deviates widely from it on one side or the other. As the square of x has been neglected the exponential form of (6) only gives *the initial stage of the motion* and ceases to be correct when x has become so large that its square cannot be neglected. It follows from this that *the motion of a particle in a circle about a centre of force in the centre is unstable if* $n > 3$.

368. *Ex.* If the law of force is $F = u^2 f(u)$, and the orbit is nearly circular, prove that a first approximation to the path is

$$u = c\{1 + M\cos(p\theta + \alpha)\}, \qquad p^2 = 1 - \frac{cf'(c)}{f(c)}.$$

Thence it follows that *the apsidal angle is independent of the mean reciprocal radius*, viz. c, only when $F = \mu u^n$, i.e., *when the law of force is some power of the distance.*

369. A second approximation. The solution (6) is in any case only a first approximation to the motion, and it may happen that, when we proceed to a second or third approximation, the value of p is altered by terms which contain M as a factor. Besides this, we shall have x expressed in a series of several trigonometrical terms whose general form is $N\cos(q\theta + \beta)$, where N contains the square or cube of M as a factor together with some divisor κ introduced by the integration, Arts. 139, 303.

Representing the corrected value of p by $p + \Delta$, the error in $p\theta + \alpha$, i.e. $\theta\Delta$, increases by $2\pi\Delta$ after each successive revolution of the particle round the centre of force. The expression (6) will therefore cease to be even a first approximation as soon as $\theta\Delta$ has become too large to be neglected. On the other hand the additional term to the value of u may be comparatively unimportant. The magnitude of the specimen term is never greater than N and, unless κ is also small, we can generally neglect such terms.

In proceeding to a higher approximation we should first seek for those terms in the differential equation which contain $\cos(p\theta + \alpha)$; these being added to the terms of the same form in equation (5) will modify the first approximate value of p.

We should also enquire if any term in the differential equation acquires by integration a small divisor κ and thus becomes comparatively large in the solution.

370. To obtain a second approximation we substitute the first approximation (6) in the *small* terms of the differential equation (3). Writing (3), for brevity, in the form

$$\frac{d^2x}{d\theta^2}=(n-3)\{x+\beta x^2+\gamma x^3+\dots\} \quad\dots\dots\dots\dots (7),$$

where $\beta=\frac{1}{2}(n-2)$, $\gamma=\frac{1}{6}(n-2)(n-4)$, &c., we find after rejecting the cubes of M

$$\frac{d^2x}{d\theta^2}=(n-3)\{x+\tfrac{1}{2}\beta M^2(1+\cos 2p\theta)\} \quad\dots\dots\dots\dots (8),$$

where $p\theta$ has been written for $p\theta+a$ for the sake of brevity. This equation shows (Art. 303) that the second approximate value of x has the form

$$x=M\cos p\theta+M^2(G+A\cos 2p\theta) \quad\dots\dots\dots\dots (9),$$

where G and A are two constants whose values may be found by substitution, and p has the same value as before.

To obtain a third approximation, we retain the term γx^3 in (7) and assume

$$x=M\cos p\theta+M^2(G+A\cos 2p\theta)+M^3B\cos 3p\theta \quad\dots\dots\dots\dots (10).$$

To find the values of p, G, A and B we substitute in (7), express all the powers of the trigonometrical functions in multiple angles and neglect all terms of the order M^4. Equating the coefficients of $\cos p\theta$, $\cos 2p\theta$, $\cos 3p\theta$ and the constants on each side, we find

$$\begin{aligned}
-Mp^2&=(n-3)\{M+2M^3G\beta+M^3A\beta+\tfrac{3}{4}M^3\gamma\},\\
-4M^2p^2A&=(n-3)\{M^2A+\tfrac{1}{2}M^2\beta\},\\
-9M^3p^2B&=(n-3)\{M^3B+M^3A\beta+\tfrac{1}{4}M^3\gamma\},\\
0&=M^2G+\tfrac{1}{2}M^2\beta.
\end{aligned}$$

Solving these equations, and remembering that p^2 differs from $3-n$ by terms of the order M^2, we find

$$G=-\tfrac{1}{4}(n-2),\qquad A=\tfrac{1}{12}(n-2),\qquad B=\tfrac{1}{96}(n-2)(n-3),$$

$$p^2=(3-n)\{1-\tfrac{1}{12}(n-2)(n+1)M^2\} \quad\dots\dots\dots\dots (11).$$

The three first are correct when M^2 is neglected and the last when M^4 is neglected.

We notice that up to and including the third order of approximation the terms G, A, B in equation (10) do not contain any small denominators, so that if M be small enough all these terms may be neglected. The motion is then represented very nearly by

$$u=c\{1+M\cos(p\theta+a)\} \quad\dots\dots\dots\dots (12),$$

$$p=\sqrt{(3-n)}\{1-\tfrac{1}{24}(n-2)(n+1)M^2\} \quad\dots\dots\dots\dots (13),$$

and this approximation holds until θ gets so large that $M^4\theta$ cannot be neglected. We notice also that *the additional term in the value of p vanishes only when the law of force is either the inverse square or the direct distance.*

Disturbed Elliptic Motion.

371. Impulsive disturbance. When a particle is describing an orbit about a centre of force it may happen that at some particular point of that orbit the particle receives an impulse and begins to describe another orbit. We have to determine how the new orbit differs from the old, for example how the major axis has been changed in position and magnitude, and in general to express the elements of the new orbit in terms of those of the undisturbed orbit.

Let the unaccented letters a, e, l, &c. represent the elements of the undisturbed orbit, while the accented letters a', e', l', &c. represent corresponding quantities for the new. We first express the velocity v and the angle β at the given point of the orbit in terms of the undisturbed elements. Thus v and β are given by

$$v^2 = \mu\left(\frac{2}{r} - \frac{1}{a}\right), \quad \sin\beta = \frac{p}{r} = \frac{\sqrt{\mu l}}{vr} \quad \text{.................(1)},$$

when the undisturbed orbit is an ellipse described about the focus.

We next consider the circumstances of the blow. Let m be the mass of the particle, mB the blow. The particle, after the impulse is concluded, is animated with the velocity B in the given direction of the blow, together with the velocity v along the tangent to the original path. Compounding these the particle has a resultant velocity v' and is moving in a known direction. Since the position of the radius vector is not changed by the blow we may conveniently refer the changes of motion to that line. If P, Q are the components of B along and perpendicular to the radius vector and β' is the angle the direction of motion makes with the radius vector, we have

$$v'\cos\beta' = v\cos\beta + P, \qquad v'\sin\beta' = v\sin\beta + Q \text{ ......(2)}.$$

Having now obtained v', β', the formulæ (1), writing accented letters for the old elements, determine the new semi-major axis a' and the new semi-latus rectum l'. The position in space of the major axis follows from Art. 336.

372. We may sometimes advantageously replace the second of the equations (1) by another formula. We notice that mh is the moment of the momentum of the particle about the centre of force. Since just after the impulse the velocity v' is the resultant of v and B, *the moment of v' is equal to that of v together with the moment of B.* Hence

$$h' = h + Bq \text{ ....(3)},$$

where q is the perpendicular on the line of action of the blow. Since $h^2 = \mu l$, when the law of force follows the Newtonian law, this equation leads to

$$\sqrt{l'} = \sqrt{l} + Bq/\sqrt{\mu} \quad\text{.....................(4)}.$$

Thus the change in the latus rectum is very easily found.

As a corollary, we may notice that *when the blow acts along the radius vector, the angular momentum mh and therefore the latus rectum of the orbit are unchanged.* We also observe that if the magnitude of the attracting force or its law of action were abruptly changed, the value of h is unaltered.

373. *Ex.* 1. Two particles, describing orbits about the same centre of force, impinge on each other. Prove

$$m_1h_1' + m_2h_2' = m_1h_1 + m_2h_2,$$

where m_1h_1, m_2h_2; m_1h_1', m_2h_2' are their angular momenta before and after impact.

Ex. 2. A particle P of unit mass is describing an ellipse about the focus S. A circle is described to touch the normal to the conic at P whose radius PC represents the velocity at P in direction and magnitude. Prove that if the particle is acted on by an impulse represented in direction and magnitude by any chord MP of the circle, the length of the major axis is unaltered by the blow.

Since $B = 2v\cos\theta$, the velocity in the direction of the blow is simply reversed. Hence $v' = v$ and $a' = a$ by Art. 335.

374. If the direction of the blow does not lie in the plane of motion, the plane of the new orbit is also changed. For the sake of the perspective, let the radius vector SP be the axis of x and let the plane of xy be the plane of the old orbit; then $v\cos\beta$, $v\sin\beta$ are the components of velocity parallel to the axes of x and y. Let the components of the blow be mX, mY, mZ; then just after the blow is concluded the components of velocity parallel to the axes are $v\cos\beta + X$, $v\sin\beta + Y$, and Z. The inclination i of the planes of the two orbits is therefore given by $\tan i = \dfrac{Z}{v\sin\beta + Y}$. The particle begins to move in its new orbit with a velocity v' in a direction making an angle β' with the radius vector SP given by

$$v'\cos\beta' = v\cos\beta + X, \qquad (v'\sin\beta')^2 = (v\sin\beta + Y)^2 + Z^2.$$

The problem is now reduced to the case already considered.

If mh' is the angular momentum in the new orbit, its components about the axes of x, y, z are 0, $-mh'\sin i$, $mh'\cos i$. Hence

$$h'\cos i = h + Yr, \qquad h'\sin i = Zr,$$

where $r = SP$.

375. Examples. *Ex.* 1. A particle is describing a given ellipse about a centre of force in the focus, and when at the farther apse A', its velocity is suddenly increased in the ratio 1 : n. Find the changes in the elements.

The direction of motion is unaltered by the blow and since this direction is at right angles to the radius vector from the centre of force, the point A' is one of the apses of the new orbit.

Let a, e; a', e' be the semi-major axes and eccentricities of the orbits. Then since SA' is unaltered in length

$$r = a'(1+e') = a(1+e) \quad\text{..................................(1)}.$$

We have here chosen as the standard figure for the new orbit an ellipse having A' for the further apse. A negative value of the eccentricity e' therefore means that A' is the nearer apse.

Also since $v' = nv$, we have

$$\mu\left(\frac{2}{r} - \frac{1}{a'}\right) = n^2\mu\left(\frac{2}{r} - \frac{1}{a}\right) \quad\text{.......................... (2),}$$

where a' must be regarded as negative if the new orbit is a hyperbola, Art. 333.

From these equations we find

$$\frac{a'}{a} = \frac{1+e}{2 - n^2(1-e)}, \qquad e' = 1 - n^2(1-e).$$

The point A' is therefore the farther or nearer apse according as $n^2(1-e)$ is $<$ or >1; if equal to unity the new orbit is a circle, if equal to -1, a parabola. The new orbit is an ellipse or hyperbola according as $n^2(1-e) <$ or >2.

Ex. 2. A particle describes an ellipse under a force tending to a focus. On arriving at the extremity of the minor axis, the force has its law changed, so that it varies as the distance, the magnitude at that point remaining the same. Prove that the periodic time is unaltered and that the sum of the new axes is to their difference as the sum of the old axes to the distance between the foci.

[Math. Tripos, 1860.]

By Art. 325 the new orbit is an ellipse having the centre of force S in the centre. Let the new law of force be $\mu' r$.

Then when $r = a$, the forces are equal, hence

$$\mu' a = \mu / a^2 \text{ .... (1).}$$

Measure a length SD parallel to the direction of motion at B, such that the velocity v at B is $\sqrt{\mu'} \, . \, SD$. Then SD is the semi-conjugate of SB in the new orbit. Equating the velocities at B in the old and new orbits, we have when $r = a$

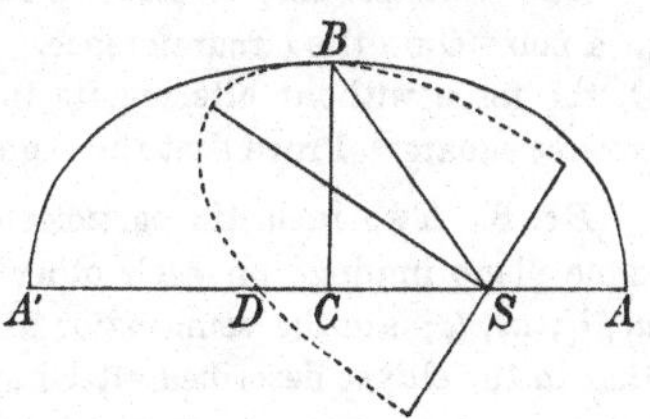

$$\mu\left(\frac{2}{r} - \frac{1}{a}\right) = \mu' \, . \, SD^2, \qquad \therefore SD = a \text{ .... (2).}$$

The conjugates SB, SD are equal diameters, the major and minor axes are therefore the internal and external bisectors of the angle BSD. Representing the semi-axes by a', b', we have

$$a'^2 + b'^2 = SB^2 + SD^2 = 2a^2, \qquad a'b' = SB \, . \, SD \sin BSD = ab \text{ .... (3).}$$

The internal bisector of the angle BSD is clearly the major axis.

If the change in the velocity had been made at any point of the ellipse, we proceed in the same way. By drawing SD parallel to the direction of motion we arrive at the known problem in conics, given two conjugate diameters in position and magnitude, construct the ellipse.

The periodic times in the two orbits are respectively $2\pi/\sqrt{\mu'}$ and $2\pi\sqrt{a^3/\mu}$. The equality of these follows from the equation (1). The rest of the question follows from (3).

Ex. 3. A particle is describing an ellipse under a force μ/r^2 to a focus: when the particle is at the extremity of the latus rectum through the focus this centre of force is removed and is replaced by a force $\mu' r'$ at the centre of the ellipse. Prove that if the particle continue to describe the same ellipse $\mu' b^4 = \mu a$.

[Coll. Exam. 1895.]

Ex. 4. A planet moving round the sun in an ellipse receives at a point of its orbit a sudden velocity in the direction of the normal outwards which transforms the orbit into a parabola, prove that this added velocity is the same for all points of the orbit, and if it be added at the end of the minor axis, the axis of the parabola will make with the major axis of the ellipse an angle whose sine is equal to the eccentricity. [Coll. Exam. 1892.]

Ex. 5. A particle describes a given ellipse about a centre of force of given intensity in the focus S. Supposing the particle to start from the further extremity of the major axis, find the time T of arriving at the extremity of the minor axis. At the end of this time the centre of force is transferred without altering its intensity from S to the other focus H, and the particle moves for a second interval T equal to the former under the influence of the central force in H. Find the position of the particle, and show that, if the centre of force were then transferred back to its original position, the particle would begin to describe an ellipse whose eccentricity is $(3e-e^2)/(1+e)$. [Math. Tripos, 1893.]

Ex. 6. A body is describing an ellipse round a force in its focus S, and HZ is the perpendicular on the tangent to the path from the other focus H. When the body is at its mean distance the intensity of the force is doubled, show that SZ is the new line of apses. [Coll. Ex.]

Ex. 7. A particle describes a circle of radius c about a centre of force situated at a point O on the circumference. When P is at the distance of a quadrant from O, the force without altering its instantaneous magnitude begins to vary as the inverse square. Prove that the semi-axes of the new orbit are $\frac{2}{3}c\sqrt{2}$ and $\frac{1}{3}c\sqrt{3}$.

Ex. 8. Two inelastic particles of masses m_1, m_2, describing ellipses in the same plane impinge on each other at a distance r from the centre of force. If a_1, l_1; a_2, l_2; are the semi-major axes and semi-latera recta before impact, prove that in the ellipse described after impact

$$(m_1+m_2)\,l^{\frac{1}{2}}=m_1 l_1^{\frac{1}{2}}+m_2 l_2^{\frac{1}{2}},$$

$$(m_1+m_2)\left(2r-l-\frac{r^2}{a}\right)^{\frac{1}{2}}=m_1\left(2r-l_1-\frac{r^2}{a_1}\right)^{\frac{1}{2}}+m_2\left(2r-l_2-\frac{r^2}{a_2}\right)^{\frac{1}{2}}.$$

Ex. 9. A planet, mass M, revolving in a circular orbit of radius a, is struck by a comet, mass m, approaching its perihelion; the directions of motion of the comet and planet being inclined at an angle of 60°. The bodies coalesce and proceed to describe an ellipse whose semi-major axis is $\dfrac{(M+m)^2 a}{M\{M+(4-\sqrt{2})\,m\}}$. Prove that the original orbit of the comet was a parabola; and if the ratio of m to M is small, show that the eccentricity of the new orbit is $(7\frac{1}{2}-4\sqrt{2})^{\frac{1}{2}}(m/M)$. [Coll. Ex. 1895.]

376. Continuous forces. We may apply the method of Art. 371 to find the effects of continuous forces on the particle. Let f, g be the tangential and normal accelerating components of any disturbing force, the first being taken positively when increasing the velocity and the second when acting inwards.

We divide the time into intervals each equal to δt and consider the effect of the forces on the elements of the ellipse at the end of each interval. We treat the forces, in Newton's manner, as small impulses generating velocities $f\delta t$ and $g\delta t$ along the tangent and normal respectively. The effect of the tangential force is to increase the velocity at any point P from v to $v + \delta v$, where $\delta v = f\delta t$, the direction of motion not being altered. To find the effect of the normal force we observe that after the interval δt the particle has a velocity $g\delta t$ along the normal, while the velocity v along the tangent is not altered. The direction of motion has therefore been turned round through an angle $\delta\beta = g\delta t/v$.

If the disturbing force were now to cease to act, the particle would move in a conic whose elements could be deduced from these two facts, (1) the velocity at P is changed to $v + \delta v$, (2) the angle of projection is $\beta + \delta\beta$. *The conic which the particle would describe if at any instant the disturbing forces were to cease to act is called the instantaneous conic at that instant.*

377. *To find the effect on the major axis,* we use the formula

$$v^2 = \mu\left(\frac{2}{r} - \frac{1}{a}\right) \quad\text{.......................... (1).}$$

Since v is increased to $v + \delta v$, we see by simple differentiation

$$2v\,\delta v = \frac{\mu}{a^2}\,\delta a, \qquad \therefore\ \delta a = \frac{2a^2 v}{\mu} f\delta t \quad\text{..............(2).}$$

In differentiating the formula for v^2 we are not to suppose that δv represents the whole change of the velocity in the time δt. The particle moves along the ellipse and experiences a change of velocity dv in the time dt given by

$$v\,dv = -\frac{\mu}{r^2}\,dr \quad\text{..(3).}$$

Taking $dt = \delta t$, the change of velocity in the time δt is $\delta v + dv$, the part δv being due to the disturbing forces and the part dv to the action of the central force.

378. *To find the changes in the eccentricity and line of apses.* We may effect this by differentiating the formulæ

$$l = a(1 - e^2), \qquad h^2 = \mu l, \qquad \frac{l}{r} = 1 + e\cos\theta \quad\text{..............(4).}$$

Since mh is the angular momentum, the increase of mh, viz. $m\delta h$, is equal to the moment of the disturbing forces about the origin (Art. 372). Let β be the angle the direction of motion at P makes with the radius vector,

$$\therefore\ \tfrac{1}{2}\sqrt{\mu}\frac{\delta l}{\sqrt{l}} = \delta h = fr\sin\beta + gr\cos\beta.$$

We deduce from equations (4)

$$\delta l = (1 - e^2)\,\delta a - 2ae\delta e, \quad \frac{\delta l}{r} = \cos\theta\,\delta e - e\sin\theta\,\delta\theta,$$

and the values of δe and $\delta\theta$ follow at once.

379. Herschel has suggested a geometrical method of finding the changes of the eccentricity and the line of apses in his *Outlines of Astronomy**. He considers the effect of the disturbing forces f, g on the position of the empty focus.

The effect of the tangential force f is to alter the velocity v and therefore to alter a. Since $SP + PH = 2a$, the empty focus H is moved, during each interval δt, along the straight line PH a distance $HH' = 2\delta a$, where δa is given by (2).

The effect of the normal force g is to turn the tangent at P through an angle $\delta\beta = g\delta t/v$. Since SP, HP make equal angles with the tangent, the empty focus H is moved perpendicularly to PH, a distance $HH'' = 2PH \,.\, \delta\beta$.

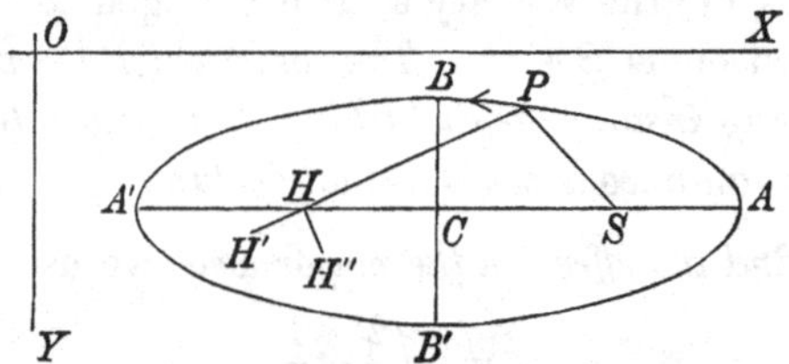

Consider first the tangential force f, we have $SH = 2ae$, $SH' = 2(ae + \delta ae)$. Hence projecting on the major axis

$$2\delta(ae) = HH' \,.\, \cos PHS = 2\delta a\,\frac{x + ae}{r'},$$

where $r' = HP = a + ex$, and x is measured from the centre;

$$\therefore\ \delta e = \frac{x - e^2 x}{r'}\,\frac{\delta a}{a} = \frac{2a(1 - e^2)}{\mu}\,\frac{xv}{r'}\,f\delta t.$$

Let ϖ be the longitude of the apse line HS measured from some fixed line through S,

$$\therefore\ 2ae\delta\varpi = HH' \sin PHS = 2\delta a\,\frac{y}{r'},$$

$$\therefore\ e\delta\varpi = \frac{y}{r'}\,\frac{\delta a}{a} = \frac{2a}{\mu}\,\frac{yv}{r'}\,f\delta t.$$

Consider secondly the normal force g. We have

$$SH = 2ae, \quad SH'' = 2(ae + \delta ae), \quad \delta a = 0;$$

$$\left.\begin{aligned} \therefore\ 2\delta(ae) &= -HH'' \sin PHS = -2r'\delta\beta\,\frac{y}{r'} \\ 2ae\delta\varpi &= \frac{HH'' \cos PHS}{r'} = 2r'\delta\beta\,\frac{x + ae}{r'} \end{aligned}\right\};$$

$$\therefore\ \delta e = -\frac{1}{a}\,\frac{y}{v}\,g\delta t, \quad e\delta\varpi = \frac{1}{a}\,\frac{x + ae}{v}\,g\delta t.$$

---

* See also some remarks by the author in the *Quarterly Journal*, 1861, vol. IV. It should be noticed that Herschel measures the eccentricity by half the distance between the foci, a change from the ordinary definition which has not been followed here.

380. The expressions for δe, $\delta\varpi$ should be put into different forms according to the use we intend to make of them. Let ψ be the angle the tangent at P makes with the major axis, then $\tan\psi = \dfrac{b^2x}{a^2y}$. We easily find by elementary conics

$$\sin\psi = \frac{b}{a}\frac{x}{\sqrt{(rr')}}, \qquad \cos\psi = \frac{a}{b}\frac{y}{\sqrt{(rr')}}.$$

Also $v^2 = \mu\left(\dfrac{2}{r} - \dfrac{1}{a}\right) = \dfrac{\mu r'}{ar}$. It immediately follows that

$$\delta e = \frac{2b}{\sqrt{(\mu a)}} f \sin\psi\,\delta t, \qquad e\,\delta\varpi = \frac{2b}{\sqrt{(\mu a)}} f\cos\psi\,\delta t,$$

$$\delta e = -\frac{br}{a\sqrt{(\mu a)}} g\cos\psi\,\delta t, \qquad e\,\delta\varpi = \frac{ar}{b\sqrt{(\mu a)}}\,\frac{x+ae}{x}\,g\sin\psi\,\delta t.$$

These formulæ give the changes of e and ϖ produced by any tangential or normal force.

381. Draw two straight lines OX, OY parallel to the principal diameters situated as shown in the figure. Since $f\cos\psi$, $f\sin\psi$ are the components of the tangential disturbing force parallel to the principal diameters, we see that *when the force acts towards OX the eccentricity is increased, and when towards OY the apse line is advanced*; the contrary effects taking place when the force tends from these lines.

The same rule applies to the normal disturbing force so far as the eccentricity is concerned. It applies also to the motion of the apse except when the particle lies between the minor axis and the latus rectum through the empty focus, and the rule is then reversed. When the eccentricity is small, $\dfrac{e\sin\psi}{x} = \dfrac{be}{a^2}$ very nearly when the particle is near the minor axis; so that the effects of the tangential force in this part of the orbit may be neglected and the rule applied generally.

382. Examples. *Ex.* 1. The path of a comet is within the orbit of Jupiter, approaching it at the aphelion. Show that each time the comet comes near Jupiter the apse line is advanced. This theorem is due to Callandreau, 1892.

The comet being near the aphelion and Jupiter just beyond, both the normal and tangential disturbing forces act towards OY; the apse therefore advances.

Ex. 2. A particle is describing an elliptic orbit about the focus and at a certain point the velocity is increased by $1/n$th, n being large. Prove that, if the direction of the major axis be unaltered, the point must be at an apse, and the change in the eccentricity is $2(1 \pm e)/n$. [Coll. Ex. 1897.]

Ex. 3. An ellipse of eccentricity e and latus rectum l is described freely about the focus by a particle of mass m, the angular momentum being mh. A small impulse mu is given to the particle, when at P, in the direction of its motion; prove that the apsidal line is turned through an angle which is proportional to the intercept made by the auxiliary circle of the ellipse on the tangent at P, and which cannot exceed lu/eh. [Math. Tripos, 1893.]

Ex. 4. A body describes an ellipse about a centre of force S in the focus. If A be the nearer apse, P the body, and a small impulse which generates a velocity T act on the body at right angles to SP, prove that the change of direction of the apse line is given approximately by $\frac{T}{h}\left(\frac{2}{e}+\cos ASP\right) SP \sin ASP$, where e is the eccentricity of the orbit and h twice the rate of description of area about S.

[Math. Tripos.]

Ex. 5. A particle describes an ellipse about a centre of force in the focus S. When the particle has reached any position P the centre of force is suddenly moved parallel to the tangent at P through a short distance x, prove that the major axis of the orbit is turned through the angle $\frac{x}{SG}\sin\phi\sin(\theta-\phi)$ where G is the point at which the normal at P meets the original major axis, θ the angle SGP and ϕ the angle the tangent makes with SP.

[Coll. Ex. 1895.]

Ex. 6. A particle describes an ellipse about a centre of force μ/r^2 and is besides acted on by a disturbing force κr^n tending to the same point. Prove that as the particle moves from a distance r_0 to r, the major axis and eccentricity change according to the law

$$\mu\left(\frac{1}{a}-\frac{1}{a_0}\right)=\frac{2\kappa}{n+1}(r^{n+1}-r_0^{n+1}),\qquad \frac{1-e^2}{1-e_0^2}=\frac{a_0}{a}.$$

Thence deduce the changes in a and e when κ is very small.

383. A resisting medium. We may also use the formulæ of Art. 380 to find the quantitative effect of a resisting medium on the motion of a particle describing an ellipse about a centre of force in the focus.

The velocity of the particle being v, let the resistance be κv. Then $g=0$ and $f=-\kappa\, ds/dt$, and the equations of motion become

$$\frac{de}{dt}=-\frac{2b\kappa}{\sqrt{(\mu a)}}\frac{dy}{dt},\qquad e\frac{d\varpi}{dt}=-\frac{2b\kappa}{\sqrt{(\mu a)}}\frac{dx}{dt}.$$

Usually f and g are so small that their squares can be neglected. Now the changes of the elements a, e, &c. are of the order of f and g, being produced by these forces. Hence in using these equations *we may regard the elements of the ellipse, when multiplied by the coefficient κ of resistance, as constants.*

Supposing then that we reject the squares of κ, we have by an easy integration

$$e=-\frac{2b\kappa}{\sqrt{(\mu a)}}y+A,\qquad e\varpi=-\frac{2b\kappa}{\sqrt{(\mu a)}}x+B,$$

where A, B are two undetermined constants. Since after a complete revolution, the coordinates x, y return to their original values, both the eccentricity and the position of the line of apses must also be the same as before. *There can therefore be no permanent change in either.* The greatest change of the eccentricity from its mean value is $2\kappa b^2/na^2$, while the apse oscillates about its mean position through an angle $2\kappa b/nea$, where $\mu = n^2a^3$, Art. 341.

384. *Ex.* A comet moves in a resisting medium whose resistance is $f = -\kappa V^p \left(\frac{a}{r}\right)^q$ where V is the velocity, r the distance from the sun and p, q are positive quantities. When the true anomaly θ is taken as the independent variable (instead of t as in Art. 380), prove that

$$\frac{1}{a}\frac{da}{d\theta} = \frac{-2A}{1-e^2}(1+2e\cos\theta+e^2)^{\frac{p+1}{2}}(1+e\cos\theta)^{q-2},$$

$$\frac{de}{d\theta} = -2A(\cos\theta+e)(1+2e\cos\theta+e^2)^{\frac{p-1}{2}}(1+e\cos\theta)^{q-2},$$

$$e\frac{d\varpi}{d\theta} = -2A\sin\theta(1+2e\cos\theta+e^2)^{\frac{p-1}{2}}(1+e\cos\theta)^{q-2},$$

where $$A = \kappa n^{p-2}a^{p-1}.(1-e^2)^{2-\frac{p}{2}-q} \text{ and } \mu = n^2a^3.$$

When the right-hand sides of these equations are expanded in series of the form

$$A + B\cos\theta + C\cos 2\theta + \ldots$$

it is obvious that the only permanent changes are derived from the non-periodical terms. Prove (1) that the longitude of the apse has no permanent changes, (2) that the eccentricity at the time t is $e - Aent(p+q-1)$, (3) the semi-major axis is $a - 2Aant$. These results are given by Tisserand, *Méc. Céleste*, 1896.

When the law of resistance is such that $p+q=1$, it follows that *neither the eccentricity nor the line of apses have any permanent change.* For any values of p and q not satisfying this relation the eccentricity will gradually change and continue to change in the same direction. When the changes of any of the elements have become so great that their products by the coefficient κ of resistance can no longer be neglected, the equations given above must be integrated in a different way.

385. Encke's Comet. The general effect of a resisting medium on the motion of a comet is to diminish its velocity and therefore also the major axis of its orbit, Art. 377. The ellipse which the comet describes is therefore continually growing smaller and the periodic time, which varies as $a^{3/2}$, continually decreases.

Encke was the first who thoroughly investigated the effect of a resisting medium on the motion of a comet. This comet has since then been called after his name. After making allowance for the disturbance due to the attraction of the sun and the planets, he found by observation that its period, viz. 1200 days, was diminished by about two hours and a half in each revolution. This he ascribed to the presence of a medium whose resistance varied as $(v/r)^2$ where v is the velocity of the comet and r its distance from the sun.

The importance and interest of Encke's result caused much attention to be given to this comet. The astronomers Von Asten of Pulkowa and afterwards Backlund* studied its motions at each successive appearance with the greatest

* In the *Bulletin Astronomique*, 1894, page 473, there is a short account of the work of Backlund by himself. He speaks of the continued decrease of the acceleration, the law of resistance, and gives references to his memoirs and particularly to

attention. The acceleration of the comet's mean motion appears to have been uniform from 1819, when Encke first took up the subject, to 1858. It then began to decrease and continued to decrease until the revolution of 1868—1871 when its magnitude was about half its former value. From 1871 to 1891 the acceleration was again nearly constant.

Assuming the law of resistance to be represented by $\kappa v^m/r^n$, Backlund found that n is essentially negative. This would make the density of the resisting medium increase according to a positive power of the distance from the sun; a result which he considered very improbable. He afterwards arrived at the conclusion that we must replace $1/r^n$ by some function $f(r)$ having maxima and minima at definite distances from the sun. In Laplace's nebular theory the planets are formed by condensations from rings of the solar nebula. In this formation all the substance of each ring would not be used up and some of it might travel along the orbit as a cloud of light material. It is suggested that Encke's comet passes through nebulous clouds of this kind and that the resistance they offer causes the observed acceleration.

It is known that comets contract on approaching the sun, sometimes to a very great extent. Tisserand remarks that when the size of the comet decreases the resistance should also decrease, and that this may help us to understand how the resistance to any comet might vary as a positive power of the distance from the sun. The size of Encke's comet also is not the same at every appearance and this again may have an effect on the law of resistance.

It is clear that if Encke's comet does meet with a resistance, every comet of short period which approaches closely to the sun must show the effect of the same influence. In 1880 Oppolzer thought he had discovered an acceleration in the motion of another comet. This was the comet Winnecke having a period of 2052 days. Further investigation showed that this was illusory, so that at present the evidence for the existence of a resisting medium rests on Encke's comet alone.

386. *Does the evidence afforded by Encke's comet prove a resisting medium?* Sir G. Stokes in a lecture* on the luminiferous medium says he asked the highest astronomical authority in the country this question. Prof. Adams replied that there might be attracting matter within the orbit of Mercury which would account for it in a different way. Sir G. Stokes then goes on to say that the comet throws out a tail near the sun and that this is equivalent to a reaction on the head towards

---

the eighth volume of his *Calculs et Recherches sur la comète d'Encke.* In the *Comptes Rendus*, 1894, page 545, Callandreau gives a summary of the results of Backlund. In the *Traité de Mécanique Céleste*, vol. IV. 1896, Tisserand discusses the influence of a resisting medium. In the *History of Astronomy* by A. M. Clerke, 1885, examples of the contraction of comets near the sun are given. M. Valz in a letter to M. Arago quoted in the *Comptes Rendus*, vol. VIII. 1838, speaks of the great contraction of a comet as it approached the sun. He remarks that as it was approaching the earth at that time, it should have appeared larger. See also Newcombe's *Popular Astronomy*, 1883.

* Presidential address at the anniversary meeting of the *Victoria Institute*, June 29, 1893: reported in *Nature*, July 27, page 307.

the sun. There is therefore an additional force towards the sun. The effect of this would be to shorten the period even if there were no resisting medium. In the course of his lecture he discusses the question, "*must the ether retard a comet*," and decides that we cannot with safety infer that the motion of a solid through it necessarily implies resistance.

Kepler's Laws and the law of gravitation.

387. Kepler's laws. The following theorems were discovered by the astronomer Kepler after thirty years of study.

(1) The orbits of the planets are ellipses, the sun being in one focus.

(2) As a planet moves in its orbit, the radius vector from the sun describes equal areas in equal times.

(3) The squares of the periodic times of the several planets are proportional to the cubes of their major axes.

The last of these laws was published in 1619 in his *Harmonice Mundi* and the first two in 1609 in his work on *the motions of Mars.*

388. From the second of these laws, it follows that *the resultant force on each planet tends towards the sun;* Art. 307.

From the first we deduce that *the accelerating force on each planet is equal to* μ/r^2, where r is the instantaneous distance of that planet from the sun, and μ is a constant; Art. 332.

It is proved in Art. 341 that when the central force is μu^2, the periodic time in an ellipse is $T = 2\pi a^{\frac{3}{2}}/\sqrt{\mu}$, where a is the semi-major axis. Now Kepler's third law asserts that for all the planets T^2 is proportional to a^3; it follows that μ *is the same for all the planets.*

Laws corresponding to those of Kepler have been found to hold for the systems of planets and their satellites. Each satellite is therefore acted on by a force tending to the primary and that force follows the law of the inverse square.

It has been possible to trace out the paths of some of the comets and all these have been found to be conics having the sun in one focus. These bodies therefore move under the same law of force as the planets.

389. The laws of Kepler, being founded on observations, are not to be regarded as strictly true. They are approximations, whose errors, though small, are still perceptible. We learn from them that the sun, planets and satellites are so constituted that the sun may be regarded as attracting the planets, and the planets the satellites, according to the law of the inverse square. We now extend this law and make the *hypothesis* that the planets and satellites also attract the sun and attract each other according to the same law. Let us consider how this hypothesis may be tested.

Let m_1, m_2, &c. be certain constants, called the masses of the bodies, such that the accelerating attraction of the first on any other body distant r_1 is m_1/r_1^2, the attraction of the second is m_2/r_2^2, and so on. Let μ be the corresponding constant for the sun.

Assuming these accelerations, we can write down the differential equations of motion of the several bodies, regarded as particles. For example, the equations of motion of the particle m_1 may be obtained by equating d^2x/dt^2, &c., to the resolved accelerating attractions of the other bodies. The equations thus formed can only be solved by the method of continued approximation. Kepler's laws give us the first approximation; as a second approximation we take account of the attractions of the planets, but suppose that m_1, m_2, &c. are so small that the squares of their ratio to μ may be neglected. This problem is usually discussed in treatises on the Planetary theory. The solution of the problem enables us to calculate the positions of the planets and satellites at any given time and the results may be compared with their actual positions at that time. The comparison confirms the hypothesis in so extraordinary a way that we may consider its truth to be established as far as the solar system is concerned.

390. Extension to other systems. The law of gravitation being established for the solar system, its extension to other systems of stars may be only a fair inference. But we should notice that this extension is not founded on observation in the same sense that the truth of the law for the solar system is established*. The constituents of some double stars move round

* Villarceau, *Connaissance des temps* for the year 1852 published in 1849; A. Hall, *Gould's Astronomical Journal*, Boston, 1888.

each other in a periodic time sufficiently short to enable us to trace the changes in their distance and angular position. We may thus, partially at least, hope to verify the law of gravitation. What we see, however, is not the real path of either constituent, but its projection on the sphere of the heavens. We can determine if the relative path is a conic and can verify approximately the equable description of areas; but since the focus of the true path does not in general project into the focus of the visible path, an element of uncertainty as to the actual position of the centre of force is introduced.

We cannot therefore use Kepler's first law to deduce from these observations alone that the law of force is the inverse square.

391. Besides this, there are two practical difficulties. First, there is the delicacy of the observations, because the errors of observations bear a larger ratio to the quantities observed than in the solar system. Secondly, a considerable number of observations on each double star is necessary. Five conditions are required to fix the position of a conic, and the mean motion and epoch of the particle are also unknown. Unless therefore more than seven *distinct* observations have been made, we cannot verify that the path is a conic. These difficulties are gradually disappearing as observations accumulate and instruments are improved.

392. Besides the motions of the double stars we can only look to the proper motions of the stars in space for information on the law of gravitation. Some of these velocities are comparable to that of a comet in close proximity to the sun and yet there is no visible object in their neighbourhood to which we could ascribe the necessary attracting forces. At present no deductions can be made, we must wait till future observations have made clear the causes of the motions.

393. Other reasons. The law of gravitation is generally deduced from Kepler's laws, partly for historical reasons and partly because the proof is at once simple and complete. It is however useful and interesting to enquire what we may learn about the law of gravitation by considering other observed facts.

Ex. 1. It is given that for all initial conditions the path of a particle is a plane curve: deduce that the force is central.

Consider an orbit in a plane P, then at every point of that orbit the resultant force must lie in the plane. Taking any point A on the orbit project particles in all directions in that plane with arbitrary velocities, then since the plane of motion of each must contain the initial tangent at A and the direction of the force at A, each particle moves in the plane P. It follows that at every point of the plane P traversed by these orbits the resultant force lies in the plane. If these orbits do not cover the whole plane we take a new point B on the boundary of the area covered, and again project particles in all directions in that plane with arbitrary velocities. By continually repeating this process we can traverse every point of the plane, provided no points are separated from A by a line along which the force is infinite. It follows that at every point of the plane P the force lies in that plane.

Next let us pass planes through any point A of one of these orbits and the direction AC of the force at A. Then by the same reasoning as before the direction of the force at points in each plane must lie in that plane and must therefore intersect AC. Thus the force at every point intersects the force at every other point. It follows that the force is central.

An observer placed at the sun, who noticed that all the planets described great circles in the heavens, would know from that one fact that the force acting on each was directed to the sun. Halphen, *Comptes Rendus*, vol. 84, Darboux's Notes to Despeyrous' *Mécanique*.

Ex. 2. If all the orbits in a given plane are conics, prove that the force is central.

If a particle P be projected from any point A in the direction of the force at A, the radius of curvature of the path is infinite at A. Since the only conic in which the radius of curvature is infinite is a straight line, the path of the particle P is a straight line and therefore the force at every point of this straight line acts along the straight line. The lines of force are therefore straight lines.

These straight lines could not have an envelope, for (unless the force at every point of that curve is infinite) we could project the particles along the tangents to the envelope past the point of contact so as to intersect other lines of force. The directions of the force would not then be the same at the same point for all paths. Bertrand, *Comptes Rendus*, vol. 84.

Ex. 3. If the orbits of all the double stars which have been observed are found to be closed curves, show that the Newtonian law of attraction may be extended to such bodies.

Bertrand has proved that all the orbits described about a centre of force (for all initial conditions within certain limits) cannot be closed unless the law of force is either the inverse square or the direct distance. By examining many cases of double stars we may include all varieties of initial conditions, and if all these orbits are closed the law of the inverse square may be rendered very probable. See Arts. 370, 426. Bertrand when giving this theorem in *Comptes Rendus*, vol. 77, 1873, quotes Tchebychef.

The Hodograph.

394. A straight line OQ is drawn from the origin O parallel to the instantaneous direction of motion and its length is proportional to the velocity of a particle P, say $OQ = kv$. The locus of Q has been called by Sir W. R. Hamilton the hodograph of the path of P. Its use is to exhibit to the eye the varying velocity and direction of motion of the particle. See Art. 29.

By giving k different values we have an infinite number of similar curves, any one of which may be used as a hodograph.

It follows from Art. 29 that, *if s' be the arc of the hodograph, ds'/dt represents in direction and magnitude the acceleration of P.*

395. If the force on the particle P is central and tends to the origin O, it is sometimes more convenient to draw OQ perpendicularly instead of parallel to the tangent. If OY be a perpendicular to the tangent, the velocity v of P is h/OY; hence if $OQ = kv$, we see that *the hodograph is then the polar reciprocal of the path with regard to the centre of force*, the radius of the auxiliary circle being $\sqrt{(hk)}$. If F be the central force at P, *the point Q travels along the hodograph with a velocity kF.*

396. Examples. *Ex.* 1. The path being an ellipse described about the centre C, and OQ being drawn parallel to the tangent, prove that the hodographs are similar ellipses.

Let CQ be the semi-conjugate of CP, then $v = \sqrt{\mu} \cdot CQ$, Art. 326. Hence if $k = 1/\sqrt{\mu}$, *the hodograph is the ellipse itself.* The point Q then travels with a velocity $\sqrt{\mu} \cdot CP$.

Ex. 2. The path being an ellipse described about the focus S, prove that a hodograph is the auxiliary circle, the other focus H being the origin and HQ drawn perpendicularly to the tangent at P.

Let SY, HZ be the two perpendiculars on the tangent, then $v = h/SY = HQ/k$, also $SY \cdot HZ = b^2$, $\therefore HQ = HZ$ if $k = b^2/h$. Since the locus of Z is the auxiliary circle the result follows at once.

Ex. 3. The path being a parabola described about the near focus S, prove that a hodograph is the circle described on AS as diameter, where A is the vertex and SQ is drawn perpendicularly to the tangent.

Ex. 4. The hodograph of the path of a projectile is a vertical straight line, the radius vector OQ being drawn parallel to the tangent.

If the tangent at P make an angle ψ with the horizon, the abscissa of Q is $kv \cos \psi$. This is constant because the horizontal velocity of P is constant. The point Q travels along this straight line with a uniform velocity kg.

Ex. 5. An equiangular spiral is described about the pole, show that a hodograph is an equiangular spiral having the same pole and a supplementary angle. See Art. 30.

Ex. 6. A bead moves under the action of gravity along a smooth vertical circle starting from rest indefinitely near to the highest point. Show that a polar equation of a hodograph is $r' = b \sin \frac{1}{2}\theta'$, the origin being at the centre.

Ex. 7. The hodograph of the path of a particle P is given, show that if the path of P is a central orbit, the auxiliary point Q must travel along the hodograph with a velocity $v' = \lambda p'^2 \rho'$, where p' is the perpendicular from the centre of force on the tangent to the hodograph and ρ' is the radius of curvature. Show also that the central force $F = v'/k$ and the angular momentum $h = 1/\lambda k^3$.

The condition that the path is a central orbit is $v^2/\rho = Fp/r$. Writing $p = c^2/r'$ and $r = c^2/p'$, we find F and thence v'.

Ex. 8. The hodograph of the path of P is a parabola with its focus at O, and the radius vector $OQ=r'$ rotates with an angular velocity proportional to r'. Prove that the path of P is a circle passing through O, described about a centre of force situated at O.

Since the angular velocity of OQ is nr', we find by resolving v' perpendicularly to OQ that $v'=nr'^3/p'$. In a parabola $lr'=2p'^2$, and since $\rho'=r'dr'/dp'$ we see that $v'=\lambda p'^2\rho'$ where $\lambda=n/l$. The path is therefore a central orbit. But the polar reciprocal of $lr'=2p'^2$ (obtained by writing $p'=c^2/r$, and $r'=c^2/p$) is $r^2=p\,(2c^2/l)$, and this is a circle passing through O.

Ex. 9. A particle describes a curve under a constant acceleration which makes a constant angle with the tangent to the path; the motion takes place in a medium resisting as the nth power of the velocity. Show that the hodograph of the curve described is of the form $b^{-n}e^{-n\theta\cot\alpha}=r^{-n}-a^{-n}$. [Coll. Ex.]

Ex. 10. A particle, moving freely under the action of a force whose direction is always parallel to a fixed plane, describes a curve which lies on a right circular cone and crosses the generating lines at a constant angle. Prove that the hodograph is a conic section. [Coll. Ex.]

397. Elliptic velocity. Since the velocity is represented in direction and magnitude by the radius vector of the hodograph we may use the triangle of velocities to resolve the velocity into convenient directions.

Thus when the path is an ellipse described about the focus S, the velocity is represented perpendicularly by HZ/k, where $k=b^2/h$ and H is the other focus. If C be the centre this may be resolved into the constant lengths HC, CZ, the former being a part of the major axis and the latter being parallel to the radius vector SP. Hence *the velocity in an ellipse described about the focus S can be resolved into two constant velocities one equal to ae/k in a fixed direction, viz. perpendicular to the major axis, and the other equal to a/k in a direction perpendicular to the radius vector SP of the particle, where $k=b^2/h$.* [Frost's *Newton*, 1854.]

398. The hodograph an orbit. We have seen that when the force is central a hodograph of the path of P is a polar reciprocal. It follows that if the hodograph is the path of a second particle P', each curve is one hodograph of the other.

Ex. 1. Let r, r' be the radii vectores of any two corresponding points P, Q of a curve and its polar reciprocal, the radius of the auxiliary circle being c. If these curves be described by two particles P, P' with angular momenta h, h', prove that the central forces at the two points P, Q are connected by $FF'=\dfrac{h^2h'^2}{c^8}rr'$.

Ex. 2. Prove that the two particles will not continue to be at points which correspond geometrically in taking the polar reciprocal, unless the orbit of each is an ellipse described about the centre. [The necessary condition is that the velocity $v'=kF$ in the hodograph should be equal to the velocity $v'=h'/p'$ in the orbit. Since $p'=c^2/r$, this proves that F varies as r.]

Motion of two or more attracting Particles.

399. Motion of two attracting particles. This is the problem of finding the motion of the sun and a single planet which mutually attract each other. To include the case of two suns revolving round each other, as some double stars are seen to do, we shall make no restriction as to the relative masses of the two particles. The problem can be discussed in two ways according as we require the relative motion of the two particles or the motion of each in space.

Let M, m be the masses of the sun and the planet, r their instantaneous distance. The accelerating attraction of the sun on the planet is M/r^2, that of the planet on the sun m/r^2.

Initially the sun and the planet have definite velocities. Let us apply to each an initial velocity (in addition to its own) equal and opposite to that of the sun; let us also continually apply to each an acceleration equal and opposite to that produced in the sun by the planet's attraction. The sun will then be placed initially at rest, and will remain at rest, while *the relative motion of the planet will be unaltered.* See Art. 39.

The planet being now acted on by the two forces M/r^2 and m/r^2, both tending towards the sun, the whole force is $(M+m)/r^2$. The planet therefore, as seen from the sun, moves in an ellipse having the sun in one focus. The period is

$$\frac{2\pi}{\sqrt{(M+m)}}a^{\frac{3}{2}},$$

where a is the semi-major axis of the relative orbit. In the same way the sun, as seen from the planet, appears to describe an ellipse of the same size in the same time.

400. We notice that *the periodic time of a double star does not depend on the mass of either constituent, but on the sum of the masses.* The time in the same orbit is the same for the same total mass however that mass is distributed over the two bodies.

401. Consider next *the actual motion in space of the two particles.* We know by Art. 92 that the centre of gravity of the two bodies is either at rest or moves in a straight line with uniform velocity. It is sufficient to investigate the motion relatively to the centre of gravity, for, when this is known, the actual motion may be constructed by imposing on each member of the system an additional velocity equal and parallel to that of the centre of gravity.

Let S and P be the sun and planet, G the centre of gravity, then $M \,.\, SP = (M+m)\, GP$. The attraction of the sun on the planet is

$$\frac{M}{SP^2} = \frac{M^3}{(M+m)^2}\frac{1}{GP^2} = \frac{M'}{GP^2}.$$

The attraction of the sun on the planet therefore tends to a point G fixed in space and follows the law of the inverse square. The planet therefore describes an ellipse in space with the centre of gravity in one focus, and the period is $\frac{2\pi}{\sqrt{M'}}\,\alpha^{\frac{3}{2}}$, where α is the semi-major axis of its actual orbit in space.

The actual orbits described by the sun and planet in space are obviously similar to each other and to the relative orbit of each about the other. If α, α' be the semi-major axes of the actual orbits of the planet and sun, a that of the relative orbit, we have by obvious properties of the centre of gravity,

$$\alpha/M = \alpha'/m = a/(M+m).$$

402. *To find the mass of a planet which has a satellite.* Since the mean accelerating attractions of the sun on the two bodies are nearly equal, their relative motion is also nearly the same as if the sun were away. Taking the relative orbit to be an ellipse, let a' be its semi-major axis. If m, m' are the masses of the planet and satellite, T' the period, we have $T'^2 = \frac{4\pi^2}{m+m'}\,a'^3$. When T' and a' have been found by observation, this formula gives the sum of the masses. The masses in this equation are measured in astronomical units, i.e. they are measured by the attractions of the bodies on a given supposititious particle placed at a given distance. It is therefore necessary to discover this unit by finding the attraction of some known body.

Consider the orbit described by the planet round the sun. Since we can neglect the disturbing attraction of the satellite, we have, if a is the semi-major axis of the relative orbit and T the period, $T^2 = \frac{4\pi^2}{M+m} a^3$.

Dividing one of these equations by the other, we find

$$\frac{m+m'}{M+m} = \left(\frac{T}{T'}\right)^2 \left(\frac{a'}{a}\right)^3.$$

This formula contains only a ratio of masses, a ratio of times and a ratio of lengths. Whatever units these quantities are respectively measured in, the equation remains unaltered. Since m is small compared with the mass M of the sun, and m' small compared with the mass m of the primary, we may take as a near approximation $\frac{m}{M} = \left(\frac{T}{T'}\right)^2 \left(\frac{a'}{a}\right)^3$. In this way the ratio of the mass of any planet with a satellite to that of the sun can be found.

403. The determination of the mass of a planet without a satellite is very difficult, as it must be deduced from the perturbations of the neighbouring planets. Before the discovery of the satellites of Mars, Leverrier had been making the perturbations due to that planet his study for many years. It was only after a laborious and intricate calculation that he arrived at a determination of the mass. After Asaph Hall had discovered Deimos and Phobos the calculation could be shortly and effectively made. According to Asaph Hall the mass of Mars is 1/3,093,500 of the sun, while Leverrier made it about one three-millionth. This close agreement between two such different lines of investigation is very remarkable; see Art. 57. The minuteness of either satellite enables us to neglect the unknown ratio m'/m in Art. 402 and thus to determine the mass of Mars with great accuracy.

404. Examples. *Ex.* 1. Supposing the period of the earth round the sun and that of the moon round the earth to be roughly $365\frac{1}{4}$ and $27\frac{1}{3}$ days and the ratio of the mean distances to be 385, find the ratio of the sum of the masses of the earth and moon to that of the sun. The actual ratio given in the *Nautical Almanac* for 1899 is 1/328129.

Ex. 2. The constituents of a double star describe circles about each other in a time T. If they were deprived of velocity and allowed to drop into each other, prove that they will meet after a time $T/4\sqrt{2}$.

Ex. 3. The relative path of two mutually attracting particles is a circle of radius b. Prove that if the velocity of each is halved, the eccentricity of the subsequent relative path is 3/4 and the semi-major axis is $4b/7$.

Ex. 4. Two particles of masses m, m', which attract each other according to the Newtonian law, are describing relatively to each other elliptic orbits of major axis $2a$ and eccentricity e, and are at a distance r when one of them, viz. m, is suddenly fixed. Prove that the other will describe a conic of eccentricity e' such that

$$(m+m')\left\{\frac{2}{r}-\frac{(m+m')(1-e'^2)}{am(1-e^2)}\right\}=m\left(\frac{2}{r}-\frac{1}{a}\right).$$

It is supposed that the centre of gravity had no velocity at the instant before the particle m became fixed. [Coll. Ex. 1895.]

Ex. 5. Two particles move under the influence of gravity and of their mutual attractions: prove that their centre of gravity will describe a parabola and that each particle will describe relatively to that point areas proportional to the time. [Math. Tripos, 1860.]

Ex. 6. The coordinates of the simultaneous positions of two equal particles are given by the equations

$$x=a\theta-2a\sin\theta,\quad y=a-a\cos\theta;\qquad x_1=a\theta,\quad y_1=-a+a\cos\theta.$$

Prove that if they move under their mutual attractions, the law of force will be that of the inverse fifth power of the distance. [Math. Tripos.]

Ex. 7. Two homogeneous imperfectly elastic smooth spheres, which attract one another with a force in the line of their centres inversely proportional to the square of the distance between their centres, move under their mutual attraction, and a succession of oblique impacts takes place between them; prove that the tangents of the halves of the angles through which the line of centres turns between successive impacts diminish in geometrical progression. [Math. T. 1895.]

Consider the relative motion. The blow at each impact acts along the line joining the centres, hence the latera recta of all the ellipses described between successive impacts are equal. The normal relative velocity is multiplied by the coefficient of elasticity at each impact. The radius vector of the relative ellipse is the same at each impact, being the sum of the radii of the spheres. The result follows immediately from Ex. 1, Art. 337.

405. *Ex.* 1. Herschel says that the star Algol is usually visible as a star of the second magnitude and continues such for the space of 2 days 13½ hours. It then suddenly begins to diminish in splendour and in 3½ hours is reduced to the fourth magnitude, at which it continues for about 15 minutes. It then begins to increase again and in 3½ hours more is restored to its usual brightness, going through all its changes in 2 d. 20 hr. 48 min. 54·7 sec. This is supposed to be due to the revolution round it of some opaque body which, when interposed between us and Algol, cuts off a portion of the light. Supposing the brilliancy of a star of the second magnitude to be to that of the fourth as 40 to 6·3 and that the relative orbit of the bodies is nearly circular and has the earth in its plane, prove that the radii of the two constituents of Algol are as 100 : 92 and that the ratios of their radii to that of their relative orbit are equal to ·171 and ·160. If the radius of the sun be 430000 miles and its density be 1·444, taking water as the unit, prove that the density of either constituent of Algol (taking them to be of equal densities) is one-fourth that of water. The numbers are only approximate.

[Maxwell Hall, *Observatory*, 1886.]

Ex. 2. The brightness of a variable star undergoes a periodic series of changes in a period of T years. The brightness remains constant for mT years, then gradually diminishes to a minimum value, equal to $1-k^2$ of the maximum, at which minimum it remains constant for nT years and then gradually rises to the original maximum. Show that these changes can be explained on the hypothesis that a dark satellite revolves round the star. Prove also that, if the relative orbit is circular, and the two stars are spherical, the ratio of the mean density of the double star to that of the sun is

$$\frac{\sin^3 \frac{1}{2} D}{T^2(1+k^3)}\left[\frac{(1+k)^2\cos^2 n\pi-(1-k)^2\cos^2 m\pi}{\cos^2 n\pi-\cos^2 m\pi}\right]^{\frac{3}{2}},$$

where D is the apparent diameter of the sun at its mean distance. [Math. T. 1893.]

406. Three attracting Particles. The problem of determining the relative motions of three or more attracting particles has not been generally solved. The various solutions in series which have as yet been obtained usually form the subjects of separate treatises, and are called the Lunar and Planetary theories. Laplace has however shown that there are some cases in which the problem can be accurately solved in finite terms*.

407. Let the several particles be so arranged in a plane that the resultant accelerating force on each passes through the common centre of gravity O of the system and that each resultant is proportional to the distance of the particle from that centre. It is then evident that if the proper common angular velocity be given to the system about O, the centrifugal force on each particle may be made to balance the attraction on that particle. The particles of the system will then move in circles round O with equal angular velocities, the lines joining them forming a figure always equal and similar to itself. Each particle also will describe a circle relatively to any other particle.

Let us next enquire what conditions are necessary that the particles may so move that the figure formed by them is always similar to its original shape, but *of varying size.* Let the distances

* Laplace's discussion may be found in the sixth chapter of the tenth book of the *Mécanique Céleste.* The proposition that the motion when the particles are in a straight line is unstable was first established by Liouville, *Académie des Sciences*, 1842, and *Connaissance des Temps* for 1845 published in 1842. His proof is different from that given in the text. The motion when the particles are at the corners of an equilateral triangle is discussed in the *Proceedings of the London Mathematical Society*, Feb. 1875. See also the author's *Rigid Dynamics*, vol. I. Art. 286, and vol. II. Art. 108. There is also a paper by A. G. Wythoff, *On the Dynamical stability of a system of particles, Amsterdam Math. Soc.* 1896.

of the particles from the centre of gravity O be r_1, r_2, &c. We then have for each particle the equations

$$\frac{d^2r}{dt^2} - r\left(\frac{d\theta}{dt}\right)^2 = -F, \quad \frac{1}{r}\frac{d}{dt}\left(r^2\frac{d\theta}{dt}\right) = G.$$

Since the figure is always similar, these equations are to be satisfied when $d\theta/dt$ is the same for every particle, and r_1, r_2, &c. have the ratios α_1, α_2, &c., where α_1, α_2, &c., are some positive finite constant quantities. It immediately follows that the arrangement must be such that the F's are in the same positive ratios and also the G's.

Since the mutual attractions of the particles form a system of forces in equilibrium, the equivalent system m_1F_1, m_2F_2, &c. and m_1G_1, m_2G_2, &c. is also in equilibrium. The sum of the moments of the G's about O must therefore be zero, which (since they are in the ratios α_1, &c.) is impossible unless each G is zero.

If also the initial conditions are such that both the radial velocities dr_1/dt, &c. and the transverse velocities $r_1d\theta/dt$, &c., have the ratios α_1, &c., all the equations will be satisfied by assuming r_1, r_2, &c. to have the constant ratios α_1, α_2, &c. The motion of some one particle, say m_1, is determined by the two polar equations of that particle.

The result is, that if the particles move so as to be always at the corners of a similar figure, that figure must be such that the resultant accelerating forces on the particles act towards the common centre of gravity O and are proportional to the distances from O. This being true initially, the particles must be projected in directions making equal angles in the same sense with their distances from O, with velocities proportional to those distances.

408. The two arrangements. *To determine how three particles must be arranged so that the force on any one may pass through the common centre of gravity; the law of force being the inverse κth power of the distance.*

It is evident that the condition is satisfied when the three particles are arranged in a straight line. We have now to enquire if any other arrangement is possible.

It is a known theorem in attraction that if two given particles of masses M, m attract a third m', placed at distances ρ, r from them, with accelerating forces $M\rho$, mr, the resultant passes through the centre of gravity of M, m and therefore through that of all three. In order that the resultant of M/ρ^κ and m/r^κ may also pass through the centre of gravity of M, m, it is evident that the ratio of M/ρ^κ to m/r^κ must be equal to the ratio $M\rho$ to mr. It immediately follows (except $\kappa = -1$) that $\rho = r$. The three particles must therefore be at equal distances; see also Art. 304.

The result is that for three attracting particles there are only two possible arrangements; (1) that in which the particles, however unequal their masses may be, are at the corners of an equilateral triangle, (2) that in which they are in the same straight line.

It may also be shown that when the law of attraction is the inverse κth, *the arrangement at the corners of an equilateral triangle is stable when* $\dfrac{(\Sigma m)^2}{\Sigma mm'} > 3\left(\dfrac{1+\kappa}{3-\kappa}\right)^2$.

409. The line arrangement. *Three mutually attracting particles whose masses are M, m', m are placed in a straight line. It is required to determine the conditions that throughout their subsequent motion they may remain in a straight line.*

Let the law of attraction be the inverse κth power of the distance. Let M, m, be the two extreme particles, m' being between the other two. Let a, b, c be the distances Mm, Mm', $m'm$; then $a = b + c$.

A necessary condition is that the resultant accelerating forces on the particles must be proportional to their distances from the centre of gravity O (Art. 407). We therefore have

$$\frac{M/a^\kappa + m'/c^\kappa}{Ma + m'c} = \frac{M/b^\kappa - m/c^\kappa}{Mb - mc} = \frac{m/a^\kappa + m'/b^\kappa}{ma + m'b} \quad \text{........(1)},$$

where the numerators express the accelerating forces on the particles and the denominators are proportional to the distances from O.

The equalities (1) are equivalent to only one equation, for if we multiply the numerators and denominators of the three fractions by m, m', $-M$ respectively, the sum of the numerators and also that of the denominators are zero. Putting $a = b(1+p)$, $c = bp$, we arrive at

$$Mp^\kappa\{(1+p)^{\kappa+1} - 1\} - m'(1+p)^\kappa(1-p^{\kappa+1}) - m\{(1+p)^{\kappa+1} - p^{\kappa+1}\} = 0 \text{...(2)}.$$

The left-hand side is negative when $p = 0$ and positive when p is infinitely large, *the equation therefore has one real positive root, whatever positive values M, m', m may have.* Putting $p = 1$, the left side becomes $(M - m)(2^{\kappa+1} - 1)$; since we may take M as the greater of the two extreme particles we see that *the real positive value of p is less than unity, provided $\kappa + 1$ is positive.* If $\kappa + 1$ were negative the root would be greater than unity.

Whatever the masses of the particles may be it follows that if they are so placed that their distances have the ratios given by this value of p, and their parallel velocities are proportional to their distances from O, they will throughout their subsequent motion remain in a straight line.

When the attraction follows the Newtonian law, the equation (2) becomes the quintic

$$(M + m')p^5 + (3M + 2m')p^4 + (3M + m')p^3 - (m' + 3m)p^2 \\ - (2m' + 3m)p - (m + m') = 0 \ldots(3).$$

The terms of this equation exhibit but one variation of sign, and there is therefore but one positive root.

It may be shown in exactly the same way that in the general case, when κ has any positive integral value, the equation (2) has only one positive root; all the terms from $p^{2\kappa+1}$ to $p^{\kappa+1}$ being positive, while those from p^{κ} to p^0 are negative.

410. When the positions of two of the masses are given, there are three possible cases; according as the third is between the other two or on either side. Since the analytical expression for the law of the inverse square does not represent the attraction when the attracted particle passes through the centre of force, Art. 135; these three cases cannot be included in the same equation. We thus have three equations of the form (3), one for each arrangement.

411. In the case of the sun, earth, and moon, M is very much greater than either m or m'. Since p vanishes when m and m' are zero, we infer that p is very small when m/M and m'/M are small. The equation (3) therefore gives $3p^3 = (m + m')/M$, or, using the numerical values of m, m' and M, $p = 1/100$ nearly.

If the moon were therefore placed at a distance from the earth one hundredth part of that of the sun, the three bodies might be projected so that they would always remain in a straight line. The moon would then be always full, but at that distance its light would be much diminished. This configuration of the sun, earth and moon however could not occur in nature because this state of steady motion is unstable. On the slightest disturbance the whole system would change and the particles would widely deviate from their former paths.

412. *Three mutually attracting particles whose masses are M, m', m describe circles round their common centre of gravity and are always in a straight line. Prove that if the force vary as any inverse power of the distance this state of motion is unstable.*

Reducing the particle M to rest we take that point as the origin of coordinates. Let (r, θ) be the coordinates of m, (r', θ') those of m'. The particle m is acted on by $(M+m)/r^\kappa$ along the straight line mM, and m'/r'^κ in a direction parallel to $m'M$. The polar equations of the motion of m are

$$\left.\begin{aligned} \frac{d^2r}{dt^2} - r\left(\frac{d\theta}{dt}\right)^2 &= -\frac{M+m}{r^\kappa} - \frac{m'}{r'^\kappa}\cos\omega - \frac{m'}{R^\kappa}\cos\phi \\ \frac{1}{r}\frac{d}{dt}\left(r^2\frac{d\theta}{dt}\right) &= -\frac{m'}{r'^\kappa}\sin\omega + \frac{m'}{R^\kappa}\frac{r'\sin\omega}{R} \end{aligned}\right\} \ldots\ldots\ldots\ldots\ldots (1),$$

where ω, ϕ are the angles at M, m of the triangle formed by joining the particles and R is the side mm'. In the same way the polar equations of the motion of m' are

$$\left.\begin{aligned} \frac{d^2r'}{dt^2} - r'\left(\frac{d\theta'}{dt}\right)^2 &= -\frac{M+m'}{r'^\kappa} - \frac{m}{r^\kappa}\cos\omega + \frac{m}{R^\kappa}\cos\phi' \\ \frac{1}{r'}\frac{d}{dt}\left(r'^2\frac{d\theta'}{dt}\right) &= \frac{m}{r^\kappa}\sin\omega - \frac{m}{R^\kappa}\sin\phi' \end{aligned}\right\} \ldots\ldots\ldots\ldots (2),$$

where ϕ' is the external angle of the triangle at m'. In forming these equations the standard case is that in which $\theta' > \theta$ and $r' < r$.

We shall now substitute in these equations $r = a(1+x)$, $\theta = nt + y$; $r' = b(1+y)$, $\theta' = nt + \eta$, and reject all powers beyond the first of the small quantities x, y, ξ, η. Remembering that $\sin\phi/r' = \sin\phi'/r = \sin\omega/R$ we find after some reduction

$$\begin{aligned} (\delta^2 - n^2 - \kappa E)\,x - 2n\delta y + m'\kappa B\xi + 0\,.\,\eta &= 0, \\ 2n\delta x + (\delta^2 + m'B)\,y + 0\,.\,\xi - m'B\eta &= 0, \\ m\kappa Ax + 0\,.\,y + (\delta^2 - n^2 - \kappa F)\,\xi - 2n\delta\eta &= 0, \\ 0\,.\,x - mAy + 2n\delta\xi + (\delta^2 + mA)\,\eta &= 0, \end{aligned}$$

where for brevity we have written δ for d/dt, and $c = a - b$,

$$A = \frac{a}{b}\left(\frac{1}{c^{\kappa+1}} - \frac{1}{a^{\kappa+1}}\right), \quad B = \frac{b}{a}\left(\frac{1}{c^{\kappa+1}} - \frac{1}{b^{\kappa+1}}\right),$$

$$E = \frac{M+m}{a^{\kappa+1}} + \frac{m'}{c^{\kappa+1}}, \quad F = \frac{M+m'}{b^{\kappa+1}} + \frac{m}{c^{\kappa+1}}.$$

The steady motion has been already found in Art. 409, but it may also be deduced from the first and third of the equations (1) and (2) by equating the constants. We thus find $n^2 = E - m'B$, $n^2 = F - mA$.

We notice that the constants E, F are positive. When $\kappa+1$ is positive, it has been shown in Art. 409 that $a>b>c$, and therefore A, B and $E+F-2n^2$ are positive. Lastly whatever κ may be $E+F-n^2$ is positive

To solve the four equations, we put $x=Ge^{\lambda t}$, $y=He^{\lambda t}$, $\xi=Ke^{\lambda t}$, $\eta=Le^{\lambda t}$. Substituting and eliminating the ratios G, H, K, L we obtain a determinantal equation whose constituents are the coefficients of x, y, ξ, η, with λ written for δ. This determinant is of the eighth degree in λ. To find its factors we must before expansion make some necessary simplifications which we can only indicate here. We first add the ξ column to the x column and the η column to the y column. The second column may now be divided by λ. Multiplying the second column by $2n$ and subtracting from the first, we see that $\lambda^2-(\kappa-3)n^2$ is another factor which we divide out. Subtracting the first row from the third and the second from the fourth, the first column acquires three zeros and the second column two. The determinant is now easily expanded and we have

$$\lambda^2\{\lambda^2-(\kappa-3)n^2\}\{(\lambda^2+C)(\lambda^2-C\kappa-(\kappa+1)n^2)+4n^2\lambda^2\}=0,$$

where $C=E+F-2n^2$. If $\kappa>3$, this equation gives a real positive value of λ and the motion is therefore unstable. If κ have any positive value C is positive, and the third factor has the product of its roots negative; one value of λ^2 is real and positive and the other real and negative. *The motion is therefore unstable for all positive values of* κ.

413. *Ex.* 1. Three mutually attracting particles are placed at rest in a straight line. Show that they will simultaneously impinge on each other if the initial distances apart are given by the value of p in the equation of the $(2\kappa+1)$th degree of Art. 409. [This equation expresses the condition that the distances between the particles are always in a constant ratio.]

Ex. 2. Three unequal mutually attracting particles are placed at rest at the corners of an equilateral triangle and attract each other according to the inverse κth power of the distances. Prove that they will arrive simultaneously at the common centre of gravity. If the law of attraction is the inverse square, the time of transit is $\frac{1}{2}\pi\,(a^3/2\mu)^{\frac{1}{2}}$ where μ is the sum of the masses and a the side of the initial triangle, Art. 131.

414. A swarm of particles. Let us suppose that a comet is an aggregation of particles whose centre of gravity describes an elliptic orbit round the sun. The question arises, what are the conditions that such a swarm could keep together*? Similar conditions must be satisfied in the case of a swarm consolidating

* The disintegration of comets was first suggested by Schiaparelli who proved that the disturbing force of the sun on a particle might be greater than the attraction of the comet. He thus obtained as a necessary condition of stability $m/b^3>2M/a^3$. The subject was dynamically treated by Charlier and Luc Picart on the supposition of a circular trajectory. They arrived at the condition $m/b^3>3M/a^3$; *Bulletin de l'Académie de S. Pétersbourg*, *Annales de l'Observatoire de Bordeaux*, Tisserand, *Méc. Céleste*, IV. The condition of stability was extended to the case of an elliptic trajectory by M. O. Callandreau in the *Bulletin Astronomique*, 1896. The brief solutions here given of these problems are simplifications of their methods.

into a planet in obedience to the Nebular theory. The following example will illustrate the method of proceeding.

We shall suppose the sun A to be fixed in space, Art. 399. Let B be the centre of the swarm, C any particle. Let r, θ be the polar coordinates of B referred to A, and ξ, η the coordinates of C referred to B as origin, the axis of ξ being the prolongation of AB. Let M be the mass of the sun. Supposing, as a first approximation, that the swarm is homogeneous and spherical, its attraction at an *internal point* C is $\mu\rho$, where $\rho = BC$. If m be the mass and b the radius of the swarm, $\mu b = m/b^2$.

The equations of motion are, by Art. 227,

$$\left.\begin{aligned}\frac{d^2(r+\xi)}{dt^2} - (r+\xi)\left(\frac{d\theta}{dt}\right)^2 - \frac{1}{\eta}\frac{d}{dt}\left(\eta^2\frac{d\theta}{dt}\right) &= \frac{-M}{(r+\xi)^2} - \mu\xi \\ \frac{d^2\eta}{dt^2} - \eta\left(\frac{d\theta}{dt}\right)^2 + \frac{1}{r+\xi}\frac{d}{dt}\left\{(r+\xi)^2\frac{d\theta}{dt}\right\} &= \frac{-M\eta}{(r+\xi)^3} - \mu\eta\end{aligned}\right\}\quad\ldots(1).$$

These equations also apply to the motion of the particle at B, where $\xi = 0$, $\eta = 0$. Hence when we expand in powers of ξ, η, all the terms independent of ξ, η must cancel out. We thus have

$$\left.\begin{aligned}\frac{d^2\xi}{dt^2} - 2\frac{d\eta}{dt}\frac{d\theta}{dt} - \eta\frac{d^2\theta}{dt^2} - \xi\left(\frac{d\theta}{dt}\right)^2 &= \frac{2M\xi}{r^3} - \mu\xi \\ \frac{d^2\eta}{dt^2} + 2\frac{d\xi}{dt}\frac{d\theta}{dt} + \xi\frac{d^2\theta}{dt^2} - \eta\left(\frac{d\theta}{dt}\right)^2 &= \frac{-M\eta}{r^3} - \mu\eta\end{aligned}\right\}\quad\ldots\ldots(2).$$

If the centre of gravity of the swarm describe a circle about the sun, we write $r = a$, $d\theta/dt = n$. The equations then become

$$\left.\begin{aligned}\frac{d^2\xi}{dt^2} - 2n\frac{d\eta}{dt} + (\mu - 3n^2)\,\xi &= 0 \\ \frac{d^2\eta}{dt^2} + 2n\frac{d\xi}{dt} + \mu\eta \qquad &= 0\end{aligned}\right\}\quad\ldots\ldots\ldots\ldots\ldots(3).$$

Putting $\xi = A\cos(pt+\alpha)$, $\eta = B\sin(pt+\alpha)$, we immediately obtain the determinantal equation

$$(p^2 - \mu + 3n^2)(p^2 - \mu) - 4p^2n^2 = 0 \ldots\ldots\ldots\ldots\ldots(4).$$

The condition that the particles of the swarm should keep together is the same as the condition that the roots of this quadratic should be real and positive. The left-hand side is positive when $p^2 = \pm\infty$, and negative when $p^2 = \mu$ and $p^2 = \mu - 3n^2$. The required condition is therefore $\mu > 3n^2$, Art. 288. *The condition that the swarm is stable is therefore* $\dfrac{m}{b^3} > 3\dfrac{M}{a^3}$.

Unless therefore the density of the swarm exceed a certain quantity the swarm cannot be stable. If the mass of the sun were distributed throughout the sphere whose radius is such that the swarm is on the surface, the density of the swarm must be at least three times that of the sphere.

The path of the particle C when describing either principal oscillation is (relatively to the axes $B\xi$, $B\eta$) an ellipse with its centre at B. Substituting the values of ξ, η in the equations of motion and using the quadratic, we find

$$\frac{A}{B}=\frac{p^2-\mu}{-2np},\quad \frac{A^2}{B^2}=1-\frac{3}{4}\frac{p^2-\mu}{p^2},\quad \frac{A_1A_2}{B_1B_2}=-\sqrt{\frac{\mu}{\mu-3n^2}}.$$

Since μ lies between the values of p^2, the first equation shows that A_1/B_1 and A_2/B_2 have opposite signs, and accordingly the radical is negative.

It follows that the oscillation which corresponds to the smaller value of p has the major axis directed along $B\xi$, while in the other that axis is along $B\eta$. The particle also describes the ellipses in opposite directions, in the former case the direction is the same as that of the swarm round the sun, in the latter, the opposite.

If the centre of gravity of the swarm describe an ellipse of small eccentricity, we may obtain an approximate solution of the equations of motion. Assuming the expansions $\theta=nt+2e\sin nt+\frac{5}{4}e^2\sin 2nt$,

$$\frac{h}{r^2}=\frac{d\theta}{dt};\quad \therefore \left(\frac{a}{r}\right)^3=1+3e\cos nt+\tfrac{3}{2}e^2+\tfrac{9}{2}e^2\cos 2nt,$$

it is evident that all the coefficients of the differential equations (2) can be at once expressed in terms of t, including all terms which contain e^2. It is however unnecessary for our present purpose to write these at length. It is easy to see that the equations become

$$\left.\begin{aligned}\frac{d^2\xi}{dt^2}-2n\frac{d\eta}{dt}+\{\mu-n^2(3+5e^2)\}\,\xi&=eX\\ \frac{d^2\eta}{dt^2}+2n\frac{d\xi}{dt}+\{\mu-\tfrac{1}{2}n^2e^2\}\,\eta&=eY\end{aligned}\right\}\ \ldots\ldots\ldots\ldots\ (5),$$

$$eX=\ \ 4en\cos nt\frac{d\eta}{dt}-2en^2\sin nt\,\eta+10en^2\cos nt\,\xi+\&c.,$$

$$eY=-4en\cos nt\frac{d\xi}{dt}+2en^2\sin nt\,\xi+\ en^2\cos nt\,\eta+\&c.$$

As a first approximation we neglect eX, eY. Comparing the equations (5) and (3) we see at once that we shall have the quadratic

$$\{p^2-\mu+n^2(3+5e^2)\}\{p^2-\mu+\tfrac{1}{2}n^2e^2\}-4p^2n^2=0\ \ldots\ldots\ldots\ldots(6).$$

The condition that the swarm is stable is then $\mu>n^2(3+5e^2)$; $\therefore\ \dfrac{m}{b^3}>\dfrac{M}{a^3}(3+5e^2)$.

It appears therefore that *the gradual dissipation of a comet is more probable when the trajectory is elliptical than when it is circular.*

As a second approximation, we substitute $\xi = A\cos(pt+\alpha)$, $\eta = B\sin(pt+\alpha)$ in the expressions X and Y. By Art. 303 the only important terms are those which become magnified by the process of solution. These terms are of the form $P\cos(\lambda t+L)$ where $\lambda = p\pm n$ or $p\pm 2n$. Unless therefore the roots p, p' of the quadratic (6) or (4) are such that $p\pm p'$ is nearly equal to n or $2n$, the terms derived from X, Y remain respectively of the order e or e^2. This relation between the roots cannot occur when e is small.

415. Tisserand's criterion*. When a comet describing a conic round the sun passes very near to a planet, such as Jupiter, its course is much disturbed. When it emerges from the sphere of perceptible influence of the planet, it may again be supposed to describe a conic round the sun, but the elements of the new path may be very different from those of the old.

Since Jacobi's integral (Art. 255) holds throughout the motion, the elements of both the conics must satisfy that equation.

Let (a_0, l_0), (a_1, l_1) be the semi-major axis and semi-latus rectum before and after passing through the sphere of influence of the planet. Let i_0, i_1 be the inclinations of the planes of the comet's orbit to the plane of the planet's motion.

Let the sun O be taken as the origin of coordinates, and let the axis of ξ pass through the planet P. Let r, ρ be the distances of the comet Q from O and P respectively and $c = OP$. Let M, m be the masses of the sun and planet, then, reducing the sun to rest (Art. 399), we regard the comet as acted on by the resultant attraction of the sun and planet together with a force m/c^2 acting parallel to PO. The field of force is therefore defined by

$$U = \frac{M}{r} + \frac{m}{\rho} - \frac{m\xi}{c^2}.$$

We suppose that the planet P describes a circular orbit relatively to O with a constant angular velocity n, where $n^2 = (M+m)/c^3$. The Jacobian integral takes the form

$$\tfrac{1}{2}V^2 - nA - \frac{M}{r} - \frac{m}{\rho} + \frac{m\xi}{c^2} = C,$$

* Tisserand's criterion may be found in his Note sur l'integrale de Jacobi, et sur son application à la théorie des comètes, *Bulletin Astronomique*, Tome VI. 1889, also in his *Mécanique Céleste*, Tome IV. 1896. M. O. Callandreau's addition is given in the second chapter of his Étude sur la théorie des comètes périodiques, *Annales de l'Observatoire de Paris, Mémoires*, 1892, Tome XX. There are also some investigations by H. A. Newton on the capture of comets by planets, especially Jupiter, *American Journal of Science*, vol. XLII. pages 183 and 482, 1891.

where V is the space velocity of the comet and A its angular momentum referred to a unit of mass. Since (Art. 333)

$$V^2 = M\left(\frac{2}{r} - \frac{1}{a}\right), \quad A = \cos i \sqrt{(Ml)},$$

the integral becomes

$$\frac{1}{2a_0} + n\cos i_0 \sqrt{\frac{l_0}{M}} + \frac{m}{M}\left(\frac{1}{\rho_0} - \frac{\xi_0}{c^2}\right) = \frac{1}{2a_1} + n\cos i_1 \sqrt{\frac{l_1}{M}} + \frac{m}{M}\left(\frac{1}{\rho_1} - \frac{\xi_1}{c^2}\right),$$

where ξ_0, ρ_0; ξ_1, ρ_1, are the values of ξ, ρ when the comet is respectively entering and leaving the sphere of influence of the planet. We obviously have $\rho_0 = \rho_1$, and since the comet does not stay long within the sphere, we may neglect $\xi_0 - \xi_1$ when multiplied by the very small quantity m/M. Writing then $n^2 = M/c^3$ as a close approximation, Art. 341, we obtain the criterion

$$\frac{1}{2a_0} + \frac{\cos i_0 \sqrt{l_0}}{c\sqrt{c}} = \frac{1}{2a_1} + \frac{\cos i_1 \sqrt{l_1}}{c\sqrt{c}}.$$

416. Tisserand uses this criterion to determine whether two comets both of which are known to have passed near Jupiter could be the same body. If the criterion is not satisfied by the known elements of the two comets, they cannot be the same body. If it is satisfied it is then worth while to examine more thoroughly how much the elements of either body have been altered by the attraction of Jupiter. This must be done by using the method of the planetary theory and is generally a laborious process.

In Tisserand's criterion the orbit of Jupiter is considered to be circular, which is not strictly correct. This defect has been corrected by M. O. Callandreau. Taking account only of the first power of the eccentricity he adds a small term containing that eccentricity as a factor. This term, unlike those in Tisserand's criterion, depends on the manner in which the comet approaches Jupiter.

417. Stability deduced from Vis Viva. The Jacobian integral has been used by G. W. Hill* to determine whether the moon could be indefinitely pulled away from the earth by the disturbing attraction of the sun. In such a problem as this, it is convenient to take the origin at the earth P and the moving axis of ξ directed towards the sun O. Reducing the earth to rest, the moon Q is acted on by $(m+m')/\rho^2$ along QP and M/c^2 parallel to OP. The Jacobian equation for relative motion, Art. 255 (3), takes the form

$$\tfrac{1}{2}v^2 = \tfrac{1}{2}n^2\rho^2 + \frac{\mu}{\rho} + \frac{M}{r} - \frac{M}{c^2}\xi + C,$$

* G. W. Hill's researches in the Lunar theory may be found in the *American Journal of Mathematics*, vol. I. 1878.

where $\rho = PQ$, $r = OQ$, $c = OP$ and μ is the sum of the masses m, m' of the earth and moon. We treat the sun's orbit as circular and put as a near approximation $M/c^3 = n^2$. Since $\rho^2 = \xi^2 + \eta^2$, this equation becomes

$$\tfrac{1}{2}v^2 = \frac{\mu}{\rho} + \frac{c^3n^2}{r} + \frac{n^2}{2}\{(c-\xi)^2 + \eta^2\} - C'.$$

Since the left-hand side is essentially positive it is clear that *the moving particle Q can never cross the surface defined by equating the right-hand side to zero, and can only move in those parts of space in which the right-hand side is positive.* Art. 299.

If the initial circumstances of the motion make C' negative, the right-hand side is always positive and the equation supplies no limits to the position of Q.

The form of the surface when C' is positive has been discussed by Hill. When C' exceeds a certain quantity the surface has in general three separate sheets. The inner of these is smaller than the other two and surrounds the earth. The second is also closed but surrounds the sun, the third is not closed. When the constants are adapted to the case of the moon, that satellite is found to be within the first sheet. It must therefore always remain there, and its distance from the earth can never exceed 110 equatorial radii. Thus *the eccentricity of the earth's orbit being neglected, we have a rigorous demonstration of a superior limit to the radius vector of the moon.*

418. *Ex.* 1. If the moon Q move in the plane of motion of the earth P and if also the sun is so remote that we may put $\frac{c^3}{r} + \frac{1}{2}r^2 = \frac{3}{2}c^2\left(1 + \frac{\xi^2}{c^2}\right)$ when the left-hand side is expanded in powers of ξ/c and η/c, the bounding surface degenerates into the curve $\frac{\mu}{\rho} + \frac{3}{2}n^2\xi^2 = C''$. It is required to trace the forms of this curve for different positive values of C''.

The curve has two infinite branches tending to the asymptotes $\frac{3}{2}n^2\xi^2 = C''$. If C'' is greater than the minimum value of $\mu/\xi + \frac{3}{2}n^2\xi^2$ there is also an oval round the body S. If the particle Q is within the oval, it cannot escape thence and its radius vector will have a superior limit. If the particle is beyond either of the infinite branches, it cannot cross them and the radius vector will have an inferior limit. The velocity at any point of the space between the oval and the infinite branches is imaginary. [Hill.]

Ex. 2. A double star is formed by two equal constituents S, P whose orbits are circles. A third particle Q whose mass is infinitely small moves in the same plane and initially is at a distance from P on SP produced equal to half SP, starting with such velocity that it would have described a circular orbit about P if S had been absent. Show that the curve of no relative velocity is closed, and that the particle being initially within that curve cannot recede indefinitely from the attracting bodies S and P.

This example is discussed by Coculesco in the *Comptes Rendus*, 1892. He also refers to a memoir of M. de Haerdtl, 1890, where the revolution of Q round P is traced during two revolutions and it is shown that at the end of the third the particle is receding from A *.

---

* Since writing the above the author has received Darwin's memoir on Periodic Orbits, *Acta Mathematica*, XXI. in which the motion of a planet about a binary star

Theory of Apses.

419. *When the law of force is a one-valued function of the distance, every apsidal radius vector must divide the orbit symmetrically.*

Let O be the centre of force, A an apse (Art. 314). The argument rests on two propositions.

(1) If two particles are projected from A with equal velocities, both perpendicularly to OA but in opposite directions, it is clear that (the force being always the same at the same distance from O) the paths described must be symmetrical about OA.

(2) If at any point of its path, the velocity of the particle were reversed in direction (without changing its magnitude), the particle would describe the same path but in a reverse direction.

If then a particle describing an orbit arrive at an apse A, its subsequent path when reversed must be the same as its previous path. Hence OA divides the whole orbit symmetrically.

We may notice that if the law of force were not one-valued, say $F = \mu\{u \pm \sqrt{(u^2 - a^2)}\}$, where the apsidal distance $OA = a$, the first proposition is not true, unless it is also given that the radical keeps one sign.

420. *There can be only two apsidal distances though there may be any number of apses.*

Let the particle after passing an apse A arrive at another apse B. Then since OB divides the orbit symmetrically, there must be a third apse C beyond B such that the angles AOB, BOC are equal and $OC = OA$. Since OC divides the orbit symmetrically, there is a fourth apse at D, where $OD = OB$ and the angles BOC, COD are equal. The apsidal distances are therefore alternately equal, and the angle contained at O by any two consecutive apsidal distances is always the same.

has been more thoroughly studied. Taking a variety of initial conditions he has traced the subsequent paths of a particle of insignificant mass. Some of the paths thus presented to the eye have such unexpected and remarkable forms that the paper is full of interest.

421. Examples. *Ex.* 1. Show that an ellipse cannot be described about a centre of force whose attraction is a one-valued function of the distance unless that centre is situated on a principal diameter and is outside the evolute.

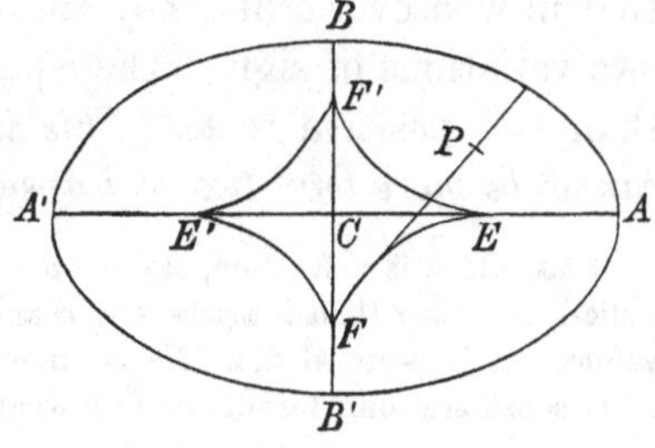

By drawing all the tangents to one arc EF of the evolute we see that they cover the whole area of the quadrant ACB of the ellipse. It follows that a normal to the ellipse can be drawn through any point P situated in this quadrant, and this normal does not divide the ellipse symmetrically, unless P lies between E and A or between F' and B.

Ex. 2. If the path is an equiangular spiral and the central force a one-valued function of the distance, prove that the centre of force must be situated in the pole.

Ex. 3. If a particle of mass m be attached to a fine elastic string of natural length a and modulus λ, and lie with the string unstretched and one extremity fixed on a smooth horizontal plane; prove that, if projected at right angles to the string with velocity v, the string will just be doubled in length at its greatest extension if $3mv^2 = 4a\lambda$. [Coll. Ex.]

Ex. 4. A particle is projected from an apse with a velocity v, prove that the apse will be an apocentre or a pericentre according as the velocity v is less or greater than that in a circle at the same distance.

422. The apsidal distances. *To find the apsidal distances when $F = \mu u^n$, and n is an integer.*

The equation of vis viva, viz. $v^2 = C - 2\int F dr$, gives

$$v^2 = h^2 \left\{ \left(\frac{du}{d\theta}\right)^2 + u^2 \right\} = C + \frac{2\mu}{n-1} u^{n-1} \quad \ldots\ldots\ldots\ldots(1).$$

Let V be the velocity at the initial distance R, β the angle of projection, then

$$V^2 = C + \frac{2\mu}{n-1}\left(\frac{1}{R}\right)^{n-1}, \quad h = VR\sin\beta \ \ldots\ldots\ldots(2).$$

Thus both h and C are known quantities, at an apse u is a max-min, and therefore $du/d\theta = 0$. The apsidal distances are therefore given by

$$\left(\frac{dr}{dt}\right)^2 = h^2\left(\frac{du}{d\theta}\right)^2 = \frac{2\mu}{n-1} u^{n-1} - h^2u^2 + C = 0 \ \ldots\ldots \text{(A)}.$$

If an equation is arranged in descending powers of the unknown quantity, we know by Descartes' theorem that there cannot be more positive roots than variations of sign. The arrangement of the terms of equation (A) will depend on whether $n-1$ is greater or less than 2; but, since there are only three terms, it is clear that in whatever order they are placed there cannot be more than two variations of sign. The equation cannot therefore have more than two positive roots. *This is an analytical proof that there cannot be more than two real apsidal distances.*

423. If n is a fraction, say $n=p/q$ in its lowest terms, we write $u=w^q$; the indices of w are then integers and w and therefore u can have only two positive values. It is assumed that if q is an even integer the sign of F is given by some other considerations, for otherwise F would not be a one-valued function of u.

424. The propositions proved in Arts. 420 and 422 are not altogether the same. The complete curve found by integrating (A) may have several branches separated from each other so that the particle cannot pass from one to the other. In 420 it is proved that *the actual branch described* cannot have more than two unequal apsidal distances. In 422 it is proved that when $F=\mu u^n$ *all the branches together* cannot have more than two unequal apsidal distances.

If the force be some other one-valued function of the distance the complete curve may have more than two unequal apsidal distances.

425. *Ex.* 1. If $\left(\frac{du}{d\theta}\right)^2=A(u-a)(u-b)(u-c)$ be the differential equation of an orbit, prove that the central force is a one-valued function of the distance. Prove also that the curve has two branches and three unequal apsidal distances, and that either branch may be described if the initial conditions are suitable. See Arts. 309, 441.

Ex. 2. If the central force is $F=\mu u^n$, where $n>3$ and the velocity is greater than that from infinity, prove that the apsidal distances lie between p and q, where $2\mu=h^2(n-1)p^{n-3}$ and $h^2=Cq^2$. [This follows from a theorem in the theory of equations applied to equation (A) of Art. 422.]

426. The apsidal angle. *To find the apsidal angle when $F=\mu u^n$, where $n<3$, and the orbit is nearly circular.*

The equation of the path with these conditions has been found by continued approximation in Arts. 367 to 370.

Taking the first approximation, we see by referring to the equation (6) of those articles that $du/d\theta$ is zero only when $p\theta+\alpha=i\pi$, where i is any integer. These values of θ therefore determine the apses and the reciprocals of the two corresponding apsidal distances are $c(1\pm M)$. The apsidal angle described between two consecutive apses is therefore π/p, where $p^2=3-n$.

Taking the higher approximations, we use the equations (12) and (13) in the same way. The apsidal angle is therefore π/p, where

$$p = \sqrt{(3-n)}\{1 - \tfrac{1}{24}(n-2)(n+1)M^2\}.$$

The reciprocals of the apsidal distances are very nearly $c(1 \pm M)$.

427. There is another method of finding the apsidal angle which is founded on a direct integration of the equations of motion*. Beginning with

$$\frac{d^2u}{d\theta^2} + u = \frac{\mu u^{n-2}}{h^2},$$

we have, as in Art. 422,

$$h^2\left(\frac{du}{d\theta}\right)^2 = \frac{2\mu}{n-1}u^{n-1} - h^2u^2 + C;$$

let $u=a$, $u=b$ be the reciprocals of the inner and outer apsidal distances. Since the right-hand side of the equation must vanish for each of these values of u, we have

$$\frac{2\mu}{n-1}a^{n-1} - h^2a^2 + C = 0, \qquad \frac{2\mu}{n-1}b^{n-1} - h^2b^2 + C = 0.$$

Eliminating h^2 and C we find

$$\left(\frac{d\theta}{du}\right)^2 = \frac{a^{n-1} - b^{n-1}}{\Delta}, \qquad \Delta = \begin{vmatrix} u^{n-1}, & u^2, & 1 \\ a^{n-1}, & a^2, & 1 \\ b^{n-1}, & b^2, & 1 \end{vmatrix}.$$

To find the apsidal angle we have to integrate the value of $d\theta$ from $u=b$ to a.

To simplify the limits we put $a=c(1+M)$, $b=c(1-M)$ and $u=c(1+Mx)$; the limits of integration are then $x=-1$ to $+1$. Also since the orbit is nearly circular, we suppose M to be a small quantity.

It now becomes necessary to expand Δ in powers of M. This may be effected by using some simple properties of determinants. If we subtract the upper row from each of the other two, the determinant is practically reduced to a determinant of two rows. Noticing that

$$(1 \pm M)^{n-1} - (1+Mx)^{n-1} = -(n-1)M(x \mp 1)\{1 + CM(x \pm 1) + DM^2(x^2 \pm x + 1) + EM^3(x^3 \pm x^2 + x \pm 1) + \&c.\},$$

where $C = \frac{1}{2}(n-2)$, $D = \frac{1}{6}(n-2)(n-3)$, $E = \frac{1}{24}(n-2)(n-3)(n-4)$, we see that the new determinant is

$$\Delta = c^{n+1}M^2(n-1)(x^2-1)\begin{vmatrix} 1 + CM(x+1) + \&c., & 2 + M(x+1) \\ 1 + CM(x-1) + \&c., & 2 + M(x-1) \end{vmatrix}.$$

Subtracting one row from the other and performing some evident simplifications, we find

$$\Delta = E^2(x^2-1)\{1 + \tfrac{1}{3}(n-2)Mx + \tfrac{1}{12}(n-2)M^2((n-4)x^2 + n - 6)\},$$

where $E^2 = 2c^{n+1}M^3(n-1)(n-3)$. We thence deduce

$$\frac{E}{\sqrt{\Delta}} = \frac{1}{\sqrt{(x^2-1)}}\{1 - \tfrac{1}{6}(n-2)Mx + \tfrac{1}{24}(n-2)M^2(2x^2 - n + 6)\}.$$

---

* The method of finding the apsidal angle by a direct integration of the apsidal equation was first used by Bertrand, *Comptes Rendus*, vol. 77, 1873. An improved version was afterwards given by Darboux in his notes to the *Cours de Mécanique* by Despeyrous, 1886.

In the same way we find after some reductions

$$(a^{n-1}-b^{n-1})^{\frac{1}{2}}=\{2c^{n-1}M(n-1)\}^{\frac{1}{2}}\{1+\tfrac{1}{12}(n-2)(n-3)M^2\}.$$

Remembering that $du=cMdx$, these give

$$\frac{d\theta}{dx}=\frac{1}{\sqrt{(3-n)}}\frac{1}{\sqrt{(1-x^2)}}\left\{1-\frac{n-2}{6}Mx+\frac{n-2}{24}M^2(2x^2+n)\right\}.$$

The integrations can be effected at sight by putting $x=\sin\phi$. Taking the limits to be $\phi=\pm\frac{1}{2}\pi$ to make the apses adjacent, we find that the apsidal angle is

$$\frac{\pi}{\sqrt{(3-n)}}\left\{1+\frac{(n-2)(n+1)}{24}M^2\right\}.$$

428. Closed orbits. *An orbit is described about a centre of force whose attraction is a one-valued function of the distance. Prove that if the orbit is closed, for all initial conditions within certain defined limits, the law of force must be the inverse square or the direct distance.* [Bertrand, *Comptes Rendus*, vol. 77, 1873.]

If the path is closed and re-entering it must admit of both a maximum and a minimum radius vector. The orbit therefore has two apsidal distances and must lie between the two circles which have these for radii and their centres at the centre of force. By varying the initial conditions we may widen or diminish the space between the circles, yet by the question the orbit is always to be closed so long as the radii of the circles remain finite.

Representing the first approximation to the reciprocals of the radii by $c(1\pm M)$ the apsidal angle will be π/p, where p can be expressed in some series of ascending powers of M. The orbit cannot be closed unless the apsidal angle is such that, after some multiple of it has been described, the particle is again at the same point of space and moving in the same way. Hence *p must be a rational fraction for all values of M whether rational or not.* The coefficients of all the powers of M must therefore be zero, while the term independent of M must be a rational fraction.

When $F=\mu u^n$ the series for p is (Art. 426)

$$p=\sqrt{(3-n)}\{1-\tfrac{1}{24}(n-2)(n+1)M^2+\&c.\}.$$

Since the coefficient of M^2 must be zero we see that $n=2$ or -1, i.e. the law of force must be the inverse square or the direct distance. In either case the condition that $\sqrt{(3-n)}$ should be a rational fraction is satisfied.

If we take the most general form for the force, we have $F=u^2f(u)$. We know by Art. 368 that the first term of the series for p is, in general, a function of c, i.e. of the reciprocal of the mean radius. Since this can be varied arbitrarily the apsidal angle cannot be commensurable with π unless this first term, viz. $cf'(c)/f(c)$, is independent of c. Putting this equal to a constant m we find by an easy integration that $f(c) = \mu c^m$. Hence $F = \mu u^{m+2}$. The general case is therefore reduced to the special case already considered.

429. Classification of orbits. *The force being $F = \mu u^n$ it is required to classify the various forms of the orbit according to the number of the apsidal distances*. We suppose μ to be positive and h not to be zero.*

Arranging the apsidal equation (A) (Art. 422) in descending powers of u, it takes one or other of the three following forms

$$\left(\frac{dr}{dt}\right)^2 = h^2\left(\frac{du}{d\theta}\right)^2 = \frac{2\mu}{n-1}u^{n-1} - h^2u^2 + C \quad\text{.......................(A)},$$

$$= -h^2u^2 + \frac{2\mu}{n-1}u^{n-1} + C,$$

$$= \left\{-h^2u^{3-n} + Cu^{1-n} - \frac{2\mu}{1-n}\right\}u^{n-1},$$

according as $n > 3$, n lies between 3 and 1, and $n < 1$.

The two constants $\frac{1}{2}C$ and h determine the *energy* and *angular momentum* of the particle, Art. 313. When these are given, we arrive, by integrating (A), at an equation of the form $\theta + \alpha = f(u)$. By varying the constant α we turn the curve round the origin without altering its form. It follows that *when C and h are known, the orbit is determined in form but not in position.* The curve thus found may have several branches which are not connected with each other. *One point on the orbit must therefore also be given to determine the value of α and to distinguish the branch actually described by the particle.*

Any point on the curve being taken as the point of projection, we may regard v as the initial velocity. We thus have $C = v^2 - V_1^2$ or $C = v^2 + V_0^2$, where V_1 is the velocity from infinity, and V_0 the velocity to the origin. The first equation is to be used when V_1 is finite, i.e. when $n > 1$; the second when V_0 is finite, i.e. when $n < 1$. See Art. 313.

430. Case I. *Let the curve have but one apsidal distance.* The right-hand side of the apsidal equation (A) must change sign once as u varies from zero to infinity. Hence, when $n > 3$, C is negative or zero, i.e. the velocity v is less than or equal to that from infinity; when n lies between 3 and 1, C must be positive or zero, i.e. the velocity v is greater than or equal to that from infinity. Lastly we see from the third form of the equation (A) that when $n < 1$ the curve cannot have only one apsidal distance.

---

* Korteweg, *Sur les trajectoires décrites sous l'influence d'une force centrale*, *Archives Néerlandaises*, vol. XIX. 1884, discusses the forms of the orbits, the conditions of stability and the asymptotic circles. Greenhill, *On the stability of orbits*, *Proc. Lond. Math. Soc.* vol. XXII. 1888, treats of the asymptotic circles which can be described when $F = \mu u^n$ for various values of n.

These conditions being satisfied, let $u=a$ be the reciprocal of the apsidal distance, found by solving the equation (A). We then have

$$\left(\frac{dr}{dt}\right)^2 = h^2\left(\frac{du}{d\theta}\right)^2 = (u-a)\,\phi(u),$$

where $\phi(u)$ cannot change sign as u varies from 0 to ∞. Since $\phi(u)$ must have the same sign as the highest power of u, its sign is positive or negative according as $n >$ or < 3.

We notice that if n is a fraction, say $n=p/q$, we replace the factor $u-a$ by $w-b$ where $u=w^q$, $a=b^q$; Art. 423. As in most cases the force F varies as some integral power of the distance, it will be more convenient to retain the form given above.

Since the left-hand side of (2) is necessarily positive, the whole of the curve must lie inside the circle $u=a$ if $n>3$, and must lie outside that circle if $n<3$. Suppose the particle, as it moves round the centre of force, to have arrived at the apse. It will then begin to recede from the circle and must always continue to recede because $du/d\theta$ is not again zero. The orbit has therefore two branches extending from the apse to the centre of force or to infinity according as $n >$ or < 3. The apse is an apocentre in the first case and a pericentre (as in a hyperbola described about the inner focus) in the second case.

The motion in the neighbourhood of the apse may be found by writing $u=a+x$ and retaining only the lowest powers of x. We then have

$$(dx/d\theta)^2 = 4Ax; \quad \therefore\ u-a=A\theta^2,$$

where $4Ah^2=\phi(a)$. The path is therefore such that the particle describes a *finite angle* θ while it moves from $u=u$ to $u=a$. Since $d\theta/dt=hu^2$ is finite, the time of describing this finite angle is also finite.

431. Cases II. and III. *To find the conditions that there may be either two apsidal distances or none.* The apsidal equation must have two positive roots or none. The condition for this is that the right-hand side of (A) must have the same sign when $u=0$ and $u=\infty$.

First. Let $n>3$, this condition requires that C should be positive and not zero. *The velocity at every point must therefore be greater than that from infinity.*

To distinguish the cases we find the max-min value M of the right-hand side by equating to zero its differential coefficient. We thus find

$$M = -\frac{\mu}{\kappa}\left(\frac{h^2}{\mu}\right)^{\kappa} + C, \qquad \kappa = \frac{n-1}{n-3}.$$

Taking the second differential coefficient we find that M is a minimum when $n>3$ and a maximum when $n<3$.

We notice that when $n>3$, the two terms of M have opposite signs and that we can make either predominate by giving h or C small values. Thus M may have any sign if the initial conditions are suitably chosen. *The path may therefore have either two apsidal distances or none; there will be two if M is negative and none if M is positive. If $M=0$ the apsidal distances are equal.*

Secondly, let $3>n>1$. The right-hand side of (A) cannot have the same sign when $u=0$ and $u=\infty$ unless C is negative. *The velocity at every point must therefore be less than that from infinity.*

Writing as before

$$M=\frac{\mu}{\kappa'}\left(\frac{\mu}{h^2}\right)^{\kappa'}+C, \qquad \kappa'=\frac{n-1}{3-n};$$

we shall prove that M is necessarily positive and has zero for its least value. Then since the right-hand side of (A) is negative when $u=0$ and $u=\infty$ and is equal to the positive quantity M for some intermediate value, *there must be two apsidal distances* which can be equal only when $M=0$.

To prove that M is positive, we notice that M is least when h is greatest. Since $h=vr\sin\beta$ (Art. 313) this occurs when $h=vr$, i.e. when the particle is projected perpendicularly to the radius vector. Substituting this value of h and remembering that $C=v^2-V_1^2$, we can see by a simple differentiation that M is again least when $v^2=\mu/r^{n-1}$, that is, when the velocity is equal to that in a circle. This value of v is less than the velocity from infinity (n being <3), and is therefore admissible here. Substituting this value of v we find that the minimum value of M is zero. The value of M is therefore positive and is zero only when the path is a circle.

We may also prove that the orbit has two apsidal distances by observing that since the velocity is insufficient to carry the particle to infinity, the orbit must have either an apocentre or must approach an asymptotic circle. In either case the apsidal equation has one positive root and therefore has another.

Thirdly, let $1>n$. Since $C=v^2+V_0^2$ we notice that C must be positive. We now have

$$M=-\frac{\mu}{\kappa}\left(\frac{h^2}{\mu}\right)^{\kappa}+v^2+V_0^2, \qquad \kappa=\frac{1-n}{3-n};$$

we may prove in the same way as before that M is least when $h=vr$ and $v^2=\mu/r^{n-1}$ and that then $M=-\frac{2\mu}{1-n}r^{1-n}+V_0^2=0$ by Art. 312. Thus M is always positive and *the curve has two apsidal distances* which can be equal only in a circle.

We verify this result by noticing that since an infinite velocity is required to carry the particle to infinity (n being <1, Art. 312), the orbit must have an apocentre or approach an asymptotic circle. The apsidal equation must therefore have two positive roots.

432. It follows from what precedes that the curve defined by the apsidal equation (A) can be without an apse only when $n>3$. In that case the orbit extends from the centre of force to infinity.

We arrive at the same result by noticing that if there is no apse, *the velocity must be sufficient to carry the particle to infinity.* If $1>n$ this condition cannot be satisfied (Art. 312). If $n>1$ this condition requires C to be positive and it is evident that the second form of the apsidal equation has then a positive root.

It also follows that there can be an asymptotic circle only when $n>3$. For if the orbit be ultimately circular the constant M must be zero, and this cannot happen when $n<3$ unless the orbit is circular throughout. See also Art. 447.

433. *To find the motion when the orbit has two apsidal distances.* If a, b be the reciprocals of these distances, the apsidal equation (A) takes the form

$$h^2\left(\frac{du}{d\theta}\right)^2=(u-a)(u-b)\,\phi(u),$$

where $\phi(u)$ is positive or negative according as $n >$ or < 3. Since the left-hand side is necessarily positive we see that u cannot lie between the limits a and b if $\phi(u)$ is positive but must lie between them if $\phi(u)$ is negative. The whole curve must therefore lie outside the annulus defined by the circles $u=a$, $u=b$ if $n>3$, and must lie within that annulus if $n<3$.

It appears that when $n>3$ the full curve defined by the differential equation (A) contains two distinct branches, either of which can be described by the particle with the given energy $\frac{1}{2}C$ and the given angular momentum h. These, being separated by the empty annulus, do not intersect, so that when the point of projection is given the particular branch described by the particle is determined. We notice also that this branch has only one apsidal distance though the complete curve has two.

When $n<3$ the path of the particle undulates between the two circles $u=a$, $u=b$, touching each alternately and being always concave to the centre of force.

434. Case IV. *To find the motion when the apsidal distances are equal.* The apsidal equation now takes the form

$$h^2 (du/d\theta)^2 = (u-a)^2 \phi(u).$$

The motion as the particle approaches the circle $u=a$ may be found by putting $u=a+x$ and retaining only the lowest powers of x. We then have

$$h^2 (dx/d\theta)^2 = \phi(a)\, x^2, \quad \therefore\ u-a = Ae^{-m\theta},$$

where $m^2 = \phi(a)/h^2$. The particle therefore approaches the limiting circle in an asymptotic path and arrives at the circle only when $\theta=\infty$. Since $d\theta/dt$ (being ultimately equal to ha^2) is finite, the time of describing an infinite number of revolutions round the centre of force is infinite.

The conditions that the right-hand side of the apsidal equation (A) may have a square factor and be positive are (1) the coefficients of the highest and lowest powers must be positive, and (2) we must have $M=0$, Art. 431. If $n>3$, C must be positive, i.e. the *velocity at every point must be greater than that from infinity.* If $n<3$ the coefficient of the highest power of u is negative, and *there can be no asymptotic circle.* (See also Art. 432.)

435. When $n>3$ and it is known that the path has an apse, we may prove that *that apse is a pericentre or apocentre according as the velocity of projection is greater or less than the velocity in a circle at the same distance.* Let v be the velocity of the particle, V_2 the velocity in a circle at the same distance r, V_1 the velocity from infinity; then (Art. 313)

$$V_1^2 = \frac{2\mu}{n-1}\frac{1}{r^{n-1}}, \quad V_2^2 = \frac{\mu}{r^{n-1}}, \quad v^2 = V_1^2 + C \text{.................}(1),$$

$$\therefore\ v^2 - V_2^2 = -\tfrac{1}{2}(n-3)\,V_1^2 + C \text{............................}(2).$$

If $r=r_1$ represent any apsidal distance, we have at that apse $v^2/\rho = F$, $V_2^2/r_1 = F$. At a pericentre the orbit lies outside the circle of radius r_1, hence $\rho > r_1$ and $\therefore\ v^2 > V_2^2$. At an apocentre the orbit lies inside the circle and $v^2 < V_2^2$.

It follows by inspection of (2) that *at a pericentre* both sides of that equation are positive, and, since V_1 decreases when r increases, both sides must continue to be positive as the particle recedes from the origin. The particle also cannot arrive at a second apse, for this requires the left side to become negative. In the same way *at an apocentre* the two sides of (2) are negative and must continue to be negative as the particle approaches the origin. The conclusion is that *the velocity at any point is greater or less than that in a circle at the same distance according as the path has a pericentre or apocentre.*

It follows also that *the path described cannot have both a pericentre and an apocentre.*

436. The following table sums up the possible orbits when $F=\mu u^n$

$n>3$, $v\leqq V_1$	{one apsidal distance, path inside the circle.
$v>V_1$, M negative	{two apsidal distances, path inside or outside both circles according as v is $<$ or $>V_2$.
$v>V_1$, M positive	{no apsidal distance, the path extends from the centre of force to infinity.
$v>V_1$, $M=0$	{an asymptotic circle, approached from within or from without according as v is $<$ or $>V_2$.
$3>n>1$, $v\geqq V_1$	{one apsidal distance, path outside the circle.
$v<V_1$	{two apsidal distances, path between the circles.
$1>n$, $v<V_1$	{two apsidal distances, path between the circles.

Here V_2 is the velocity in a circle at the distance of the point of projection.

Ex. When the force $F=\mu u^n$ is repulsive show that the path, if not rectilinear, has a pericentre with branches stretching to infinity.

437. *The motion in the neighbourhood of the origin* is found by retaining the highest powers only of u. We thus have by (A), Art. 429,

$$\left(\frac{dr}{dt}\right)^2=h^2\left(\frac{du}{d\theta}\right)^2=B^2u^{n-1} \text{ or } -h^2u^2,$$

according as $n>3$ or <3, where $(n-1)B^2=2\mu$. The first alternative gives after integration, supposing the particle to be approaching the origin,

$$r^p-r_0{}^p=-\frac{Bp}{h}\theta, \qquad r^q-r_0{}^q=-Bqt,$$

where $p=\frac{1}{2}(n-3)$, $q=\frac{1}{2}(n+1)$; showing that the particle (except when $n=3$) describes a finite angle in a finite time when the radial distance decreases from $r=r_0$ to zero.

The negative sign in the second alternative shows that, when $n<3$, the particle cannot reach the origin unless $h=0$, i.e. unless the path is a radius vector.

438. *The motion at an infinite distance from the origin* is found by retaining the lowest powers only of u. We then have

$$\left(\frac{dr}{dt}\right)^2=h^2\left(\frac{du}{d\theta}\right)^2=C \text{ or } -\frac{2\mu}{1-n}u^{n-1},$$

according as $n>$ or <1. The negative sign in the second alternative shows that when $n<1$ the curve can have no branches which extend to infinity.

When C is positive, i.e. when the velocity v of projection is greater than that from infinity, the first alternative leads to

$$h(u-u_0)=-\theta\sqrt{C}, \qquad r-r_0=t\sqrt{C},$$

showing that when the particle travels from $r=r_0$ to infinity it describes a finite angle θ round the origin, and that the time is infinite. The path therefore tends to a rectilinear asymptote whose distance from the origin is $-d\theta/du=h/\sqrt{C}$.

If however $C=0$, i.e. the velocity v of projection is equal to that from infinity, the lowest existing power of u in the apsidal equation (A) is u^2 or u^{n-1}. We

then have $$\left(\frac{dr}{dt}\right)^2 = h^2\left(\frac{du}{d\theta}\right)^2 = -h^2u^2, \text{ or } +\frac{2\mu}{n-1}u^{n-1},$$
according as $n>3$ or $n<3$ but >1. The first alternative shows that (except when $h=0$) there are no branches leading to infinity. The second alternative, i.e. $n<3$, gives, supposing the particle to recede from the origin,

$$r^p - r_0{}^p = \frac{Bp}{h}\theta, \qquad r^q - r_0{}^q = Bqt,$$

where $(n-1)B^2=2\mu$, $p=-\frac{1}{2}(3-n)$, $q=\frac{1}{2}(n+1)$. These equations show that as the particle proceeds from $r=r_0$ to infinity it describes a finite angle in an infinite time. The path tends to a rectilinear asymptote at an infinite distance from the origin.

439. Stability of the orbits. Referring to Art. 436 we see that when $n>3$ the orbit extends to the origin or to infinity except when the particle is approaching an asymptotic circle. The existence of such a circle depends on the equality of the factors of the right-hand side of the apsidal equation, and a slight change in the constants C, h may render the factors unequal or imaginary. In either case the new path will lead the particle either to the centre of force or to infinity. *Such orbits may be called unstable.*

When $n<3$ and the velocity of projection less than that from infinity, the path is restricted to lie between the two circles $u=a$, $u=b$, and the values of a and b depend on the constants C and h. Any slight disturbance will alter the values of these constants, but the orbit will still be restricted to lie between two circles though the radii will not be exactly the same as before. *Such orbits may be called stable.*

440. *Ex.* Prove that any small decrease of the angular momentum h or increase of the energy $\frac{1}{2}C$ will widen the annulus within which the particle moves; that is, will increase the oscillation of the particle on each side of the central line.

441. Apsidal boundaries when $F=f(u)$. When the law of force contains several terms the argument becomes more complicated. Let $F=\Sigma A_n u^n$, then

$$v^2 = h^2\left\{\left(\frac{du}{d\theta}\right)^2 + u^2\right\} = C + 2\Sigma\frac{A_n}{n-1}u^{n-1}.$$

Transposing the terms, the apsidal equation is

$$\left(\frac{dr}{dt}\right)^2 = h^2\left(\frac{du}{d\theta}\right)^2 = 2\Sigma\frac{A_n}{n-1}u^{n-1} - h^2u^2 + C \quad\ldots\ldots\ldots\ldots\ldots\ldots \text{(B)},$$
$$= (u-a_1)(u-a_2)\ldots(u-a_\kappa)\,\phi(u),$$

where a_1, a_2, ... are positive quantities arranged in descending order, and $\phi(u)$ contains all the factors which do not vanish between $u=0$ and $u=\infty$. The factor $\phi(u)$ keeps one sign, viz. that of the highest power of u.

Let us divide the plane of motion into annular portions by circles whose common centre is at the centre of force and whose radii are the reciprocals of a_1, a_2, &c. Then since $(du/d\theta)^2$ changes sign when u passes any one of these boundaries, it is clear that the curve defined by the differential equation (B) can have branches only in the alternate annuli, the intervening ones being vacant. The space between $u=a_1$ and $u=\infty$ being occupied or vacant according as $\phi(u)$ is positive or negative.

If the initial position of the particle lie between any two contiguous circles, the subsequent path is restricted to lie between these circles and touches each alternately. If the initial position lie outside the greatest circle or inside the least, the subsequent path must also lie outside or inside these circles and must therefore extend to infinity or to the centre of force.

442. Next, let some of the factors of the apsidal equation be equal, say

$$\left(\frac{dr}{dt}\right)^2 = h^2\left(\frac{du}{d\theta}\right)^2 = (u-a)^m f(u)\,\phi(u),$$

where $f(u)$ has been written for the remaining factors. To determine the motion in the neighbourhood of the circle $u=a$, we write $u=a+x$ and retain only the lowest powers of x. We then have, supposing $m>2$,

$$\left(\frac{dx}{d\theta}\right)^2 = \frac{B^2}{h^2}x^m, \qquad \frac{1}{x^\kappa} - \frac{1}{{x_0}^\kappa} = \pm\frac{B\kappa}{h}\theta,$$

where $B^2 = f(a)\,\phi(a)$, and $\kappa = \frac{1}{2}(m-2)$. The case in which $m=2$ is discussed in Art. 434. *We see that the circle $u=a$ is asymptotic.* The particle arrives at the circle after describing an infinite number of revolutions round the centre of force and at the end of an infinite time.

443. Let us trace the surface of revolution whose abscissa is r and ordinate $z=Fr^3$, and let the ordinate z be perpendicular to the plane of motion of the particle. We notice that *this surface is independent of the initial conditions and that its form depends solely on the law of force.*

It is easy to see that the ordinate z corresponding to any value of r represents the square of the angular momentum in a circular orbit described with radius r. It will therefore be useful also to trace the plane whose ordinate is $z=h^2$, where h is the angular momentum of the path described.

By describing circles whose radii are the abscissæ of the maximum and minimum ordinates of the surface, we may divide the plane of motion into annular portions in which the function $z=Fr^3$ is alternately increasing or decreasing *outwards* from the centre of force. These we may call the *ascending or descending portions of the surface.*

444. If r represent any apsidal distance, we have at the corresponding apse $v^2/\rho = F$ and $v=h/r$; hence $h^2 = F\rho r^2$. At a pericentre the orbit lies outside the circle of radius r, hence $\rho > r$, and the angular momentum h of the path must be greater than that in a circle of radius r. In the same way, at an apocentre the orbit lies inside the circle, and the angular momentum h is less than that in a circle of radius r.

Referring to the surface $z=Fr^3$, we see that *a pericentral distance $r=OA$ must have an ordinate AA' less than that of the plane $z=h^2$,* and *an apocentral distance OB must have an ordinate BB' greater than that of the plane.* It immediately follows that if A, B are the pericentre and apocentre of *the same path, both the points A', B', cannot lie on the same descending portion of the surface.* This conclusion does not apply if A, B are the pericentre and apocentre of different branches of the complete curve; (Art. 441).

We infer from this result that an annular space on the plane of motion (Art. 443) in which Fr^3 decreases outwards has this element of instability, viz. that *a path having both a pericentre and an apocentre cannot be described within the space.* If the path have a pericentre the particle will leave the space on its outer margin;

if an apocentre it will move out of the space on its inner boundary. We see also that when the particle has left the annular space it must proceed to infinity or to the centre of force, unless it come into some other external annular space in which Fr^3 has increased sufficiently to exceed the h^2 of its own path or into some internal space in which Fr^3 has become less than h^2.

445. We may also deduce this result very simply from the radial resolution.

We have $$\frac{d^2r}{dt^2}=r\left(\frac{d\theta}{dt}\right)^2-F=\frac{1}{r^3}(h^2-Fr^3).$$

As the particle approaches and passes an apocentre r increases to a maximum and decreases, hence dr/dt changes sign from positive to negative and d^2r/dt^2 is negative. In the same way, when the particle passes a pericentre, d^2r/dt^2 is positive. It immediately follows that at an apocentre $Fr^3>h^2$ and at a pericentre $Fr^3<h^2$.

446. If the orbit have an asymptotic circle $r=a$, the angular momentum h must be equal to that in a circle of that radius. Hence *the asymptotic circle must be the projection of some one of the intersections of the surface $z=Fr^3$ with the plane $z=h^2$*; (Art. 443).

As the asymptotic circle is itself an apocentre or pericentre, it follows, as in Art. 445, that when the particle is approaching the circle from within h^2-Fr^3 is negative and ultimately zero. Hence Fr^3 is decreasing *outwards*. When the particle is approaching the circle from without h^2-Fr^3 is positive and ultimately zero, hence Fr^3 is increasing *inwards*. In either case it follows that *only those intersections which lie on a descending portion of the surface $z=Fr^3$ can correspond to asymptotic circles.*

As each descending portion of the surface can have only one intersection with the plane $z=h^2$, there cannot be more asymptotic circles than descending branches.

There may be fewer asymptotic circles than descending branches because two conditions are necessary that an asymptotic circle of given radius $r=a$ should exist; (1) the angular momentum must be equal to that in the circle, and (2) the constant C must be such that the velocity at a distance $r=a$ is equal to that in the circle, i.e. $v^2/a=F$.

447. As an example, consider the force $F=\mu u^n$. If $n>3$, the surface $z=Fr^3$ has only a descending portion, there can therefore be one and only one asymptotic circle. Also the path described cannot have both an apocentre and a pericentre, though different branches of the same curve may have one an apocentre and another a pericentre. See Arts. 444, 446, 436. If $n<3$, the surface $z=Fr^3$ has only an ascending portion. Hence there cannot be an asymptotic circle, but the path can have both an apocentre and a pericentre.

448. *Ex.* Discuss the properties of the surface $Z=Fr-v^2$, where the velocity v is a known function of r given in Art. 441. Prove that (1) the abscissæ of its max-min ordinates are the same as those of the surface $z=Fr^3$, so that the ascending and descending portions of each correspond (Art. 443); (2) each asymptotic circle must be one of the intersections of the surface with the plane of motion; (3) conversely, if at any intersection we also have $z=h^2$, that intersection is an asymptotic circle.

The first result follows from $\dfrac{dZ}{dr}=\dfrac{1}{r^2}\dfrac{dz}{dr}$. To prove the second and third we notice that when $Z=0$, the velocity is equal to that in a circle; and when $z=h^2$, the angular momentum is equal to that in a circle.

449. Examples. *Ex.* 1. Find the law of force with the lowest index of u such that an orbit can be described having two given asymptotic circles whose radii are the reciprocals of a and b, and find the path. Find also the conditions of projection that the path may be described.

Referring to Art. 441 we see that the right-hand side of the apsidal equation (B) must be $\mu(u-a)^2(u-b)^2$. We then find

$$F=\mu u^2(u-a)(u-b)(2u-a-b)+\mu' u^3,$$

and the angular momentum at projection must be $\sqrt{\mu'}$.

Ex. 2. Let $F=\mu u^2\{(u-a)(3u-a-b)+cu\}$, where F is the central force. If the conditions of projection are such that $h^2=\mu c$ and the velocity v when $u=a$ is $v^2=\mu ca^2$, show that the path is $\dfrac{a-b}{u-b}=(\tanh\theta)^2$, where $c\kappa^2=2(a-b)$. Show also that the curve has two infinite branches tending to the same asymptotic circle $u=a$, with an apse at a distance $1/b$.

Ex. 3. A particle arrives at an apse distant r from the centre of force with a velocity v equal to that in a circle at the distance r. If the velocity be reversed in direction, will the particle describe the same path in a reverse order or will it travel along the circle? See Art. 419.

At such an apse the radius of curvature ρ of the path must be equal to r. But since $\dfrac{1}{\rho}=u+\dfrac{d^2u}{d\theta^2}$ at any apse this requires that $d^2u/d\theta^2=0$. The apsidal equation (B) of Art. 441 must therefore have equal roots, and the apse is at the extremity of a path with an asymptotic circle. The particle therefore can never arrive at such an apse in any finite time (Art. 442).

If the particle be projected from a point on the asymptotic circle with the given values of v and h it may be said to describe either orbit, for the deviation of one from the other is indefinitely small at the end of any finite time.

Boussinesq, *Comptes Rendus*, vol. 84, 1877, considers the circular motion to be a singular integral of the differential equation. Korteweg and Greenhill have also discussed this problem.

On the law of force by which a conic is described.

450. Newton's theorem*. *An orbit is described by a particle about a centre of force C whose law is known: it is required to find the law of force by which the same orbit can be described about another centre of force O.*

* Newton's theorem is given in Prop. VII. Cor. 3 of the second section of the first book of the *Principia*. The application to the motion of a particle in a circle acted on by a force parallel to a fixed direction follows in the next proposition. Sir W. R. Hamilton's paper, giving the law $F=\mu r/p^3$, is in the third volume of the *Proceedings of the Irish Academy*, 1846. Villarceau in the *Connaissance des Temps*

Let F, F' be the forces of attraction tending respectively to C and O. Let CY, OZ be the perpendiculars on the tangents at any point P; $CP = r$, $OP = r'$. Then since $\sin CPY = CY/r$, we have

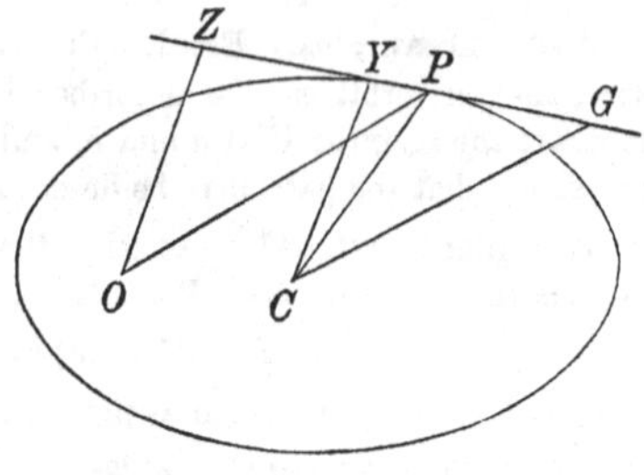

$$\frac{v^2}{\rho} = F\frac{CY}{r}, \quad v = \frac{h}{CY},$$

$$\therefore\ F = \frac{h^2 r}{\rho \, . \, CY^3}.$$

Similarly $$F' = \frac{h'^2 r'}{\rho \, . \, OZ^3}, \qquad \therefore\ \frac{F'}{F} = \frac{h'^2}{h^2}\frac{r'}{r}\left(\frac{CY}{OZ}\right)^3.$$

If we draw CG parallel to OP, the triangles OPZ and CGY are similar, and

$$\frac{OZ}{OP} = \frac{CY}{CG}; \qquad \therefore\ \frac{F'}{F} = \frac{h'^2}{h^2}\frac{CG^3}{r'^2 r}.$$

If then F is given as a function of r, the law of force F' tending to any assumed point O is also known, when we have deduced CG as a function of r and r' from the geometrical properties of the curve.

Remembering that the area $A = \frac{1}{2}ht$, we see that the periodic times in which the whole curve is described about C and O respectively are inversely as the arbitrary constants h and h'. By choosing these properly we can make *the ratio of the periodic times have any ratio we please.*

We also notice that if the time of describing any arc PQ is known when the central force tends to C, the area PCQ is known. Now the area POQ differs from this by a *rectilinear*

for 1852, using Cartesian coordinates, arrived at two possible laws of force. Afterwards Darboux and Halphen investigated two laws equivalent to these, and proved that there is no other law in which the central force is a function only of the coordinates of its point of application. Their results may be found in vol. 84 of the *Comptes Rendus*, 1877. The investigations of Darboux were reproduced by him at somewhat greater length in his notes to the *Cours de Mécanique* by Despeyrous, 1884. There is a third paper by Glaisher in vol. 39 of the *Monthly Notices of the Astronomical Society*, 1878, who also gives the expression $(2\pi/\sqrt{\mu})\,\varpi^{\frac{3}{2}}$ for the periodic time. Darboux uses chiefly polar coordinates, while Halphen employs Cartesian, beginning with the general differential equation of all conics: Glaisher simplifies the arguments by frequently using geometrical methods. There is also a paper by S. Hirayama of Tokyo in *Gould's Astronomical Journal*, 1889.

figure whose area can therefore be found. Hence the area POQ and therefore the time of describing the same arc PQ when the central force tends to O can be found.

451. Suppose *the orbit is a conic,* then the force tending to the centre C is $F = \mu r$, and $h = \sqrt{\mu} \,.\, ab$. It immediately follows that *the force tending to any point O is* $F' = \dfrac{h'^2}{a^2b^2} \cdot \dfrac{CG^3}{r'^2}$. If, for example, O is a focus, it is a known geometrical property of a conic that G lies on the auxiliary circle and that therefore $CG = a$. We then have $F' = \mu'/r'^2$, where $h'^2 = \mu' b^2/a$.

452. Parallel forces. *To find the force parallel to a given straight line by which a conic can be described.* See Art. 323.

Let the point O be at an infinite distance, then in Newton's formula PO and CG remain parallel to the given straight line throughout the motion. Also the length $r' = OP$ is constant. The required law of force is therefore $F' = \mu \,.\, CG^3$, where μ is some constant.

If the direction PO of the force at P cut the diameter conjugate to CG in N, we have $CG \,.\, PN = b'^2$, where b' is the semi-diameter parallel to CG. The law of force may therefore also be written $F' = A/PN^3$, where $A = \mu b'^6$.

To find the constant μ, we notice that in any central orbit, the velocity being $v = h/p$, the component of the velocity perpendicular to the radius vector r' is h/r'. In our case when the force acts parallel to a given straight line this component is constant. Representing this transverse velocity by V, the Newtonian formula of Art. 451 becomes $F' = \dfrac{V^2}{a^2b^2} CG^3$.

453. Hamilton's formula. *A particle describes a conic about a centre of force situated at any point O. It is required to find the law of force.* Taking the same notation as in Newton's theorem, we let F, F' be the forces tending respectively to the centre C and the point O. Then (Art. 450)

$$\frac{F'}{F} = \frac{h'^2}{h^2} \frac{OP}{CP} \left(\frac{CY}{OZ}\right)^3, \qquad F = \mu \,.\, CP, \qquad h = \sqrt{\mu} \,.\, ab.$$

It is a geometrical property of a conic that, if p and ϖ are the perpendiculars drawn from P and the centre C on the polar line of O, $\frac{p}{\varpi} = \frac{OZ}{CY}$*. It follows that *the law of force tending to O is* $F' = \frac{h'^2}{a^2b^2}\left(\frac{\varpi}{p}\right)^3 r'$, where p and r' vary from point to point of the curve and h', a, b and ϖ are constant.

If we write the Hamiltonian expression for the force in the form $F' = \mu' r'/p^3$, we see that the angular momentum $h' = \sqrt{\mu'} . ab/\varpi^{\frac{3}{2}}$, where as before ϖ is the perpendicular from the centre on the polar line.

From this we easily deduce the periodic time in an elliptic orbit. Remembering that the whole area is πab, the formula $A = \frac{1}{2}h't$ gives as the time of describing a complete ellipse $T = \frac{2\pi}{\sqrt{\mu}} \varpi^{\frac{3}{2}}$.

454. *To find the time of describing any portion of the ellipse with Hamilton's law of force.* The coordinates of any point P referred to an origin at the centre of force O with axes parallel to the principal diameters are

$$x = a\cos\phi - f, \qquad y = b\sin\phi - g,$$

where ϕ is the eccentric angle of P and f, g the coordinates of the centre of force referred to the centre of the curve. Then, if h be the angular momentum,

$$hdt = xdy - ydx = (ab - fb\cos\phi - ga\sin\phi)\,d\phi,$$

$$\therefore\ ht = ab\phi - fb\sin\phi + ga\cos\phi - ga,$$

where the time is measured from the passage through the apse from which ϕ is measured. This, if required, can be expressed in terms of x and y,

$$ht = ab\phi - fy + gx - ga.$$

This result can be deduced at once from the formula $A = \frac{1}{2}ht$, by equating A to the excess of the area of the sector ACP (viz. $\frac{1}{2}ab\phi$) over the sum of the triangles ACO, OCP.

* The following is a short analytical proof: Let the conic be $Ax^2 + By^2 = 1$ and let f, g be the coordinates of O. The polar line of O and the tangent at P are respectively $\qquad Af\xi + Bg\eta = 1, \qquad Ax\xi + By\eta = 1.$

The perpendiculars from P and O, viz. p and OZ, are therefore

$$p = \frac{1 - Afx - Bgy}{\sqrt{(A^2f^2 + B^2g^2)}}, \qquad OZ = \frac{1 - Afx - Bgy}{\sqrt{(Ax^2 + By^2)}}.$$

The perpendiculars from the centre, viz. ϖ and CY, are found by replacing the numerators by unity. It follows that $p/OZ = \varpi/CY$.

The time of describing an arc of a hyperbola or parabola may be found by proceeding as in Arts. 348, 349.

455. Examples. *Ex.* 1. Deduce from Hamilton's expression (1) the central force to the focus of a conic, and (2) that to the centre. [In the latter case ϖ and p are both infinite but their ratio is unity.]

Ex. 2. A particle describes an ellipse whose centre is C under the action of a centre of force F situated at a point R in the major axis. If the tangent at P cut the major axis in T, prove that the force F varies as $RP\,.\,(CT/RT)^3$.

456. The Hamiltonian expression for the force may be put into two different forms.

First, we have the form $F=\mu r/p^3$ (Art. 453).

Secondly. Let OA, OB be two tangents drawn to the conic from the centre of force O, and let $PL=\alpha$, $PM=\beta$, $PN=\gamma$; these being the three perpendiculars drawn from any point P on the sides of the triangle OAB. By a property of conics we have $\alpha\beta=\kappa\gamma^2$, where κ is a constant for the same conic. The central force may therefore be expressed in either of the forms

$$F=\mu\frac{OP}{PN^3}=\nu\frac{OP}{(PL\,.\,PM)^{\frac{3}{2}}}. \qquad [\nu=\mu\kappa^{\frac{3}{2}}.]$$

Each of these expressions is a one-valued function of the position of P though their values are not necessarily equal except at points on the orbit.

We may suppose either of these laws to be extended to all points of the plane of motion and enquire what would be the path for any given conditions of projection. These problems will be considered in turn.

457. The conic being given in its general form referred to any rectangular axes, viz.,

$$Ax^2+2Cxy+By^2+2Dx+2Ey+G=0,$$

the two Hamiltonian expressions for the force to the origin may be put into the forms

$$F=\frac{h^2\Delta r}{(Dx+Ey+G)^3}, \qquad F=\frac{h^2\Delta r}{(\alpha x^2+2\gamma xy+\beta y^2)^{\frac{3}{2}}},$$

where $\alpha=D^2-AG$, $\gamma=DE-CG$, $\beta=E^2-BG$, and Δ is the discriminant.

To prove this we notice that the polar line of the origin is $Dx+Ey+G=0$, so that the ratio of the perpendiculars from the centre $\bar{x}$, $\bar{y}$ and from the point P is

$$\frac{\varpi}{p}=\frac{D\bar{x}+E\bar{y}+G}{Dx+Ey+G}.$$

If we refer the equation of the conic to the centre as origin, it becomes

$$Ax^2+2Cxy+By^2=-D\bar{x}-E\bar{y}-G=-\frac{\Delta}{AB-C^2}.$$

Turning the axes round the origin let this become

$$A'x^2+B'y^2=-D\bar{x}-E\bar{y}-G,$$

where by the theory of invariants $A'B'=AB-C^2$ and $A'+B'=A+B$. Since the conic is now referred to its principal diameters, we have $a^2b^2=\dfrac{(D\bar{x}+E\bar{y}+G)^2}{A'B'}$.

It immediately follows by substituting in Art. 453 that

$$F=\frac{h^2}{a^2b^2}\left(\frac{\varpi}{p}\right)^3 r=h^2\Delta\,.\,\frac{r}{(Dx+Ey+G)^3}\,.$$

Since the equation of the conic may be written in the form

$$G\,(Ax^2+2Cxy+By^2)+(Dx+Ey+G)^2=(Dx+Ey)^2,$$

the expression just obtained for the force F may be put by a simple substitution into the second form.

The straight lines $ax^2+2\gamma xy+\beta y^2=0$, when real, pass through the origin and make $Dx+Ey+G=0$. They therefore meet the curve at the points where the polar line of the origin cuts it, i.e. *these straight lines are the tangents drawn from the centre of force to the conic.*

458. In the same way we may express F as a function of the coordinates x, y in a variety of different forms each of which gives the same magnitude for the force when the particle lies on the given conic. When these expressions for the force are generalized and supposed to hold at all points of space, they are not always one-valued functions of the coordinates. A law which gives several different values for the force at the same point may be set aside as altogether improbable.

For example, we might deduce from Hamilton's law an expression for F in terms of r alone. To do this we find the distance p of any point P on the orbit from the polar line of the origin O in terms of the distance r of P from O. But there are four points on the conic at the same distance r from the origin and each of these is, in general, at a different distance from the polar line. The expression for the central force F as a function of r only will therefore have four values for each value of r.

459. The First law of force. Supposing the first form of the Hamiltonian law of force to be extended to all points of the plane, *we put* $F=\dfrac{\mu r}{p^3}$, *where r is the distance of any point P from a fixed centre of force O, and p is the perpendicular from P on an arbitrary straight line fixed in space.* It is supposed that p is positive when P and the origin are on the same side of the given straight line.

We shall now prove that, *if a particle be projected from any point P in any direction PT, with any velocity V, the path is a conic having O, and the given straight line, for pole and polar.*

This follows from the results of Art. 453. It is obvious that we can describe a conic to satisfy (1) the three conditions that it shall pass through P, touch PT and have such a radius of curvature that V^2/ρ is equal to the normal force at P, (2) the two conditions that the polar line of O shall be the given straight line. We may also prove that this conic is a *real conic.* This being so, the conic must be the path.

We may however obtain a proof independent of Art. 453 by integrating the equation of motion. Let the origin be at the centre of force, and the given straight line be parallel to the axis of x at a distance c, then $p=c-r\sin\theta$.

We have
$$\frac{d^2u}{d\theta^2}+u=\frac{F}{h^2u^2}=\frac{\mu}{h^2}\frac{1}{(up)^3}=\frac{\mu}{h^2}\frac{1}{(cu-\sin\theta)^3}.$$

To integrate this, put $cu=\sin\theta+cu'$;
$$\therefore\ \frac{d^2u'}{d\theta^2}+u'=\frac{\mu}{h^2c^3u'^3}.$$

This is the differential equation of the path of a particle acted on by a central force $F=\mu r/c^3$. This path is known to be a conic having its centre at the origin, Art. 325;
$$\therefore\ c^2u'^2=A'\cos^2\theta+2C'\cos\theta\sin\theta+B'\sin^2\theta \quad\text{.....................(1).}$$

The polar equation of the required orbit is therefore
$$(cu-\sin\theta)^2=A'\cos^2\theta+2C'\cos\theta\sin\theta+B'\sin^2\theta,$$
which when written in Cartesian coordinates becomes
$$(c-y)^2=A'x^2+2C'xy+B'y^2 \quad\text{.............................(2).}$$

Writing this equation in the form $\kappa\gamma^2=\alpha\beta$ where α, β are the factors of the right-hand side, it is obvious that the polar line of the origin is the given straight line $y=c$.

When the conic is given in the form (2), *the constant h is given by* $\dfrac{\mu c}{h^2}=A'B'-C'^2$.

To prove this we notice that h represents the angular momentum of both orbits. We have therefore by Art. 326 $h^2c^3/\mu=a'^2b'^2$, where a', b' are the semi-axes of the conic (1). We know by the theory of conics that $A'B'-C'^2=c^4/a'^2b'^2$, the result therefore follows at once.

When the conic is given in the general form of Art. 457, we find $\dfrac{\mu}{h^2}=\Delta\left(\dfrac{c}{G}\right)^3$.

Since the central force is not a function of r only, *it is not conservative* and the velocity cannot be found without a knowledge of the path. In such cases we use the formula $v=h/(OZ)$, see the figure of Art. 450.

460. *To classify the paths according to the sign of μ, the law of force being $F=\mu r/p^3$.*

Let μ be positive; the force is attractive and the orbit concave to O at all points on the side of the given straight line nearest to the centre of force and the contrary at all points on the far side. When a conic cuts the polar line of a point O, the part of the curve nearest to O is convex; hence *the orbit does not cut the polar line.* It also follows that the orbit may be an ellipse or hyperbola on the side near O, but must be a hyperbola on the far side.

Let μ be negative; the force is repulsive and the orbit convex to O on the near side of the polar line while the contrary holds on the far side. The conic may be an ellipse or a hyperbola. By drawing a figure we see that the polar line must cut the conic though, in the case of a hyperbola, the path may be the other branch.

461. Examples. *Ex.* 1. The conic $Ax^2+2Cxy+By^2+2cy-c^2=0$ is described by a particle under the action of a central force $F=\dfrac{\mu r}{p^3}$ tending to the origin, where $p=c-y$ is the distance of the particle from the given straight line

$y=c$. The conic must have the form given if the polar line of the origin is to be $y=c$. Prove that

(1) $\{A(B+1)-C^2\}h^2=\mu c$, (2) $\{AB-C^2\}\dfrac{h^2}{c^2}=\mu\dfrac{2p-c}{p^2}-\left(\dfrac{dy}{dt}\right)^2$,

(3) $Ah^2=\dfrac{\mu c}{p^2}y^2+\left(c\dfrac{dy}{dt}\right)^2$, (4) $(B+1)h^2=\dfrac{\mu c}{p^2}x^2+\left(c\dfrac{dx}{dt}+h\right)^2$,

(5) $-Ch^2=\dfrac{\mu c}{p^2}xy+\left(c\dfrac{dx}{dt}+h\right)c\dfrac{dy}{dt}$.

From these equations, when the path is known, we can find the angular momentum h and the two components of velocity; conversely we can deduce the path when the circumstances of projection are given.

These equations follow from the preceding propositions. An independent proof may be obtained by differentiating the equation of the conic twice and writing for d^2x/dt^2, d^2y/dt^2 their values $-\mu x/p^3$, $-\mu y/p^3$. We thus obtain three equations which may be transformed into those given above by simple processes.

Ex. 2. Prove that the conic described is an ellipse, parabola or hyperbola according as $\mu(2p-c)/p^2-p'^2$ is positive, zero or negative, where p is the distance of the point of projection from the polar line and p' the *resolved* initial velocity.

Ex. 3. If $Ax^2+2Cxy+By^2+2Dx+2Ey+G=0$ is the conic described, show that the periodic time in an ellipse is $T=\dfrac{2\pi}{\sqrt{\mu}}\left\{\dfrac{\Delta c}{(AB-C^2)G}\right\}^{\frac{3}{2}}$.

Ex. 4. A particle is acted on by a central force $F=\mu\dfrac{r^{n-2}}{p^n}$ tending to the origin where r is the radius vector and p the distance from a fixed straight line. Prove that the equation of the path is $c/r=\sin\theta+f(\theta)$, where $c/r=f(\theta)$ is the polar equation of the path when the force tending to the origin is $F=\mu r^{n-2}/c^n$, both orbits being described with the same angular momentum h.

462. The second law of force. Supposing the second form of the Hamiltonian law of force to be extended to all points of the plane of motion, we put

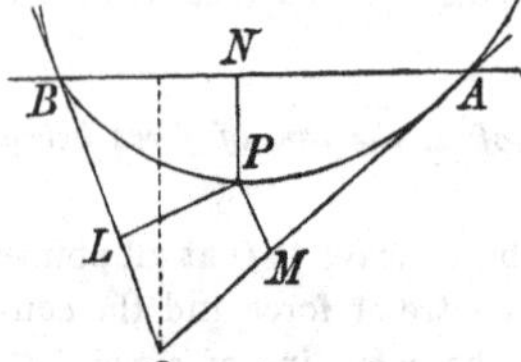

$$F=\nu\frac{OP}{(PL\,.\,PM)^{\frac{3}{2}}},$$

where PL, PM are the perpendiculars from any point P on two fixed straight lines OA, OB, drawn through the centre of force O; Art. 456.

The form of the path may be obtained by following either of the methods described in Art. 459. The result is that the path is always a conic touching the given straight lines OA, OB.

If the force at any point P given by this formula is to be a function of the position of P only, it should be supposed to keep one sign throughout each of the triangular spaces formed by the given straight lines OA, OB (supposed to be real), though that sign may be different in different triangles. In any triangle in which the sign is negative only the convex portions of the conic can be described, while the concave portions are alone possible when the sign is positive. The force is infinite when the particle arrives at either of the straight lines OA, OB and the path becomes discontinuous.

If we suppose the magnitude alone of the force to be given by the formula, the sign being taken at pleasure, arcs of both parts of each conic could be described by giving F the proper sign.

463. Examples. *Ex.* 1. If the trilinear equation of the conic is $\alpha\beta=\kappa\gamma^2$, prove that $h^2=-4\nu c\kappa^{\frac{1}{2}}\operatorname{cosec}^2\theta$ where $\nu=\mu\kappa^{\frac{3}{2}}$, θ is the angle at the corner of the triangle occupied by the central force, and c is the perpendicular from the centre of force O on the polar line AB. The negative sign shows (what is indeed obvious from the figure) that the force is repulsive on the side of the polar line nearest to the centre of force, i.e. μ is negative.

Ex. 2. A particle is projected from the point P with a velocity V and the tangent GPH intersects the given straight lines OA, OB in G and H. Prove that the areal equation of the path, referred to the triangle OGH, is

$$\sqrt{(lx)}+\sqrt{(my)}+\sqrt{\left(\frac{\rho\Delta}{em}z\right)}=0,$$

where $l=GP$, $m=HP$, Δ is the area of the triangle, and the radius of curvature ρ of the path at P is given by $V^2/\rho=F\sin GPO$. It follows that the conic is inscribed or escribed according as F is positive or negative, i.e. according as the force is attractive or repulsive.

464. There are no other laws of force besides

$$F=\mu\frac{OP}{PN^3}=\nu\frac{OP}{(PL\,.\,PM)^{\frac{3}{2}}},$$

which, being a one-valued function of the coordinates (except as regards sign), are such that a conic will be described with *any* initial conditions.

To prove this consider two conics intersecting in the four points A, B, C, D, which it is convenient to take as real. It follows from Hamilton's theorem that for points on any one conic the force to a given point O must be $F=\mu r/p^3$. Hence if the force is to be one-valued, i.e. the same at the same point of space for all paths through that point, we must have at each of the four points A, B, C, D, $p^3/\mu=\pm p'^3/\mu'$, where p, p' are the perpendiculars on the two polar lines of O.

We now require the following geometrical theorem*. If two conics intersect in four points A, B, C, D and the ratios of the perpendiculars from each of these points on the polar lines of a point O are equal, then either the polar lines are coincident or two common tangents (real or imaginary) can be drawn from O.

In the former case the common law of force for the two conics is given by the first form of F, in the latter case by the second form.

---

* Let the conics be, see Art. 457,

$$ax^2+2\gamma xy+\beta y^2=(Dx+Ey+G)^2,$$
$$a'x^2+2\gamma' xy+\beta' y^2=(D'x+E'y+G')^2.$$

Since $Dx+Ey+G=0$, $D'x+E'y+G'=0$ are the polar lines of the origin, we must have at the points of intersection

$$ax^2+2\gamma xy+\beta y^2=m\,(a'x^2+2\gamma' xy+\beta' y^2).$$

This quadratic equation gives only two values of y/x for the same value of m. The equation cannot therefore be satisfied at four points unless either α, β, γ are respectively proportional to α', β', γ', or the four points lie on two straight lines (say OAB, OCD) passing through O. In the former case the two conics have a pair of common tangents, in the latter the polar line of O is common to the two conics. This common polar line can be constructed by dividing OAB, OCD harmonically in E, F and then joining EF.

Singular Points in Central Orbits.

465. Singular Points. It has already been pointed out in Art. 100 that cases present themselves in our mathematical processes in which either the force, the velocity or both become infinite. Such infinite quantities do not occur in nature and if we limit ourselves to problems which have a direct application to natural phenomena these are only matters of curiosity. Nevertheless it is useful to consider them because they call our attention to peculiarities in the analysis which we might otherwise pass over. The utility of such a discussion is perhaps shown by *the differences of opinion* which exist regarding the subsequent path of a particle on arriving at a singular point*.

466. Points of infinite Force. Let us suppose that a particle P, describing an orbit about a centre of force O, arrives at a point B where the tangent passes through the centre of force and therefore coincides with the radius vector. At first sight we might suppose that the particle would move along the straight line BO and proceed in a direct line to the centre of force. But this is not necessarily the case.

Supposing B to be at a finite distance from O and the curvature to be finite, we see from the equations (Art. 306)

$$\frac{v^2}{\rho}=F\frac{p}{r}, \qquad v=\frac{h}{p}, \qquad r\frac{d\theta}{dt}=\frac{h}{r},$$

that both v and F are infinite at the point B. We shall also suppose that when the particle passes on *the force changes its direction* and reduces the velocity again to a finite quantity.

At the same time the component of the velocity perpendicular to the radius vector OP, viz. $rd\theta/dt$, remains finite however near the particle approaches B. Since there is no force to destroy this transverse velocity, the particle must cross the straight line OB and proceed to describe an arc on the opposite side.

* The singularity of the motion when the particle describes a circle about an external centre of force is discussed in Frost's *Newton*, 1854 and 1863. The same result is independently arrived at by Sylvester in the *Phil. Mag.* 1866. Other cases are considered by Asaph Hall in the *Messenger of Mathematics*, 1874. There are several papers also in the *Bulletin de la Société Mathématique de France*, such as Gascheau in vol. x. 1881, and Lecornu in vol. XXII.

467. To simplify the argument, let us suppose that the particle describes a circle about a centre of force O external to the circumference. By Art. 321, the circumstances of the motion are given by

$$F=\frac{\mu r}{(r^2-b^2)^3},\quad v^2=\frac{\mu}{2}\frac{1}{(r^2-b^2)^2},\quad h^2=\frac{\mu}{8a^2},$$

where b is the length of each of the tangents OB, OB' drawn from O to the circle.

Describe a second circle having a radius equal to that of the given circle and touching OB at B on the opposite side. If a second particle, properly projected along the second circle, arrive at B simultaneously with the given particle P, but moving in the opposite direction, both the velocity v and the transverse velocity h/r of the two particles will be equal and opposite each to each.

If the velocity of the second particle be reversed, Art. 419, it will retrace its former path in a reverse order and this must be also the subsequent path of the particle P.

The particle will therefore describe in succession a series of arcs of equal circles. The points of discontinuity at which the particle changes from one circle to the next lie on a circle whose centre is O and radius $OB=b$, and the successive arcs are alternately concave and convex to the centre of force. The particle will thus continually move round the centre of force *in the same direction* in an undulating orbit, but the curve will not be re-entering after one circuit unless the angle BOB' is a submultiple of four right angles.

The same arguments will apply to other orbits. When a conic is described about an external centre of force O as explained in Art. 462, the particle by a proper projection can be made to describe either of the arcs contained between the tangents drawn from O. On arriving at the point of contact B, it will cross the tangent and describe an arc of a conic equal to the undescribed arc of the original conic.

468. The particle arrives at the centre of force. When the particle P arrives at the centre of force in a finite time, the determination of the subsequent path presents some other peculiarities.

Taking first the Newtonian case in which the particle describes a circle about a centre of force O on its circumference, we notice that the transverse velocity h/r (as well as the velocity v) becomes infinite at O. To understand how the particle can have an infinite velocity in a direction perpendicular to what is ultimately a tangent to the path, we observe that, since $2ap=r^2$, the transverse velocity h/r is infinitely less than the tangential velocity h/p.

When the particle has passed through the origin, the central force, changing its direction, reduces the velocity again to a finite quantity. Meantime the transverse velocity carries the particle across the tangent to the circle. By the same reasoning as before, the subsequent path is an equal circle which touches the original circle at the centre of force. On arriving a second time at the centre of force, the particle returns to the original circle, and so on continually.

469. One peculiarity of this case is that the radius vector of the particle while describing the second circle moves round the centre of force in the opposite direction to that in the first circle. Let P, P' be two positions of the particle, equidistant from the centre of force, just before and just after passing through that point. The transverse velocity being unaltered the moments of the velocity at P and P' taken in the same direction round O are equal and opposite. Since this moment is $r^2 d\theta/dt$, it follows that at the point of discontinuity h changes its sign.

470. When the particle moves in an equiangular spiral about a centre of force whose law is the inverse cube, it describes an infinite number of continually decreasing circuits and arrives at the centre of force at the end of a finite time, Art. 319. The subsequent path is another equiangular spiral, Art. 357, having the same angle. To determine its position we consider the conditions of motion at the point of junction.

Let us construct a second equiangular spiral obtained from the first by producing each radius vector PO backwards through the origin O to an equal distance OP'. If two particles P, P' describe these spirals so as to arrive simultaneously at the centre of force O, the particles are always in the same straight line with O, and at equal distances from it. Their radial and transverse velocities are also always equal and opposite each to each. If the velocity of P' be reversed, it will retrace its former path in a reverse order, and this must therefore be the subsequent path of P.

On passing the centre of force the particle will recede from the origin and describe the spiral above constructed. We notice also that the radius vector of the particle moves round the centre of force in the opposite direction to that in the first spiral.

471. Limiting Problems. We may sometimes simplify the discussion of some singularities by replacing the dynamical problem by another more general one of which the given problem is a limiting case. But the use of the method requires some discrimination. For example the motion of a particle attracted by a centre of force at a point O whose law of force is the inverse cube, may in some cases be regarded as a limit of the motion when the particle is constrained to move in a smooth fixed plane and is attracted by an equal centre of force situated at a point C outside the plane, where CO is perpendicular to the plane and is equal to some small quantity c. The method requires that the limiting motion should be the same whether we put the radius vector $r=0$ first and then $c=0$, or $c=0$ first and then $r=0$. We know by the principles of the differential calculus that the order in which the variables r and c assume their limiting values is not always a matter of indifference.

The component of force in the direction of the radius vector PO is $\mu r/(r^2+c^2)^2$ when the centre of force is at C, and is μ/r^3 when the centre is at O. As long as the particle is at a finite distance from the origin, these components are substantially the same, but when the particle is in the immediate neighbourhood of O, the former is $\mu r/c^4$ and therefore zero when the particle passes through O, while the latter is infinite.

In the former case, though the orbit at a distance from O is very nearly an equiangular spiral, it becomes elliptical in the neighbourhood of O. The force is not sufficient to draw the particle into the centre; the path has a pericentre and the particle retires again to an infinite distance. See also Art. 322.

472. Examples. *Ex.* 1. A particle describes one branch of the spiral $r\theta=a$ under the action of a centre of force in the origin (Art. 358). Show that after passing through the centre of force it will describe another spiral of the same kind, obtained from the first by producing each radius vector backwards through the origin to an equal distance.

Since the tangent to the curve is ultimately perpendicular to the radius vector, the two branches of the spiral may have a common tangent, and it might therefore be supposed that the particle would describe the second branch. But this argument requires that the particle should not pass through the origin, so that the radial velocity dr/dt (which is known to be constant) has its direction altered without any change in the direction of the force.

Ex. 2. A particle describes an epicycloid with the centre of force in the centre of the fixed circle (Art. 322). Supposing the force to become repulsive when the particle enters that circle, show that the path on passing the cusp is a hypocycloid.

Kepler's Problem.

473. *A particle describes an ellipse about a centre of force in one focus, it is required to express in series the two anomalies and the radius vector in terms of the time.*

If we require only the first few terms of the series it is convenient to start from the equations

$$r^2\frac{d\theta}{dt}=\sqrt{\{\mu a(1-e^2)\}},\quad \frac{a(1-e^2)}{r}=1+e\cos v\ \ldots\ldots(1),$$

where v is the true anomaly. Eliminating r, we have

$$\begin{aligned}\sqrt{\frac{\mu}{a^3}}\frac{dt}{d\theta}&=(1-e^2)^{\frac{3}{2}}(1+e\cos v)^{-2}\\&=(1-\tfrac{3}{2}e^2+\&c.)(1-2e\cos v+3e^2\cos^2 v-\&c.)\\&=1-2e\cos v+\tfrac{3}{2}e^2\cos 2v+\&c.\end{aligned}$$

Remembering that $v=\theta-\alpha$, where α is the longitude of the apse nearest to the centre of force, we have

$$nt+\epsilon=\theta-2e\sin(\theta-\alpha)+\tfrac{3}{4}e^2\sin 2(\theta-\alpha)+\&c.\ldots\ldots(2),$$

where $$n^2=\mu/a^3.$$

We notice that when the planet makes a complete revolution, θ increases by 2π and that the corresponding increment of t is $2\pi/n$. It follows immediately that n represents the mean angular velocity, the mean being taken with regard to the time; see Art. 341.

The equation (2) may be extended to higher powers of e, and therefore when e is small it may be used to determine the time of describing any angle θ.

474. To find θ in terms of t, we reverse the series. Writing it in the form

$$\theta = nt + \epsilon + 2e \sin(\theta - \alpha) - \tfrac{3}{4}e^2 \sin 2(\theta - \alpha),$$

we have as a first approximation

$$\theta = nt + \epsilon;$$

a second approximation gives

$$\theta = nt + \epsilon + 2e \sin(nt + \epsilon - \alpha).$$

Writing $v_0 = nt + \epsilon - \alpha$, a third approximation gives

$$\theta - \alpha = v_0 + 2e \sin(v_0 + 2e \sin v_0) - \tfrac{3}{4}e^2 \sin 2v_0;$$

$$\therefore\ \theta = nt + \epsilon + 2e \sin(nt + \epsilon - \alpha) + \tfrac{5}{4}e^2 \sin 2(nt + \epsilon - \alpha) \ldots (3),$$

and so on, the labour of effecting the successive approximations increasing at each step. As the eccentricity of the earth's orbit is about 1/60th it is obvious however that the terms become rapidly evanescent.

475. For the sake of clearness we recapitulate the meaning of the letters in the important equation we have just investigated; θ is the true longitude of the planet measured from any axis of x in the plane of the orbit; α is the longitude of the apse nearest the centre of force or origin; n is the mean angular velocity, the mean being taken with regard to time for one complete revolution; ϵ is a constant whose magnitude depends on the instant from which the time t is measured.

To define the epoch ϵ. Let a particle P_0 move round the centre of force in such a manner that its longitude is given by the equation $\theta_0 = nt + \epsilon$. It follows that this planet moves with a uniform angular velocity n and has therefore the same periodic time as the true planet P. When the radius vector of the particle P_0 passes through an apse $\theta_0 - \alpha$ and therefore $nt + \epsilon - \alpha$ is an integral multiple of π. It immediately follows from (2) that $\theta = nt + \epsilon$. Hence the radii vectores of the two planets coincide when the true planet passes through either apse. The definition of P_0 may be shortly summed up thus.

Let an imaginary planet move round the centre of force with a uniform angular velocity in the same period as the true planet and let their radii vectores coincide at one apse and therefore at the other. This planet is called the Dynamical Mean Planet. Its longitude at the time $t=0$ is the constant ϵ and is called the epoch.

476. *To express the mean anomaly and radius vector in terms of the time.*

Since both the mean and true planets cross the nearer apse at the time given by $nt + \epsilon = \alpha$, the mean anomaly may be represented by $m = nt + \epsilon$. If u be the eccentric anomaly we have by Art. 342,

$$u = m + e \sin u.$$

Proceeding as before we have for the three first approximations,

$$u = m, \qquad u = m + e \sin m,$$

$$\begin{aligned} u &= m + e \sin (m + e \sin m) \\ &= m + e \sin m + \tfrac{1}{2} e^2 \sin 2m \text{..................... (4)}. \end{aligned}$$

Again, as in Art. 343,

$$\begin{aligned} r &= a - ex = a - ae \cos u \\ &= a - ae \cos (m + e \sin m) \\ &= a \{1 - e \cos m + \tfrac{1}{2} e^2 (1 - \cos 2m)\} \text{ .... (5)}. \end{aligned}$$

The series for the longitude and radius vector are given here only to the second power of the eccentricity. Laplace in the *Mécanique Céleste* (page 207) and Delaunay in his *Théorie de la Lune* (vol. I. pages 19 and 55) give the series up to the sixth power. Stone has continued the expansion up to the seventh power in the *Astronomical Notices*, 1896 (vol. LVI. page 110). Glaisher has given the expansion of the eccentric anomaly up to the eighth power in the *Astronomical Notices*, 1877 (vol. XXXVII. page 445).

477. *When the eccentricity e is very nearly equal to unity*, as in the case of some comets, the formulæ giving the relations between t and v must be modified. Starting as before (Art. 473) from the equations

$$r^2 \frac{d\theta}{dt} = \sqrt{\{\mu a (1 - e^2)\}}, \qquad \frac{a(1-e^2)}{r} = 1 + e \cos v,$$

we put the perihelion distance $a(1-e) = p$.

$$\therefore t \sqrt{\frac{\mu}{p^3}} = \int \frac{(1+e)^{\frac{3}{2}} dv}{(1 + e \cos v)^2}.$$

Let $(1-e)/(1+e)=f$ and put $\tan\frac{1}{2}v=x$ for the sake of brevity;

$$\therefore \tfrac{1}{2}t\sqrt{\frac{\mu(1+e)}{p^3}}=\int\frac{(1+x^2)\,dx}{(1+fx^2)^2}$$
$$=x+\frac{x^3}{3}-2f\left(\frac{x^3}{3}+\frac{x^5}{5}\right)+3f^2\left(\frac{x^5}{5}+\frac{x^7}{7}\right)-\&c.$$

When v is given this formula determines the time t measured from perihelion. If f is small the term independent of f is the one requiring the most arithmetical calculation and this can be abbreviated by using the tables constructed for that purpose; see Art. 349. Conversely when t is given and v is required the same tables give a first approximate value of x. Representing this by $\tan\frac{1}{2}\omega$, it is usual to expand the correction $v-\omega$ in terms of ω in a series ascending in powers of f. For these formulæ we refer the reader to Watson's *Astronomy* and Gauss, *Theoria*, &c.

478. When the eccentric anomaly is given, the true and mean anomalies and the radius vector are expressed by the equations

$$m=u-e\sin u,\qquad \tan\frac{v}{2}=\sqrt{\frac{1+e}{1-e}}\tan\frac{u}{2}\ \ldots\ldots\ldots\ldots(1),$$
$$r=a-ex=a(1-e\cos u)\ \ldots\ldots\ldots\ldots(2).$$

When any one of the other quantities is taken as the independent variable, the corresponding equations can be deduced from these in the form of series. *Two methods are used to find the general term of these series.* First we may have recourse to Lagrange's theorem, viz., when

$$y=z+x\phi(y),\qquad f(y)=f(z)+\Sigma\frac{x^i}{Li}\frac{d^{i-1}}{dz^{i-1}}\{(\phi(z))^i f'(z)\},$$

where $Li=1.2.3\ldots i$, and the Σ implies summation from $i=1$ to ∞. By the second method the general term is expressed by a definite integral which is usually a Bessel's function.

479. Lagrange's theorem. *To express the eccentric anomaly u and the radius vector r in terms of the time.*

Since $u=m+e\sin u$, we have by Lagrange's theorem

$$u=m+\Sigma\frac{e^i}{Li}\frac{d^{i-1}}{dm^{i-1}}(\sin m)^i.$$

The expansion of $(\sin m)$ in cosines of multiple angles when i is even and in sines when i is odd is given in books on trigonometry; (see Hobson's *Trigonometry*, Art. 52). The $(i-1)$th differential is always a series of sines and is easily seen to be

$$2^{i-1}\frac{d^{i-1}}{dm^{i-1}}(\sin m)^i=i^{i-1}\sin im-i(i-2)^{i-1}\sin(i-2)m+\frac{i(i-1)}{2}(i-4)^{i-1}\sin(i-4)m-\&c.$$

In the same way, expanding $\cos u$ by Lagrange's theorem, i.e. writing $f(y)=\cos y$, we find

$$\frac{r}{a}-1=-e\cos u=-e\cos m+\Sigma\frac{e^{i+1}}{Li}\frac{d^{i-1}}{dm^{i-1}}(\sin m)^{i+1},$$

where as before Σ implies summation from $i=1$ to ∞.

480. Bessel's functions. We shall now briefly examine the second method by which we express the general term in a definite integral. We know by Fourier's theorem that we can expand any function $\phi(m)$ in a series of the form

$$\phi(m)=A_0+A_1\cos m+\ldots+A_i\cos im+\ldots$$
$$+B_1\sin m+\ldots+B_i\sin im+\ldots,$$

which holds for all values of m from $-\pi$ to $+\pi$. If also $\phi(m)$ is a periodic function having the period 2π, the expansion will hold for all values of m. If $\phi(m)$ does not change sign with m we may omit the second line of the expansion, while if it does change sign with m, we omit the first line.

To find A_i we use Fourier's rule; multiply both sides by $\cos im$ and integrate from $m=-\pi$ to $+\pi$. Remembering that

$$\int \cos im \cos i'm\, dm = 0, \qquad \int \cos im \sin i'm\, dm = 0,$$

$$\int \cos^2 im\, dm = \int \sin^2 im\, dm = \pi,$$

we find

$$\int \phi(m) \cos im\, dm = \pi A_i, \qquad \int \phi(m)\, dm = 2\pi A_0.$$

Similarly multiplying by $\sin im$ and integrating between the same limits, we find

$$\int \phi(m) \sin im\, dm = \pi B_i.$$

481. *To expand $u-m=e\sin u$ in a series of sines of multiples of m.* We put

$$u-m=\Sigma B_i \sin im;$$

$$\therefore\ \pi B_i = \int (u-m)\sin im\, dm,$$

the limits being $m=-\pi$ to π. Integrating by parts,

$$\pi i B_i = -(u-m)\cos im + \int \cos im\,(du-dm).$$

The integrated part is zero, for u and m are equal when $u=\pm\pi$. We thus have

$$\pi i B_i = \int \cos im\, du - \int \cos im\, dm.$$

The second integral is zero; substituting for m its value in terms of u,

$$\pi i B_i = \int \cos i\,(u - e\sin u)\, du.$$

This definite integral when taken between the limits 0 and π is written $\pi J_i(ie)$. We have

$$u = m + \Sigma B_i \sin im, \qquad iB_i = 2J_i(ie).$$

482. The series thus obtained is convergent, for

$$\pi i^2 B_i = \int \frac{du}{dm}\, d\sin im = \frac{du}{dm}\sin im - \int \frac{d^2u}{dm^2}\sin im\, dm.$$

The integrated part vanishes at both the limits $m=\pm\pi$. Also

$$u = m + e\sin u, \qquad \therefore\ \frac{d^2u}{dm^2} = \frac{-e\sin u}{(1-e\cos u)^3},$$

and since $e<1$, it is clear that d^2u/dm^2 has a numerical maximum value; let this be k. Since $\sin im<1$, it follows that $\pi i^2 B_i$ is numerically $<2k\pi$. The series is therefore at least as convergent as $\Sigma 1/i^2$.

483. *To compare the two expansions of $u-m$.* In the Lagrangian series the terms are collected according to the powers of e, the coefficient of e^i being a series of the sines of multiple angles. In the series with Bessel's functions the terms are arranged according to the multiple angles, the coefficient of $\sin im$ being a series of powers of e.

The series for $u-m$ is really a double series containing both trigonometrical terms of the form $\sin im$ and also powers of e. If the terms are collected and arranged according to the multiple angles, it follows from what precedes, that each coefficient B_i is a convergent series, and that the series of coefficients B_1, B_2, &c. also form a convergent series, provided the eccentricity e is less than unity.

But if the series is arranged according to the powers of e, the positive and negative terms are added together in a different way. It may then be that the series of coefficients of e, e^2, &c. are only made convergent by more limited values of e. The condition of convergency is given in Art. 488.

484. The expression for B_i may be written

$$\pi i B_i = \int \cos iu \,.\, \{1 - \tfrac{1}{2} i^2 e^2 \sin^2 u + \&c.\}\, du + \int \sin iu \,.\, \{ie \sin u - \&c.\}\, du.$$

If we expand $\sin^2 u$, $\sin^4 u$, &c. in cosines of multiple angles and remember that $\int \cos iu \cos i'u\, du = 0$, we see that every term in the first integral will be zero in which the power of e is less than i. A similar remark applies to the second integral. Hence *the lowest power of e which accompanies the term sin im is* e^i.

485. *To express* $r/a = 1 - e \cos u$ *in a series of cosines of multiples of m, we put*

$$-e \cos u = A_0 + \Sigma A_i \cos im;$$

$$\therefore\ \pi A_i = -e \int \cos u \cos im\, dm,$$

where the limits of integration are $m = -\pi$ to π. Integrating by parts to change dm into du, we have

$$\pi i A_i = -e \cos u \sin im - e \int \sin im \sin u\, du.$$

The integrated part vanishes between the limits. Writing $m = u - e \sin u$, the integral becomes

$$\begin{aligned} \pi i A_i &= -e \int \sin i (u - e \sin u) \sin u\, du \\ &= \tfrac{1}{2} e \int \cos \{(i+1) u - ie \sin u\}\, du - \tfrac{1}{2} e \int \cos \{(i-1) u - ie \sin u\}\, du; \end{aligned}$$

$$\therefore\ iA_i = e \{J_{i+1}(ie) - J_{i-1}(ie)\}.$$

Similarly $\qquad 2\pi A_0 = -e \int \cos u\, dm = -e \int \cos u \,.\, (1 - e \cos u)\, du.$

Integrating between limits $u = -\pi$ to π, we find $A_0 = \frac{1}{2} e^2$:

$$\therefore\ r/a = 1 + \tfrac{1}{2} e^2 + \Sigma A_i \cos im.$$

486. That this series is convergent may be proved in the same way as before. We have

$$\pi i^2 A_i = -e \int \frac{d \cos u}{dm}\, d \cos im = e \int \frac{d^2 \cos u}{dm^2} \cos im\, dm,$$

by integrating by parts. Since $u = m + e \sin u$, we find by differentiation $\dfrac{d^2 \cos u}{dm^2} = \dfrac{e - \cos u}{(1 - e \cos u)^2}$. This has obviously a maximum value, say k. Then since $\cos im < 1$, $\pi i^2 A_i$ is numerically less than $2\pi k e$, and the series is at least as convergent as $\Sigma 1/i^2$.

487. Examples. *Ex.* 1. Prove $\cos \kappa u = \Sigma A_i \cos im$, $\sin \kappa u = \Sigma B_i \sin im$, where $iA_i = \kappa \{J_{i-\kappa}(ie) - J_{i+\kappa}(ie)\}$, $iB_i = \kappa \{J_{i-\kappa}(ie) + J_{i+\kappa}(ie)\}$ and κ is not equal to unity, and the summations extend from $i = 1$ to ∞. Also $J_{-n}(x) = (-1)^n J_n(x)$.

Since $J_{-n}(-x) = J_n(x)$, these series may be written

$$\frac{1}{\kappa} \cos \kappa u = \Sigma J_{i-\kappa}(ie) \frac{\cos im}{i}, \qquad \frac{1}{\kappa} \sin \kappa u = \Sigma J_{i-\kappa}(ie) \frac{\sin im}{i},$$

where Σ implies summation from $i = -\infty$ to $+\infty$, and the term $J_{i-\kappa}(ie)/i$, when $i = 0$, is $-\frac{1}{2} e$ or 0 according as κ is equal or unequal to unity (Art. 485).

Since the Cartesian coordinates, referred to the centre of the ellipse, are $x = a \cos u$, $y = b \sin u$, we deduce the expansions of these in terms of the mean anomaly by putting $\kappa = 1$.

Ex. 2. Prove that $a/r = 1 + 2\Sigma J_i(ie) \cos im$, where the summation extends from $i = 1$ to ∞.

This follows from $a/r = du/dm$; see Arts. 343, 481.

Ex. 3. Prove that $v=m+\Sigma C_i \sin im$, where

$$C_i=\frac{2\sqrt{(1-e^2)}}{\pi i}\int_0^{\pi}\frac{\cos i(u-e\sin u)}{1-e\cos u}du.$$

Proceeding as before we find $\pi i C_i=\int_{-\pi}^{+\pi}\cos im\,(dv-dm)$; substituting for dv/du, the result follows. Also by integrating again by parts, we can prove that this series is at least as convergent as $\Sigma 1/i^2$. This integral is given by Poisson in the *Connaissance des Temps*, 1825, 1836. See also Laplace, vol. v. and Lefort, *Liouville's Journal*, 1846. See also Art. 343.

Ex. 4. Prove the expansions

$$\tfrac{1}{2}(v-u)=\ \ \lambda\sin u+\tfrac{1}{2}\lambda^2\sin 2u+\tfrac{1}{3}\lambda^3\sin 3u+\dots$$
$$\tfrac{1}{2}(u-v)=-\lambda\sin v+\tfrac{1}{2}\lambda^2\sin 2v-\tfrac{1}{3}\lambda^3\sin 3v-\dots$$

where $$\lambda=\frac{e}{1+\sqrt{(1-e^2)}}.$$ [Laplace.]

In $\tan\frac{1}{2}v=\mu\tan\frac{1}{2}u$, where $\mu^2=(1+e)/(1-e)$, substitute the exponential values of the tangents, solve for $e^{(v-u)\sqrt{-1}}$ and take logarithms; the results follow easily.

Ex. 5. Show that $m=v+2\Sigma\frac{(-\lambda)^i}{i}\{1+i\sqrt{(1-e^2)}\}\sin iv$ where Σ implies summation from $i=1$ to ∞.

We have from the geometrical meaning of u, $r\sin v=b\sin u$ (Art. 342),

$$\therefore \sin u=\frac{\sqrt{(1-e^2)}\sin v}{1+e\cos v}=-\sqrt{(1-e^2)}\frac{d}{edv}\log(1+e\cos v)$$
$$=-\sqrt{(1-e^2)}\frac{d}{edv}\log\frac{(1+\lambda e^{v\sqrt{-1}})(1+\lambda e^{-v\sqrt{-1}})}{1+\lambda^2}.$$

Expand, substitute in $m=u-e\sin u$, remembering the theorem in Ex. 3, the result follows. This is Tisserand's proof of Laplace's theorem, *Méc. Céleste*, page 223.

488. Convergency of the series for r and θ. Laplace was the first to prove that the expansions of the radius vector and true anomaly in terms of the time and in powers of the eccentricity are not convergent for all values of the eccentricity less than unity (see Arts. 474, 476). He showed by a difficult and long process that the condition necessary for the convergence of both series is that the eccentricity should be less than ·66195. *Méc. Céleste*, Tome v. Supplément, p. 516.

This important result was afterwards confirmed by Cauchy, *Exercises d'Analyse*, &c. An account is also given by Moigno in his *Differential Calculus*. The whole argument was put on a better foundation by Rouché in a memoir on Lagrange's series in the *Journal Polytechnique*, Tome XXII. The process was afterwards further simplified by Hermite in his *Cours à la Faculté des Sciences*, Paris 1886. In these investigations the test of convergency requires the use of the complex variable. The latter part of the method of Rouché may be found in Tisserand, *Méc. Céleste*, Art. 100, and is also given here.

489. The theorem arrived at may be briefly stated. Having given the equation $z=m+x\phi(z)$ we have (1) to distinguish which root we expand in powers of x, (2) to determine the test of convergency. It is shown that if a contour exist enclosing the complex point $z=m$, such that at every point of the boundary the modulus of $\dfrac{x\phi(z)}{z-m}$ is less than unity, the given equation has but one root within the area and the Lagrangian expansion for that root is convergent*.

To apply this theorem to Kepler's problem we put $\phi(z)=\sin z$ and let x represent the eccentricity of the ellipse, Art. 478.

We measure a real length $OA=m$ from an assumed origin O, and with A for centre describe a circle with an arbitrary radius r. Representing the complex line OP by z, the Lagrangian series will be convergent if r can be so chosen that the modulus of $\dfrac{x \sin z}{z-m}$ is less than unity for all positions of P on the circle. Since

$$(\text{mod})^2 \text{ of } \phi(\xi+\eta i)=\phi(\xi+\eta i)\,.\,\phi(\xi-\eta i),$$

$$z=m+re^{\theta i},$$

where e is the base of Napier's logarithms, we have

$$(\text{mod})^2 \text{ of } \frac{x\sin z}{z-m}=\left(\frac{x}{r}\right)^2\frac{\sin(m+re^{\theta i})\sin(m+re^{-\theta i})}{e^{\theta i}e^{-\theta i}}$$

$$=\frac{1}{2}\left(\frac{x}{r}\right)^2\{\cos(2ri\sin\theta)-\cos(2m+2r\cos\theta)\}$$

$$=\left(\frac{x}{r}\right)^2\{\tfrac{1}{4}(e^{r\sin\theta}+e^{-r\sin\theta})^2-\cos^2(m+r\cos\theta)\}.$$

---

* If $f(x)$ be a continuous one-valued function over the area of a circular contour whose centre is $x=a$, then Cauchy's theorem asserts that $f(x)$ can be expanded by Taylor's theorem in a convergent series of powers of $x-a$ for all points within the contour; (see Forsyth's *Theory of Functions*, Art. 26).

When $z=m+x\phi(z)$, the Lagrangian expansion of z, or $\psi(z)$, in powers of x is a transformation, term for term, of Taylor's, and we may use Cauchy's theorem, provided z, or $\psi(z)$, is one-valued.

If z have two values for the same value of x, the equation $F(z)=z-m-x\phi(z)=0$ (regarded as an equation to find z when x is given) has two roots. To determine whether this is so, we use another theorem of Cauchy's (see Burnside and Panton, *Theory of Equations*).

We measure $OA=m$ from the assumed origin O and with A for centre describe a circle of radius r. Let a point P describe this circle once, then by Cauchy's theorem if $\log F(z)$ is increased by $2n\pi i$, the equation $F(z)$ has n roots within the contour. Hermite writes

$$\log F(z)=\log(z-m)+\log\left(1-\frac{x\phi(z)}{z-m}\right).$$

(1) The equation $z-m=0$ has but one root and that root lies within the contour, hence as P moves round, $\log(z-m)$ is increased by $2\pi i$.

(2) If the modulus of $u=\dfrac{x\phi(z)}{z-m}$ is less than unity at all points of the circle, the value of $\log(1-u)$, (being the same on departing from and arriving again at any point of the contour) increases by zero when P moves round the contour.

It follows that $\log F(z)$ increases by $2\pi i$ when P makes one circuit, that is *the equation $z=m+x\phi(z)$ has but one root within the contour if the modulus of $\dfrac{x\phi(z)}{z-m}$ is less than unity at all points on the circumference.*

Now, putting $e^{r\sin\theta}+e^{-r\sin\theta}=v+\frac{1}{v}$, we see that the first term of this expression continually increases from $v=1$, or $\theta=0$ to $v=\infty$, and is therefore greatest when $\theta=\frac{1}{2}\pi$. The least value of the second term is zero. The modulus is therefore less than $\frac{1}{2}\frac{x}{r}(e^r+e^{-r})$. The Lagrangian series is therefore convergent for all values of the eccentricity x less than $2r/(e^r+e^{-r})$.

To find the maximum value of this function of r, we equate its differential coefficient to zero. This gives

$$V=e^r(r-1)-e^{-r}(r+1)=0.$$

Since dV/dr is positive for all values of r this equation has but one positive root, and this root lies between 1 and 2. Using the value of e^r given by the equation $V=0$, we find that the maximum value of the eccentricity is $\sqrt{(r^2-1)}$, which reduces to ·66.

CHAPTER VII.

MOTION IN THREE DIMENSIONS.

The four elementary resolutions and moving axes.

490. The Cartesian equations. The equations of motion of a particle in three dimensions may be written in a variety of forms all of which are much used.

The Cartesian forms of these equations are

$$\frac{d^2x}{dt^2}=X,\quad \frac{d^2y}{dt^2}=Y,\quad \frac{d^2z}{dt^2}=Z\text{.................(A)},$$

where x, y, z are the coordinates of the particle and X, Y, Z the components of the accelerating forces on the particle. These equations are commonly used with rectangular axes, but it is obvious that they hold for oblique axes also, provided X, Y, Z are obtained by oblique resolution.

491. The Cylindrical equations. From these we may deduce *the cylindrical or semi-polar forms of the equations.* Let the coordinates of the particle P be ρ, ϕ, z, where ρ, ϕ are the polar coordinates in the plane of xy of the projection N of the particle P on that plane, and $z=PN$. By referring to Art. 35, we see that the first two of the equations (A) change by resolution into the first two of the following equations (B), while the third remains unaltered. We have

$$\frac{d^2\rho}{dt^2}-\rho\left(\frac{d\phi}{dt}\right)^2=P,\quad \frac{1}{\rho}\frac{d}{dt}\left(\rho^2\frac{d\phi}{dt}\right)=Q,\quad \frac{d^2z}{dt^2}=Z\text{ ....(B)},$$

where P, Q are the components of the accelerating forces respectively along and perpendicular to the radius vector ρ.

492. Principle of angular momentum. Since the moments of the components P and Z about the axis of z are zero, the moment of the whole acceleration about the axis of z is equal to $Q\rho$. In the same way the moment of the velocity about Oz is equal to the moment of its component perpendicular to the plane POz, and this is $\rho^2 d\phi/dt$. Introducing the mass m of the particle as a factor, the second of the equations (B) may be written in the form

$$\frac{d}{dt}\begin{pmatrix}\text{moment of}\\ \text{momentum}\end{pmatrix} = \begin{pmatrix}\text{moment of}\\ \text{forces}\end{pmatrix}.$$

The moments may be taken about any straight line which is fixed in space, such a line being here represented by the axis of z. The moment of the momentum is also called the angular momentum of the particle (Arts. 79, 260).

When the forces have no moment about a fixed straight line the angular momentum about that straight line is constant throughout the motion.

493. The polar equations. *We may immediately deduce from the semi-polar form* (B), *the polar equations* (C). Let r, θ, ϕ be the polar coordinates of P, where $r = OP$, θ is the angle OP makes with the axis Oz, and ϕ the angle the plane POz makes with the plane xOz.

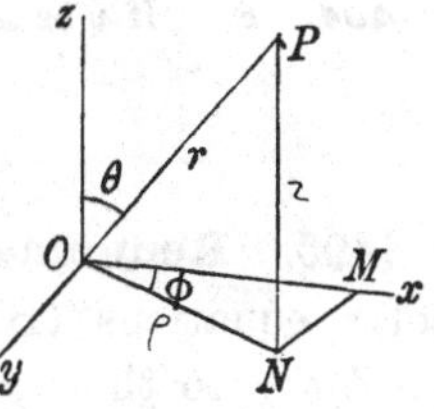

Since $OP = r$ is the radius vector corresponding to the coordinates $ON = \rho$, $NP = z$, we see by Art. 35 that the accelerations

$$\frac{d^2\rho}{dt^2} \text{ and } \frac{d^2z}{dt^2} \text{ are equal to } \frac{d^2r}{dt^2} - r\left(\frac{d\theta}{dt}\right)^2 \text{ and } \frac{1}{r}\frac{d}{dt}\left(r^2\frac{d\theta}{dt}\right).$$

Hence the whole acceleration of P is the resultant of

(1) $\dfrac{d^2r}{dt^2} - r\left(\dfrac{d\theta}{dt}\right)^2$ along OP in the direction in which r is measured;

(2) $\dfrac{1}{r}\dfrac{d}{dt}\left(r^2\dfrac{d\theta}{dt}\right)$ perpendicular to OP, in the plane zOP, taken positively in the direction in which θ is measured;

(3) $\rho\left(\dfrac{d\phi}{dt}\right)^2$ in the direction of the perpendicular drawn from P on Oz, i.e. parallel to NO;

(4) $\dfrac{1}{\rho}\dfrac{d}{dt}\left(\rho^2\dfrac{d\phi}{dt}\right)$ perpendicular to the plane zOP in the direction in which ϕ increases.

If R, S, T are the components of the acceleration of the particle respectively in the directions of (1) the radius vector OP, (2) the perpendicular to OP in the plane of zOP, and (3) the perpendicular to the plane zOP, taken positively when they act in the directions in which r, θ, ϕ are respectively increasing, we have

$$\left.\begin{aligned} \frac{d^2r}{dt^2} - r\left(\frac{d\theta}{dt}\right)^2 - \rho\left(\frac{d\phi}{dt}\right)^2 \sin\theta &= R \\ \frac{1}{r}\frac{d}{dt}\left(r^2\frac{d\theta}{dt}\right) - \rho\left(\frac{d\phi}{dt}\right)^2 \cos\theta &= S \\ \frac{1}{\rho}\frac{d}{dt}\left(\rho^2\frac{d\phi}{dt}\right) \qquad\qquad &= T \end{aligned}\right\}\text{...........(C).}$$

We notice that $\rho = r\sin\theta$.

494. *Ex.* If v be the velocity, show that the radial acceleration is

$$P = \frac{d^2r}{dt^2} + \frac{1}{r}\left\{\left(\frac{dr}{dt}\right)^2 - v^2\right\}.$$

495. Reducing a plane to rest. Referring to the semi-polar equations (B), we notice that if we transfer the term $\rho\,(d\phi/dt)^2$ to the right-hand side of the first equation and include it among the impressed accelerating forces, the first and third equations become the same as the Cartesian equations of motion of a particle moving in a fixed plane zOP (Art. 31), while the second equation determines the motion perpendicular to that plane. *We may therefore replace the first and third resolutions by any of the other forms which have been proved to be equivalent to them.* Art. 38.

For example, if we replace these two resolutions by their polar forms (Art. 35) we obtain at once the equations (C).

The process of regarding $\rho\,(d\phi/dt)^2$ as an impressed accelerating force acting at P and tending *from* the axis of z is sometimes called *reducing the plane zOP to rest.* See Arts. 197, 257.

496. The intrinsic equations. *To find the intrinsic equations of motion, due to the tangential and normal resolutions.*

Let P, P' be the positions of the particle at the times $t, t+dt$; $v, v+dv$ the velocities in those positions, $d\psi$ the angle between the tangents.

In the time dt, the component of velocity along the tangent at P has increased from v to $(v+dv)\cos d\psi$. Writing unity for $\cos d\psi$, the acceleration along the tangent, i.e. the rate of increase of the velocity, is dv/dt.

The component of velocity along the radius of curvature at P has increased from zero to $(v+dv)\sin d\psi$, which in the limit is $v d\psi$. The acceleration along the radius of curvature is therefore $v d\psi/dt$, or which is the same thing v^2/ρ.

The osculating plane by definition contains two consecutive tangents. The component of velocity perpendicular to that plane is zero and remains zero. The acceleration along the perpendicular to the osculating plane, i.e. the binormal, is therefore zero.

If F and G are the component accelerations measured positively in the directions of the arc s, the radius of curvature ρ and H the component perpendicular to the osculating plane, the equations of motion are

$$v\frac{dv}{ds}=F, \quad \frac{v^2}{\rho}=G, \quad 0=H \text{..............(D)}.$$

497. *Show that the solution of the equations of motion of a particle in polar coordinates can be reduced to integrations when the work function has the form*

$$U=f_1(r)+\frac{f_2(\theta)}{r^2}+\frac{f_3(\phi)}{r^2\sin^2\theta},$$

where $f_1(r)$, $f_2(\theta)$ and $f_3(\phi)$ are arbitrary functions.

The third of the equations (C) gives, with this form of U, the mass being unity,

$$\frac{1}{r\sin\theta}\frac{d}{dt}\left(r^2\sin^2\theta\frac{d\phi}{dt}\right)=\frac{df_3(\phi)}{r\sin\theta\,d\phi}\frac{1}{r^2\sin^2\theta};$$

$$\therefore \frac{1}{2}\left(r^2\sin^2\theta\frac{d\phi}{dt}\right)^2=f_3(\phi)+A \text{ .... (1)}.$$

The second of the equations (C) gives

$$\frac{1}{r}\frac{d}{dt}\left(r^2\frac{d\theta}{dt}\right)-r\sin\theta\cos\theta\left(\frac{d\phi}{dt}\right)^2=\frac{df_2(\theta)}{r\,d\theta}\frac{1}{r^2}-\frac{2f_3(\phi)\cos\theta}{r^3\sin^3\theta}.$$

Substituting for $d\phi/dt$, we obtain

$$\frac{1}{2}\left(r^2\frac{d\theta}{dt}\right)^2=-\frac{A}{\sin^2\theta}+f_2(\theta)+B \text{ .... (2)}.$$

The equation of vis viva is

$$\left(\frac{dr}{dt}\right)^2 + r^2\left(\frac{d\theta}{dt}\right)^2 + r^2\sin^2\theta\left(\frac{d\phi}{dt}\right)^2 = 2U + 2C \quad\text{.................}\quad (3).$$

After substituting from (1) and (2) this becomes

$$\left(\frac{dr}{dt}\right)^2 + \frac{2B}{r^2} = 2f_1(r) + 2C \quad\text{..........................}\quad (4).$$

These are the first integrals of the equations of motion. Since the variables are separable in all the equations, they can be reduced to integrations. Substituting for dt from (4) in (2), that equation gives θ in terms of r. Substituting again in (1), we find ϕ in terms of r. Lastly (4) determines t in terms of r.

498. **Moving axes.** *To find the equations of motion of a particle referred to rectangular axes which move about the origin O in an arbitrary manner.*

Let us suppose that the moving axes Ox, Oy, Oz are turning round some instantaneous axis OI with an angular velocity which we may call θ. Let θ_1, θ_2, θ_3 be the components of θ about the instantaneous positions of Ox, Oy, Oz. Then in the figure θ_1 represents the rate at which any point in the circular arc yOz is moving along that arc, θ_2 is the rate at which any point of the circular arc zOx is moving along the arc, and so on.

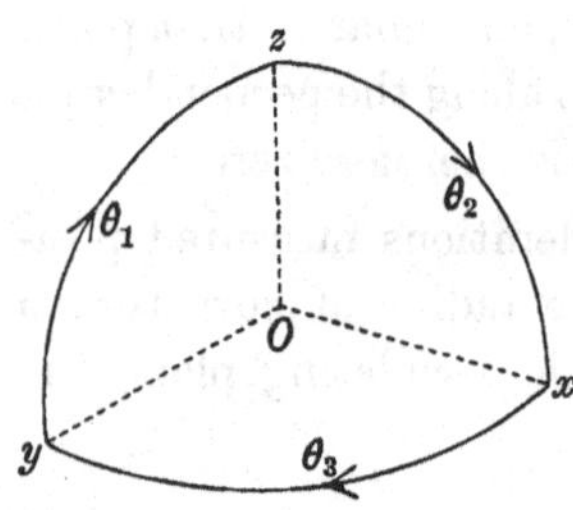

Let us represent by the symbol V any directed quantity or vector such as a force, a velocity, or an acceleration. Let V_x, V_y, V_z be its components with regard to the moving axes.

Let $O\xi$, $O\eta$, $O\zeta$ be three rectangular axes fixed in space and let V_1, V_2, V_3 be the components of the same vector along these axes. Let α, β, γ be the angles the axis $O\zeta$ makes with Ox, Oy, Oz. Then

$$V_3 = V_x\cos\alpha + V_y\cos\beta + V_z\cos\gamma,$$

$$\therefore\ \frac{dV_3}{dt} = \frac{dV_x}{dt}\cos\alpha + \frac{dV_y}{dt}\cos\beta + \frac{dV_z}{dt}\cos\gamma$$

$$- V_x\sin\alpha\frac{d\alpha}{dt} - V_y\sin\beta\frac{d\beta}{dt} - V_z\sin\gamma\frac{d\gamma}{dt}.$$

Let the arbitrary axis of ζ coincide with Oz at the time t, i.e. let the moving axis be passing through the fixed axis. Then $\alpha = \frac{1}{2}\pi$,

$\beta = \frac{1}{2}\pi$, $\gamma = 0$. Hence

$$\frac{dV_3}{dt} = \frac{dV_z}{dt} - V_x \frac{d\alpha}{dt} - V_y \frac{d\beta}{dt}.$$

Now $d\alpha/dt$ is the angular rate at which the axis Ox is separating from a fixed line $O\zeta$ momentarily coincident with Oz, hence $d\alpha/dt = \theta_2$. Similarly $d\beta/dt = -\theta_1$. Substituting

$$\frac{dV_3}{dt} = \frac{dV_z}{dt} - V_x\theta_2 + V_y\theta_1.$$

Similarly
$$\frac{dV_1}{dt} = \frac{dV_x}{dt} - V_y\theta_3 + V_z\theta_2,$$

$$\frac{dV_2}{dt} = \frac{dV_y}{dt} - V_z\theta_1 + V_x\theta_3.$$

When the moving axes momentarily coincide with the fixed axes, the components of the vector V are equal, each to each, i.e. $V_x = V_1$, $V_y = V_2$, $V_z = V_3$. As the moving axes pass on, this equality ceases to exist. The rates of increase of the components relatively to the moving axes are dV_x/dt, dV_y/dt, dV_z/dt; while the rates of increase relative to the fixed axes are dV_1/dt, dV_2/dt, dV_3/dt. The relations which exist between these rates of increase are given by the equations just investigated.

499. If the vector V is the radius vector of a moving point P, the components V_x, V_y, V_z are the Cartesian coordinates of P, and the rates of increase are the component velocities. If the vector V is the velocity of P, the rates of increase are the component accelerations.

Let then x, y, z be the coordinates of a point P; u, v, w the components of its velocity in space; X, Y, Z the components of its accelerations. Then

$$u = \frac{dx}{dt} - y\theta_3 + z\theta_2, \qquad X = \frac{du}{dt} - v\theta_3 + w\theta_2,$$

$$v = \frac{dy}{dt} - z\theta_1 + x\theta_3, \qquad Y = \frac{dv}{dt} - w\theta_1 + u\theta_3,$$

$$w = \frac{dz}{dt} - x\theta_2 + y\theta_1, \qquad Z = \frac{dw}{dt} - u\theta_2 + v\theta_1.$$

500. If the origin of coordinates is also in motion these equations require some slight modification. Let p, q, r be the resolved parts of the velocity of the origin in the directions of the axes. In order that u, v, w may represent the resolved velocities of the particle P in space (i.e. referred to an origin fixed in space), we must add p, q, r respectively to the expressions given for u, v, w in Art. 499. These additions having been made, u, v, w represent the component space velocities of P, and the expressions for the space accelerations X, Y, Z are the same as those given above. See Art. 227.

The theory of moving axes is more fully given in the author's treatise on *Rigid Dynamics*. The demonstration here given of the fundamental theorem is founded on a method used by Prof. Slesser in the *Quarterly Journal*, 1858. Another simple proof is given in the chapter on moving axes at the beginning of vol. II. of the treatise just referred to.

501. Moving field of force. *When the field of force is fixed relatively to axes moving about a fixed origin we may obtain the equation corresponding to that of vis viva.*

If T be the semi vis viva, we know that dT/dt is equal to the sum of the virtual moments of the forces divided by dt. Hence, the mass being unity,

$$\frac{dT}{dt} = Xu + Yv + Zw$$
$$= Xx' + Yy' + Zz' + \theta_3 (xY - yX) + \ldots\ldots$$

If A_1, A_2, A_3 are the angular momenta about the axes (Art. 492),

$$A_1 = yw - zv, \qquad A_2 = zu - xw, \qquad A_3 = xv - yu,$$

and, taking moments about the axes,

$$dA_1/dt = yZ - zY, \qquad dA_2/dt = zX - xZ, \qquad dA_3/dt = xY - yX.$$

The equation of vis viva therefore becomes

$$\frac{dT}{dt} - \theta_1 \frac{dA_1}{dt} - \theta_2 \frac{dA_2}{dt} - \theta_3 \frac{dA_3}{dt} = \frac{dU}{dt},$$

where U is a function of the coordinates x, y, z only. If θ_1, θ_2, θ_3 are constant, this, when integrated, reduces to the equation of Art. 256.

502. *Ex.* 1. Show how to deduce the polar forms (C), Art. 493, from the equations for moving axes.

Let the moving axes be represented by $O\xi$, $O\eta$, $O\zeta$. Let the axis of ξ move so as always to coincide with the radius vector OP; let $O\eta$ be always perpendicular to the plane zOP. The angular velocity $d\theta/dt$ of the radius vector may therefore be represented by $\theta_2 = d\theta/dt$ about $O\eta$. The plane zOP has an angular velocity $d\phi/dt$ about Oz, and this may be resolved into $\theta_1 = \cos\theta\, d\phi/dt$ and $\theta_3 = \sin\theta\, d\phi/dt$. Also the coordinates of P are $\xi = r$, $\eta = 0$, $\zeta = 0$.

It immediately follows from the equations of moving axes that $u = dr/dt$, $v = \theta_3 r$, $w = -\theta_2 r$. Substituting these in the expressions for X, Y, Z we obtain the components of acceleration already written at length in Art. 493.

Ex. 2. If $(\alpha_1\beta_1\gamma_1)$, $(\alpha_2\beta_2\gamma_2)$, $(\alpha_3\beta_3\gamma_3)$ are the direction cosines of a system of orthogonal axes moving about the origin, prove that

$$\theta_3 = \frac{d\alpha_1}{dt}\alpha_2 + \frac{d\beta_1}{dt}\beta_2 + \frac{d\gamma_1}{dt}\gamma_2,$$

where θ_3 is positive when the rotation is from the first axis to the second.

To prove this we notice that θ_3 measures the rate at which the axis of y is separating from the position of the axis of x at the time t. Hence $-\theta_3 dt$ is the cosine of the angle the new axis of y makes with the old axis of x.

Ex. 3. A particle is describing an orbit about a centre of force which varies as any function of the distance, and is acted on by a disturbing force which is always perpendicular to the plane of the instantaneous orbit and is inversely proportional to the distance of the particle from the centre of force. Prove that the plane of the instantaneous orbit revolves uniformly round its instantaneous axis.
[Math. Tripos, 1860.]

Lagrange's Equations.

503. Lagrange has given a general theorem by which we can form the equations of motion of a particle, or of a system of particles, in any kind of coordinates*.

The expression "coordinate" is here used in a generalized sense. *Any quantities are called the coordinates of a particle, or of a system of particles, which determine the position of that particle or system in space.*

In using Lagrange's equations, it will be found convenient to represent by some special symbols, such as accents, all total differential coefficients with regard to the time; thus x', x'' represent respectively dx/dt and d^2x/dt^2.

504. Lemma. *Let L be a function of any variables x, y, &c., their velocities x', y', &c., and the time t. If we express x, y, &c. as functions of some independent variables θ, ϕ, &c. and the time t, say*

$$x = f(t, \theta, \phi, \&c.), \quad y = F(t, \theta, \phi, \&c.), \quad z = \&c. \ldots\ldots(1),$$

then will

$$\frac{d}{dt}\frac{dL}{d\theta'} - \frac{dL}{d\theta} = \left(\frac{d}{dt}\frac{dL}{dx'} - \frac{dL}{dx}\right)\frac{dx}{d\theta} + \left(\frac{d}{dt}\frac{dL}{dy'} - \frac{dL}{dy}\right)\frac{dy}{d\theta} + \&c.$$

Representing partial differential coefficients by suffixes, we have by differentiating (1),

$$x' = f_t + f_\theta \theta' + f_\phi \phi' + \&c. \ldots\ldots\ldots\ldots\ldots\ldots(2).$$

Since θ enters into the expression L through both x, y, &c. and their velocities x', y', &c. while θ' enters only through x', y', &c.,

* The Lagrangian equations are of the greatest importance in the higher dynamics and are usually studied as a part of *Rigid Dynamics*. We give here only such theorems as may be of use in the rest of this treatise. The application of the method to impulses, to the cases in which the geometrical equations contain the differential coefficients of the coordinates, the use of indeterminate multipliers, the Hamiltonian function, &c., are regarded as a part of the higher dynamics.

we have the partial differential coefficients

$$\frac{dL}{d\theta} = \frac{dL}{dx}\frac{dx}{d\theta} + \frac{dL}{dx'}\frac{dx'}{d\theta} + \&c. \quad\quad (3),$$

$$\frac{dL}{d\theta'} = \frac{dL}{dx'}\frac{dx'}{d\theta'} + \&c. \quad\quad (4),$$

where in each case the &c. represents the corresponding terms for y, z, &c.

By differentiating (2) we see that $\dfrac{dx'}{d\theta'} = f_\theta = \dfrac{dx}{d\theta}$. Hence

$$\frac{d}{dt}\frac{dL}{d\theta'} - \frac{dL}{d\theta} = \left(\frac{d}{dt}\frac{dL}{dx'} - \frac{dL}{dx}\right)\frac{dx}{d\theta} + \&c.$$

$$+ \frac{dL}{dx'}\left(\frac{d}{dt}f_\theta - \frac{dx'}{d\theta}\right) + \&c. \quad\quad (5).$$

By differentiating f_θ totally with regard to t, we have

$$\frac{d}{dt}f_\theta = f_{\theta t} + f_{\theta\theta}\theta' + \&c. \quad\quad (6).$$

The right-hand side of this equation is seen by differentiating (2) to be equal to $\dfrac{dx'}{d\theta}$. It therefore follows that all the terms in the second line of (5) vanish. The lemma has therefore been proved.

505. By using this lemma we may deduce Lagrange's equations from the Cartesian equations of motion. For the sake of generality, let there be any number of particles, of any masses m_1, m_2, &c., and let their coordinates be (x_1, y_1, z_1), (x_2, y_2, z_2), &c. Let T be the semi vis viva of the system, then

$$2T = \Sigma m\,(x'^2 + y'^2 + z'^2) \quad\quad (7).$$

Let U be the work function of the impressed forces, then U is a function of the coordinates only. Let R_x, R_y, R_z be the components of any forces of constraint which act on the typical particle m. We have as many Cartesian equations of motion of the form

$$mx'' - \frac{dU}{dx} = R_x, \quad my'' - \frac{dU}{dy} = R_y, \quad mz'' - \frac{dU}{dz} = R_z,$$

as there are particles.

The particles may be free or connected together, or constrained by curves and surfaces, but after using all the given geometrical relations, the position of the system may be made to depend on some *independent* auxiliary quantities or coordinates. Let these be θ, ϕ, &c.; then writing $L = T + U$, we have for the particle m,

$$\frac{d}{dt}\frac{dL}{dx'} - \frac{dL}{dx} = \frac{d}{dt} mx' - \frac{dU}{dx} = R_x,$$

with similar forms for y and z. Hence using the lemma,

$$\frac{d}{dt}\frac{dL}{d\theta'} - \frac{dL}{d\theta} = \Sigma \left(R_x \frac{dx}{d\theta} + R_y \frac{dy}{d\theta} + R_z \frac{dz}{d\theta} \right) \text{ ....(8)},$$

where Σ implies summation for all the particles.

The right-hand side of this equation (after multiplication by $\delta\theta$) is the virtual moment of the forces of constraint for a geometrical displacement $\delta\theta$. This by the principle of virtual work is known to be zero.

Since the variations of the coordinates x, y, &c. due to the displacement $\delta\theta$ are deduced from the partial differential coefficients $dx/d\theta$, $dy/d\theta$, &c., t not varying, the displacement given to the system is one consistent with the geometrical relations as they exist at the instant of time t.

Taking the various kinds of forces of constraint it has been proved in Art. 248 that the virtual moment of each for such a displacement is zero. Consider the case of a particle constrained to rest on a curve or surface, the virtual moment is zero for any displacement tangential to the *instantaneous* position of the curve or surface. *The restriction that the geometrical equations must not contain the time explicitly is not necessary in Lagrange's equations.*

If some of the particles are connected together so as to form a rigid body, the mutual actions and reactions of the molecules are equal. Their virtual moments destroy each other because each pair of particles remain at a constant distance from each other. The Lagrangian equations may therefore be applied to rigid bodies.

506. *The Lagrangian equations of motion* are therefore

$$\frac{d}{dt}\frac{dL}{d\theta'} - \frac{dL}{d\theta} = 0, \quad \frac{d}{dt}\frac{dL}{d\phi'} - \frac{dL}{d\phi} = 0, \quad \text{\&c.} = 0 \text{ ...... (9)}.$$

The function $L = T + U$ and is therefore *the sum of the kinetic energy and the work function.* If we use the function V to represent the potential energy, we have, by definition, $U + V$ equal to a constant. We then put $L = T - V$, so that L *is the difference between the kinetic and potential energies.* Substituting these values for L, and remembering that U and V are functions of the coordinates and not of their velocities, we may also write the Lagrangian equations in the two typical forms

$$\frac{d}{dt}\frac{dT}{d\theta'} - \frac{dT}{d\theta} = \frac{dU}{d\theta}; \quad \frac{d}{dt}\frac{dT}{d\theta'} - \frac{dT}{d\theta} + \frac{dV}{d\theta} = 0 \text{.....(10)},$$

where θ stands for any one of the coordinates. It should be noticed that in these equations, all the differential coefficients are partial, except those with regard to t.

The function L is sometimes called *the Lagrangian function*. We see that *when once it has been found, all the dynamical equations, free from all unknown reactions, can be deduced by simple differentiation.*

507. Virtual moment of the effective forces. If we substitute for L in the lemma of Art. 504 the value of T given by (7) we have

$$\frac{d}{dt}\frac{dT}{d\theta'} - \frac{dT}{d\theta} = \Sigma\left\{mx''\frac{dx}{d\theta} + my''\frac{dy}{d\theta} + \&c.\right\} \quad \text{.................. (11).}$$

The right-hand side (after multiplication by $\delta\theta$) is the sum of the virtual moments of the effective forces mx'', my'', &c. It follows therefore that *the Lagrangian expression on the left-hand side (after multiplication by $\delta\theta$) represents the sum of the virtual moments of the effective forces, when expressed in terms of the generalized coordinates θ, ϕ, &c.*

In the same way writing T for the arbitrary function L in (4), we have by (7)

$$\frac{dT}{d\theta'} = \Sigma\left\{mx'\frac{dx}{d\theta} + my'\frac{dy}{d\theta} + \&c.\right\}.$$

The left-hand side (after multiplication by $\delta\theta$) therefore represents the sum of the virtual moments of the momenta of the several particles of the system for the displacement $\delta\theta$. It is often called *the generalized θ component* of the momentum.

508. Meaning of the lemma. The fundamental equation represented by the lemma has been deduced from the principles of the differential calculus without reference to any mechanical theorem.

Analytically, it expresses the fact that the Lagrangian operator symbolized by

$$\Delta_\theta = \frac{d}{d\theta} - \frac{d}{dt}\frac{d}{d\theta'}$$

follows the same law as the differential coefficient $d/d\theta$, i.e.

$$\Delta_\theta L = \Delta_x L \cdot \frac{dx}{d\theta} + \Delta_y L \cdot \frac{dy}{d\theta} + \ldots\ldots,$$

which may also be written

$$\Delta_\theta L \cdot \delta\theta = \Delta_x L \cdot \delta x + \Delta_y L \cdot \delta y + \ldots\ldots,$$

where $\delta\theta$, δx, δy, &c. are any small arbitrary variations consistent with the geometrical relations which hold at the time t.

If we interpret the lemma *dynamically* (Art. 506), the equation asserts that the sum of the virtual moments of the effective and impressed forces for a displacement $\delta\theta$ has the same value whatever changes are made in the coordinates.

509. Working rule. When we solve a dynamical problem we begin by writing down the equation of vis viva, viz. $T = U + C$.

It appears that when we have done this, Lagrange's method enables us to write down all the equations of motion of the second order by performing certain differentiations on the quantities on each side of the equation (Art. 506).

We shall presently show that before performing these differentiations, we may remove certain factors from one side to the other by making a change in the independent variable t; Art. 524.

510. The function T. We have assumed that the Cartesian coordinates x, y, z of every particle of the system can be expressed in terms of the generalized coordinates θ, ϕ, &c. by means of equations of the form

$$x = f(t, \theta, \phi, \text{\&c.}) \ldots\ldots\ldots\ldots\ldots\ldots\ldots\ldots(1);$$

these equations may contain t, but not θ', ϕ', &c. (Art. 504). *In choosing therefore the Lagrangian coordinates, we see that they must be such that the Cartesian coordinates of every particle could be expressed if required in terms of them by means of equations which may contain the time, but do not contain differential coefficients with regard to the time.*

Differentiating the geometrical equations (1) as in Art. 504

$$x' = f_t + f_\theta\theta' + f_\phi\phi' + \text{\&c.}, \quad y' = \text{\&c.} \ldots\ldots\ldots\ldots(2),$$

and substituting in the expression for the vis viva

$$2T = \Sigma m(x'^2 + y'^2 + z'^2) \ldots\ldots\ldots\ldots\ldots\ldots(7),$$

given in Art. 505, we observe that $2T$ takes the form

$$2T = A_{11}\theta'^2 + 2A_{12}\theta'\phi' + \ldots + B_1\theta' + B_2\phi' + \ldots + C,$$

where the coefficients A_{11}, &c., B_1, B_2, &c., and C are functions of t, θ, ϕ, &c.

In most dynamical problems, *the geometrical equations do not contain the time explicitly*, i.e. t does not enter into the equations (1) except implicitly through θ, ϕ, &c. The term f_t will therefore be absent from the equation (2), Art. 504. Hence x', y', z' are homogeneous functions of θ', ϕ', &c. of the first order. When substituted in (7), we find that $2T$ is a *homogeneous function of θ', ϕ', &c. of the second order, viz.*

$$2T = A_{11}\theta'^2 + 2A_{12}\theta'\phi' + \ldots,$$

where A_{11}, A_{12}, &c. are functions of the coordinates θ, ϕ, &c. but not of t.

511. Examples of Lagrange's equations. *Ex.* Two particles, of masses M, m, are connected by a light rod, of length l. The first A is constrained to move along a smooth fixed horizontal wire, while the other B is free to oscillate in the vertical plane under the action of gravity. It is required to find the motion.

To fix the positions of both the particles in space, we require two coordinates, say, the distance ξ of the point A from some origin O and the inclination θ of AB

to the vertical. The Cartesian coordinates of B are then $x=\xi+l\sin\theta$ and $y=l\cos\theta$. The semi vis viva and work functions are then

$$T=\tfrac{1}{2}M\xi'^2+\tfrac{1}{2}m\{(\xi'+l\cos\theta\theta')^2+(l\sin\theta\theta')^2\}$$
$$=\tfrac{1}{2}(M+m)\xi'^2+ml\cos\theta\xi'\theta'+\tfrac{1}{2}ml^2\theta'^2 \quad\text{.......... (1)},$$
$$U=mgl\cos\theta \quad\text{..........(2)}.$$

Substituting in the Lagrangian equations,

$$\frac{d}{dt}\frac{dT}{d\xi'}-\frac{dT}{d\xi}=\frac{dU}{d\xi},\qquad \frac{d}{dt}\frac{dT}{d\theta'}-\frac{dT}{d\theta}=\frac{dU}{d\theta},$$

we have

$$\left.\begin{array}{l}\dfrac{d}{dt}\{(M+m)\xi'+ml\cos\theta\theta'\}=0\\ \dfrac{d}{dt}\{ml\cos\theta\xi'+ml^2\theta'\}+ml\sin\theta\xi'\theta'=-mgl\sin\theta\end{array}\right\}.$$

These give

$$(M+m)\xi'+ml\cos\theta\theta'=A,\qquad \cos\theta\xi''+l\theta''=-g\sin\theta \quad\text{.......... (3)},$$

where A is a constant of integration. Eliminating ξ, we have

$$(M+m\sin^2\theta)\theta'\theta''+m\sin\theta\cos\theta\theta'\,.\,\theta'^2=-\frac{g}{l}(M+m)\sin\theta\theta'.$$

This gives by integration

$$(M+m\sin^2\theta)\theta'^2=C+\frac{2g}{l}(M+m)\cos\theta \quad\text{.......... (4)}.$$

In this way the velocities ξ' and θ' have been found in terms of the coordinates ξ, θ.

We have here used both the Lagrangian equations, but we might have replaced the second by the equation of vis viva, viz. $T=U+C$. Eliminating ξ' by the help of the first of equations (3), we should then have arrived at the result (4) without any further integrations.

512. *Ex.* 1. *The four elementary forms for the acceleration of a point follow at once from Lagrange's equations.* For example, let us deduce the polar form given in Art. 493.

We notice that the components of velocity of P along the radius vector and perpendicular to it, are respectively r' and $r\theta'$, while that perpendicular to the plane zOP is $r\sin\theta\phi'$. Since these three directions are orthogonal, we have

$$2T=m(r'^2+r^2\theta'^2+r^2\sin^2\theta\phi'^2).$$

Substituting in the Lagrangian equation

$$\frac{d}{dt}\frac{dT}{d\xi'}-\frac{dT}{d\xi}=\frac{dU}{d\xi},$$

where ξ in turn stands for r, θ, ϕ, we obtain

$$\frac{d}{dt}(mr')-m(r\theta'^2+r\sin^2\theta\phi'^2)=\frac{dU}{dr},$$
$$\frac{d}{dt}(mr^2\theta')-mr^2\sin\theta\cos\theta\phi'^2=\frac{dU}{d\theta},$$
$$\frac{d}{dt}(mr^2\sin^2\theta\phi')=\frac{dU}{d\phi},$$

which evidently reduce to the forms given in Art. 493.

Ex. 2. *To deduce the accelerations for moving axes from Lagrange's equations when the component velocities are known.*

We have given by Art. 499,

$$u=x'-y\theta_3+z\theta_2,\qquad v=y'-z\theta_1+x\theta_3,\qquad w=z'-x\theta_2+y\theta_1.$$

Also $$T=\tfrac{1}{2}(u^2+v^2+w^2),$$

the mass of the particle being unity. Since x' enters into the expression for T only through u, while x enters through both v and w, we have

$$\frac{dT}{dx'}=\frac{dT}{du}\frac{du}{dx'}=u,\qquad \frac{dT}{dx}=\frac{dT}{dv}\frac{dv}{dx}+\frac{dT}{dw}\frac{dw}{dx}=v\theta_3-w\theta_2.$$

The Lagrangian equation $$\frac{d}{dt}\frac{dT}{dx'}-\frac{dT}{dx}=\frac{dU}{dx}$$

becomes $$\frac{du}{dt}-v\theta_3+w\theta_2=X.$$

Ex. 3. *To deduce the equation of vis viva from Lagrange's equations.*

Multiplying the Lagrangian equations

$$\frac{d}{dt}\frac{dT}{d\theta'}-\frac{dT}{d\theta}=\frac{dU}{d\theta},\qquad \frac{d}{dt}\frac{dT}{d\phi'}-\frac{dT}{d\phi}=\frac{dU}{d\phi},\ \&c.,$$

by θ', ϕ', &c. respectively and adding the results, we have

$$\Sigma\left\{\frac{d}{dt}\left(\theta'\frac{dT}{d\theta'}\right)-\theta''\frac{dT}{d\theta'}\right\}-\Sigma\theta'\frac{dT}{d\theta}=\Sigma\theta'\frac{dU}{d\theta},$$

where Σ implies summation for all the coordinates.

If the geometrical equations do not contain the time explicitly, T is a homogeneous function of θ', ϕ', &c., Art. 510, and by Euler's theorem $\Sigma\theta'\dfrac{dT}{d\theta'}=2T$. Also since T and U are not functions of t,

$$\frac{dT}{dt}=\Sigma\left(\theta'\frac{dT}{d\theta}+\theta''\frac{dT}{d\theta'}\right),\qquad \frac{dU}{dt}=\Sigma\theta'\frac{dU}{d\theta}.$$

Substituting in the expression given above, we have

$$2\frac{dT}{dt}-\frac{dT}{dt}=\frac{dU}{dt};\qquad \therefore\ T=U+C,$$

where C is an arbitrary constant, usually called the constant of vis viva.

Ex. 4. The position of a moving point is determined by the radii $1/\xi$, $1/\eta$, $1/\zeta$ of the three spheres which pass through it and touch three fixed rectangular coordinate planes at the origin. Find the component velocities u, v, w of the point in the directions of the outward normals of the spheres, and prove that the component accelerations in the same directions are $du/dt+v(\eta u-\xi v)-w(\xi w-\zeta u)$, and two similar expressions. [Coll. Ex. 1896.]

Writing $D=\xi^2+\eta^2+\zeta^2$ we deduce from the equations of the spheres that $x=2\xi/D$, &c. Noticing that the spheres are orthogonal, we find, by resolving the velocities x', y', z' along them, $u=-x\xi'/\xi$, $v=-y\eta'/\eta$, $w=-z\zeta'/\zeta$. Hence

$$T=\tfrac{1}{2}(u^2+v^2+w^2)=2(\xi'^2+\eta'^2+\zeta'^2)/D^2.$$

Also the acceleration along the ξ axis is dU/udt or $-\tfrac{1}{2}D\,dU/d\xi$. Substituting in the Lagrangian formula $\dfrac{dU}{d\xi}=\dfrac{d}{dt}\dfrac{dT}{d\xi'}-\dfrac{dT}{d\xi}$, we obtain the required result. It may also be deduced from the formulæ of Arts. 499, 502, Ex. 2.

513. *To apply the Lagrangian equations to determine the small oscillations of a system of particles about a position of equilibrium, when the geometrical equations do not contain the time explicitly.*

Let the system have n coordinates and let these be θ, ϕ, &c. Let their values in the position of equilibrium be α, β, &c., and at any time t, let $\theta = \alpha + x$, $\phi = \beta + y$, &c.

The vis viva being a homogeneous function of θ', ϕ', &c. (Art. 510), we have

$$2T = P\theta'^2 + 2Q\theta'\phi' + R\phi'^2 + \&c.,$$

where P, Q, &c. are functions of θ, ϕ, &c. When we substitute $\theta = \alpha + x$, &c. *and reject all powers of the small quantities above the second*, this reduces to an expression of the form

$$2T = A_{11}x'^2 + 2A_{12}x'y' + A_{22}y'^2 + \&c. \ldots\ldots\ldots\ldots(1),$$

where the coefficients are constant, and are known functions of α, β, &c.

The work function U is a function of θ, ϕ, &c. and when expanded takes the form

$$2U = 2U_0 + 2B_1x + 2B_2y + \&c. + B_{11}x^2 + 2B_{12}xy + \&c. \ldots(2).$$

We assume that these expansions are possible.

Since the system is in equilibrium in the position defined by $x = 0$, $y = 0$, &c., we have by the principle of virtual work,

$$\frac{dU}{dx} = 0, \quad \frac{dU}{dy} = 0, \ \&c.; \ \therefore \ B_1 = 0, \ B_2 = 0, \ \&c. \ldots\ldots(3).$$

If the position of equilibrium is not known beforehand, the values of α, β, &c. may be obtained by solving the n equations (3).

To find the equations of motion we substitute in the n Lagrangian equations typified by

$$\frac{d}{dt}\frac{dT}{dx'} - \frac{dT}{dx} = \frac{dU}{dx} \ldots\ldots\ldots\ldots\ldots\ldots\ldots\ldots(4).$$

Since the expansion for T does not contain the coordinates x, y, &c., we have $dT/dx = 0$, $dT/dy = 0$, &c. The equation (4) therefore becomes

$$\left.\begin{array}{c} A_{11}x'' + A_{12}y'' + A_{13}z'' + \&c. = B_{11}x + B_{12}y + B_{13}z + \&c. \\ A_{12}x'' + A_{22}y'' + A_{23}z'' + \&c. = B_{12}x + B_{22}y + B_{23}z + \&c. \\ \&c. = \&c. \end{array}\right\} \ldots(5).$$

To solve the equations (5) we follow the rules given in Art. 292. Let any *principal oscillation* be represented by

$$x = G\sin(pt+\alpha), \qquad y = H\sin(pt+\alpha), \text{ \&c.} \quad \text{........(6)},$$

where G, H, &c. are constants. We find by an easy substitution

$$\left.\begin{aligned}(A_{11}p^2+B_{11})G+(A_{12}p^2+B_{12})H+\ldots&=0\\(A_{12}p^2+B_{12})G+(A_{22}p^2+B_{22})H+\ldots&=0\\\text{\&c.}&=0\end{aligned}\right\}\quad\text{........(7)}.$$

Eliminating the ratios $G : H$: &c., the n values of p^2 are given by the Lagrangian equation

$$\begin{vmatrix}A_{11}p^2+B_{11}, & A_{12}p^2+B_{12}, & \text{\&c.}\\A_{12}p^2+B_{12}, & A_{22}p^2+B_{22}, & \text{\&c.}\\\text{\&c.} & \text{\&c.} & \text{\&c.}\end{vmatrix}=0\quad\text{..........(8)}.$$

514. It is shown in the higher dynamics that, because the vis viva $2T$ is necessarily positive for all real values of x', y', &c., the values of p^2 given by this determinantal equation are real. If all the roots are positive the values of p are real, and the system of particles then oscillates about the position of equilibrium. If any or all the values of p^2 are negative, some or all the values of p take the form $\pm q\sqrt{-1}$. The corresponding trigonometrical terms in (6) become exponential and the system does not oscillate. See Art. 120.

515. If a value of p^2 is zero p has two *equal* zero values, and the corresponding term in (6) takes the form $A+Bt$. In such a case the coordinate may become large and the system will then depart so far from the position of equilibrium that it will be necessary to take account of the small terms in (1) and (2) of higher orders than the second.

516. Rule. When applying Lagrange's equations to any special case of oscillation *about a position of equilibrium* we begin by writing down the expressions for the vis viva and work function for the system in *its displaced position*, and express these in the *quadratic forms* (1) and (2) (Art. 513). If the whole motion is required we follow in each special case the process described in the general investigation. But if, as usually happens, only the periods are required, we omit the intervening steps and deduce the determinant (8) immediately from the expansions (1), (2).

To help the memory, we notice that, *if we drop the accents in the expression for T, the determinant* (8) *is the discriminant of the quadric $Tp^2 + U$.*

517. *To apply Lagrange's equations to determine the initial motion of a system.*

The method has been already explained in Art. 282. The Lagrangian equations give the values of θ'', ϕ'', &c. in the initial position without introducing the unknown reactions. Differentiating the Lagrangian equations of Art. 506 we obtain θ''', ϕ''', &c., and any higher differential coefficients.

If x, y, z are the Cartesian coordinates of any point P of the system, we have by Art. 510,

$$x = f_1(\theta, \phi, \&c.), \quad y = f_2(\theta, \phi, \&c.), \quad z = \&c.,$$

and therefore by differentiation the initial values of x', x'', &c., y', y'', &c., z', &c. may be found. The initial radius of curvature follows from the formulæ of the differential calculus, Art. 280.

518. *Let, for example, the initial accelerations be required when the system starts from rest.* The initial position being $\theta = \alpha$, $\phi = \beta$, &c. we put, as in Art. 513, $\theta = \alpha + x$, $\phi = \beta + y$, &c. Since the system starts from rest, the velocities x', y', &c. are small and we can make the expansions (1) and (2) as before. Since the initial position is not one of equilibrium, we no longer have $B_1 = 0$, $B_2 = 0$, &c. Retaining only the lowest powers of x, y, &c. which occur in the equations of motion, we have

$$\left.\begin{aligned} A_{11}x'' + A_{12}y'' + \&c. &= B_1 \\ A_{12}x'' + A_{22}y'' + \&c. &= B_2 \\ \&c. &= \&c. \end{aligned}\right\}.$$

These determine the initial accelerations of the coordinates and therefore the component accelerations of every point of the system.

519. *Ex.* 1. Let us apply the Lagrangian equations to find the small oscillations of the two particles described in Art. 511.

The quantities ξ, θ represent the deviations of the rod from its position of equilibrium. The vis viva and work function expressed in quadratic forms are

$$T = \tfrac{1}{2}(M+m)\,\xi'^2 + ml\xi'\theta' + \tfrac{1}{2}ml^2\theta'^2, \qquad U = mgl\,(1 - \tfrac{1}{2}\theta^2).$$

The determinant is the discriminant of

$$Tp^2 + U = \tfrac{1}{2}(M+m)\,p^2\xi^2 + mlp^2\xi\theta + \tfrac{1}{2}ml\,(lp^2 - g)\,\theta^2;$$

$$\therefore \begin{vmatrix} (M+m)\,p^2, & mlp^2 \\ mlp^2, & ml\,(lp^2 - g) \end{vmatrix} = 0.$$

One principal motion is given by

$$p^2=\frac{g}{l}\frac{M+m}{M},\qquad \xi=G\sin(pt+\alpha),\qquad \theta=H\sin(pt+\alpha).$$

The other is determined by $p^2=0$; this implies that one coordinate takes the form $A+Bt$. It is evident that the rod could be so projected along the horizontal wire that ξ has this form while $\theta=0$.

The student should apply Lagrange's equations to the problems on small oscillations and initial motions already considered in the chapter on motion in two dimensions. He will thus be able to form a comparison of the advantages of the different methods.

Ex. 2. Three uniform rods AB, BC, CD have lengths $2a$, $2b$, $2a$ and masses m, m', m. They are hinged together at B and C, and at A, D are small smooth rings which are free to move along a fixed fine horizontal bar. The rods hang in equilibrium, forming with the bar a vertical rectangle. When a slight symmetrical displacement is given, the period of a small oscillation is given by $4map^2=3g(m+m')$. Find also the periods when the displacement is unsymmetrical. [Coll. Ex. 1897.]

Ex. 3. Two equal strings AC, BC have their ends at the fixed points A, B, on the same horizontal line, and at C a heavy particle is attached. From C a string CD hangs down with a second heavy particle at D. Find the periods of the three small oscillations. [The two periods of the oscillations perpendicular to the vertical plane through A and B are given in Art. 300, Ex. 1.]

520. Solution of Lagrange's Equations. Our success in obtaining the first integrals of the Lagrangian equations will greatly depend on the choice of coordinates. When the position of the system is determined by only one coordinate, the equation of vis viva is the first integral, and this is sufficient to determine the motion.

When there are two or more coordinates, integrals can be found only in special cases. The general problem of the solution of the Lagrangian equations is too great a subject to be attempted here. It is sufficient to state a few elementary rules which may assist the student.

521. We should, if possible, *so choose the coordinates that some one of them is absent from the expression for the work function U.* For example, if there be any direction such that the component of the impressed forces is zero throughout the motion, we should take the axis of z in that direction and let z be one of the coordinates. Again if the moment of the forces about some straight line fixed in space, say Oz, is always zero, the angle ϕ which the plane POz makes with xOz will be a suitable coordinate. In that case $dU/d\phi=0$ and U is independent of ϕ. These, or similar, mechanical considerations generally enable us to make a proper choice.

Let θ be the coordinate absent from the work function, then *if θ is also absent from the expression for T*, though the differential coefficient θ' is present, the Lagrangian equation

$$\frac{d}{dt}\frac{dT}{d\theta'}-\frac{dT}{d\theta}=\frac{dU}{d\theta} \text{ becomes } \frac{dT}{d\theta'}=A,$$

where A is the constant of integration. Thus a first integral, *different from that of vis viva*, has been found.

522. Liouville's integral. Liouville has given an integral of Lagrange's equations which has the advantage of great simplicity when it can be applied. This may be found in vol. XI. of his *Journal*, 1846; the following is a slight modification of his method.

Let us suppose that the vis viva has the form

$$2T=M(P\theta'^2+Q\phi'^2+R\psi'^2+\&c.)\text{..........................(1)},$$

where the products $\theta'\phi'$, $\phi'\psi'$, &c. are absent. *The method requires that the coefficient P should be a function of θ only, while Q, R, &c., are not functions of θ.* We notice that M may be a function of all or any of the coordinates, and Q, R, &c. functions of any except θ. *It is also necessary that* the impressed forces should be such that *the work function U has the form*

$$M(U+C)=F_1(\theta)+F(\phi,\psi,\&c.)\text{..........................(2)},$$

where C is the constant in the equation of vis viva,

$$T=U+C\text{..(3)}.$$

We shall now prove that *when these conditions are satisfied, a first integral is*

$$\tfrac{1}{2}M^2P\theta'^2=F_1(\theta)+A\text{ .... (4)}.$$

We first put $P\theta'^2=\xi'^2$, then ξ is a function of θ only and we may temporarily take ξ, ϕ, ψ, &c. as the coordinates. We now have

$$T=\tfrac{1}{2}M(\xi'^2+Q\phi'^2+\&c.)=U+C,$$

and the Lagrangian equation for ξ is

$$\frac{d}{dt}(M\xi')-\frac{1}{2}\frac{dM}{d\xi}(\xi'^2+Q\phi'^2+\&c.)=\frac{dU}{d\xi}.$$

Using the equation of vis viva, this takes the form

$$M\frac{d}{dt}(M\xi')=(U+C)\frac{dM}{d\xi}+M\frac{dU}{d\xi}=\frac{d}{d\xi}M(U+C).$$

Substituting on the right-hand side from (2) and multiplying by ξ', we have

$$M\xi'\frac{d}{dt}(M\xi')=\frac{dF_1(\theta)}{d\xi}\xi'.$$

Since $F_1(\theta)$ is a function of ξ and not of any of the other coordinates, this gives by an easy integration

$$\tfrac{1}{2}M^2\xi'^2=F_1(\theta)+A.$$

Returning to the coordinate θ, we have the integral (4).

When the initial conditions are given, the value of C can be found by introducing these conditions into the equation of vis viva. If a solution is required for all initial conditions C is arbitrary and in that case the condition (2) requires that both MU and M should have the general form indicated on the right-hand of that equation. If

$$M(U+C)=F_1(\theta)+F_2(\phi)+\&c.,$$

and Q, R, &c. are respectively functions of ϕ, ψ, &c. only, it is evident that the method supplies *all* the first integrals.

Ex. If $T=M(P\theta'^2+Q\phi'^2)$, $M=f_1(\theta)+f_2(\phi)$, $MU=F_1(\theta)+F_2(\phi)$, *integrate the Lagrangian equations by Liouville's method.* The integrals are

$$\tfrac{1}{2}M^2P\theta'^2=F_1(\theta)+Cf_1(\theta)+A_1, \qquad \tfrac{1}{2}M^2Q\phi'^2=F_2(\phi)+Cf_2(\phi)+A_2,$$

adding these and using the equation of vis viva we see that $A_1+A_2=0$. The paths are then given by

$$\frac{\sqrt{P}\,d\theta}{\sqrt{(F_1+Cf_1+A_1)}}=\frac{\sqrt{Q}\,d\phi}{\sqrt{(F_2+Cf_2+A_2)}}=\frac{\sqrt{2}\,dt}{M}.$$

Multiplying these by f_1, f_2 and adding, the time is found by

$$\frac{f_1\sqrt{P}\,d\theta}{\sqrt{(F_1+Cf_1+A_1)}}+\frac{f_2\sqrt{Q}\,d\phi}{\sqrt{(F_2+Cf_2+A_2)}}=\sqrt{2}\,dt,$$

where all the variables have been separated.

523. Jacobi's integral. If T be a homogeneous function of the coordinates θ, ϕ, &c. of n dimensions and U a homogeneous function of the same coordinates of $-(n+2)$ dimensions, then one integral is

$$\theta\frac{dT}{d\theta'}+\phi\frac{dT}{d\phi'}+\&c.=(n+2)\,Ct+A,$$

where C is the constant of vis viva and A an arbitrary constant.

To prove this, we multiply the Lagrangian equations by θ, ϕ, &c. and add the products. Remembering Euler's theorem on homogeneous functions, we have

$$\theta\frac{d}{dt}\frac{dT}{d\theta'}+\&c.=nT-(n+2)\,U.$$

The left-hand side is the same as

$$\frac{d}{dt}\left(\theta\frac{dT}{d\theta'}+\&c.\right)-\left\{\theta'\frac{dT}{d\theta'}+\&c.\right\}=\frac{d}{dt}\left(\theta\frac{dT}{d\theta'}+\&c.\right)-2T,$$

since T is a homogeneous function of θ', ϕ', &c. of two dimensions. Remembering that $T-U=C$, we have $\dfrac{d}{dt}\left\{\theta\dfrac{dT}{d\theta'}+\&c.\right\}=(n+2)\,C.$

Ex. A free system of particles moves under the influence of their mutual attractions, the law of force being the inverse cube: show that $\Sigma mr^2=A+Bt+Ct^2$ where r is the distance of the particle m from the origin.

[*Vorlesungen über Dynamik.*]

Some developments of these results are given in the first volume of the author's treatise on *Rigid Dynamics.*

524. Change of the independent variable. *It is sometimes useful to be able to change the independent variable in Lagrange's equations from t to some other quantity τ so that $d\tau=P\,dt$, where P is any function of the coordinates.*

We suppose that the geometrical equations do not contain the time explicitly, so that T is a homogeneous function of the form

$$T=\tfrac{1}{2}A_{11}\theta'^2+A_{12}\theta'\phi'+\tfrac{1}{2}A_{22}\phi'^2+\ldots\ldots\ldots\ldots\ldots\ldots\ldots\ldots\text{(1)}.$$

Let suffixes applied to the coordinates mean differentiations with regard to τ just as accents denote differentiations with regard to t. Then

$$\theta' = P\theta_1, \quad \phi' = P\phi_1, \text{ \&c.}$$

Consider how any one of Lagrange's equations, say,

$$\frac{d}{dt}\frac{dT}{d\phi'} - \frac{dT}{d\phi} = \frac{dU}{d\phi} \quad\text{.............................. (2),}$$

is affected by the change of t. Let us write

$$T_2 = (\tfrac{1}{2}A_{11}\theta_1^2 + A_{12}\theta_1\phi_1 + \ldots\ldots)\,P \quad\text{........................ (3).}$$

Supposing that P is a function of the coordinates only, not of θ', ϕ', &c., we have

$$\frac{dT}{d\phi'} = A_{12}\theta' + A_{22}\phi' + \ldots = (A_{12}\theta_1 + A_{22}\phi_1 + \ldots)\,P = \frac{dT_2}{d\phi_1},$$

$$\frac{dT}{d\phi} = \frac{1}{2}\frac{dA_{11}}{d\phi}\theta'^2 + \ldots = \left(\frac{1}{2}\frac{dA_{11}}{d\phi}\theta_1^2 + \ldots\right)P^2 = P^2\frac{d}{d\phi}\left(\frac{T_2}{P}\right).$$

The Lagrangian equation therefore becomes after a slight reduction

$$\frac{d}{d\tau}\frac{dT_2}{d\phi_1} - \frac{dT_2}{d\phi} = -\frac{T_2}{P}\frac{dP}{d\phi} + \frac{1}{P}\frac{dU}{d\phi} \quad\text{........................ (4).}$$

If we use the equation of vis viva, viz. $T = U + C$, and notice that $T = PT_2$, the right-hand side of this equation becomes $\dfrac{d}{d\phi}\dfrac{U+C}{P}$. The typical Lagrangian form therefore takes the form

$$\frac{d}{d\tau}\frac{dT_2}{d\phi_1} - \frac{dT_2}{d\phi} = \frac{d}{d\phi}\frac{U+C}{P} \quad\text{.......................... (5).}$$

We notice that though $T = PT_2$, they are differently expressed. To obtain the partial differential coefficients of T_2, the quantities θ', ϕ', &c. must be replaced by $P\theta_1$, $P\phi_1$, &c. before differentiation.

Suppose for example that the equation of vis viva (Art. 509) is

$$T = M\{\tfrac{1}{2}A\theta'^2 + \text{\&c.}\} = U + C,$$

and that we wish to remove the factor M before deducing the Lagrangian equations. Changing the independent variable so that $d\tau = Pdt$, we deduce the Lagrangian equations by operating on

$$T_2 = MP\{\tfrac{1}{2}A\theta_1^2 + \text{\&c.}\}, \qquad U_2 = \frac{U+C}{P}.$$

Choosing $MP = 1$, we have

$$T_2 = \tfrac{1}{2}A\theta_1^2 + \text{\&c.}, \qquad U_2 = M(U + C).$$

The factor M has thus been transferred from the expression for the vis viva to the work function. Here M is a function of the coordinates only.

We may now change the suffixes into accents if we remember that the differentiations are to be taken with regard to τ instead of t. This difference is of no importance if we require only the paths of the particles and not their positions at any time. If the time also be required, we add the equation $dt = Md\tau$.

525. Orthogonal Coordinates. The Lagrangian equations are much simplified when the expression for T can be put into the form

$$T = \tfrac{1}{2}(P\theta'^2 + Q\phi'^2 + \text{\&c.}),$$

where the products $\theta'\phi'$, &c. are absent. We shall now prove that this will be the case when the coordinates of the particle are the parameters of systems of curves or surfaces at right angles.

Let the equations of three systems of surfaces which intersect at right angles be $f_1(x, y, z)=\rho_1, \quad f_2(x, y, z)=\rho_2, \quad f_3(x, y, z)=\rho_3$ ....(1), where ρ_1, ρ_2, ρ_3 are three constants or parameters whose values determine which surface of each system is taken. These parameters may be regarded as the coordinates of the point of intersection of the three surfaces.

Such coordinates are called sometimes *orthogonal coordinates* and sometimes *curvilinear coordinates.* Their theory was given by Lamé in his *Leçons sur les coordonnées curvilignes*, 1859. In what follows we adopt his notation as far as possible.

As an example of orthogonal coordinates we call to mind a *system of confocal ellipsoids and hyperboloids of one and two sheets*, the lengths of the major axes being usually taken as the parameters. These are called elliptic coordinates. We may also notice that *all the coordinates in common use, whether Cartesian, cylindrical or polar, are orthogonal.* In the first the point is defined as the intersection of three orthogonal planes, in the second we use a cylinder cut by two planes, and in the third a sphere cut by a right cone and a plane.

Let (a_1, b_1, c_1) be the direction cosines of a normal to the surface whose parameter is ρ_1, then

$$a_1=\frac{1}{h_1}\frac{d\rho_1}{dx}, \quad b_1=\frac{1}{h_1}\frac{d\rho_1}{dy}, \quad c_1=\frac{1}{h_1}\frac{d\rho_1}{dz} \text{ ....(2)},$$

where

$$h_1^2=\left(\frac{d\rho_1}{dx}\right)^2+\left(\frac{d\rho_1}{dy}\right)^2+\left(\frac{d\rho_1}{dz}\right)^2 \text{ .... (3)}.$$

Let ds_1 be an elementary arc of the intersection of the two surfaces ρ_2, ρ_3; then ds_1 is also an elementary length measured along the normal to the surface ρ_1. As we travel along this arc x, y, z and ρ_1 vary, while ρ_2, ρ_3 are constant. Hence

$$\frac{d\rho_1}{dx}dx+\frac{d\rho_1}{dy}dy+\frac{d\rho_1}{dz}dz=d\rho_1;$$

$$\therefore\ a_1dx+b_1dy+c_1dz=d\rho_1/h_1 \text{ .... (4)}.$$

But the left side is the sum of the projections of dx, dy, dz on the normal and is therefore ds_1; hence $ds_1=d\rho_1/h_1$. It follows that the component v_1 of velocity along the normal to the surface ρ_1 is $v_1=\dfrac{1}{h_1}\dfrac{d\rho_1}{dt}$. In the same way the components of velocity normal to the other two surfaces may be found, and since these are at right angles,

$$v^2=\frac{1}{h_1^2}\rho_1'^2+\frac{1}{h_2^2}\rho_2'^2+\frac{1}{h_3^2}\rho_3'^2 \text{ .... (5)},$$

where accents denote differential coefficients.

In order to use this expression, it will be necessary to express h_1, h_2, h_3, in terms of the new coordinates ρ_1, ρ_2, ρ_3. To effect this we solve the equations (1) and determine x, y, z as functions of ρ_1, ρ_2, ρ_3; finally substituting these values in the expressions (3) for h_1, h_2, h_3. This is sometimes a lengthy process.

Motion on a Curve.

526. Fixed Curves. *To find the motion of a particle on a smooth curve fixed in space.*

To find the velocity, we resolve the forces along the tangent to the curve. If F be the component of the impressed forces

X, Y, Z, this gives as in Art. 181,

$$mv\frac{dv}{ds} = F = X\frac{dx}{ds} + Y\frac{dy}{ds} + Z\frac{dz}{ds}.$$

If U be the work function, $F = dU/ds$, and we have

$$\tfrac{1}{2}mv^2 = U + C,$$

which is the equation of vis viva.

To find the pressure, we resolve in any two directions which may suit the problem under consideration. Supposing that we choose the radius of curvature and binormal, we have

$$\frac{mv^2}{\rho} = G + R_1, \qquad 0 = H + R_2,$$

where G, H are the components of the impressed forces; R_1, R_2 the corresponding components of the pressure on the particle.

These equations show that the pressure of the particle on the curve is the resultant of two forces, (1) the statical pressure due to the forces urging the particle against the curve, (2) the centrifugal force mv^2/ρ acting in the direction opposite to that in which ρ is measured, Art. 183.

527. *Ex.* 1. A plane is drawn through the tangent at P making an angle i with the osculating plane. If ρ' be the radius of the circle of closest contact to the curve in this plane, then $\dfrac{mv^2}{\rho'} = G' + R'$ where G' and R' are the components of the impressed accelerating force and of the pressure respectively.

This follows from the theorem on curves $\rho' \cos i = \rho$, corresponding to Meunier's theorem on surfaces.

Ex. 2. A helix is placed with its axis vertical, and a bead slides on it under the action of gravity. Find the motion and pressure.

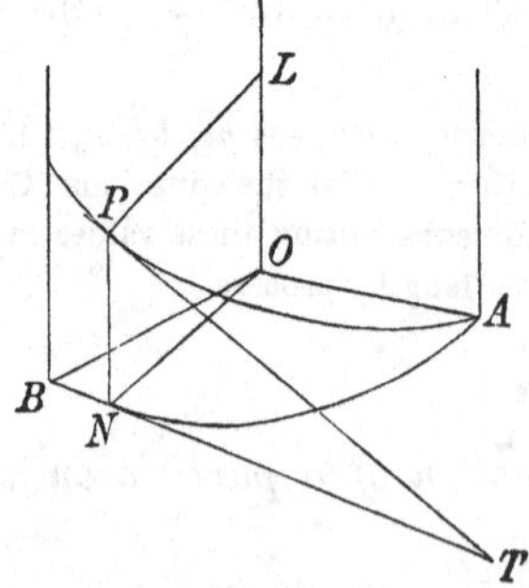

Let a be the radius of the cylinder, α the inclination of the tangent to the horizon. Drawing PL perpendicular to the axis of z, the radius of curvature is a length measured along PL equal to $a \sec^2 \alpha$. If PT is the tangent, the osculating plane is LPT. If the helix is smooth we have

$$v^2 = -2gz + C, \qquad \frac{v^2 \cos^2 \alpha}{a} = \frac{R_1}{m},$$

$$0 = g\cos\alpha + \frac{R_2}{m}.$$

If the particle start from rest at a height h, we see that $C = 2gh$. Since $v = -ds/dt$ and $ds \sin\alpha = dz$, we find that the time of descending that height is $\operatorname{cosec} \alpha \sqrt{2h/g}$.

If the helix is rough, the friction is $\mu\sqrt{(R_1^2+R_2^2)}$. Supposing that the coefficient of friction is $\mu=\tan\alpha$, the resolution along the tangent becomes

$$v\frac{dv}{ds}=-g\sin\alpha+\frac{\sin\alpha}{a}\sqrt{(v^4\cos^2\alpha+a^2g^2)};$$

writing $v^2\cos\alpha=\xi$ for brevity, we find

$$\int\frac{d\xi}{\sqrt{(\xi^2+a^2g^2)}-ag}=\frac{s\sin 2\alpha}{a}+C.$$

To integrate this we multiply the numerator and denominator of the fraction on the left-hand side by the denominator with the minus sign changed. We then find

$$\log\{v^2\cos\alpha+\sqrt{(v^4\cos^2\alpha+a^2g^2)}\}-\frac{ag+\sqrt{(v^4\cos^2\alpha+a^2g^2)}}{v^2\cos\alpha}=\frac{s\sin 2\alpha}{a}+C.$$

To find C we require the initial value of v. If this were zero the particle would remain at rest because $\mu=\tan\alpha$.

Ex. 3. A rough helical tube of pitch α and radius a is placed so as to have its axis vertical and the coefficient of friction is $\tan\alpha\cos\epsilon$. An extended flexible string which just fits the tube is placed in it: show that when the string has fallen through a vertical distance ma its velocity is $(ag\sec\alpha\sinh 2\mu)^{\frac{1}{2}}$, where μ is determined by the equation

$$\cot\tfrac{1}{2}\epsilon\tanh\mu=\tanh(\mu\sin\epsilon+\tfrac{1}{2}m\cos\alpha\sin 2\epsilon).\qquad\text{[Math. Tripos, 1886.]}$$

Ex. 4. Two small rings of masses m, m' can slide freely on two wires each of which is a helix of pitch p, the axes being coincident and the principal normals common; the rings repel one another with a force equal to $\mu mm'r$ when they are at a distance r from one another. Prove that if ϕ be the angle the plane through one ring and the axis makes with the plane through the other ring and the axis, the time in which ϕ increases from α to β is $\int_\alpha^\beta\{A\phi^2-2B\cos\phi+C\}^{-\frac{1}{2}}d\phi$, where

$$A=\mu mm'p^2\left\{\frac{1}{m(a^2+p^2)}+\frac{1}{m'(b^2+p^2)}\right\},\qquad B=\frac{ab}{p^2}A,$$

and a, b are the radii of the cylinders on which the helices are drawn.

[Coll. Ex. 1896.]

528. Moving curves. *Ex.* 1. *A particle P is constrained to move on the plane curve $z=f(x)$, which rotates about a straight line Oz in its plane with an angular velocity ω. It is required to form the equations of motion.*

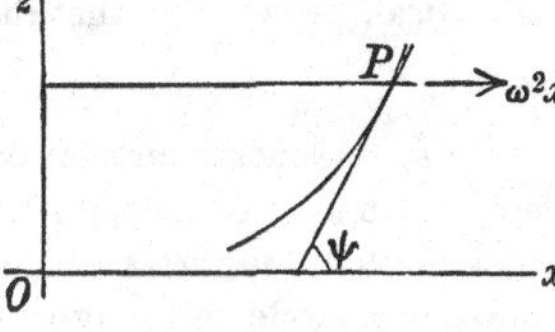

Applying to P an acceleration ω^2x tending from the axis of rotation, we treat the curve as if it were fixed, Art. 495. Taking the tangential and normal resolutions, we have

$$v\frac{dv}{ds}=\frac{F}{m}+\omega^2x\cos\psi,\qquad\frac{v^2}{\rho}=\frac{G}{m}-\omega^2x\sin\psi+\frac{R}{m},$$

where v is the velocity of the particle *relatively to the curve*, ψ the angle the tangent at P makes with the axis of x, and ρ is the radius of curvature. Also F and G are the components of the impressed forces along the tangent and radius of curvature at P.

We may replace the first of these equations by the integral of vis viva, viz.

$$\tfrac{1}{2}mv^2=\int(Fds+m\omega^2x\,dx).$$

The second equation then gives the component R of pressure in the plane of the curve. The component R' of pressure perpendicular to the plane of the curve is given by

$$m\frac{1}{x}\frac{d}{dt}(x^2\omega)=H+R',$$

where H is the corresponding component of the impressed force, and x is the distance of the particle from the axis of rotation.

Ex. 2. A circular wire is constrained to turn round a vertical tangent Oz with a uniform angular velocity ω. A heavy smooth bead, starting from the highest point A without any velocity relative to the curve, descends under the action of gravity. Find the velocity and pressure.

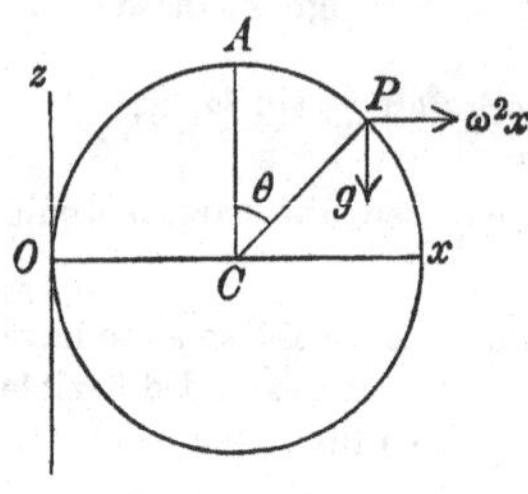

Let C be the centre, $OC=a$; let P be the particle, the angle $ACP=\theta$, $v=a\,d\theta/dt$. We reduce the plane to rest by applying to P an accelerating force measured by ω^2x, where $x=a+a\sin\theta$, and acting parallel to OC. The equation of vis viva then gives

$$\tfrac{1}{2}v^2=g\,(a-a\cos\theta)+\omega^2\int_a^x x\,dx;$$

$$\therefore\ v^2=2g\,(a-a\cos\theta)+\omega^2a^2\,(2\sin\theta+\sin^2\theta).$$

The components R, R' of the pressure on the particle respectively along PC and perpendicular to the plane are given by

$$\frac{v^2}{a}=g\cos\theta-\omega^2x\sin\theta+\frac{R}{m},\qquad \frac{1}{x}\frac{d}{dt}(x^2\omega)=\frac{R'}{m}.$$

The latter equation reduces to $R'=2m\omega v\cos\theta$.

Ex. 3. Two small rings of masses m, m', $(m>m')$ are capable of sliding on a smooth circular wire of radius a, whose vertical diameter is fixed, the rings being below the centre and connected by a light string of length $a\sqrt{2}$: prove that if the wire is made to rotate round the vertical diameter with an angular velocity $\left\{\frac{2g}{a\sqrt{3}}\frac{m\sqrt{3}-m'}{m-m'}\right\}^{\frac{1}{2}}$, the rings can be in relative equilibrium on opposite sides of the vertical diameter, the radius through the ring m being inclined at an angle 60° to the vertical. Show also that the tension of the string is $\frac{mm'}{m-m'}\frac{\sqrt{3}-1}{\sqrt{2}}g$.

[Coll. Ex. 1897.]

Ex. 4. A smooth circular cone of angle 2α has its axis vertical and its vertex, pierced with a small hole, downwards. A mass M hangs at rest by a string which passes through the vertex and a mass m attached to the upper extremity describes a horizontal circle on the inner surface of the cone. Find the time T of a complete revolution, and prove that small oscillations about the steady motion take place in the time $T\operatorname{cosec}\alpha\,\{(M+m)/3m\}^{\frac{1}{2}}$. [Coll. Ex. 1896.]

Ex. 5. A smooth plane revolves with uniform angular velocity ω about a fixed vertical axis which intersects it in the point O, at which a heavy particle is placed at rest. Show that during the subsequent motion $v^2=p^2\omega^2+2gz$; where z is the depth of the particle below O, p its distance from the axis and v the speed with which the path is traced on the plane. [Coll. Ex. 1893.]

529. A case of free motion with two centres of force. *Ex.* 1. *A particle P, of unit mass, is constrained to move along an elliptic wire without inertia which can turn freely about its major axis. The particle is acted on by two centres of force, situated in the foci S, H, which attract according to the law of the inverse square. Prove that the pressure on the curve is zero in certain cases.*

We take the major axis as the axis of z and the origin at the centre. Let ω be the angular velocity of the wire. Representing the distance of the particle P from the major axis by y, the component R' of pressure on the particle perpendicular to the plane of the curve is given by

$$\frac{1}{y}\frac{d}{dt}(y^2\omega)=R'.$$

But since the wire is without inertia, i.e. without mass, the wire moves so that the pressure R' on it is zero, Art. 267. We therefore have throughout the motion

$$y^2\omega=B,$$

where B is the constant of angular momentum about the axis of rotation.

Let the distances of the particle from the foci S, H be r_1, r_2; and let the central forces be μ_1/r_1^2, μ_2/r_2^2.

To find the motion in the plane zOP, we apply to P an acceleration $\omega^2y=B^2/y^3$, tending *from* the major axis, and then treat the curve as if it were fixed. We notice that the particle could freely describe the ellipse under any one of the forces μ_1/r_1^2, μ_2/r_2^2, B^2/y^3 if properly projected; see Arts. 333, 323. It immediately follows that if all the three forces act simultaneously, the pressure on the particle will be a constant multiple of the curvature, Art. 272.

The pressure will be zero, if the square of the velocity of projection is equal to the sum of the squares of the velocities when the particle describes the curve freely under each force separately; Art. 273. We find therefore that if v_1 be the velocity relatively to the curve, the pressure is zero, if

$$v_1^2=\mu_1\left(\frac{2}{r_1}-\frac{1}{a}\right)+\mu_2\left(\frac{2}{r_2}-\frac{1}{a}\right)-B^2\left(\frac{1}{y^2}+\frac{a^2-b^2}{b^4}\right).$$

If v be the resultant velocity of the particle in space, we have $v^2=v_1^2+\omega^2y^2$. Hence

$$v^2=\mu_1\left(\frac{2}{r_1}-\frac{1}{a}\right)+\mu_2\left(\frac{2}{r_2}-\frac{1}{a}\right)-B^2\frac{a^2-b^2}{b^4}.$$

When the pressure is zero, the wire may be removed and the particle describes its path freely in space under the action of the two given centres of force. The general path under all circumstances of projection has not been found. *If the particle is projected along the tangent to any ellipse having S, H for foci with a velocity whose component in the plane of the ellipse is v_1, and whose component perpendicular to the plane is $v'=\omega y=B/y$, it will continue to describe the ellipse freely, while the ellipse itself moves round the straight line SH with a variable angular velocity $\omega=B/y^2$.*

Ex. 2. If the particle is also acted on by a third centre of force situated at the centre and attracting according to the direct distance, prove that the pressure on the revolving wire is zero in certain cases.

530. *Ex. A particle P of unit mass moves on a smooth curve which is constrained to turn about a fixed axis with an angular velocity ω. It is required to find the relative motion.*

Let the axis of rotation be the axis of z and let the axes of x, y be fixed to the curve and rotate round Oz with the angular velocity ω. Let us refer the motion to these moving axes. Since $\theta_1=0$, $\theta_2=0$, $\theta_3=\omega$, the equations of Art. 499 become

$$\left.\begin{aligned} X+R_1&=\frac{du}{dt}-v\omega, & u&=\frac{dx}{dt}-y\omega\\ Y+R_2&=\frac{dv}{dt}+u\omega, & v&=\frac{dy}{dt}+x\omega\\ Z+R_3&=\frac{dw}{dt}, & w&=\frac{dz}{dt} \end{aligned}\right\}\ \ldots\ldots\ldots\ldots\ldots\ldots\ldots (1),$$

where R_1, R_2, R_3 are the components of the pressure on the particle. Eliminating u, v, w,

$$\left.\begin{aligned} \frac{d^2x}{dt^2}&=X+R_1+\omega^2x+\frac{d\omega}{dt}y+2\omega\frac{dy}{dt}\\ \frac{d^2y}{dt^2}&=Y+R_2+\omega^2y-\frac{d\omega}{dt}x-2\omega\frac{dx}{dt}\\ \frac{d^2z}{dt^2}&=Z+R_3 \end{aligned}\right\}\ \ldots\ldots\ldots\ldots\ldots\ldots\ldots (2).$$

The resultant of the two accelerating forces $X_1=\omega^2x$, $Y_1=\omega^2y$ is a force tending directly from the axis of rotation and whose magnitude is $F_1=\omega^2r$, where r is the distance of the particle P from the axis.

The resultant F_2 of the two forces $X_2=y\,d\omega/dt$, $Y_2=-x\,d\omega/dt$ is $F_2=-r\,d\omega/dt$, and it acts perpendicularly to the plane containing the axis of rotation and the particle in the direction in which the angular velocity ω is measured.

To find the resultant F_3 of the forces $X_3=2\omega dy/dt$, $Y_3=-2\omega dx/dt$, we notice that the component along the tangent to the curve, viz. $X_3dx/ds+Y_3dy/ds$, is zero. The resultant acts perpendicularly to the given curve, and may be compounded with and included in the reaction. When only the motion of the particle is required, the force F_3 may be omitted.

Reasoning as in Art. 197, we see that the equations of motion (2) become the same as if the particle were moving on a fixed curve, provided we impress on the particle (in addition to given forces X, Y, Z) *two* accelerating forces, viz. (1) a force $F_1=\omega^2r$ and (2) a force $F_2=-r\,d\omega/dt$.

The process of including the two forces F_1, F_2 among the impressed forces is sometimes called *reducing the curve to rest.*

The curve having been reduced to rest, the velocity of the particle relatively to the curve is found either by the equation of vis viva or by resolving along the tangent. We find

$$\tfrac{1}{2}v^2=C+U+\int\omega^2r\,dr-\int r\frac{d\omega}{dt}\,.\,rd\phi,$$

where U represents the work function. If the angular velocity is uniform, this reduces to

$$\tfrac{1}{2}v^2=C+U+\tfrac{1}{2}\omega^2r^2.$$

The velocity thus found is the velocity relative to the curve. The actual velocity in space is the resultant of velocity v and the velocity ωr of the point of the curve instantaneously occupied by the particle.

531. The pressure of the fixed curve on the particle is not the same as the actual pressure of the moving curve. Representing the first by R' and the second by R, we see that R' is the resultant of R and the two forces $X_3=2\omega dy/dt$, $Y_3=-2\omega dx/dt$. We may compound these two forces into a single force F_3. We project the moving curve on a plane perpendicular to the axis of rotation. If P' be the projection of P, dx/dt and dy/dt are the component velocities of P. The resultant is then evidently $F_3=2\omega v'$ where v' is the velocity of P' relatively both to the curve and its projection. The direction of F_3 is perpendicular not only to the given curve but also to its projection. The components along and perpendicular to the radius vector are $+2\omega r d\theta/dt$ and $-2\omega dr/dt$.

532. *Ex.* A small bead slides on a smooth circular ring of radius a which is made to revolve about a vertical axis passing through its centre with uniform angular velocity ω, the plane of the ring being inclined at a constant angle α to the horizontal plane. Show that the law of angular motion of the bead on the ring is the same as that of a bead on the ring of radius $a/\sin\alpha$ revolving round a vertical diameter with angular velocity $\omega\sin\alpha$. [Coll. Ex.]

533. A changing curve. *A bead of unit mass moves on a smooth curve whose form is changing in any given manner. It is required to find the motion.*

Let the equations of the curve be written in the form

$$x=f_1(\theta,\ t),\qquad y=f_2(\theta,\ t),\qquad z=f_3(\theta,\ t)\ \ldots\ldots\ldots\ldots\ (1),$$

where θ is an auxiliary variable. We may regard the position of the particle at any given time t as defined by some value of θ. Our object is to find θ in terms of the time.

Let us use Lagrange's equations. We have

$$T=\tfrac{1}{2}\Sigma(f_\theta\theta'+f_t)^2\ \ldots\ldots\ldots\ldots\ (2),$$

where Σ implies summation for all the coordinates, and partial differential coefficients are indicated by suffixes. The Lagrangian equation is

$$\frac{d}{dt}\frac{dT}{d\theta'}-\frac{dT}{d\theta}=\frac{dU}{d\theta}\ \ldots\ldots\ldots\ldots\ (3);$$

$$\therefore\ \Sigma\frac{d}{dt}(f_\theta\theta'+f_t)f_\theta-\Sigma(f_\theta\theta'+f_t)(f_{\theta\theta}\theta'+f_{\theta t})=\frac{dU}{d\theta}\ \ldots\ldots\ldots\ldots\ (4).$$

This is a differential equation of the second order from which θ may be found.

The three components of the pressure on the particle in the directions of the axes may be found by differentiating the equations (1). If X, Y, Z, be the components of the impressed forces; R_1, R_2, R_3 those of the pressure, the Cartesian equations of motion are

$$\frac{d^2x}{dt^2}=X+R_1,\qquad \frac{d^2y}{dt^2}=Y+R_2,\qquad \frac{d^2z}{dt^2}=Z+R_3.$$

Since the pressure must be perpendicular to the tangent to the instantaneous position of the curve, we do not necessarily require all these equations, though it may be convenient to use them.

534. *Ex.* A helix is constrained to turn about its axis Oz, which is vertical, with a uniform angular velocity ω. Find the motion of a particle of unit mass descending on it under the action of gravity.

Let the axes OA, OB move with the curve and let OA make an angle ωt with some axis of x fixed in space. Let the angle $AON=\theta$. See the figure of Art. 527.

The equations of the helix referred to axes fixed in space are

$$x=a\cos(\theta+\omega t),\qquad y=a\sin(\theta+\omega t),\qquad z=a\theta\tan\alpha;$$

$$\therefore\ 2T=x'^2+y'^2+z'^2=a^2\{(\theta'+\omega)^2+\tan^2\alpha\theta'^2\}.$$

Substituting in Lagrange's equation, we find after a little reduction

$$a\theta''=-g\sin\alpha\cos\alpha,$$

which admits of easy integration. It should be noticed that this result is independent of the angular velocity of the guiding curve, provided only it is constant. A similar result holds for any curve on a right circular cylinder turning uniformly about its axis.

To find the pressure of the helix on the particle we use cylindrical coordinates, Art. 491. Let P, Q, R be the components of the pressure, then since in the helix $\rho=a$, $\phi=\theta+\omega t$, we find by substitution

$$P=-a(\theta'+\omega)^2,\qquad Q=a\theta'',\qquad Z-g=a\tan\alpha\theta''.$$

These show that the pressure on the particle is equivalent to a sustaining force $g\cos\alpha$ acting perpendicularly to the osculating plane together with the radial pressure P.

Motion on a Surface.

535. Any Surface. *To find the motion of a particle on a fixed surface.*

Let P be the position of the particle at the time t, $P\eta$ a tangent to the path, $P\zeta$ a normal to the surface, and $P\xi$ that tangent to the surface which is perpendicular to the path. Let PC be the radius of curvature of the path, PA the binormal, then PA, PC lie in the plane $\xi\zeta$. Let χ be the angle $CP\zeta$.

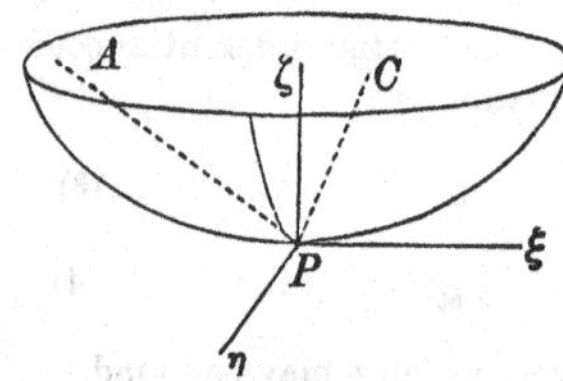

Let X, Y, Z be the resolved impressed forces parallel to any axes x, y, z fixed in space. Let the equation of the surface be $f(x, y, z)=0$.

The resolved accelerations of the particle in the directions PA, $P\eta$, PC are known to be zero, vdv/ds and v^2/ρ respectively, Art. 496. Hence resolving in the direction $P\eta$,

$$mv\frac{dv}{ds}=X\frac{dx}{ds}+Y\frac{dy}{ds}+Z\frac{dz}{ds},$$

which if U be the work function at once reduces to

$$\tfrac{1}{2}mv^2=U+C \quad \ldots\ldots\ldots\ldots\ldots\ldots\ldots\ldots(1).$$

This is the equation of vis viva.

Let R be the pressure of the constraining surface on the particle measured positively inwards. Then resolving along the normal,

$$\frac{mv^2}{\rho}\cos\chi = H + R,$$

where H is the component of the impressed forces. If ρ' be the radius of curvature of the normal section $\eta P\zeta$ of the surface made by a plane through the tangent to the path, it is proved in solid geometry that $\rho = \rho' \cos\chi$. We therefore have

$$\frac{mv^2}{\rho'} = H + R \qquad\qquad (2).$$

536. If a, b are the radii of curvature of the principal sections of the surface at P, ϕ the angle the tangent to the path makes with the section a, we have by Euler's theorem

$$\frac{1}{\rho'} = \frac{\cos^2\phi}{a} + \frac{\sin^2\phi}{b}.$$

Let v_1, v_2 be the resolved velocities of the particle along the tangents to the principal sections, then $v_1 = v\cos\phi$ and $v_2 = v\sin\phi$. The equation (2) then takes the form

$$m\left(\frac{v_1^2}{a} + \frac{v_2^2}{b}\right) = H + R.$$

537. If the forces are conservative, *the velocity of the particle is given by the equation* (1) *in terms of its coordinates at any instant and of the initial conditions.* To determine the velocity at any point we do not require to know the path by which the particle arrived at that point (Art. 181).

The pressure R is given by (2) in terms of the velocity at that point, the normal component of force and the radius of curvature of the normal section of the surface through the tangent. *The pressure is therefore also independent of the path. The whole energy C being given, the pressure depends on the position of the particle and the direction of motion.*

The equation (2) shows that the acceleration of the particle normal to the surface is v^2/ρ'. It is therefore independent of the position of the osculating plane but depends on the direction of motion.

538. To find the path of the particle we resolve in some third direction. Choosing the direction $P\xi$, we have

$$\frac{mv^2}{\rho}\sin\chi = F \qquad\qquad (3),$$

where F is the component of the impressed force along that tangent to the surface which is perpendicular to the path. This may also be written in the forms

$$\frac{mv^2}{\rho'}\tan\chi = F, \quad \frac{mv^2}{\rho''} = F,$$

where ρ'' is the radius of curvature of the projection of the path on the tangent plane. It is also called the geodesic radius of curvature.

539. **Geodesic path.** If the only impressed forces acting on the particle are normal to the surface, $F = 0$, and the third equation shows that either $\sin\chi = 0$ or that the path is a straight line. The path is therefore necessarily a geodesic line.

If the surface is rough, the friction acts opposite to the direction of motion, and F would still be zero. So also if the particle moves in a resisting medium the resistance is opposite to the direction of motion. Generally we conclude that *the path of a particle on a rough surface in a resisting medium when acted on by forces normal to the surface is a geodesic.*

Conversely, *if the path is a geodesic line* we must have $\sin\chi = 0$ and therefore $F = 0$. *The component of the impressed force tangential to the surface must then also be tangential to the path.*

540. *To find the radius of curvature of the path and the position of the osculating plane when the position and direction of motion of the particle are given.*

To effect this we use the two equations

$$\frac{mv^2}{\rho}\sin\chi = F, \quad \frac{1}{\rho'} = \frac{\cos^2\phi}{a} + \frac{\sin^2\phi}{b} = \frac{\cos\chi}{\rho}.$$

The particle being in a given position, v^2, a and b are known. Since ϕ is the angle the direction of motion makes with the principal section whose radius of curvature is a, we have

$$F = A\cos\phi + B\sin\phi,$$

where A and B are the given components of impressed force along the tangents to the principal sections. Thus the values of both $\sin\chi/\rho$ and $\cos\chi/\rho$ follow at once.

541. Motion on a surface of revolution. *When the surface on which the particle moves is one of revolution, it is generally more convenient to use cylindrical coordinates.*

Let the axis of figure be the axis of z and let ξ be the distance of the particle P from that axis. Let the equation of the surface be $z=f(\xi)$. Let U be the work function, and let the mass be unity. The equation of motion obtained by resolving perpendicularly to the plane zOP is

$$\frac{1}{\xi}\frac{d}{dt}(\xi^2\phi')=\frac{dU}{\xi\, d\phi}\text{ ....(1).}$$

We have also the equation of vis viva

$$T=\tfrac{1}{2}\{\xi'^2+z'^2+\xi^2\phi'^2\}=U+C\text{ ....(2),}$$

which, by using the equation of the surface, may be written in the form

$$\tfrac{1}{2}\xi'^2\left\{1+\left(\frac{dz}{d\xi}\right)^2\right\}+\tfrac{1}{2}\xi^2\phi'^2=U+C\text{ ....(3).}$$

Here accents denote differentiations with regard to the time.

By solving (1) and (3) we determine the two coordinates ξ, ϕ in terms of the time.

In certain cases the solution can be effected. The equation (1) gives

$$\xi^2\phi'\frac{d}{dt}(\xi^2\phi')=\xi^2\frac{dU}{d\phi}\phi'.$$

Let the impressed forces be such that

$$\xi^2U=F_1(\phi)+F_2(\xi, z)\text{ ....(4),}$$

where F_1, F_2 are arbitrary functions. We then have

$$\xi^2\phi'\frac{d}{dt}(\xi^2\phi')=\frac{dF_1(\phi)}{d\phi}\phi'\,;\quad\therefore\ \tfrac{1}{2}\xi^4\phi'^2=F_1(\phi)+A\text{ ...(5).}$$

Substituting this value of ϕ' in (3) we find

$$\tfrac{1}{2}\xi^2\xi'^2\left\{1+\left(\frac{dz}{d\xi}\right)^2\right\}=F_2(\xi, z)+C\xi^2-A\text{ ....(6).}$$

Since z is a known function of ξ, the variables in this equation are separable. The determination of ξ as a function of t has therefore been reduced to integration. The differential equation of the path is found by dividing (5) by (6). It is evident that here also the variables are separable.

Since the expression for the vis viva, given in (3), can be written in the form

$$T = \tfrac{1}{2}\xi^2\{P\xi'^2 + \phi'^2\},$$

where P is a function of ξ only, this solution is an example of Liouville's method of solving Lagrange's equations; Art. 522.

542. Motion on a sphere. When the surface on which the particle moves is a sphere, we may use polar coordinates, the centre being the origin. The equations corresponding to (1) and (3) of Art. 541 are found by putting $\xi = l\sin\theta$, where l is the radius; we then have

$$l^2\frac{d}{dt}(\sin^2\theta\phi') = \frac{dU}{d\phi}, \quad \tfrac{1}{2}l^2\{\theta'^2 + \sin^2\theta\phi'^2\} = U + C.$$

These admit of integration when U, expressed in polar coordinates, has the form

$$\sin^2\theta\, U = F_1(r, \phi) + F_2(r, \theta).$$

The resulting integrals are

$$\left.\begin{aligned} \tfrac{1}{2}l^2\sin^4\theta\phi'^2 &= F_1(l, \phi) + A \\ \tfrac{1}{2}l^2\sin^2\theta\theta'^2 &= F_2(l, \theta) + C\sin^2\theta - A \end{aligned}\right\}.$$

543. Examples. *Ex.* 1. A particle of mass m moves on the inner surface of a cone of revolution, whose semi-vertical angle is α, under the action of a repulsive force $m\mu/r^3$ from the axis; the angular momentum of the particle about the axis being $m\sqrt{\mu}\tan\alpha$; prove that its path is an arc of a hyperbola whose eccentricity is $\sec\alpha$. [Math. Tripos, 1897.]

Resolve along the generator and take moments about the axis, thus avoiding the reaction, Art. 541. These prove by integration that the path lies on a plane parallel to the axis. The angle between the asymptotes is therefore equal to the angle of the cone.

Ex. 2. A particle P moves on a sphere of radius l under the action both of gravity and a force $X = \mu/x^3$ tending directly from a vertical diametral plane taken as the plane of yz. Show that the determination of the motion can be reduced to integration. If the particle is projected horizontally from the extremity of the axis of x, show that when next moving horizontally, it is in a lower position.

Ex. 3. A particle is acted on by a force the direction of which meets an infinite straight line AB at right angles and the intensity of which is inversely proportional to the cube of the distance from AB. The particle is projected with the velocity from infinity from a point P at a distance a from the nearest point O of the line in a direction perpendicular to OP and inclined at an angle α to the plane AOP. Prove that the particle is always on the sphere the centre of which is O, that it meets every meridian line through AB at the angle α, and that it reaches the line AB in the time $a^2\sec\alpha/\sqrt{\mu}$, where μ is the absolute force. [Math. Tripos, 1860.]

Ex. 4. A particle moves on a spherical surface of unit radius, its position being determined by its polar distance θ and its longitude ϕ. If the tangential acceleration is always in the meridian, and $\sin^2\theta\, d\phi/dt = h$, $\cot\theta = u$, prove that its value is $h^2(1+u^2)\left(u+\dfrac{d^2u}{d\phi^2}\right)$.

Prove also that the law of force perpendicular to the equatorial plane under which the sphero-conic $\dfrac{1}{\sin^2\theta} = \dfrac{\cos^2\phi}{\sin^2 a} + \dfrac{\sin^2\phi}{\sin^2 b}$ can be described is that of the inverse cube of the distance. [Math. Tripos, 1893.]

Ex. 5. A particle moves on a smooth helicoid, $z = a\phi$, under the action of a force μr per unit mass directed at each point along the generator inwards, r being the distance from the axis of z. The particle is projected along the surface perpendicularly to the generator at a point where the tangent plane makes an angle α with the plane of xy, its velocity of projection being $a\sqrt{\mu}$. Prove that the equation of the projection of its path on the plane of xy is

$$1 + a^2/r^2 = \sec^2\alpha\,\{\cosh(\phi/\cos\alpha)\}^2. \qquad \text{[Math. Tripos, 1896.]}$$

544. Cylinders. *Ex.* 1. A particle moves on a rough circular cylinder under the action of no external forces. Prove that the space described in time t is $\dfrac{a\sec^2\alpha}{\mu}\log\left(1+\dfrac{\mu V\cos^2\alpha t}{a}\right)$ where the particle has initially a velocity V in a direction making an angle α with the transverse plane of the cylinder.

[Math. Tripos, 1888.]

Ex. 2. A heavy particle moves on a rough vertical circular cylinder and is projected horizontally with a velocity V. Prove that at the point where the path cuts the generator at an angle ϕ, the velocity v is given by

$$ag/v^2 \sin^2\phi = ag/V^2 + 2\mu\log(\cot\phi + \operatorname{cosec}\phi),$$

and that the azimuthal angle θ and vertical descent z are $ag\theta = \int v^2 d\phi$ and $gz = \int v^2 \cot\phi\, d\phi$, the limits being $\phi = \frac{1}{2}\pi$ to ϕ. [Math. Tripos, 1888.]

The cylindrical equations of motion give

$$\frac{d}{dt}(v\sin\phi) = -\frac{\mu}{a}v^2\sin^3\phi, \qquad \frac{d}{dt}(v\cos\phi) = g - \frac{\mu}{a}v^2\sin^2\phi\cos\phi.$$

First eliminating dt and putting $v = 1/w$ we obtain the first result. Secondly eliminating μ we obtain the others.

Ex. 3. A smooth cylinder whose cross section is a cardioid is placed with its generators inclined at an angle α to the vertical and having the generator through the cusp in its highest position, and a particle is projected from the cusp line with velocity V along the inner surface of the cylinder inclined at an angle β to the generator; show that it will leave the surface if $V^2 < \dfrac{6}{5}\dfrac{ag\sin\alpha}{\sin^2\beta}$, where $2a$ is the breadth of any section through the cusp. [Math. Tripos, 1887.]

545. String on a surface. *Ex.* 1. A string, one end of which is fastened at a point of the surface of a smooth circular cylinder whose axis is vertical, winds round the cylinder for part of its length, and terminates in a straight portion of length c at the end of which a particle is tied. Show that when the particle is projected in the direction horizontal and perpendicular to the string it begins to rise or fall according as the velocity is greater or less than $\sin\alpha\,(gc\sec\alpha)^{\frac{1}{2}}$; α being the angle at which the string cuts the generators.

Prove also that during the ensuing motion $\frac{1}{r}\frac{d}{dt}(r^2\omega)+a\omega^2=0$; r being at any time the length of the projection of the straight portion of the string on a horizontal plane, ω the angular velocity of the vertical plane drawn through the string and a the radius of the cylinder. [Coll. Ex. 1895.]

Ex. 2. A string is wound round a vertical cylinder of radius a in the form of a given helix, the inclination to the horizon being i. The upper end is attached to a fixed point on the cylinder, and the lower, a portion of the string of length $l \sec i$ having been unwound, has a material particle attached to it which is also in contact with a rough horizontal plane, the coefficient of friction being μ. Supposing a horizontal velocity V perpendicular to the free portion of the string to be applied to the particle so as to tend to wind the string on the cylinder, determine the motion and prove that the particle will leave the plane after the projection of the unwound portion of the string upon the plane has described an angle

$$\frac{1}{2\mu\tan i}\log\frac{ga}{2\mu V^2\tan^2 i-2\mu gl\tan i+ga}. \qquad \text{[Math. T. 1860.]}$$

Ex. 3. A fine string of length l is fastened to a point A of a smooth cylinder of radius a, and, being wound round the cylinder, has a particle of given mass attached to the free end. Show that, if the particle is projected in any direction, it will, so long as the string is tight and some portion of it remains wound on the cylinder, describe a geodesic line on the surface

$$x\cos\frac{1}{a}\left(\sqrt{l^2-z^2}-\sqrt{x^2+y^2-a^2}\right)+y\sin\frac{1}{a}\left(\sqrt{l^2-z^2}-\sqrt{x^2+y^2-a^2}\right)=a,$$

where the axis of the cylinder is the axis of z, and the axis of x is the radius through A.

Show also that the particle cannot be so projected that the string shall not slip on the cylinder, except when the path lies in the plane of the circular section of the cylinder drawn through A. [Math. Tripos, 1893.]

546. Gauss' coordinates. The motion of a particle on a surface may also be investigated by using the geodesic polar coordinates of Gauss. In this method every surface has a geometry of its own, in which all the lines under consideration are drawn on the surface. The geodesics on the surface correspond to straight lines on a plane, and the properties of the figures are discussed by reasoning analogous to that of two dimensions.

Let O be any origin, ρ the length of the geodesic drawn from O to any moving point P. Let ω be the angle OP makes with some fixed geodesic Ox. Let OP' be a neighbouring geodesic, PL the perpendicular to OP'. Then in the limit $LP'=d\rho$, $PL=Pd\omega$. The theorem that $OP=OL$ is proved in Salmon's *Solid Geometry*, Art. 394, edition of 1882. The quantity P is a function of ρ and ω, whose form depends on the particular surface under consideration. On a plane $P=\rho$, and on a sphere of radius a, $P=a\sin\rho/a$. On an ellipsoid when the origin O is at an umbilicus, $P=y\operatorname{cosec}\omega$, where ω is the angle the geodesic OP makes with the arc containing the four umbilici. The difficulty of finding the value of P for any surface prevents this method from coming into general use.

The vis viva $2T$ of a particle of unit mass is given by

$$T=\tfrac{1}{2}(\rho'^2+P^2\omega'^2),$$

where accents as usual denote differential coefficients with regard to the time.

Let U be the work function; F, G the accelerations at P along and perpendicular to the geodesic radius vector OP. We have by Lagrange's theorems,

$$\frac{d}{dt}\frac{dT}{d\rho'} - \frac{dT}{d\rho} = \frac{dU}{d\rho} = F; \qquad \therefore F = \rho'' - P\frac{dP}{d\rho}\omega'^2 \ldots\ldots\ldots\ldots\ldots(1).$$

$$\frac{d}{dt}\frac{dT}{d\omega'} - \frac{dT}{d\omega} = \frac{dU}{d\omega} = PG; \qquad \therefore G = \frac{1}{P}\frac{d}{dt}(P^2\omega') - \frac{dP}{d\omega}\omega'^2.$$

Since $\dfrac{dP}{dt} = \dfrac{dP}{d\rho}\rho' + \dfrac{dP}{d\omega}\omega'$, this reduces to

$$G = P\omega'' + \frac{dP}{d\omega}\omega'^2 + 2\frac{dP}{d\rho}\omega'\rho' \ldots\ldots\ldots\ldots\ldots\ldots(2).$$

547. *We may also arrive at these results without using Lagrange's equations.* Let u, v be the component velocities of P along and perpendicular to the tangent PT at P to the geodesic OP. Let $P'T'$ be the projection of the tangent to OP' on the tangent plane at P. Since the tangent planes at P, P' make an indefinitely small angle with each other the component velocities at P along and perpendicular to $P'T'$ are $u+du$ and $v+dv$. If $d\theta$ be the angle PT makes with $P'T'$, the accelerations along and perpendicular to PT are (as in Art. 225),

$$F = \frac{du}{dt} - v\frac{d\theta}{dt}, \qquad G = \frac{dv}{dt} + u\frac{d\theta}{dt}.$$

Now $u=\rho'$, $v=P\omega'$, and by a theorem proved in Salmon's *Solid Geometry*, Art. 392, $d\theta = \dfrac{dP}{d\rho}d\omega$. We therefore have

$$F = \rho'' - P\omega'^2\frac{dP}{d\rho}, \qquad G = \frac{d}{dt}(P\omega') + \rho'\omega'\frac{dP}{d\rho}.$$

These reduce to the same forms as before.

548. *Ex.* A particle P, constrained to move on an ellipsoid, is attached to an umbilicus by a string of given length, which also lies on the surface. Prove that the particle describes a geodesic circle with a uniform velocity V, and that the angular velocity of the string about the umbilicus is $V\sin\omega/y$. Prove also that the accelerating tension is $V^2\cos\beta/y$, where β is the angle the tangent at P to the string makes with the axis of y.

549. Developable surfaces. When the surface on which the particle moves is developable, we may sometimes fix the position of the particle by using the edge as a curve of reference. Let s be the arc of the edge measured from some fixed point A to a point Q such that the tangent at Q passes through P. Let $QP=u$ measured positively in the same direction as s. We then have

$$v^2 = \frac{u^2}{\rho^2}s'^2 + (u'+s')^2.$$

The form of the surface being given, the radius of curvature ρ of the edge at Q is known as a function of s. When U is given as a function of u and s the Lagrangian method supplies two equations to find the coordinates u and s.

Ex. A heavy particle moves on a developable surface whose edge is a helix with its axis vertical. Obtain two integrals by which s' and u' may always be found in terms of u and s. Show also that if the particle is projected along a tangent to the helix, it will continue to describe that tangent.

Motion of a heavy particle on a surface of revolution.

550. *To find the motion of a heavy particle on a surface of revolution the axis of which is vertical.*

Let the axis of z be the axis of the surface and let z be measured upwards. The velocity v is then given by

$$v^2 = 2g(h - z)\text{.........................}(1),$$

where h is a constant depending on the initial conditions. Let the plane $z = h$ be called *the level of no velocity.*

Let ξ be the distance of the particle P from the axis of figure, and ϕ the angle the plane zOP makes with the plane zOx. Then

$$\xi^2 \frac{d\phi}{dt} = A\text{.............................}(2),$$

where mA is the constant angular momentum and its value is known when the initial values of ξ and $d\phi/dt$ are given; Art. 492. The velocity v at any point being given by (1), the angular momentum A must lie between zero and $v\xi$. It is the former when the particle is moving in the plane zOP and the latter when moving horizontally. The particle therefore can occupy only those points of the surface at which $v\xi > A$, i.e. those points at which $2g(h - z)\xi^2 > A^2$. If then we describe the cubic surface

$$(h - z)\xi^2 = A^2/2g\text{.......................}(3),$$

the ξ of the particle for any value of z must be greater than the corresponding ξ of the cubic surface.

This cubic divides the given surface of revolution into zones, separated by horizontal circles, and the particle can move only in those zones which are more remote from the axis of figure than the corresponding portions of the cubic. The zone actually moved in is determined by the point of projection. The particle moves round the axis of figure and must continue to ascend or to descend until it arrives at a point at which the vertical velocity can be zero, that is, until it reaches one of the boundaries of the zone.

If the particle is projected horizontally it is on the boundary of two zones. It will move on that neighbouring zone which is the more remote from the axis than the corresponding portion of the cubic. If the cubic touch the surface of revolution, the particle is situated on an evanescent zone and will then describe a horizontal circle. The path is stable or unstable according as the neighbouring zones are less or more remote from the axis of figure than the cubic surface.

551. *Ex. A particle is projected horizontally with a velocity V at a point whose coordinates are* ξ, z. *Will it rise or fall?*

If mR be the pressure on the particle, ψ the angle the radius of curvature makes with the vertical, we see by resolving vertically, that the particle if inside and $\psi < \frac{1}{2}\pi$ will rise or fall according as $R\cos\psi$ is greater or less than g.

To find R we resolve along the normal to the surface. Since the particle is moving along that principal section whose radius of curvature is the normal n, we have $V^2/n = R - g\cos\psi$, Art. 536. Since $n\sin\psi = \xi$, we see that *the particle will rise, fall, or describe a horizontal circle according as* V^2 *is greater, less, or equal to* $g\xi\tan\psi$. If $z = f(\xi)$ be the equation of the surface of revolution, $\tan\psi = dz/d\xi$.

To find the level to which the particle will rise or fall we use the cubic surface described in Art. 550, the constants A and h being known from the equations $V\xi = A$, $V^2 = 2g(h-z)$. The intermediate motion may be deduced from the equations (1), (2) of the same article.

552. *Ex. To find the pressure on the particle when in any position.*

We use the formula given in Art. 536. The principal radii of curvature of the surface are the radius of curvature ρ of the meridian and the normal n. The velocity perpendicular to the meridian being $v_2 = \xi d\phi/dt$, the velocity v_1 along the meridian is given by $v^2 = v_1{}^2 + v_2{}^2$. The formula

$$\frac{v_1{}^2}{\rho} + \frac{v_2{}^2}{n} = R - g\cos\psi,$$

shows that

$$R = g\cos\psi + \frac{2g(h-z)}{\rho} + \frac{A^2}{\xi^2}\left(\frac{1}{n} - \frac{1}{\rho}\right).$$

This problem has a special interest because we can use it to represent experimentally the path of a particle under the action of a centre of force. If Q be the projection of the particle on a horizontal plane, the motion of Q is the same as that of a particle moving under the action of a central force whose magnitude is $R\sin\psi$. If then a surface is so constructed that the generating curve satisfies the differential equation $R\sin\psi = \mu/\xi^2$, where R has the value given above, the path of Q should be a conic with a focus at the origin.

The experiment cannot be properly tried with a particle, for the surface must then be very smooth. It is better to replace the particle by a small sphere which is made to roll on a rough surface, but in that case, the theory must be modified to allow for the size of the particle. *Nature*, 1897.

553. Small oscillation. *Ex. A heavy particle P, describing a horizontal circle on a surface of revolution, is slightly disturbed. It is required to find the oscillations to a first approximation.*

The plane zOP may be reduced to rest if we apply to the particle a horizontal acceleration $\xi(d\phi/dt)^2$, Art. 495. Since $\xi^2 d\phi/dt = A$, this acceleration is equal to A^2/ξ^3. Resolving along the meridian, we have

$$\frac{d^2s}{dt^2} = \frac{A^2}{\xi^3}\cos\psi - g\sin\psi,$$

where ψ is the angle PGO which the normal to the surface makes with the axis.

Let the radius of the mean circle be $N_1P_1=c$ and let the normal to the surface at any point of its circumference make an angle $P_1G_1O=\gamma$ with the vertical.

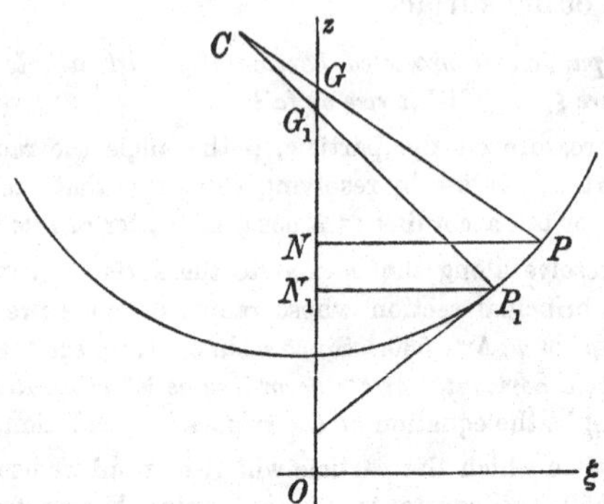

Since s may be taken to be the arc of the meridian between the particle and the mean circle, we have

$$\xi=c+s\cos\gamma, \qquad \psi=\gamma+s/\rho,$$

where ρ is the radius of curvature of the meridian at its intersection with the mean circle.

Substituting, we find by Taylor's theorem $\dfrac{d^2s}{dt^2}=F-p^2s$,

$$F=\frac{A^2}{c^3}\cos\gamma-g\sin\gamma, \qquad p^2=\frac{A^2\sin\gamma}{c^3\rho}+\frac{3A^2\cos^2\gamma}{c^4}+\frac{g\cos\gamma}{\rho}.$$

The position of the circle of reference is as yet arbitrary except that the deviation s must be small. Let it be so chosen that the mean value of s (taken for any long time) is zero; we then have $F=0$. The mean circle and the angular momentum mA are so related that $A^2=c^3g\tan\gamma$, while the oscillatory motion is given by $s=L\sin(pt+M)$ where L, M are the constants of integration.

To find the motion round the axis of figure we use the equation $\xi^2 d\phi/dt=A$;

$$\therefore \frac{d\phi}{dt}=\frac{A}{\xi^2}=\frac{A}{c^2}\left(1-\frac{2s}{c}\cos\gamma\right);$$

$$\therefore \phi=\frac{At}{c^2}+\frac{2A\cos\gamma}{c^3}\cdot\frac{L}{p}\cos(pt+M)+N,$$

where N is the constant of integration.

If we write ω for the mean value of $d\phi/dt$, we have $A=c^2\omega$. We then find

$$c\omega^2=g\tan\gamma, \qquad p^2=\omega^2\left(\frac{c\sin\gamma}{\rho}+3\cos^2\gamma\right)+\frac{g\cos\gamma}{\rho}.$$

The time the particle takes to travel from the highest position to the lowest or the reverse is π/p.

554. The Paraboloid. *Ex.* 1. A smooth paraboloid is placed with its axis vertical and vertex downwards, and its equation is $\xi^2=4az$. A heavy particle is projected horizontally with velocity V, the initial altitude being $z=b$, show that the particle is again moving horizontally at an altitude $z=V^2/2g$. Show also that the pressure on the surface at any point of the path is inversely proportional to the radius of curvature of the parabola.

To prove the first, we notice that the angular momentum $A=V\xi$ where $\xi^2=4ab$. The cubic $\xi^2(h-z)=A^2/2g$ becomes $z^2-hz+V^2b/2g=0$, one root of the quadratic being $z=b$, the other b' is given either by $b+b'=h$ or $b'=V^2/2g$. The second part follows from Art. 552.

If the time T of passing from one limit to the other be required, we first notice that

$$v^2=\left\{\left(\frac{d\xi}{dz}\right)^2+1\right\}\left(\frac{dz}{dt}\right)^2+\frac{A^2}{\xi^2}=2g(h-z);$$

$$\therefore \sqrt{2g}\,T=\int\frac{\sqrt{(a+z)}\,dz}{\sqrt{(z-b)(b'-z)}},$$

the limits being b and b'. This integral can be reduced to elliptic forms by putting

$$a+z=(b'+a)\cos^2\theta.$$

Ex. 2. A particle moves under the action of gravity on a smooth paraboloid whose axis is vertical, vertex downwards and latus rectum $4a$. If the particle be projected along the surface in the horizontal plane through the focus with a velocity $\sqrt{(2nag)}$, prove that the initial radius of curvature ρ of the path, and the angle θ which the radius of curvature makes with the axis, are given by

$$\sqrt{(n^2+1)}\,\rho=2na\sqrt{2},\qquad (1-n)\tan\theta=1+n.\qquad \text{[Math. T. 1871.]}$$

Ex. 3. A heavy particle moves on a paraboloid with its axis vertical, the equation of the surface being $x^2/a+y^2/\beta=4z$. Show that the particle when moving horizontally must lie on the quartic surface $\frac{a\beta}{4}\left(\frac{1}{p^2}-4\right)\left(\frac{B}{2g}-\frac{1}{p^2}\right)=\frac{z(h-z)}{p^2}$, where $\frac{1}{p^2}=\frac{x^2}{a^2}+\frac{y^2}{\beta^2}+4$, and B is the initial value of $\frac{1}{p^2}\left(\frac{x'^2}{a}+\frac{y'^2}{\beta}+2g\right)$. Show also that when the paraboloid is a surface of revolution, the intersection reduces to two horizontal planes and two coincident planes at the vertex.

555. The Conical Pendulum. *To find the motion of a heavy particle P on a smooth sphere*.*

It will be convenient in this problem to take the origin of coordinates at the centre O of the sphere and to measure Oz vertically *downwards.* Let l be the length of the string OP and θ the angle it makes with Oz. Let ϕ be the angle the vertical plane zOP makes with some fixed plane zOx. Let r be the

* The problem of the conical pendulum has been considered by Lagrange in the second volume of his *Mécanique Analytique.* He deduces equations equivalent to (1) and (3) of Art. 555 from his generalized equations, and notices that the cubic has three real roots. He reduces the determination of t and ϕ to integrals, and makes approximations when the bounding planes are close together. He refers also to a memoir of Clairaut in 1735. There is an elaborate memoir by Tissot in *Liouville's Journal*, vol. XVII. 1852. He expresses t, z, ϕ and the arc s in elliptic integrals in terms of u. A long communication by Chailan may be found in the *Bulletin de Soc. Math. de France*, 1889, vol. XVII. There is a brief discussion of this problem in Greenhill's *Applications of Elliptic Functions*, 1892, Art. 208.

distance of P from Oz. Let h be the altitude above O of the level of zero velocity. We now proceed as in Art. 550.

By the principles of angular momentum and vis viva,

$$r^2 \frac{d\phi}{dt} = A, \quad v^2 = l^2 \left(\frac{d\theta}{dt}\right)^2 + r^2 \left(\frac{d\phi}{dt}\right)^2 = 2g(h + l\cos\theta) \ldots(1).$$

Eliminating $d\phi/dt$ and writing $r = l\sin\theta$,

$$l^2 \sin^2\theta \left(\frac{d\theta}{dt}\right)^2 = 2g(h + l\cos\theta)\sin^2\theta - \frac{A^2}{l^2} \ldots\ldots\ldots(2).$$

Putting $z = l\cos\theta$, this may also be written in the form

$$l^2 \left(\frac{dz}{dt}\right)^2 = 2g(h+z)(l^2 - z^2) - A^2 \ldots\ldots\ldots\ldots\ldots(3).$$

To find the positions of the horizontal sections between which the particle oscillates (Art. 550), we put $dz/dt = 0$. We thus have the cubic

$$(h+z)(l^2 - z^2) - A^2/2g = 0 \ldots\ldots\ldots\ldots\ldots\ldots(4).$$

Since the initial value of z must make $(dz/dt)^2$ positive, the left-hand side of the cubic (4) is positive for some value of z lying between $z = \pm l$. When $z = \pm l$ the left-hand side is negative, hence the cubic has two real roots lying between $\pm l$ and separated by the initial value of z. Let these roots be $z = a$ and $z = b$. Lastly when z is very large and negative the left-hand side is positive, the third root of the cubic is therefore negative and numerically greater than l. Let this root be $z = -c$. *The particle oscillates between the two horizontal planes defined by $z = a$, $z = b$.*

Since the cubic can be written in the form

$$z^3 + hz^2 - l^2 z + (A^2/2g - l^2 h) = 0,$$

we have the obvious relations

$$a + b - c = -h, \quad (a+b)c - ab = l^2, \quad abc = A^2/2g - l^2 h.$$

Conversely, when the depths a and b of the two boundaries of the motion are given, the values of the other constants of the motion, viz. c, h, and A, follow at once. We have

$$c = \frac{l^2 + ab}{a+b}, \quad \frac{A^2}{2g} = \frac{(l^2 - a^2)(l^2 - b^2)}{a+b}, \quad h = c - a - b.$$

556. *Ex.* Prove (1) that one of the two horizontal planes bounding the motion lies below the centre; (2) that the plane equidistant from the two bounding planes also lies below the centre; (3) that both the bounding planes lie below the centre if $2ghl^2 < A^2$; (4) if a length $OC = c$ be measured upwards from the centre O, the point C is not only above the top of the sphere, but above the level of zero velocity.

To prove (1), we notice that if all the roots were negative, every coefficient of the cubic (4) would be positive, which is not the case. To prove (2); since both a and b are numerically less than l, it follows from the value of c that $a+b$ is positive. (3) The two roots a and b will have the same or different signs according as the left-hand side of the cubic when $z=0$ and $z=l$ has the same or different signs. The fourth result follows from the fact that $c-h$, i.e. $a+b$, is positive.

The first and third results follow also from Descartes' rule of signs; for since all the roots of the cubic are real, there are as many positive roots as changes of sign, and as many negative roots as continuations.

557. *Ex.* To find *the tension of the string* we produce the radius vector OP outwards to a point Q so that PQ is half the length of the string. Let z' be the depth of Q below the level of zero velocity. Prove that the tension mR is given by $lR=2gz'$. Thence show that *the string can become slack only when Q crosses the level of zero velocity.* It may be noticed that the tension or pressure on a sphere is independent of the angular momentum mA.

558. *Ex.* 1. *A particle P is projected horizontally with a velocity V. Determine whether it will rise or fall, and find the position of the other boundary to the motion.*

Let the initial radius OP make an angle α with the vertical. Resolving along the normal, we find that the initial tension mR is given by $R=g\cos\alpha+V^2/l$. The particle will rise or fall according as $R\cos\alpha$ is $>$ or $<g$, that is, according as $V^2\cos\alpha$ is $>$ or $<lg\sin^2\alpha$. If these are equal the particle describes a horizontal circle. See Art. 551.

To determine how far it will rise or fall, we notice that one root of the cubic in Art. 555 is known, viz. $z=l\cos\alpha$; the cubic may therefore be reduced to a quadratic. But *it is more easy to repeat the reasoning.* We have by the principles of angular momentum and vis viva

$$r^2\frac{d\phi}{dt}=Vl\sin\alpha,\qquad l^2\left(\frac{d\theta}{dt}\right)^2+r^2\left(\frac{d\phi}{dt}\right)^2=V^2+2gl(\cos\theta-\cos\alpha).$$

Eliminating $d\phi/dt$ and putting zero for $d\theta/dt$, the limiting values of θ are found from

$$V^2\frac{\sin^2\alpha}{\sin^2\theta}=V^2+2gl(\cos\theta-\cos\alpha);$$

$$\therefore\ V^2(\cos\theta+\cos\alpha)=2gl\sin^2\theta.$$

Putting $V^2/2gl=2n$ for brevity, we find

$$\cos\theta=-n+\sqrt{(1-2n\cos\alpha+n^2)},$$

where the positive sign is given to the radical because $\cos\theta$ must be less than unity. This value of $\cos\theta$ and $\cos\theta=\cos\alpha$ determine the positions of the bounding planes of the motion.

Ex. 2. A heavy particle, constrained to move on the surface of a smooth sphere of radius a, is projected horizontally with a velocity V from a point on the surface whose depth below the centre is x. Prove that, when next moving horizontally, the depth x' of the particle below the same point is given by

$$2g(x'^2-a^2)+V^2(x'+x)=0.$$

Ex. 3. In the centre of a hollow sphere resides a repulsive force. A heavy particle is projected horizontally along the surface of the sphere from a point distant 60° from the highest point with a velocity due to falling through the diameter by its weight only. Prove that it will be again moving horizontally at a point whose distance from the lowest point is $\tan^{-1}\sqrt[4]{\frac{3}{4}}$. [Coll. Ex.]

Ex. 4. A particle is attached by a string to the top of a hemispherical dome, and is projected horizontally along the interior surface, which is rough, with a velocity just sufficient to prevent it from at once leaving the surface. Find the velocity after describing a given arc, and show that it will always remain in contact with the surface. [Math. Tripos, 1853.]

559. *Ex.* 1. *Show that the radius of curvature of the path and the inclination χ of the osculating plane to the normal to the sphere are given by*

$$\frac{l^2}{\rho^2}=1+\frac{g^2A^2}{v^6}, \qquad \tan\chi=\frac{gA}{v^3},$$

where v is the velocity and mA the constant angular momentum.

We follow the method given in Art. 540. Let F be the component of acceleration along that tangent to the sphere which is perpendicular to the direction of motion. Then $\frac{\cos\chi}{\rho}=\frac{1}{l}$, $\frac{v^2}{\rho}\sin\chi=F$. To find F, we notice that the acceleration perpendicular to the meridian plane is zero, while that tangential is $g\sin\theta$. Hence if the direction of motion makes an angle ψ with the meridian,

$$F=g\sin\theta\sin\psi.$$

Since the components of velocity in and perpendicular to the meridian plane are $a\theta'$ and $a\sin\theta\phi'$, we have $v\cos\psi=a\theta'$, $v\sin\psi=l\sin\theta\phi'$. Choosing the latter component to find ψ and remembering that $l^2\sin^2\theta\phi'=A$, the values of $\cos\chi/\rho$ and $\sin\chi/\rho$ are evident.

Ex. 2. A particle is projected with velocity V horizontally from a point on the surface of a smooth sphere. Prove that the radius of curvature of its path is $\frac{lV^2}{\sqrt{(V^4+l^2g^2\sin^2\alpha)}}$ where l is the radius of the sphere and α the inclination to the vertical of the radius at the point. [Coll. Ex. 1881.]

Ex. 3. A particle is projected inside a smooth sphere of radius l with a velocity $\sqrt{2gl}$ along a tangent to the horizontal equator, prove that at first the radius of curvature is $2l/\sqrt{5}$. [Coll. Ex. 1897.]

560. *Ex.* Prove that *the projection of the path of the particle on a horizontal plane is a central orbit described under a force* $R\sin\theta=\frac{gr}{l^2}\{2h+3\sqrt{(l^2-r^2)}\}$, *where the radical changes sign when $r=l$.*

Show also that if the two roots a and b of the cubic in Art. 555 have the same signs, the central path is a spiral curve touching alternately two circles whose radii are $\sqrt{(l^2-a^2)}$ and $\sqrt{(l^2-b^2)}$, the curve being always concave to the centre of force. If a and b have opposite signs the central path after touching each bounding circle, touches the circle $r=l$ and then touches the other bounding circle. There will be a point of inflexion only if R vanishes and changes sign.

561. *Ex.* *If we write $\frac{1}{3}h+l\cos\theta=\kappa\cos\phi$, the general equation of motion of a conical pendulum may be reduced to the form*

$$-\sin^2\phi\left(\frac{d\phi}{dt}\right)^2=\frac{\kappa g}{2l^2}(\cos 3\phi-\cos 3\alpha),$$

by properly choosing the constants κ and α.

Show that these values are

$$\kappa=\frac{2}{3}\sqrt{(h^2+3l^2)},\qquad -\kappa^3\cos 3\alpha=\frac{2A^2}{g}+\frac{8}{27}h^3-\frac{8}{3}hl^2.$$

Find also the positions of the bounding planes when the constants κ and α of the motion are given.

562. Time of passage. *The motion of the particle as it travels from one boundary to the other may be found by an elliptic integral.*

We write the equation (3) of Art. 555 in the form

$$\frac{\sqrt{(2g)}}{l}t=\int\frac{dz}{\sqrt{(a-z)(z-b)(z+c)}},$$

where the limits are $z=a$ and $z=b$, and $a>b$. Putting $z=a-\xi^2$, the integral takes a standard form which is reduced to an elliptic integral by writing $\xi=\sin\psi\sqrt{(a-b)}$, i.e. we write

$$z=a\cos^2\psi+b\sin^2\psi;$$

$$\therefore\ \frac{\sqrt{(2g)}}{l}t=\frac{2}{\sqrt{(a+c)}}\int\frac{d\psi}{\sqrt{(1-\kappa^2\sin^2\psi)}},$$

where $$\kappa^2=\frac{a-b}{a+c},\qquad c=\frac{l^2+ab}{a+b}.$$

If the time of passage from one boundary to the other is required, the limits are 0 and $\frac{1}{2}\pi$.

If the two bounding planes are close together, κ is small. By expanding in powers of κ and effecting the integrations we find that the time from one boundary to the other is given by

$$\frac{\sqrt{(2g)}}{l}t=\frac{\pi}{\sqrt{(a+c)}}\left\{1+(\tfrac{1}{2})^2\kappa^2+\left(\frac{1\,.\,3}{2\,.\,4}\right)^2\kappa^4+\&c.\right\}.$$

If the two bounding planes are also close to the lowest point, we put

$$a=l\cos\alpha=l(1-\tfrac{1}{2}\alpha^2),\quad b=l\cos\beta=l(1-\tfrac{1}{2}\beta^2).$$

We then find that the time of passage from one boundary to the other is

$$t=\frac{\pi}{2}\sqrt{\frac{l}{g}}\left(1+\frac{\alpha^2+\beta^2}{16}\right),$$

the fourth powers of α and β being neglected. This result is given by Lagrange.

Let $u = \int_0^\psi \frac{d\psi}{\sqrt{(1-\kappa^2 \sin^2\psi)}}$ and K be the value of u when $\psi = \frac{1}{2}\pi$. Let t be the time of passage from the lower boundary to the depth z defined by any value of ψ, and T the time from one boundary to the other, then $t/T = u/K$.

563. *Ex.* 1. Prove that when half the time of passing from the lower to the upper boundary has elapsed, the particle is above the mean level between the two boundaries. Prove also that the depth of the particle is then $(\kappa' a + b)/(\kappa' + 1)$, where $\kappa'^2 = 1 - \kappa^2$. [Tissot.]

Ex. 2. Prove that when a quarter of the time has elapsed, the depth z of the particle is

$$z = \frac{a\sqrt{\kappa'}\,(\sqrt{(1+\kappa')}+1) + b\,(\sqrt{(1+\kappa')} - \sqrt{\kappa'})}{(1+\sqrt{\kappa'})\sqrt{(1+\kappa')}}.$$

564. The apsidal angle. *To find the change in the value of ϕ as the particle moves from one bounding plane to the other.*

Eliminating dt between (1) and (3) of Art. 555 we find

$$\frac{\sqrt{(2g)}}{Al}\,\phi = \int \frac{dz}{\sqrt{(a-z)}\sqrt{(z-b)}\sqrt{(z+c)}\,(l^2 - z^2)},$$

where the limits of integration are $z = b$ and $z = a$, and $a > b$. Putting $a = m + \mu$, $b = m - \mu$, $z = m + \xi$ so that m is the middle value of z and μ the extreme deviation on each side of the middle, we have

$$\frac{\sqrt{(2g)}}{Al}\,\phi = \int \frac{d\xi}{\sqrt{(\mu^2 - \xi^2)}\sqrt{(m + c + \xi)}\,\{l^2 - (m+\xi)^2\}},$$

where the limits are $\xi = -\mu$ and μ.

565. When the *bounding planes are close to each other*, the range μ of the values of ξ is small. If also the planes are not near the lowest point, the two last factors in the denominator are not small for any value of ξ. We may therefore expand these in powers of ξ and thus put the integral into the form

$$\phi = \int \frac{d\xi}{\sqrt{(\mu^2 - \xi^2)}}\,(P + Q\xi + R\xi^2) = \pi\,(P + \tfrac{1}{2}R\mu^2).$$

After calculating P and R, this gives

$$\phi = \frac{\pi l}{\sqrt{(l^2 + 3m^2)}}\left\{1 - \frac{3\,(3l^2 + 13m^2)\,m^2\mu^2}{4\,(l^2 - m^2)\,(l^2 + 3m^2)^2}\right\}.$$

566. *If both the bounding planes are near the lowest point of the sphere,* l and z are nearly equal, and the last factor in the denominator of ϕ (Art. 564), may be so small that its changes in value are considerable fractions of itself. We write the integral in the form

$$\frac{\sqrt{(2g)}}{Al}\phi=\int\frac{dz}{\sqrt{(a-z)}\sqrt{(z-b)(l-z)}}\cdot\frac{1}{\sqrt{(c+z)(l+z)}}.$$

The two factors in the denominator of the second fraction are not small and these may be expanded in powers of some small quantity properly chosen. We shall make the expansion in powers of $l-z=\eta$.

Remembering the values of A and c found in Art. 555, we have

$$\phi=\frac{1}{2}\sqrt{(l-a)}\sqrt{(l-b)}\int\frac{dz}{\sqrt{(a-z)}\sqrt{(z-b)(l-z)}}\left\{1+\frac{1}{2}\frac{\eta}{c+l}+\frac{\eta}{2l}\right\};$$

all these integrals are common forms. To find the first we put $l-z=1/u$. We have

$$\int\frac{dz}{\sqrt{(a-z)}\sqrt{(z-b)(l-z)}}=\frac{1}{\sqrt{(l-a)}\sqrt{(l-b)}}\int\frac{du}{\sqrt{(\alpha-u)}\sqrt{(u-\beta)}},$$

where α and β are two constants which we need not calculate. For since the limits of the first integral, viz. $z=a$, $z=b$, make the denominator vanish, the limits of the other must be $u=\alpha$, $u=\beta$. Putting $u=\frac{1}{2}(\alpha+\beta)+\xi$ we see at once that the value of that integral is π. Since $\eta=l-z$ the values of the remaining integrals have just been found. Hence

$$\phi=\frac{\pi}{2}\left\{1+\frac{1}{2}\sqrt{(l-a)}\sqrt{(l-b)}\left(\frac{a+b}{(l+a)(l+b)}+\frac{1}{l}\right)\right\},$$

where we have written for $c+l$ its value given in Art. 555.

If p, q be the radii of the circles which bound the oscillation, we have

$$l-a=\frac{p^2}{2l},\qquad l-b=\frac{q^2}{2l},$$

and in the small terms which contain the product pq as a factor, we can write $a=l$, $b=l$; hence (see Art. 562)

$$\phi=\frac{\pi}{2}\left\{1+\frac{3}{8}\frac{pq}{l^2}\right\},\qquad t=\frac{\pi}{2}\sqrt{\frac{l}{g}}\left\{1+\frac{1}{16}\frac{p^2+q^2}{l^2}\right\}.$$

The first of these results differs from that given by Lagrange. The correction was first made by M. Bravais in a note to the *Mécanique Analytique*.

567. *Ex.* A simple spherical pendulum of length l is drawn out to the horizontal position and is then projected horizontally with a velocity $2pl$. Show that, if θ is the angle that the string makes with the vertical, and ϕ the azimuthal angle of the vertical plane through the string, $\sin\theta\sin(\phi-pt)=\frac{p}{n}\sqrt{2\cos\theta}$, where n is equal to $\sqrt{g/l}$. [Math. Tripos, 1893.]

Motion on an Ellipsoid.

568. Cartesian coordinates. *To find the motion of a particle of unit mass on an ellipsoid*.*

Let X, Y, Z be the components of the impressed forces in the directions of the principal axes. Let R be the pressure on the particle measured positively inwards. Since the direction cosines of the normal are px/a^2, &c., the equations of motion are

$$x'' = X - Rp\frac{x}{a^2}, \quad y'' = Y - Rp\frac{y}{b^2}, \quad z'' = Z - Rp\frac{z}{c^2} \ldots(1),$$

where accents denote differential coefficients with regard to the time. We also have from the equation of the surface

$$\frac{x^2}{a^2} + \frac{y^2}{b^2} + \frac{z^2}{c^2} = 1, \quad \frac{xx'}{a^2} + \frac{yy'}{b^2} + \frac{zz'}{c^2} = 0 \ldots\ldots\ldots\ldots(2),$$

$$\frac{xx''}{a^2} + \frac{yy''}{b^2} + \frac{zz''}{c^2} + \frac{x'^2}{a^2} + \frac{y'^2}{b^2} + \frac{z'^2}{c^2} = 0 \ldots\ldots\ldots\ldots(3).$$

Multiplying the dynamical equations (1) by x', y', z', adding and integrating, we have

$$\tfrac{1}{2}(x'^2 + y'^2 + z'^2) = C + \int (Xdx + Ydy + Zdz) \ldots\ldots(4);$$

$$\therefore \tfrac{1}{2}v^2 = C + U,$$

where U is the work function and C is a constant. This is of course the equation of vis viva.

Substituting from (1) in (3), we find

$$Rp\left(\frac{x^2}{a^4} + \frac{y^2}{b^4} + \frac{z^2}{c^4}\right) = \left(\frac{x'^2}{a^2} + \frac{y'^2}{b^2} + \frac{z'^2}{c^2}\right) + \left(\frac{Xx}{a^2} + \frac{Yy}{b^2} + \frac{Zz}{c^2}\right) \ldots(5).$$

* The motion of a particle constrained to remain on an ellipsoid is discussed by Liouville in his *Journal*, vol. XI. 1846. He uses elliptic coordinates and shows that the variables can be separated when $U(\mu^2 - \nu^2) = F_1(\mu) - F_2(\nu)$. There is also a paper on the same subject by W. R. Westropp Roberts in the *Proceedings of the Mathematical Society*, 1883. He also uses elliptic coordinates and especially treats of the case in which the path is a line of curvature. The case in which the particle is attracted to the centre by a force proportional to the distance is solved in Cartesian coordinates by Painlevé, *Leçons sur l'integration des équations différentielles de la Mécanique*, 1895. He also treats separately the limiting case of a heavy particle moving on a paraboloid whose axis is vertical. There is a short paper by T. Craig in the *American Journal of Mathematics*, vol. I. 1878. He discusses the same problem as Painlevé, beginning with Cartesian coordinates, but passing quickly to Elliptic coordinates. He shows that the path is a geodesic when the central force is zero and the particle is acted on by what is equivalent to a force tangential to the path and varying as $f(t) + F(s)v$ where s is the arc described. This result follows also from Art. 539.

In an ellipsoid we have

$$\frac{1}{p^2}=\frac{x^2}{a^4}+\frac{y^2}{b^4}+\frac{z^2}{c^4},\qquad \frac{1}{D^2}=\frac{l^2}{a^2}+\frac{m^2}{b^2}+\frac{n^2}{c^2}\ \ldots\ldots\ldots(6),$$

where D is the semi-diameter of the ellipsoid whose direction cosines are (l, m, n). Also the radius of curvature of the normal section whose tangent is parallel to D is $\rho=D^2/p$. Taking D to be parallel to the tangent to the path $l=x'/v$, $m=y'/v$, $n=z'/v$. The equation (5) is therefore the Cartesian equivalent of

$$R=\frac{v^2}{\rho}-N\ \ldots\ldots\ldots\ldots\ldots\ldots\ldots\ldots(7),$$

where N is the *inward* normal component of the impressed force.

569. *In certain cases we may find another integral.* Differentiating (5) and remembering (6), we have

$$\frac{d}{dt}\left(\frac{R}{p}\right)=2\left(\frac{x'x''}{a^2}+\frac{y'y''}{b^2}+\frac{z'z''}{c^2}\right)+\frac{d}{dt}\left(\frac{Xx}{a^2}+\frac{Yy}{b^2}+\frac{Zz}{c^2}\right).$$

Substituting for x'', y'', z'' from (1) and using (6),

$$2Rp\left(-\frac{p'}{p^3}\right)+\frac{d}{dt}\left(\frac{R}{p}\right)=\frac{1}{a^2}\left(2x'X+\frac{dXx}{dt}\right)+\&c.,$$

$$\therefore\ p^2\frac{d}{dt}\left(\frac{R}{p^3}\right)=\frac{1}{a^2x^2}\frac{d}{dt}(Xx^3)+\&c.\ \ldots\ldots\ldots\ldots\ldots(8).$$

If then the forces acting on the particle are such that

$$\frac{1}{a^2x^2}\frac{d}{dt}(Xx^3)+\frac{1}{b^2y^2}\frac{d}{dt}(Yy^3)+\frac{1}{c^2z^2}\frac{d}{dt}(Zz^3)=0\ \ldots\ldots(9),$$

we have

$$R=Ap^3\ \ldots\ldots\ldots\ldots\ldots\ldots\ldots\ldots(10).$$

Substituting in (5) or (7), we have the third integral which may be written in either of the forms

$$\left.\begin{aligned}&\frac{x'^2}{a^2}+\frac{y'^2}{b^2}+\frac{z'^2}{c^2}+\frac{Xx}{a^2}+\frac{Yy}{b^2}+\frac{Zz}{c^2}=Ap^2\\&\frac{v^2}{\rho}-N=Ap^3\end{aligned}\right\}\ \ldots\ldots\ldots(11).$$

If only the direction of motion is required, we eliminate v between the equations (4) and (7). Remembering that $\rho=D^2/p$, we see that the direction of motion at any point of the path is parallel to that semi-diameter D whose length is given by

$$\frac{2(U+C)}{D^2}=Ap^2+\frac{N}{p}\ \ldots\ldots\ldots\ldots\ldots\ldots(12).$$

Supposing the condition (9) to be satisfied we notice that when the initial velocity and direction of motion are such that the equation (11) gives $A = 0$, it follows by (10) that *the pressure R is zero throughout the motion. The particle is therefore free and moves unconstrained by the ellipsoid.* Conversely, if the particle, when properly projected, can freely describe a curve on the ellipsoid, the condition (9) is satisfied. If it can describe the same curve when otherwise projected, the pressure varies as p^3.

If the components X, Y, Z do not satisfy the condition (9), we may sometimes make them do so by adding to them the components of an arbitrary normal force F and subtracting F from the reaction R. The condition (9) then becomes

$$\frac{1}{a^2x^2}\frac{d}{dt}(Xx^3)+\frac{1}{b^2y^2}\frac{d}{dt}(Yy^3)+\frac{1}{c^2z^2}\frac{d}{dt}(Zz^3)=p^2\frac{d}{dt}\left(\frac{F}{p^3}\right),$$

where F is an arbitrary function of x, y, z and p is a function of x, y, z given by (6). The equation (10) then becomes $R = F + Ap^3$.

It is only necessary that the condition (9) should hold for the path of the particle, but as this is generally unknown, the condition should be true for every arc on the ellipsoid.

570. *Ex.* A particle is acted on by a centre of attractive force situated at the centre of the ellipsoid, the force being κr. If D is the semi-diameter parallel to the tangent to the path, prove that

$$\frac{v}{D}=\sqrt{(\kappa+Ap^2)}, \qquad v^2=2C-\kappa r^2.$$

These reduce to the ordinary formulæ of central forces when $A=0$.

Since $X=-\kappa x$, &c. the condition (9) is satisfied. The first of the results to be proved then follows from (11), for $N=\kappa p$.

571. *Ex.* A particle P moves on the ellipsoid under the action of a force $Y=-\kappa/y^3$, whose direction is always parallel to the axis of y, and is projected from any point P with a velocity $v^2=\kappa/y^2$ in a direction perpendicular to the geodesic joining P to an umbilicus. Prove that the path is a geodesic circle having the umbilicus for centre, i.e. the geodesic distance of P from the umbilicus is constant*.

We see by substitution that the condition (9) is satisfied by this law of force. The path is therefore given by

$$\frac{v^2}{D^2}=Ap^2+\frac{N}{p}, \qquad v^2=\frac{\kappa}{y^2}+2C,$$

where, as before, D is the semi-diameter parallel to the tangent to the path. Since the cosine of the angle the normal makes with the axis of y is py/b^2, we have

* This result is due to W. R. W. Roberts, who gives a proof by elliptic coordinates in the *Proceedings of the London Math. Soc.* 1883.

$N=\kappa p/b^2y^2$. The conditions of projection show that $C=0$. Hence $\dfrac{1}{D^2}=\dfrac{A}{\kappa}p^2y^2+\dfrac{1}{b^2}$. If ρ, σ are the semi-axes of the diametral plane of P

$$\rho^2+\sigma^2=a^2+b^2+c^2-r^2, \qquad \rho\sigma=abc/p.$$

If also D, D' are two semi-diameters *at right angles* of the same plane

$$\frac{1}{D^2}+\frac{1}{D'^2}=\frac{1}{\rho^2}+\frac{1}{\sigma^2}=\frac{a^2+b^2+c^2-r^2}{a^2b^2c^2}p^2;$$

$$\therefore \frac{1}{p^2D'^2}=\frac{a^2+b^2+c^2-r^2}{a^2b^2c^2}-\frac{1}{b^2p^2}-\frac{Ay^2}{\kappa}.$$

Substituting for p and r their Cartesian values

$$\frac{1}{p^2D'^2}=\frac{1}{a^2c^2}+\frac{a^2+c^2}{a^2b^2c^2}\left(1-\frac{x^2}{a^2}-\frac{z^2}{c^2}\right)-\left(\frac{1}{a^2b^2c^2}+\frac{1}{b^6}+\frac{A}{\kappa}\right)y^2.$$

Using the equation to the surface, this becomes

$$\frac{1}{p^2D'^2}=\frac{1}{a^2c^2}+\left\{\frac{(a^2-b^2)(b^2-c^2)}{a^2b^2c^2}-\frac{Ab^4}{\kappa}\right\}\frac{y^2}{b^4}.$$

Since the particle is projected perpendicularly to the geodesic defined by $pD'=ac$, the coefficient of y^2 must be zero. It then follows that throughout the subsequent motion $pD'=ac$, and the path cuts all the geodesics from the umbilicus at right angles. These geodesics are therefore all of constant length.

Let ω be the angle which the geodesic joining the particle P to an umbilicus U makes with the arc joining the umbilici. If ds be an arc of the orthogonal trajectory of the geodesics, $ds=Pd\omega$, where $P=y/\sin\omega$ (Art. 546). Since $v^2=\kappa/y^2$, it follows that the angular velocity ω' of the geodesic radius vector is given by $\omega'=\dfrac{\sqrt{\kappa}}{y^2}\sin\omega$.

When the ellipsoid reduces to a disc lying in the plane xy, the geodesics become straight lines and the geodesic circle reduces to a Euclidian circle having its centre at H (Art. 576). The theorem is then identical with one given by Newton, viz. that a circle can be described under the action of a force $Y=-\kappa/y^3$.

The motion of a particle in a *geodesic circle under the action of a force*, or tension, *along the geodesic radius* is given in Art. 548, where the result is deduced from Gauss' coordinates.

572. *Ex.* 1. A particle, moving on the ellipsoid, is acted on by a centre of force situated at any given point E. If the force F is such that the condition (9) is satisfied, prove that $F=\mu r/P^3$, where r and P are the distances of the particle from E and from the polar plane of E respectively. Thence show that, if the initial conditions are such that the constant $A=0$, the path is a conic and the velocity at any point is given by $v^2=\rho N$.

To prove this we put $X=G(x-\alpha)$, $Y=G(y-\beta)$, $Z=G(z-\gamma)$, where $G=F/r$ and (α, β, γ) are the coordinates of E. Substituting in the equation (9) and remembering (2) Art. 568, we have an easy differential equation to find G. When $A=0$, the particle moves freely on the ellipsoid under the action of a central force. The path is a plane curve and is therefore a conic. The equation of vis viva fails to give the velocity, but this is determined by (11) Art. 569, when the direction of motion is known.

Ex. 2. A particle moving on a prolate spheroid is acted on by a central force tending to one focus and attracting according to the Newtonian law. Prove that the integrals of the equations of motion are

$$\left(\frac{dr}{dt}\right)^2 = \frac{2\mu}{r} - \frac{\mu b^2}{ar^2} + A\frac{b^2p^2}{a^2} + C, \qquad v^2 = \frac{2\mu}{r} + C,$$

where p is the perpendicular from the centre on the tangent plane, r the distance from the focus, and A, B the constants of integration.

573. *Ex.* 1. A particle under the action of no external forces is projected from an umbilicus of an ellipsoid, prove that the path is one of the geodesics defined by $pD = ac$.

Ex. 2. A particle is projected with a velocity v along the surface of an indefinitely thin ellipsoidal shell bounded by similar ellipsoids. Prove that when it leaves the ellipsoid the perpendicular p from the centre on the tangent plane is given by $MP^2R^2 = v^2p^2abc$, where R is the radius vector parallel to the initial direction of motion, P the perpendicular on the initial tangent plane, M the attracting mass and a, b, c the semi-axes of the ellipsoid. [Math. Trip. 1860.]

574. *Ex.* Let the forces be such that $\dfrac{1}{p^3}(Xd\lambda + Yd\mu + Zd\nu)$ is a perfect differential, say dS, for all displacements on the ellipsoid, where λ, μ, ν are the direction cosines of the normal, i.e. $\lambda = px/a^2$, &c. Prove that

$$R + N = 2p^3(S+B), \qquad \frac{v^2}{\rho} = p\left(\frac{x'^2}{a^2} + \frac{y'^2}{b^2} + \frac{z'^2}{c^2}\right) = 2p^3(S+B),$$

where B is the constant of integration.

Divide (8), Art. 569, by p^2 and integrate by parts. The integrals of the equations of motion are then obtained by using (6) and (7), remembering that $\rho = D^2/p$.

575. *In order to include in one form all the different cases of paraboloids, cones, and cylinders*, it may be useful to state the results when the quadric on which the particle moves is written in its most general form $\phi(x, y, z) = 0$.

Writing $\dfrac{4}{p^2} = \phi_x{}^2 + \phi_y{}^2 + \phi_z{}^2$, where suffixes denote partial differential coefficients, let the forces satisfy the condition

$$\frac{1}{\phi_x{}^2}\frac{d}{dt}(\phi_x{}^3X) + \frac{1}{\phi_y{}^2}\frac{d}{dt}(\phi_y{}^3Y) + \frac{1}{\phi_z{}^2}\frac{d}{dt}(\phi_z{}^3Z) = 0 \quad\ldots\ldots\ldots\ldots\ldots(9),$$

for all displacements on the quadric. We then find that the pressure $R = Ap^3$. The three components x', y', z' of the velocity may be deduced from the equations

$$\phi_x x' + \phi_y y' + \phi_z z' = 0 \ldots\ldots\ldots(2), \qquad \tfrac{1}{2}(x'^2 + y'^2 + z'^2) = U + C \ldots\ldots\ldots(4),$$

$$\phi_{xx}x'^2 + \&\text{c.} + 2\phi_{xy}x'y' + \&\text{c.} + \phi_x X + \phi_y Y + \phi_z Z = \frac{2R}{p} \ldots\ldots\ldots\ldots(5),$$

where the numbers appended to the equations correspond to those in Arts. 568, &c.

576. Elliptic coordinates. *Preliminary statement.* The position of the particle P in space is defined by the intersection of three quadrics confocal to a given quadric. In the figure ABC, $A'MM'$, $A''NN'$ are respectively the ellipsoid, hyperboloid of one sheet and that of two sheets; only that part of each being drawn which lies in the positive octant. Let their major axes $OA = \lambda$, $OA' = \mu$,

$OA''=\nu$. Let a, b, c be the three axes of any confocal. If $a^2-b^2=h^2$, $a^2-c^2=k^2$, then $OH=h$, $OK=k$ are the major axes of the focal conics.

The quantities λ, μ, ν are the elliptic coordinates of P; the first λ is always positive and greater than k; the second μ is less than k and greater than h; the

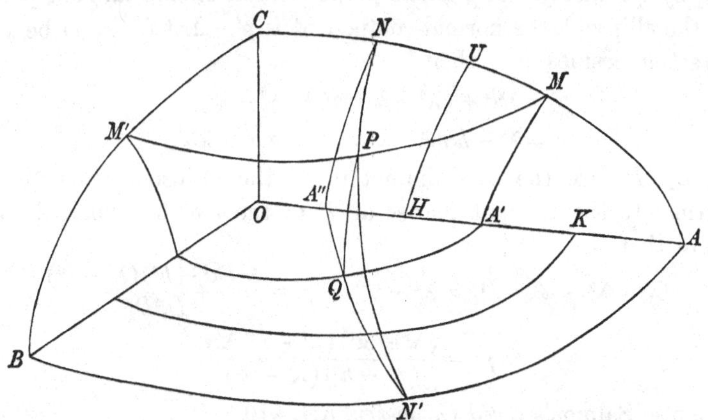

third ν is less than h, and changes sign when the particle crosses the plane of yz. The y axes of the quadrics are $\sqrt{(\lambda^2-h^2)}$, $\sqrt{(\mu^2-h^2)}$, $\sqrt{(\nu^2-h^2)}$; two of these are real and the third is imaginary. These radicals are positive when the particle lies in the positive octant, but the second or third vanishes and changes sign when the particle crosses the plane of xz, according as it travels along PN or PM. Similar remarks apply to the z axes.

The major axes of the three confocals which intersect in any point (x, y, z) are given by the cubic

$$\frac{x^2}{a^2}+\frac{y^2}{a^2-h^2}+\frac{z^2}{a^2-k^2}=1,$$

where h and k are the constants of the system. Clearing of fractions and arranging the cubic in descending powers of a^2, we see that the three roots λ^2, μ^2, ν^2 are such that

$$\left.\begin{aligned}\lambda^2+\mu^2+\nu^2&=x^2+y^2+z^2+h^2+k^2\\ \lambda^2\mu^2+\mu^2\nu^2+\nu^2\lambda^2&=h^2(x^2+z^2)+k^2(x^2+y^2)+h^2k^2\\ \lambda\mu\nu&=hkx\end{aligned}\right\}\ \ldots\ldots\ldots\ldots\ (1).$$

From the third equation we infer by symmetry

$$\left.\begin{aligned}\sqrt{(\lambda^2-h^2)}\sqrt{(\mu^2-h^2)}\sqrt{(h^2-\nu^2)}&=h\sqrt{(k^2-h^2)}\,y\\ \sqrt{(\lambda^2-k^2)}\sqrt{(k^2-\mu^2)}\sqrt{(k^2-\nu^2)}&=k\sqrt{(k^2-h^2)}\,z\end{aligned}\right\}\ \ldots\ldots\ldots\ldots\ldots\ (2).$$

577. *To prove that the velocity v of a particle in elliptic coordinates is given by*

$$v^2=\frac{(\lambda^2-\mu^2)(\lambda^2-\nu^2)\lambda'^2}{(\lambda^2-h^2)(\lambda^2-k^2)}+\frac{(\mu^2-\lambda^2)(\mu^2-\nu^2)\mu'^2}{(\mu^2-h^2)(\mu^2-k^2)}+\frac{(\nu^2-\lambda^2)(\nu^2-\mu^2)\nu'^2}{(\nu^2-h^2)(\nu^2-k^2)}\ \ldots\ (3).$$

We notice that the three quadrics confocal to a given quadric cut each other at right angles at P, so that the square of the velocity is the sum of the squares of the normal components of velocity. It is therefore sufficient to prove that the first term is the square of the component normal to the ellipsoid, the other terms following by symmetry. If p is the perpendicular on the tangent plane to the ellipsoid, the normal component is p'. Let (l, m, n) be the direction cosines of p, then

$$p^2 = \lambda^2 l^2 + (\lambda^2 - h^2) m^2 + (\lambda^2 - k^2) n^2$$
$$= \lambda^2 - h^2 m^2 - k^2 n^2; \quad \therefore\ pp' = \lambda\lambda'.$$

If D_1, D_2 are the semi-diameters of the ellipsoid respectively normal to the tangent planes at P to the two hyperboloids, we know that

$$D_1^2 = \lambda^2 - \mu^2, \quad D_2^2 = \lambda^2 - \nu^2, \qquad p^2 = \frac{\lambda^2(\lambda^2 - h^2)(\lambda^2 - k^2)}{D_1^2 D_2^2};$$
$$\therefore\ p'^2 = \frac{(\lambda^2 - \mu^2)(\lambda^2 - \nu^2)\lambda'^2}{(\lambda^2 - h^2)(\lambda^2 - k^2)}.$$

See also Salmon's *Solid Geometry*, Art. 410.

578. *To find the motion of a particle on an ellipsoid in elliptic coordinates.* Let the ellipsoid on which the particle moves be defined by a given value of λ. The mass being taken as unity the vis viva is determined by

$$2T = v^2 = (\mu^2 - \nu^2)\left\{\frac{(\mu^2 - \lambda^2)\mu'^2}{(\mu^2 - h^2)(\mu^2 - k^2)} - \frac{(\nu^2 - \lambda^2)\nu'^2}{(\nu^2 - h^2)(\nu^2 - k^2)}\right\} \ldots(4).$$

This we write for brevity in the form

$$2T = M\{P\mu'^2 + Q\nu'^2\}\ldots\ldots\ldots\ldots\ldots\ldots\ldots(5).$$

If we express the work function U in terms of (λ, μ, ν), we have (since λ is constant) the Lagrangian function $T + U$ expressed in terms of two independent coordinates μ, ν.

Comparing (5) with Liouville's form, Art. 522, we may obviously solve the Lagrangian equations by proceeding as in that article. The results are that when the forces are such that the work function takes the form

$$(\mu^2 - \nu^2)U = F_1(\mu) + F_2(\nu)\ldots\ldots\ldots\ldots\ldots\ldots(\text{A}),$$

the integrals are

$$\left.\begin{aligned} \tfrac{1}{2}(\mu^2 - \nu^2)^2 \frac{(\mu^2 - \lambda^2)\mu'^2}{(\mu^2 - h^2)(\mu^2 - k^2)} &= F_1(\mu) + C\mu^2 + A \\ -\tfrac{1}{2}(\mu^2 - \nu^2)^2 \frac{(\nu^2 - \lambda^2)\nu'^2}{(\nu^2 - h^2)(\nu^2 - k^2)} &= F_2(\nu) - C\nu^2 - A \end{aligned}\right\}\ldots(\text{B}).$$

There is also the equation of vis viva

$$\tfrac{1}{2}v^2 = U + C \text{..........................(C)}.$$

Dividing one of the equations (B) by the other, and remembering that λ is constant, the equation of the path takes the forms

$$\frac{(\mu^2-\lambda^2)(d\mu)^2}{(\mu^2-h^2)(\mu^2-k^2)\{F_1(\mu)+C\mu^2+A\}} = \frac{-(\nu^2-\lambda^2)(d\nu)^2}{(\nu^2-h^2)(\nu^2-k^2)\{F_2(\nu)-C\nu^2-A\}} \text{...(D)},$$

in which the variables are separated.

579. *Ex.* 1. Let v_1 and v_2 be the components of the velocity of the particle in the directions of the lines of curvature defined by μ=constant and ν=constant respectively. Prove that

$$\frac{1}{2}v_1^2 = \frac{F_2(\nu)-C\nu^2-A}{\mu^2-\nu^2}, \quad \frac{1}{2}v_2^2 = \frac{F_1(\mu)+C\mu^2+A}{\mu^2-\nu^2}.$$

Prove also that the pressure R on the particle is given by

$$R+N = \left\{\frac{F_1(\mu)+C\mu^2+A}{\lambda^2-\mu^2} + \frac{F_2(\nu)-C\nu^2-A}{\lambda^2-\nu^2}\right\}\frac{2p}{\mu^2-\nu^2},$$

where p is the perpendicular on the tangent plane and N the normal impressed force. The value of p in elliptic coordinates is given in Art. 577. See Art. 568.

Ex. 2. Supposing that the equation (D) of Art. 578 is written in the form $Pd\mu = Qd\nu$ in which the variables are separated, show that the time

$$t = \int P\mu^2 d\mu - \int Q\nu^2 d\nu. \qquad \text{[}\textit{Liouville}\text{, XI.]}$$

The equations (B) become

$$(\mu^2-\nu^2)Pd\mu = dt, \quad (\mu^2-\nu^2)Qd\nu = dt.$$

Multiplying these by μ^2, ν^2 respectively and subtracting we obtain the result.

580. *To translate the elliptic expressions into Cartesian geometry* we use the equations (1) and (2) of Art. 576. Let the normals at the four umbilici U_1, U_2, &c. intersect the major axis in the two points E_1, E_2, which of course are equally distant from the centre O. We easily find that

$$OE_1 = \frac{hk}{\lambda}, \quad E_1U_1 = \frac{bc}{a} = \frac{\sqrt{(\lambda^2-h^2)}\sqrt{(\lambda^2-k^2)}}{\lambda} \text{.................. (1)}.$$

The equations (1) Art. 576 give

$$(\mu \pm \nu)^2 = \left(x \pm \frac{hk}{\lambda}\right)^2 + y^2 + z^2 - \frac{(\lambda^2-h^2)(\lambda^2-k^2)}{\lambda^2}.$$

Let r_1, r_2 be the distances of the particle from the points E_1, E_2, and let m be the distance of E_1 from the umbilicus U_1; then

$$(\mu-\nu)^2 = r_1^2 - m^2, \quad (\mu+\nu)^2 = r_2^2 - m^2 \text{........................(2)}.$$

From these μ, ν may be found in terms of x, y, z and the constant λ.

581. *Ex.* Show that the equation $U(\mu^2-\nu^2) = F_1(\mu) + F_2(\nu)$ is equivalent to $\frac{d^2}{d\rho_1^2}(U\rho_1\rho_2) = \frac{d^2}{d\rho_2^2}(U\rho_1\rho_2)$, where $\rho_1 = \sqrt{(r_1^2-m^2)}$, $\rho_2 = \sqrt{(r_2^2-m^2)}$.

We have $\frac{d^2}{d\mu\, d\nu}U(\mu^2-\nu^2) = 0$, and by (2) Art. 580

$$\frac{d}{d\mu} = \frac{d}{d\rho_2} + \frac{d}{d\rho_1}, \quad \frac{d}{d\nu} = \frac{d}{d\rho_2} - \frac{d}{d\rho_1}.$$

The result follows at once.

582. *The condition* (A) *of Art.* 578, *viz.*

$$(\mu^2-\nu^2)\,U=F_1(\mu)+F_2(\nu)\ \ldots\ldots\ldots\ldots\ldots\ldots\ldots\ldots\ldots\ (A),$$

can be satisfied by several laws of force.

1. Let the force tend to the centre of the ellipsoid and vary as the distance. Representing the force by Hr, we have, by (1) Art. 576,

$$U=-\tfrac{1}{2}Hr^2=-\tfrac{1}{2}H\{\mu^2+\nu^2+(\lambda^2-h^2-k^2)\};$$

$$\therefore\ F_1(\mu)=-\tfrac{1}{2}H\{\mu^4+(\lambda^2-h^2-k^2)\mu^2\},\qquad F_2(\nu)=\tfrac{1}{2}H\{\nu^4+(\lambda^2-h^2-k^2)\nu^2\}.$$

Substituting these in the equations (B), the motion is known.

2. Let the direction of the force be parallel to the axis of x, and $X=-2H/x^3$. Then

$$U=\frac{H}{x^2}=\frac{Hh^2k^2}{\lambda^2\mu^2\nu^2},\qquad \therefore\ (\mu^2-\nu^2)\,U=\frac{Hh^2k^2}{\lambda^2}\left\{-\frac{1}{\mu^2}+\frac{1}{\nu^2}\right\}.$$

3. Let the work function $U=\dfrac{H}{\sqrt{(r_1{}^2-m^2)}}$, where r_1 is the distance of the particle from the point E_1, Art. 580. We then have

$$U=\frac{H}{\mu-\nu},\qquad \therefore\ (\mu^2-\nu^2)\,U=H(\mu+\nu).$$

To find the force we notice that since $dU/d\lambda=0$, the direction of the force is tangential to the ellipsoid. Also

$$(\mu-\nu)^2=x^2+y^2+z^2-2xhk/\lambda-\lambda^2+h^2+k^2;$$

$$\therefore\ X=\frac{dU}{dx}=-\frac{H}{(\mu-\nu)^3}\left\{x-\frac{hk}{\lambda}+\left(\frac{hkx}{\lambda^2}-\lambda\right)\frac{d\lambda}{dx}\right\},$$

with similar expressions for Y and Z. Now the equation to the ellipsoid being $\lambda=$ constant, the last term of each of the three expressions represents the component of a normal force. This normal force has no effect on the motion. *Taking only the remaining terms we see that X, Y, Z are the components of a central force tending to the point E whose magnitude is* $\dfrac{Hr_1}{(r_1{}^2-m^2)^{\frac{3}{2}}}$. When the ellipsoid is reduced to a disc, $\lambda=k$ (Art. 576), and $m=0$ (Art. 580). The point E_1 becomes a focus and the law of force is the inverse square.

583. *Ex.* 1. Show that a particle can describe the line of curvature defined by $\mu=\mu_0$ under the action of the central force $\dfrac{Hr_1}{(r_1{}^2-m^2)^{\frac{3}{2}}}$ tending to the point E_1. Show also that the velocity at any point is then given by $v^2=H\left\{\dfrac{2}{(r_1{}^2-m^2)^{\frac{1}{2}}}-\dfrac{1}{\mu_0}\right\}$.

We notice that when the ellipsoid reduces to a plane, $m=0$, and this becomes the common expression for the velocity under the action of a central force varying as the inverse square.

Referring to the general expressions marked (A) and (B) in Art. 578, we see that the particle will describe the line of curvature if both $\mu'=0$ and $\mu''=0$ when $\mu=\mu_0$. This will be the case if we choose the constants C and A so that

$$F_1(\mu)+C\mu^2+A=(\mu-\mu_0)^2\,\phi(\mu),$$

where $\phi(\mu)$ is some function of μ. Supposing this done, we have, when $\mu=\mu_0$,

(Art. 579) $$\tfrac{1}{2}v^2=U+C=\frac{F_2(\nu)-C\nu^2-A}{\mu_0{}^2-\nu^2}.$$

In the special case proposed $U=H/(\mu-\nu)$. We have therefore to make $C\mu^2+H\mu+A=(\mu-\mu_0)^2 C$. This gives $-2C\mu_0=H$, $A=C\mu_0^2$. Also $F_2(\nu)=H\nu$.

$$\therefore v^2=H\left\{\frac{2}{\mu_0-\nu}-\frac{1}{\mu_0}\right\}=H\left\{\frac{2}{(r_1^2-m^2)^{\frac{1}{2}}}-\frac{1}{\mu_0}\right\}.$$

Ex. 2. A particle is constrained to move on the surface $y=x\tan nz$. By putting $x=\mu\cos nz$, $y=\mu\sin nz$, we have

$$v^2=(\mu^2n^2+1)\left\{\frac{\mu'^2}{\mu^2n^2+1}+z'^2\right\}=M(\xi'^2+z'^2).$$

Hence show that when the forces are such that

$$(\mu^2n^2+1)\,U=F_1(\mu)+F_2(z),$$

the Lagrangian equations can be integrated. The path is given by

$$\frac{\mu'^2}{(\mu^2n^2+1)\{F_1(\mu)+C(\mu^2n^2+1)+A\}}=\frac{z'^2}{F_2(z)-A}.$$ [Liouville, 1846.]

If the particle is acted on by a force tending directly from the axis of z and varying as the distance from that axis, find the components of velocity along the lines of curvature.

584. Spheroids. When the ellipsoid on which the particle moves becomes a spheroid either prolate or oblate, the formulæ (A) and (B) of Art. 578 require some slight modifications.

Let (λ, b, c), (μ, b', c'), (ν, b'', c'') be the semi-axes of the three quadrics which intersect in P; then also $a=\lambda$, $a'=\mu$, $a''=\nu$.

In a prolate spheroid $b=c$, $h=k$, and the focal conics become coincident with

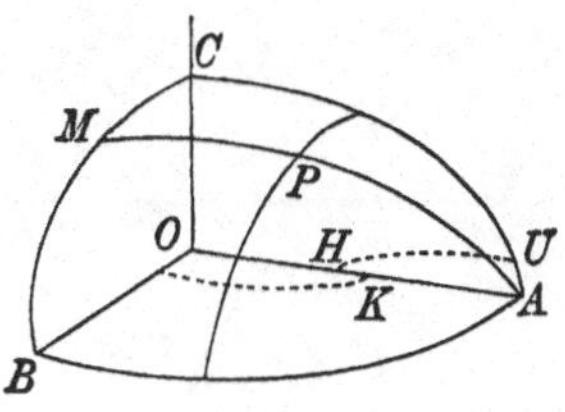

Prolate.

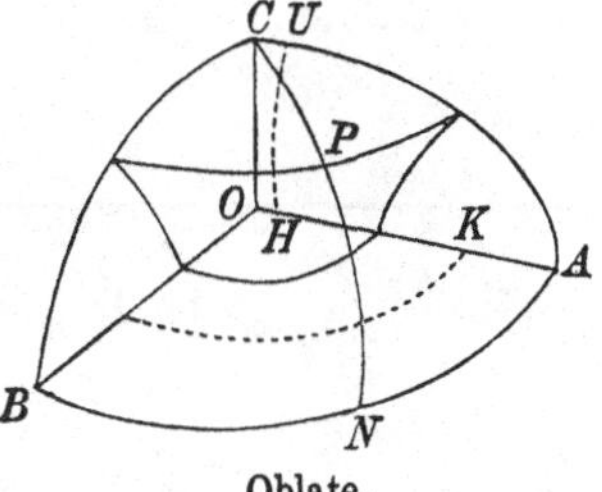

Oblate.

OH and HA. The axes of the hyperboloid of one sheet are $\mu=h$, $b'=0$, $c'=0$; it therefore reduces to the two planes $y^2/b'^2+z^2/c'^2=0$, the ratio b'/c' being indeterminate. Art. 576.

In an oblate spheroid $\lambda=b$, $h=0$; one focal conic becomes coincident with OC, while the other is a circle of radius k. The axes of the hyperboloid of two sheets are $\nu=0$, $b''=0$, $c''^2=-k^2$; it therefore reduces to the two planes $x^2/\nu^2+y^2/b''^2=0$, the limiting ratio ν/b'' being indeterminate.

In the figure the positions of the focal conics *just before* they assume their limiting positions are represented by the dotted lines, while PM or PN represents one of the planes assumed by the hyperboloid.

Before taking the limits of the equations (A) and (B) we shall make a change of variables. In the prolate spheroid we replace μ by a new variable ϕ, such that

$$\tan^2\phi=-\frac{\mu^2-k^2}{\mu^2-h^2},\qquad \therefore\ \sin^2\phi=-\frac{\mu^2-k^2}{k^2-h^2},\qquad \therefore\ -\phi'^2=\frac{\mu^2\mu'^2}{(\mu^2-h^2)(\mu^2-k^2}$$

Thus $\tan\phi$ varies between the limits 0 and ∞ as μ varies between k and h. Since $b'^2=\mu^2-h^2$, $c'^2=\mu^2-k^2$, and $y^2/b'^2+z^2/c'^2=0$, it is clear that ϕ is ultimately the angle the plane PM makes with the plane AB. Putting $\mu=h$, the formulæ (A) and (B) become

$$(h^2-\nu^2)\,U=f_1(\phi)+F_2(\nu),$$

$$-\tfrac{1}{2}(h^2-\nu^2)^2\frac{(h^2-\lambda^2)\,\phi'^2}{h^2}=f_1(\phi)+Ch^2+A,$$

$$-\tfrac{1}{2}(\nu^2-\lambda^2)\,\nu'^2=F_2(\nu)-C\nu^2-A.$$

In the oblate spheroid, we replace ν by the variable ϕ where

$$\tan^2\phi=-\frac{\nu^2-h^2}{\nu^2}, \qquad \therefore\ \nu=h\cos\phi, \qquad \therefore\ -\phi'^2=\frac{\nu'^2}{\nu^2-h^2},$$

thus $\tan\phi$ varies between 0 and ∞ as ν varies between h and 0. Also since $x^2/\nu^2+y^2/b''^2=0$, ϕ is ultimately the angle the plane PM makes with the plane AC. Putting $\nu=0$, $h=0$, the limiting forms of the equations (A), (B) are

$$\mu^2U=F_1(\mu)+f_2(\phi),$$

$$\tfrac{1}{2}\mu^2\frac{(\mu^2-\lambda^2)\,\mu'^2}{\mu^2-k^2}=F_1(\mu)+C\mu^2+A,$$

$$\tfrac{1}{2}\mu^4\frac{\lambda^2\phi'^2}{k^2}=f_2(\phi)-A.$$

CHAPTER VIII.

SOME SPECIAL PROBLEMS.

Motion under two centres of force.

585. *To find the motion of a particle of unit mass in one plane under the action of two centres of force*.*

Let the position of a point P be defined as the intersection of two confocal conics, the foci being H_1, H_2, and let $OH_1 = h$. Let the semi-major axes be $OA = \mu$, $OA' = \nu$: the semi-minor axes are therefore $\sqrt{(\mu^2 - h^2)}$, $\sqrt{(\nu^2 - h^2)}$.

Since $\dfrac{x^2}{\mu^2} + \dfrac{y^2}{\mu^2 - h^2} = 1$, we have

$$\mu^4 - (x^2 + y^2 + h^2)\mu^2 + h^2x^2 = 0 \quad\text{..............(1).}$$

The relations between the elliptic coordinates μ, ν of any point P and the Cartesian coordinates x, y are therefore

$$x = \frac{\mu\nu}{h}, \quad y = \frac{(\mu^2 - h^2)^{\frac{1}{2}}(\nu^2 - h^2)^{\frac{1}{2}}}{h\sqrt{-1}}, \quad r^2 = \mu^2 + \nu^2 - h^2,$$

where r is the distance from the centre. We also have $r_2 = \mu + \nu$, $r_1 = \mu - \nu$, where r_1, r_2 are the distances of P from the foci.

* Euler was the first who attacked the problem of the motion of a particle in one plane about two fixed centres of force, *Mémoires de l'Académie de Berlin*, 1760. Lagrange, in the *Mécanique Analytique*, page 93, begins by excusing himself for attempting a problem which has nothing corresponding to it in the system of the world, where all the centres of force are in motion. He supposes the motion to be in three dimensions and obtains a solution where the forces are $\alpha/r^2 + 2\gamma r$ and $\beta/r'^2 + 2\gamma r$. Legendre in his *Fonctions elliptiques* pointed out that the variables used by Euler were really elliptic coordinates, and Serret remarks that this is the first time these coordinates were used. Jacobi took this problem as an example of his principle of the least multiplier, *Crelle*, XXVII. and XXIX. Liouville in 1846 and 1847 gives two methods of solution, the first by Lagrange's equations and the second by the Hamiltonian equations. Serret extends Liouville's first method to three dimensions, *Liouville's Journal*, XIII. 1848, and gives a history of the problem. Liouville in the same volume gives a further communication on the subject.

Proceeding as in Art. 577, the velocity v of the particle expressed in elliptic coordinates is

$$2T = v^2 = (\mu^2 - \nu^2)\left\{\frac{\mu'^2}{\mu^2 - h^2} - \frac{\nu'^2}{\nu^2 - h^2}\right\} \text{............(2)},$$

where the accent represents d/dt. Comparing this with Liouville's form

$$2T = M(P\mu'^2 + Q\nu'^2)$$

in Art. 522, we may obviously solve the Lagrangian equations by proceeding as in that Article. The results are that when the work function has the form

$$(\mu^2 - \nu^2)U = F_1(\mu) + F_2(\nu) \text{..................(3)},$$

we have the two integrals

$$\left.\begin{aligned} \tfrac{1}{2}(\mu^2-\nu^2)^2\frac{\mu'^2}{\mu^2-h^2} &= F_1(\mu) + C\mu^2 + A \\ -\tfrac{1}{2}(\mu^2-\nu^2)^2\frac{\nu'^2}{\nu^2-h^2} &= F_2(\nu) - C\nu^2 - A \end{aligned}\right\} \text{.........(4)}.$$

There is also the equation of vis viva which may be deduced from these by simple addition, viz.

$$\tfrac{1}{2}v^2 = U + C \text{..........................(5)}.$$

586. Let the central forces tending to the foci be respectively H_1/r_1^2 and H_2/r_2^2. We then have

$$U = \frac{H_1}{r_1} + \frac{H_2}{r_2}; \quad \therefore (\mu^2 - \nu^2)U = (H_1 + H_2)\mu + (H_1 - H_2)\nu \text{...(6)}.$$

The integrals (4) then become

$$\left.\begin{aligned} \tfrac{1}{2}(\mu^2-\nu^2)^2\frac{\mu'^2}{\mu^2-h^2} &= K_1\mu + C\mu^2 + A \\ -\tfrac{1}{2}(\mu^2-\nu^2)^2\frac{\nu'^2}{\nu^2-h^2} &= K_2\nu - C\nu^2 - A \end{aligned}\right\} \text{............(7)},$$

where $K_1 = H_1 + H_2$, $K_2 = H_1 - H_2$. To find the path we eliminate t,

$$\frac{(d\mu)^2}{(\mu^2-h^2)(C\mu^2+K_1\mu+A)} = \frac{-(d\nu)^2}{(\nu^2-h^2)(-C\nu^2+K_2\nu-A)} = \frac{2(dt)^2}{(\mu^2-\nu^2)^2} \text{...... (8)}.$$

The initial values of μ, μ', ν, ν' being given, the equations (7) determine the constants A, C. Another constant is introduced by the integration of (8) which is also determined by the initial values of μ, ν. A fourth constant makes its appearance when the time is found in terms of either μ or ν.

587. *Ex.* 1. Show that the particle will describe the ellipse defined by $\mu=\mu_0$, if the particle is projected along the tangent at any point with a velocity v given by

$$v^2=H_1\left(\frac{2}{r_1}-\frac{1}{\mu_0}\right)+H_2\left(\frac{2}{r_2}-\frac{1}{\mu_0}\right).$$

To prove this we notice that if the particle describe the ellipse, μ is constant throughout the motion, and the values of μ', μ'' given by (7) must be zero. The right-hand side of that equation must take the form $C(\mu-\mu_0)^2$, and therefore $-2C\mu_0=K_1$. Substituting for C in the equation of vis viva (5) the result follows at once. See also Art. 274.

Ex. 2. A particle is projected so that both the constants A and C are zero. Show that the velocity is that due to an infinite distance and that the path is given by

$$\int\frac{d\phi}{\sqrt{(1-\frac{1}{2}\sin^2\phi)}}=\left(\frac{K_1}{K_2}\right)^{\frac{1}{2}}\int\frac{d\theta}{\sqrt{(1-\frac{1}{2}\sin^2\theta)}}+B,$$

where $\mu=h\sec^2\phi$, $\nu=h\cos^2\theta$ and B is a constant.

Ex. 3. A particle moves under the action of two equal centres of force, one attracting and the other repelling like the poles of a magnet. The particle is projected with a velocity due to an infinite distance. Show that if the direction of projection be properly chosen the particle will oscillate in a semi-ellipse, the two poles being the foci. If otherwise projected the path is given by

$$\sqrt{\frac{h}{\beta}}\log\{\mu+\sqrt{(\mu^2-h^2)}\}+B=\int\frac{d\phi}{\sqrt{(1-k\sin^2\phi)}},$$

where $\nu=h\cos^2\phi+\beta\sin^2\phi$, $2k=1-\beta/h$ and $A=2H\beta$.

Ex. 4. Prove that the lemniscate, $rr'=c^2$, can be described under the action of two centres of force each H/r^3 tending to the foci, provided the velocity at the node is $\frac{2}{c}\sqrt{\frac{2H}{3}}$. See Art. 190, Ex. 11.

588. *To find the motion of a particle of unit mass in three dimensions under the action of two centres of force attracting according to the Newtonian law.*

Let the two centres of force H_1, H_2, be situated in the axis of z and let the origin O bisect the distance H_1H_2. Let ϕ be the angle the plane zOP makes with zOx and let ρ be the distance of P from Oz.

Since the impressed forces have no moment about Oz, we have by the principle of angular momentum (Art. 492),

$$\rho^2\phi'=B\text{..............................}(1).$$

We now adopt the method explained in Art. 495. We treat the particle as if it were moving in a fixed plane zOP under the influence of the two centres of force together with an additional force $\rho\phi'^2=B^2/\rho^3$ tending from the axis of z. This problem has been partly solved in Art. 585; it only remains to consider the effect of the additional force. This force adds the term $-B^2/2\rho^2$ to the work function U.

Taking H_1, H_2 as the foci of a system of confocal conics, let μ, ν be the elliptic coordinates of P. As before, we suppose that the work function U of the impressed forces satisfies the condition

$$(\mu^2-\nu^2)\,U=F_1(\mu)+F_2(\nu)\dots\dots\dots\dots\dots\dots(2).$$

Since ρ is the ordinate of the conics [Art. 585],

$$\rho^2=\frac{(\mu^2-h^2)(\nu^2-h^2)}{-h^2};\qquad \therefore\ \frac{\mu^2-\nu^2}{\rho^2}=\frac{h^2}{\mu^2-h^2}-\frac{h^2}{\nu^2-h^2}\dots(3).$$

The term to be added to U has therefore the same form as those already existing in U and shown in (2). To obtain the integrals we have merely to add the terms given in (3), (after multiplication by $-\frac{1}{2}B^2$) to the functions F_1, F_2.

In this way, we find the integrals

$$\left.\begin{aligned}\tfrac{1}{2}(\mu^2-\nu^2)^2\frac{\mu'^2}{\mu^2-h^2}&=F_1(\mu)+C\mu^2-\tfrac{1}{2}\frac{B^2h^2}{\mu^2-h^2}+A\\ -\tfrac{1}{2}(\mu^2-\nu^2)^2\frac{\nu'^2}{\nu^2-h^2}&=F_2(\nu)-C\nu^2+\tfrac{1}{2}\frac{B^2h^2}{\nu^2-h^2}-A\end{aligned}\right\}\dots(4).$$

When the central forces follow the Newtonian law,

$$U=\frac{H_1}{r_1}+\frac{H_2}{r_2};\qquad \therefore\ (\mu^2-\nu^2)\,U=K_1\mu+K_2\nu,$$

where $K_1=H_1+H_2$, $K_2=H_1-H_2$, as in Art. 586. We therefore write in the solution (4), $F_1(\mu)=K_1\mu$, $F_2(\nu)=K_2\nu$.

If the particle is acted on by a third centre of force situated at the origin and attracting as the distance, we add to the expression for U the term $-\frac{1}{2}H_3r^2=-\frac{1}{2}H_3(\mu^2+\nu^2-h^2)$. The effect of this is to increase the functions F_1, F_2 by $-\frac{1}{2}H_3(\mu^4-h^2\mu^2)$, and $\frac{1}{2}H_3(\nu^4-h^2\nu^2)$ respectively.

In the same way if the particle is also acted on by a force tending directly from the axis of z and equal to κ/ρ^3, or a force parallel to z and equal to κ/z^3, the effect is merely to give additional terms to the functions F_1 and F_2. See Art. 582.

589. *Ex.* A particle P moves under the attraction of two centres of force at A and B. If the angles PAB, PBA be respectively θ_1, θ_2, the distances AP, BP be r_1, r_2, and the accelerations be μ_1/r_1^2, μ_2/r_2^2, prove that

$$\left(r_1^2\frac{d\theta_1}{dt}\right)\left(r_2^2\frac{d\theta_2}{dt}\right)=a\,(\mu_1\cos\theta_1+\mu_2\cos\theta_2)+C,$$

where $AB=a$, C is a constant and the motion is in one plane.

If the motion is in three dimensions, prove that

$$\left(r_1^2 \frac{d\theta_1}{dt}\right)\left(r_2^2 \frac{d\theta_2}{dt}\right) + h^2 \cot\theta_1 \cot\theta_2 = a\,(\mu_1 \cos\theta_1 + \mu_2 \cos\theta_2) + C,$$

where h is the areal description round the line of centres. [Coll. Ex. 1895.]

On Brachistochrones.

590. Preliminary Statement. Let a particle P, projected from a point A at a time t_0 with a velocity v_0, move along a smooth fixed wire under the influence of forces whose potential U is a given function of the coordinates of P, and let the particle arrive at a point B at a time t_1 with a velocity v_1. Let us suppose that the circumstances of the motion are slightly varied. Let a particle start from a neighbouring point A' at a time $t_0 + \delta t_0$ with a velocity $v_0 + \delta v_0$. Let it be constrained by a smooth wire to describe an arbitrary path nearly coincident with the former under forces whose potential is the same function of the coordinates as before, and let it arrive at a point B' near the point B at a time $t_1 + \delta t_1$ with velocity $v_1 + \delta v_1$.

According to the same notation, if x, y, z; x', y', z', are the coordinates and resolved velocities at any point P of the first path at the time t, then $x + \delta x$, &c.; $x' + \delta x'$, &c., are the coordinates and resolved velocities at any point P' of the varied path occupied by the particle at the time $t + \delta t$.

Let P, Q be any two points on the two paths simultaneously occupied at the time t. Let the coordinates of Q be $x + \Delta x$, $y + \Delta y$, &c. Then δx exceeds Δx by the space described in the time δt,

$$\therefore \Delta x = \delta x - (x' + \delta x')\,\delta t = \delta x - x'\delta t$$

when quantities of the second order are neglected.

We may regard δx, δy, δz, as any indefinitely small arbitrary functions of x, y, z, limited only by the geometrical conditions of the problem.

We here consider two independent changes of the coordinates. There are (1) the differentials dx, dy, dz when the particle travels along the undisturbed path, and (2) the variations δx, δy, δz when the particle is displaced to some neighbouring path. It follows from the independence of these two displacements that $d\delta x = \delta dx$.

591. The Brachistochrone. *A particle of unit mass moves under the action of forces so that its velocity v at any point is given by $\frac{1}{2}v^2 = U + C$, where U is a known function of the coordinates, the constant C being also known. Supposing the initial and final positions A, B to lie on two given surfaces, it is required to find the path the particle must be constrained to take that the time of transit may be a minimum*.*

* An account of the early history of this problem is given in Ball's *Short History of Mathematics*. Passing to later times, the theorem $v = Ap$ for a central force is given by Euler, *Mechanica*, vol. II. There is a memoir by Roger in *Liouville's Journal*, vol. XIII. 1848; he discusses the brachistochrone on a surface

The time t of transit being $t = \int ds/v$, we have to make this integral a minimum. Since a variation is only a kind of differential, we follow the rules of the differential calculus and make the first variation of t equal to zero. Let the curve AB be varied into a neighbouring curve $A'B'$, each element being varied into a corresponding element. Since the number of elements is not altered, the variation of the integral is the integral of the variation. Writing ϕ for $1/v$ to avoid fractions, we have

$$\delta t = \int \delta(\phi ds) = \int (\phi d\delta s + ds\delta\phi).$$

Since $(ds)^2 = (dx)^2 + (dy)^2 + (dz)^2$, we have

$$ds\delta ds = dx\delta dx + dy\delta dy + dz\delta dz;$$

$$\therefore \delta t = \int \phi \left(\frac{dx}{ds} d\delta x + \frac{dy}{ds} d\delta y + \&c.\right) + \int \left(\frac{d\phi}{dx}\delta x + \frac{d\phi}{dy}\delta y + \&c.\right) ds.$$

Integrating the first three terms by parts,

$$\delta t = \phi \left(\frac{dx}{ds}\delta x + \&c.\right) + \int \left\{\left[\frac{d\phi}{dx} - \frac{d}{ds}\left(\phi \frac{dx}{ds}\right)\right]\delta x + \&c.\right\} ds,$$

where the part outside the sign of integration is to be taken between the limits A to B.

We notice that in this variation, C has not been varied. If C were different for the different trajectories, we should have

$$\delta\phi = \frac{d\phi}{dx}\delta x + \frac{d\phi}{dy}\delta y + \frac{d\phi}{dz}\delta z + \frac{d\phi}{dC}\delta C.$$

There would then be an additional term inside the integral. It follows that v^2 *is regarded as the same function of* x, y, z *for all the trajectories.*

Since the time t is to be a minimum for all variations consistent with the given conditions, it must be a minimum when the ends A, B are fixed (Art. 144). We then have at these points $\delta x = 0$, $\delta y = 0$, $\delta z = 0$, and the part outside the integral vanishes.

The required curve must therefore be such that the integral is zero whatever small values the arbitrary functions δx, δy, δz may have. It is proved in the calculus of variations (and is

and generalises Euler's theorem that the normal force is equal to the centrifugal force. Jellett in his *Calculus of Variations*, 1850, proves these theorems and deduces from the principle of least action that the brachistochrone becomes a free path when $v = k^2/v'$. Tait has applied Hamilton's characteristic function to the problem in the *Edinburgh Transactions*, vol. XXIV. 1865, and deduces from a more general theorem the above relation to free motion. Townsend in the *Quarterly Journal*, vol. XIV. 1877, obtains the relation $v = v'$ in free motion, and gives numerous examples. There are also some theorems by Larmor in the *Proceedings of the London Mathematical Society*, 1884.

perhaps evident) that the coefficients of δx, δy, δz must separately vanish. We therefore have, writing $1/v$ for ϕ,

$$\frac{d}{dx}\left(\frac{1}{v}\right)=\frac{d}{ds}\left(\frac{1}{v}\frac{dx}{ds}\right),\quad \frac{d}{dy}\left(\frac{1}{v}\right)=\frac{d}{ds}\left(\frac{1}{v}\frac{dy}{ds}\right),\quad \frac{d}{dz}\left(\frac{1}{v}\right)=\frac{d}{ds}\left(\frac{1}{v}\frac{dz}{ds}\right).$$

These are the differential equations of the brachistochrone.

These three equations really amount to only two, for if we multiply them by $\phi\,dx/ds$, $\phi\,dy/ds$, &c. and add the products, we find

$$\phi\frac{d\phi}{ds}=\frac{1}{2}\frac{d}{ds}\left\{\phi^2\left(\frac{dx}{ds}\right)^2+\phi^2\left(\frac{dy}{ds}\right)^2+\phi^2\left(\frac{dz}{ds}\right)^2\right\},$$

which is an evident identity.

592. Supposing these differential equations to have been solved, it remains to determine the constants of integration. To effect this we resume the expression for δt, now reduced to the part outside the integral sign. We have

$$\delta t=\phi\left(\frac{dx}{ds}\delta x+\frac{dy}{ds}\delta y+\frac{dz}{ds}\delta z\right),$$

which is to be taken between the limits A to B. Since we may vary the ends A, B of the curve, one at a time, along the bounding surface (Art. 144), this expression for δt must be zero at each end. The variations δx, δy, δz are proportional to the direction cosines of the displacement of the end, and dx/ds, &c. are the direction cosines of the tangent to the brachistochrone. This equation therefore implies that *the brachistochrone meets the bounding surface at right angles.*

The expression for δt may be put into a geometrical form which is sometimes useful. Let $\delta\sigma_1$, $\delta\sigma_2$ be the displacements AA', BB' of the two ends. Let θ_1, θ_2 be the angles these displacements respectively make with the tangents at A and B to the brachistochrone AB. Let v_1, v_2 be the velocities at A, B. Then

$$\delta t=\frac{\delta\sigma_2\cos\theta_2}{v_2}-\frac{\delta\sigma_1\cos\theta_1}{v_1}.$$

593. In some problems the velocity v is a given function of the coordinates of one or both ends of the curve. This does not affect the differential equations, for in these the coordinates of the ends, when fixed, are merely constants.

The case is different when we vary the ends in that portion of the expression for δt which is outside the integral sign. We must add to that expression the terms of $\delta\phi$ due to the variation of the ends. If x_0, y_0, z_0; x_1, y_1, z_1, are the coordinates of the ends A, B, we then have

$$\delta t = \left[\phi\left(\frac{dx}{ds}\,\delta x + \&c.\right)\right]_0^1 + \delta x_0 \int \frac{d\phi}{dx_0}\,ds + \&c. + \delta x_1 \int \frac{d\phi}{dx_1}\,ds + \&c.,$$

where the &c. indicate terms with y and z respectively written for x. The conditions at the ends are then found by equating this expression to zero.

594. The equations of the brachistochrone are found by equating the first variation of the time to zero. To determine whether this curve makes the time a maximum, a minimum, or neither, it is necessary to examine the terms of the second order. For this we refer the reader to treatises on the calculus of variations. In most cases there is obviously some one path for which the time is a minimum, and if our equations lead to but one path, that path must be a true brachistochrone. In other cases we can use Jacobi's rule. Let AB be the curve from A to B given by the calculus of variations. Let a second curve *of the same kind* but with varied constants be drawn through the initial point A and make an indefinitely small angle at A, with the curve AB. If they again intersect in some point C, the curve satisfies the conditions for a true minimum only if C be beyond B.

595. **Theorem I.** When the only force on the particle acts (like gravity) in a vertical direction, $\phi = 1/v$ is a function of z only, and the first two differential equations of the curve (Art. 591) admit of an immediate integration. Remembering that $dx/ds = \cos\alpha$, $dy/ds = \cos\beta$, it follows that *the brachistochrone for a vertical force is such a curve that at every point $v = a\cos\alpha$, $v = b\cos\beta$, where α, β are the angles the tangent makes with any two horizontal straight lines, and a, b are the two constants of integration.* By equating the two values of v and integrating, we see that the brachistochrone is a plane curve.

596. **Theorem II.** Let X, Y, Z be the components of the impressed forces, the mass being unity; then since $\frac{1}{2}v^2 = U + C$, we have $X = \frac{1}{2}dv^2/dx$, &c. The differential equations of the brachistochrone therefore become

$$\frac{d}{ds}\left(\frac{1}{v}\frac{dx}{ds}\right) + \frac{X}{v^3} = 0, \quad \frac{d}{ds}\left(\frac{1}{v}\frac{dy}{ds}\right) + \frac{Y}{v^3} = 0, \quad \&c. = 0 \ldots(1).$$

Let λ, μ, ν be the direction cosines of the binormal, then since the binormal is perpendicular both to the tangent and the radius of curvature

$$\lambda\frac{dx}{ds} + \mu\frac{dy}{ds} + \nu\frac{dz}{ds} = 0, \qquad \lambda\frac{d^2x}{ds^2} + \mu\frac{d^2y}{ds^2} + \nu\frac{d^2z}{ds^2} = 0 \ldots\ldots(2).$$

Using the values of X, Y, Z given in (1) we find

$$\lambda X + \mu Y + \nu Z = 0 \quad\text{.......................(3)},$$

the resultant force is therefore perpendicular to the binormal, and its direction lies in the osculating plane.

Let $l = \rho \dfrac{d^2x}{ds^2}$, $m = \rho \dfrac{d^2y}{ds^2}$, &c. be the direction cosines of the positive direction of the radius of curvature, then

$$-\frac{lX + mY + nZ}{v^3} = \frac{1}{v}\,\frac{l^2 + m^2 + n^2}{\rho} + \frac{d}{ds}\left(\frac{1}{v}\right)\left\{l\frac{dx}{ds} + \&c.\right\}.$$

Since the radius of curvature is at right angles to the tangent, the last term is zero, and we have

$$lX + mY + nZ = -\frac{v^2}{\rho} \quad\text{....................(4)}.$$

This equation proves that *in any brachistochrone the component of the impressed forces along the radius of curvature is equal to minus the component of the effective forces in the same direction.*

597. *To find the pressure on the constraining curve.* Let F_1, F_2 be the components of the impressed forces in the directions of the radius of curvature and binormal. Let R_1, R_2 be the pressures on the particle in the same directions. Then by Art. 526

$$v^2/\rho = F_1 + R_1, \quad 0 = F_2 + R_2.$$

In a brachistochrone $F_2 = 0$ and $F_1 = -v^2/\rho$, hence $R_2 = 0$ and $R_1 = -2F_1$.

598. *To find a dynamical interpretation of Theorem II.*

We see by referring to the equations of motion in Art. 597, that if we changed the sign of F_1, the component of pressure R_1 would be zero, and the path would then be free. We also suppose the tangential component of force to remain unchanged so that the velocity is not altered. It follows immediately, that *a brachistochrone and a free path may be changed, either into the other, by making the resultant force at each point act at the same angle to the same direction of the tangent as before, but on the other side, and still in the osculating plane.* In this comparison the velocities of the particle, when free and when constrained, are equal at the same point of the path, i.e. $v' = v$.

599. Theorem III. The equations of motion of a particle P constrained to describe the brachistochrone are

$$\frac{d}{ds}\left(\frac{1}{v}\frac{dx}{ds}\right)=\frac{d}{dx}\left(\frac{1}{v}\right),\quad \frac{d}{ds}\left(\frac{1}{v}\frac{dy}{ds}\right)=\frac{d}{dy}\left(\frac{1}{v}\right),\ \&\text{c.}$$

If we now write $vv'=k^2$ or, which is the same thing $vds=k^2dt'$, where $v'=ds/dt'$, the first of these equations becomes

$$\frac{d}{ds}\left(v'\frac{dx}{ds}\right)=\frac{dv'}{dx}.$$

Now $v'dx/ds$ being the x component of the velocity, is equal to dx/dt'. Multiplying by v' or ds/dt', the equations take the form

$$\frac{d^2x}{dt'^2}=\frac{1}{2}\frac{dv'^2}{dx},\quad \frac{d^2y}{dt'^2}=\frac{1}{2}\frac{dv'^2}{dy},\ \&\text{c.}$$

These are the equations of motion of a free particle P' moving along the same path with a velocity v' and occupying the position x, y, z at the time t'. It follows that *the brachistochrone from point to point in a field $U+C$ is the same as the path of a free particle in a field $U'+C'$, provided* $U'+C'=\frac{k^4}{4}\frac{1}{U+C}$; i.e. $v'=\frac{k^2}{v}$.

To understand better the relation between the two fields of force we notice that if X, X' be the components of force in any the same direction at the same point,

$$X=\frac{dU}{dx},\quad X'=\frac{dU'}{dx},\quad \therefore\ X'=-X\left(\frac{k}{v}\right)^4.$$

We also notice that $dt'/dt=v/v'$.

600. This theorem is useful, as it enables us to apply to a brachistochrone the dynamical rules we have already studied for free motion. It also enables us to express at once the fundamental differential equations in polar or other co-ordinates.

The first theorem (Art. 595) follows at once from the third, for when the force is vertical we see by resolving horizontally that $v'\cos\alpha$ is constant. Since $v'=k^2/v$, this gives the result.

To deduce the second theorem, we notice that in the free motion $v'^2/\rho=F_1'$, where F_1' is the component of force along the radius of curvature. Using the theorems $v'=k^2/v$, $X'=-X(k/v)^4$, (where X is here F_1) this becomes $v^2/\rho=-F_1$.

601. *Ex.* 1. *To find the brachistochrone from one given curve to another, the acting force being gravity and the level of no velocity given.* The motion is supposed to be in a vertical plane.

Let the axis of x be at the level of no velocity and let y be measured downwards; then $v^2=2gy$. By Art. 595 the curve is such that $v=a\cos\alpha$. This gives $y=2b\cos^2\alpha$, where b is an undetermined constant. This is the well-known *equation of a cycloid, having its cusps at the level of no velocity.* The radius of the generating circle and the position of the cusps on the axis are determined by the conditions that the cycloid cuts each of the bounding curves at right angles; Art. 592.

Ex. 2. If in the last example the bounding curves are two straight lines which intersect the axis of no velocity in the points L, L'; and make angles β, β' with the horizon, prove that the diameter $2b$ of the generating circle is $LL'/(\beta-\beta')$ and the distance of the cusp from L is $2b\beta$. Explain the results when the lines are parallel.

602. *Ex.* Show by using Jacobi's rule that the cycloid from one given point A to another B is a real minimum, the level of zero velocity being given (Art. 594).

The cycloid found by the calculus of variations passes through A and B and there is no cusp between these points. Describe a neighbouring cycloid passing through A and having its cusps on the same horizontal line, the radii of the generating circles being b and $b+db$. Since the base of a cycloid from cusp to cusp is $2\pi b$, it is easy to prove that the next intersection of the two curves lies in a vertical which passes between the two next cusps. The cycloids therefore cannot again intersect between A and B and the time from A to B must be a minimum. See also Art. 654.

603. *Ex.* Find the brachistochrone from one given curve to another when the acting force is gravity and the particle starts from rest at the upper curve.

Fixing the ends, it follows, from Art. 601, that *the brachistochrone is a cycloid having a cusp on the higher curve.* To determine the constants of the curve, we examine the part of δt due to the variation of the two ends. Let x_0, y_0; x_1, y_1 be the coordinates of the upper and lower ends, then $v^2=2g(y-y_0)$. By Art. 593 we have

$$\delta t=\phi\left\{\frac{dx}{ds}\delta x+\frac{dy}{ds}\delta y\right\}+\delta y_0\int\frac{d\phi}{dy_0}ds=0,$$

where $\phi=1/v$ and the expression is taken between limits. Now in our problem

$$-\frac{d\phi}{dy_0}=\frac{d\phi}{dy}=\frac{d}{ds}\left(\phi\frac{dy}{ds}\right),$$

by using the differential equation of the brachistochrone in Art. 591. We therefore have

$$\left[\phi\left(\frac{dx}{ds}\delta x+\frac{dy}{ds}\delta y\right)\right]_0^1-\delta y_0\left[\phi\frac{dy}{ds}\right]_0^1=0.$$

Remembering that $\phi=1/v$ and $v=a\cos\alpha$, this takes the form

$$[\delta x+\tan\alpha\,\delta y]_0^1-\delta y_0[\tan\alpha]_0^1=0.$$

When we fix the lower end, we have, since y is measured downwards, $\delta x_1=0$, $\delta y_1=0$. Hence

$$-(\delta x_0+\tan\alpha_0\,\delta y_0)-\delta y_0(\tan\alpha_1-\tan\alpha_0)=0 \quad\text{.................}\quad (1).$$

When we fix the upper end, $\delta x_0=0$, $\delta y_0=0$;

$$\therefore\ \delta x_1+\tan\alpha_1\,\delta y_1=0 \quad\text{..............................}\quad (2).$$

The last of these two equations proves that *the brachistochrone cuts the lower curve at right angles*, while the first, giving $\delta y_0/\delta x_0=\delta y_1/\delta x_1$, proves that *the tangents to the bounding curves at the points where the brachistochrone meets them are parallel.*

604. *Ex.* 1. A particle falls from rest at a fixed point A to a fixed point C, passing through another point B; find the entire path when the time of motion is a minimum, (1) supposing B to be a fixed point, (2) supposing B constrained to lie on a given curve. [Math. Tripos, 1866.]

The paths from A to B, B to C are cycloids having their cusps on a level with the point A. It is supposed that there is no impact at B in passing from one cycloid to the next. The particle describes a small arc of a curve of great curvature and moves off along the next cycloid without loss of velocity.

We have yet to find the position of B when it is only known to lie on a given curve. Taking the origin at A, and the axis of z vertically downwards, we have $v^2=2gz$. The time is given by

$$\sqrt{(2g)}t=\int_A^B \frac{ds}{\sqrt{z}}+\int_B^C \frac{ds'}{\sqrt{z'}},$$

where accents refer to the lower cycloid.

$$\therefore \sqrt{(2g)}\,\delta t=\left[\frac{1}{\sqrt{z}}\left(\frac{dx}{ds}\delta x+\frac{dy}{ds}\delta y+\frac{dz}{ds}\delta z\right)\right]_A^B+\left[\frac{1}{\sqrt{z'}}\left(\frac{dx'}{ds'}\delta x+\frac{dy'}{ds'}\delta y+\frac{dz'}{ds'}\delta z\right)\right]_B^C=0,$$

by Art. 592. Let (α, β, γ), $(\alpha', \beta', \gamma')$, (θ, ϕ, ψ) be the direction angles of the tangents at B to the two cycloids and to the constraining curve. Then remembering that A and C are fixed points and that B is varied on the curve, we have

$$(\cos\alpha\cos\theta+\cos\beta\cos\phi+\cos\gamma\cos\psi)-(\cos\alpha'\cos\theta+\cos\beta'\cos\phi+\cos\gamma'\cos\psi)=0.$$

It follows that *the tangent to the locus of B makes equal angles with the tangents to the two cycloids AB, BC.* This determines the point B.

Ex. 2. Find the curve of quickest descent from a fixed point A to another C, supposing that a screen is interposed between A and C having a given finite aperture through which the path must pass. [So long as the curve AC can be arbitrarily varied the minimum curve is found by Arts. 591, 601. Hence if the single cycloid AC does not pass through the aperture, the minimum curve must pass through a point B on the boundary of the aperture. The curve then consists of two cycloids AB, BC, and the position of B is found by Ex. 1.] [Todhunter.]

605. *Ex.* 1. If the brachistochrone is a parabola when the force is parallel to the axis, prove that the magnitude of the force is inversely proportional to the square of the distance from the directrix. [This follows from the equation $v=a\cos\alpha$.] Prove also that the time of describing any arc PQ varies as the area contained by the focal radii, SP, SQ. [For $\cos\alpha$ varies as $1/p$, therefore dt varies as pds.] See also Art. 649.

Ex. 2. A point moves in a plane with a velocity always proportional to the curvature of the path, prove that the brachistochrone of continuous curvature between any two given points is a complete cycloid. [Math. Tripos, 1875.]

We here have $\int\rho ds=\int\phi dx$ a minimum, where $\phi=(1+y'^2)^2/y''$. The curve can be immediately found by using two rules in the calculus of variations. *First*, we have

$$\delta\int\phi dx=\phi\delta x+(Y_,-Y_{,,}')\,\omega+Y_{,,}\omega'+\int(-Y_,'+Y_{,,}'')\,\omega dx,$$

where $Y_,$, $Y_{,,}$, are the partial differential coefficients of ϕ with regard to y', y''; $\omega=\delta y-y'\delta x$, and the part outside the integral sign is to be taken between limits. Also accents denote differentiation with regard to x. The extreme points being given, $\delta x=0$, $\delta y=0$ at each end. Hence exactly as in Art. 591, 592, the differential equation of the curve is $Y_,'-Y_{,,}''=0$ and $Y_{,,}=0$ at each end. This gives $Y_,-Y_{,,}'=A$.

Secondly, the calculus of variations gives also the integral

$$\phi=(Y_{,}-Y_{,,}')\,y'+Y_{,,}y''+B.$$

Eliminating $Y_{,,}'$ between our two first integrals we find $\phi=Ay'+Y_{,,}y''+B$, which contains two arbitrary constants A, B. Substituting for ϕ and $Y_{,,}$, this leads to $(1+y'^2)^2/y''=\frac{1}{2}Ay'+\frac{1}{2}B$; $\therefore\ \rho\,ds=\frac{1}{2}A\,dy+\frac{1}{2}B\,dx$.

Taking the straight line $Ay+Bx=0$ as an axis of ξ, this is equivalent to $\rho=C\sin\psi$ where $\sin\psi=d\eta/ds$ and C is a constant. This is the known equation of a cycloid. The condition $Y_{,,}=0$ at each end gives y'' infinite and therefore $\rho=0$. The cycloid is therefore complete.

Ex. 3. Prove that the differential equation of the brachistochrone from rest at one given point A to another point B, when *the length of the curve is also given*, is $\dfrac{a}{\sqrt{y}}+b=\sqrt{\left\{1+\left(\dfrac{dy}{dx}\right)^2\right\}}$. [Airy's Tracts.]

To make $\int ds/v$ a minimum subject to the condition that $\int ds$ is a given quantity we use a rule supplied by the calculus of Variations. We make $\int(\lambda/v+1)\,ds$ a minimum without regard to the given condition and finally determine the constant λ so that the arc has the given length.

606. Central force. *Ex.* 1. Prove that the brachistochrone for a central force F is given by $v=Ap$, where $\frac{1}{2}v^2=\int F dr$ and p is the perpendicular from the centre of force on the tangent. The mass is unity, as is usual in these problems.

The brachistochrone is a free path for a particle moving about the same centre but with such a law of force that the velocity $v'=k^2/v$. Since $v'p=h$ by Art. 306, we have $v=Ap$.

When $F=\mu u^n$, and the velocity is equal to that from infinity, the differential equation $v=Ap$ can be integrated exactly as in Arts. 360, 363.

Ex. 2. Prove that the same path will be a brachistochrone for $F=\mu u^n$ and a free path for $F'=\mu' u^{n'}$ if $n+n'=2$, provided the velocity in each case varies as some power of the distance.

For the brachistochrone and the free paths respectively, we have

$$v^2=2\mu u^{n-1}/(n-1),\quad v'^2=2\mu' u^{n'-1}/(n'-1).$$

These satisfy the condition $vv'=k^2$ if $n+n'=2$, (Art. 599).

Ex. 3. Prove that the ellipse is a brachistochrone for a central force tending from the focus and equal to $\mu/(2a-r)^2$. [Townsend.]

The conic is a free path for a force μ/SP^2 tending to the focus S. Hence making the force act on the other side of the tangent as described in Art. 598, the conic is a brachistochrone for an equal force tending from the other focus H.

Ex. 4. Prove that the central repulsive force for the brachistochronism of a plane curve varies as $d\,(p^2)/dr$, the circle of zero velocity being given by the vanishing of p.

Prove that the cissoid $x\,(x^2+y^2)=2ay^2$ is brachistochronous for a central repulsive force from the point $(-a,\,0)$ which at the distance r from that point is proportional to $r/(r^2+15a^2)^2$, the particle starting from rest at the cusp.

[Math. Tripos, 1896.]

Ex. 5. Prove that the lemniscate of Bernoulli can be described as a brachistochrone in a field of potential μr^6, r being measured from the node of the lemniscate, and find the necessary velocity. [See Arts. 320, 606, Ex. 2.]

[Math. Tripos, 1893.]

Ex. 6. A particle, acted on by a central attractive force whose accelerating effect at a distance r is $\dfrac{\mu r}{(a^2+r^2)^2}$, a being a constant, is projected from a given point with the velocity from infinity. Prove that the form of the groove in which it must move in order to arrive at another given point in the shortest possible time is a hyperbola whose centre coincides with the centre of force. [Math. Tripos.]

Ex. 7. Show that the force of attraction towards the directrix of a catenary, along perpendiculars to it, for which the catenary is a brachistochrone, will vary as the inverse cube of the perpendicular. [Coll. Ex. 1897.]

607. Brachistochrone on a surface. To find the brachistochrone on a given surface we require only a slight modification in the argument of Art. 591. Proceeding as before, we find

$$\delta t = \frac{1}{v}\left(\frac{dx}{ds}\,\delta x + \text{\&c.}\right) + \int (P\,\delta x + Q\,\delta y + R\,\delta z)\,ds,$$

where $P = \dfrac{d}{dx}\dfrac{1}{v} - \dfrac{d}{ds}\left(\dfrac{1}{v}\dfrac{dx}{ds}\right)$, with similar expressions for Q and R. Since δt is zero for all variations of the curve on the surface, we must have

$$P\,\delta x + Q\,\delta y + R\,\delta z = 0.$$

If $f(x, y, z) = 0$ is the equation of the surface, the variations are connected by the one equation

$$f_x\delta x + f_y\delta y + f_z\delta z = 0,$$

where suffixes imply partial differential coefficients. We must therefore have $P/f_x = Q/f_y = R/f_z$. The equations of a brachistochrone on the surface $f(x, y, z) = 0$ are therefore given by

$$\left(\frac{d}{dx}\frac{1}{v} - \frac{d}{ds}\frac{dx}{v\,ds}\right)\bigg/ f_x = \left(\frac{d}{dy}\frac{1}{v} - \frac{d}{ds}\frac{dy}{v\,ds}\right)\bigg/ f_y = \left(\frac{d}{dz}\frac{1}{v} - \frac{d}{ds}\frac{dz}{v\,ds}\right)\bigg/ f_z.$$

If the brachistochrone is to begin and end at given bounding curves drawn on the surface, we equate to zero the integrated part of δt, taken between the limits. Fixing the ends in turn, we see that at each end the cosine of the angle between the tangents to the curve and to the boundary is zero (Art. 592). *The brachistochrone therefore cuts the boundaries at right angles.*

608. By writing $v = k^2/v'$ as in Art. 599 these equations may be put into the form

$$\left(\frac{d^2x}{dt'^2} - \frac{dU'}{dx}\right)\bigg/ f_x = \left(\frac{d^2y}{dt'^2} - \frac{dU'}{dy}\right)\bigg/ f_y = \left(\frac{d^2z}{dt'^2} - \frac{dU'}{dz}\right)\bigg/ f_z.$$

These are the equations of motion of a particle moving freely on the constraining surface. It follows that *the brachistochrone from point to point on a constraining surface in a field $U+C$ is a free path on the same surface in a field $U'+C'$, where*

$$\tfrac{1}{2}v^2 = U+C, \qquad \tfrac{1}{2}v'^2 = U'+C', \qquad vv' = k^2.$$

The relation between the component forces in any direction is $F' = -F\left(\frac{k}{v}\right)^4$.

Ex. If the particle is constrained by a smooth wire to describe the brachistochrone on the surface without a change in the field of force, prove that

$$-v^2 \sin\chi/\rho = G, \quad v^2 \cos\chi/\rho = H+R, \quad -2G = R_2,$$

where H, G are the components of the impressed forces along the normal to the surface, and that tangent to the surface which is perpendicular to the path, and R, R_2 are the components of the pressure in the same directions. Also ρ is the radius of curvature of the path, and χ the angle the osculating plane makes with the normal to the surface.

The first is obtained by transforming the equation of motion of a free particle P', viz. $v'^2 \sin\chi/\rho = G'$ by the rule given above, the others then follow from the ordinary equations of motion of the particle P.

609. We may also sometimes find the brachistochrone on a given surface by making a comparison with the brachistochrone on some other more suitable surface.

Let us derive a second surface from the given one by writing for the coordinates x, y, z of any point P some functions of ξ, η, ζ, the coordinates of a corresponding point Q. Let these functions be such that

$$(dx)^2 + (dy)^2 + (dz)^2 = \mu^2 \{(d\xi)^2 + (d\eta)^2 + (d\zeta)^2\},$$

where μ is a function of ξ, η, ζ. Geometrically this equation implies that every elementary arc ds drawn from a point P on the surface bears the same ratio to the corresponding arc $d\sigma$ drawn from Q, viz. the ratio $\mu : 1$.

The brachistochrone on the given surface is found by making t a minimum, where

$$t = \int \frac{ds}{v} = \int \frac{\mu\, d\sigma}{v},$$

and the velocity v of P is some given function of the coordinates of P.

Expressing v in terms of ξ, η, ζ, this integral implies that the corresponding curve on the derived surface is also a brachistochrone, the velocity v' being given by $v' = v/\mu$. The work functions for the motions of P and Q are respectively $v^2 = 2(U+C)$ and $U' = (U+C)/\mu^2$.

If we arrange matters so that μ/v is constant, the velocity on the second surface is constant. *The brachistochrones on the given surface then correspond to geodesics on the derived surface.*

This comparison assists us in determining the point on a brachistochrone with one end given at which the time ceases to be a minimum.

The derived surface may be obtained in many ways, for example by using the method of inversion. The theory of this surface is also used in making maps; see the United States Coast Survey, *Craig's treatise on Projections*. The application to brachistochrones is given by Darboux in his *Théorie générale des Surfaces*.

Ex. A particle P moves on a sphere under the action of a centre of repulsive force situated at a point O on the surface, and the velocity v at any point distant

r from O is $v=Ar^2$. Prove that the brachistochrone from one given point to another is a circle whose plane passes through O.

Inverting the sphere with regard to O, the diameter $2a$ being the constant of inversion, the derived surface is a tangent plane. The curve is traced out by Q, usually called *the stereographic projection* of that traced by P. The ratio of the elementary arcs described by P and Q are in the ratio $r^2 : 4a^2$. Hence if the path of P is a brachistochrone for a velocity $v=Ar^2$, that of Q is a brachistochrone for a uniform velocity. The path of Q is therefore a straight line and that of P is a circle. Another proof follows from Arts. 608, 318.

610. Bertrand's theorem. A series of brachistochrones is drawn on a given surface from a point A, and the arcs AB, AB', &c. are described in equal times, the velocity at A being given. Prove that the locus of B cuts all the brachistochrones at right angles.

The following amounts to Bertrand's proof. If possible let the angle $AB'B$ be acute. Drawing the arc BC so that the angle $CBB' > CB'B$, the sides of the triangle BCB' will *then* be elementary and the triangle may be regarded as rectilinear. It follows that the arc $CB' > CB$. The time of describing CB' is $>$ than that of describing CB because the velocity at every point in the neighbourhood of C is ultimately the same. The time of describing the line ACB is therefore less than that of describing AB' or AB. The path AB could not then be a brachistochrone. This proof is the same as that used by Salmon in his *Solid Geometry*, Art. 394, to prove the corresponding theorem for geodesics. Bertrand's theorem is now generally enunciated in a generalized form and to this we proceed in the next article.

611. A surface S_1 being given, let us draw from every point A on it that brachistochrone which starts off at right angles to the surface. Let lengths AB be taken along these lines so that the time t of transit from the surface along each is equal to a given quantity. The locus of the extremities B traces out a second surface which we may call S_2. By Art. 592, we have

$$\delta t = \delta\sigma_2 \cos\theta_2/v_2 - \delta\sigma_1 \cos\theta_1/v_1 .$$

By construction $\cos\theta_1 = 0$ for each line and, since the times of describing neighbouring lines are equal, $\delta t = 0$. *It follows that the surface S_2 also cuts the lines at right angles.*

If the surface S_1 is an infinitely small sphere all the brachistochrones diverge from a given point A. The locus of the other extremities of the arcs drawn from A and described in equal times is therefore an orthogonal surface.

This proof may be applied to brachistochrones drawn on a given surface by expressing the conditions at the limits in Art. 607 in a form similar to that in Art. 592.

This theorem though enunciated for a brachistochrone applies generally to problems in the calculus of variations. The time t may stand for any integral of the form $\int\phi\,.\,ds$ where ϕ is a given function of x, y, z, and the curve is such that the integral is a minimum between any two points taken on it.

612. *Ex.* 1. Prove that the equations of a brachistochrone on a surface of revolution for a heavy particle with a given level of zero velocity are $r^2\dfrac{d\phi}{ds} = Av$,

$v^2=2gz$, where r, ϕ, z are cylindrical coordinates, z being measured downwards from the zero level. Prove also that the brachistochrone touches the meridian at the zero level.

Ex. 2. A heavy particle is projected from a given point along a smooth groove cut on the surface of a right circular cone whose axis is vertical and vertex upwards, with a velocity due to the depth from the vertex. Prove that, if it reach another given point not more than half-way round the cone in the least possible time, the curve of the groove must be such as would, if the cone were developed, become a parabola with the point corresponding to the vertex as focus.

[Math. Tripos, 1873.]

Ex. 3. Prove that the brachistochrone on a vertical cylinder for a heavy particle with a given level of zero velocity becomes the brachistochrone on a vertical plane when the cylinder is developed on the plane. [Roger.]

Ex. 4. Find the brachistochrone when the velocity at any point of space is proportional to the distance from a given straight line. Prove that the curve lies on a sphere and cuts all the circles whose planes are perpendicular to the given straight line at a constant angle, i.e., the curve is a loxodrome. [Tait.]

Motion of a particle relative to the earth.

613. Let O be any point on the surface of the earth and let λ be its latitude. Then λ is the angle which the normal to the surface of still water at O makes with the plane of the equator. Let $OL=b$ be a perpendicular from O on the axis of rotation. Let ω be the angular velocity of the earth, then the earth turns round its axis from west to east in the time $2\pi/\omega$.

As we intend to discuss the motion of a particle P relative to axes moving with the earth and having the origin at O, it is convenient to begin by reducing O to rest. We therefore apply to the particle P an accelerating force equal to $\omega^2 b$ and acting in the direction LO. We also apply an initial velocity equal to ωb opposite to the direction of motion of O, i.e. in a direction due westwards from O.

When the particle has been projected from the earth it is acted on by the attraction of the earth and the applied force $\omega^2 b$. The force usually called gravity is not the attraction of the earth, but is the resultant of that attraction and the centrifugal force. The form of the earth is such that at every point of its surface this resultant acts perpendicularly to the surface of still water. Let g be this force at the point O, then when the particle is at O, and O has been reduced to rest, the resultant force is represented by g.

When the moving point P has ascended to a height h, the attraction of the earth is altered and is nearly equal to $g(1-2h/a)$, where a is the radius of the earth. Since h is usually not more than a few hundred feet and a is roughly 4000 miles, it is obvious that the change in the value of gravity is so small that, *for a first approximation at least, we may regard gravity as a force constant in direction and magnitude.* Since $2\pi/\omega$ is 24 hours, we find that $\omega^2 a$ is nearly equal to $g/289$. Hence if we neglect gh/a *we must also neglect $\omega^2 h$ at all points near O.* The applied force $\omega^2 b$ is not neglected because at points near the equator b is nearly as large as the radius of the earth.

614. The equations of motion of a particle referred to axes moving with the earth have been already formed in Art. 499. We have here merely to express the components θ_1, θ_2, θ_3 in terms of the angular velocity ω of the earth. We then substitute the values of the space velocities u, v, w in the equations of the second order and neglect all terms of the form $\omega^2 x$, $\omega^2 y$, $\omega^2 z$. We thus find

$$u=\frac{dx}{dt}-y\theta_3+z\theta_2,\qquad \frac{d^2x}{dt^2}-2\frac{dy}{dt}\theta_3+2\frac{dz}{dt}\theta_2=X,$$

$$v=\frac{dy}{dt}-z\theta_1+x\theta_3,\qquad \frac{d^2y}{dt^2}-2\frac{dz}{dt}\theta_1+2\frac{dx}{dt}\theta_3=Y,$$

$$w=\frac{dz}{dt}-x\theta_2+y\theta_1,\qquad \frac{d^2z}{dt^2}-2\frac{dx}{dt}\theta_2+2\frac{dy}{dt}\theta_1=-g+Z,$$

where X, Y, Z are the impressed forces other than gravity, the mass being unity.

615. It will clearly be convenient to choose as the axis of z the vertical at O. If the axis of x be directed along the meridian towards the south and the axis of y towards the west, we have

$$\theta_1=\omega\cos\lambda,\quad \theta_2=0,\quad \theta_3=-\omega\sin\lambda,$$

since λ is the latitude of the place.

It is sometimes necessary to take the axis of x inclined to the meridian at some angle β, the angle β being measured from the south towards the west. We then have

$$\theta_1=\omega\cos\lambda\cos\beta,\quad \theta_2=-\omega\cos\lambda\sin\beta,\quad \theta_3=-\omega\sin\lambda.$$

616. If we wish the axes to move round the vertical with an angular velocity p, we have $\beta=pt+\epsilon$, where ϵ is some constant.

We then have

$$\theta_1 = \omega \cos\lambda \cos\beta, \quad \theta_2 = -\omega \cos\lambda \sin\beta, \quad \theta_3 = -\omega \sin\lambda + p.$$

The components θ_1, θ_2, θ_3 are not now constants, and in making the substitutions for u, v, w in the equations of motion their differential coefficients will not disappear. But if p be any small quantity of the same order as ω, these differential coefficients are of the order ω^2. The equations of motion will then be still represented by the forms given in Art. 614.

617. As in some few cases it is necessary to examine the terms which contain ω^2, we give the results of the substitution when the axis of z is vertical, while those of x, y point respectively southward and westward:

$$\frac{d^2x}{dt^2} + 2\omega \sin\lambda \frac{dy}{dt} - \omega^2 \sin^2\lambda x - \omega^2 \sin\lambda \cos\lambda z = X,$$

$$\frac{d^2y}{dt^2} - 2\omega \cos\lambda \frac{dz}{dt} - 2\omega \sin\lambda \frac{dx}{dt} - \omega^2 y = Y,$$

$$\frac{d^2z}{dt^2} + 2\omega \cos\lambda \frac{dy}{dt} - \omega^2 \cos^2\lambda z - \omega^2 \sin\lambda \cos\lambda x = -g + Z.$$

618. *Ex. A particle P is attached to a point A at the summit of a high tower and when in relative rest the particle is allowed to fall freely. The point A being at a height h vertically above O, it is required to find the point at which the particle strikes the horizontal plane at O.*

Taking the axes of x, y to point due south and west, the equations of motion are

$$x'' - 2y'\theta_3 = 0, \qquad y'' - 2z'\theta_1 + 2x'\theta_3 = 0, \qquad z'' + 2y'\theta_1 = -g,$$

where $\theta_1 = \omega \cos\lambda$, $\theta_3 = -\omega \sin\lambda$, and the accents denote d/dt (Art. 614). *We solve these by successive approximation.*

As a first approximation, we neglect the terms which contain ω. Remembering that initially x, y, x', y', z' are each zero and $z = h$, we arrive at $x = 0$, $y = 0$, $z = h - \frac{1}{2}gt^2$.

As a second approximation we substitute these values of x, y, z in the terms of the differential equations which contain θ or ω. We obtain after an easy integration

$$x = At + B, \qquad y = Ct + D - \tfrac{1}{3}gt^3\theta_1, \qquad z = Et + F - \tfrac{1}{2}gt^2.$$

The particle being initially in *relative rest* we have $x' = 0$, $y' = 0$, $z' = 0$, hence $A = 0$, $C = 0$, $E = 0$. The initial velocities *in space* are not required here, but (after O has been reduced to rest) these are given by $u = 0$, $v = -h\theta_1$, $w = 0$. To the value of v we may add the velocity of O, viz. $-\omega b$. Also when $t = 0$, we have $x = 0$, $y = 0$, $z = h$;

$$\therefore x = 0, \qquad y = -\tfrac{1}{3}gt^3\theta_1, \qquad z = h - \tfrac{1}{2}gt^2.$$

We see from the value of z that *the vertical motion is unaffected by the rotation of the earth.* The time of falling is given by $h = \frac{1}{2}gt^2$. Since $x = 0$ throughout the motion, the particle strikes the horizontal plane on the axis of y, and *there is no southerly deviation.* Since $\theta_1 = \omega \cos\lambda$ we have $y = -\frac{1}{3} g\omega \cos\lambda t^3$; *there is therefore a deviation towards the east* which is proportional to the cube of the time of descent. This deviation is greatest at the equator.

619. *Ex.* 1. Show that the path of a particle falling from relative rest is nearly the curve $325ay^2 = \cos^2\lambda z^3$.

Ex. 2. A particle is projected vertically upwards in vacuo with a velocity V. Prove that when the particle reaches the ground there is no deviation to the south, and that the deviation to the west is $4\omega\cos\lambda V^3/3g^2$. [Laplace IV., p. 341.]

Ex. 3. A particle falls from relative rest at a point A situated at a height h above the point O. Supposing the resistance of the air to be represented by κv where v is the velocity and κ a small quantity, find the effect on the easterly deviation.

Measuring z upwards and neglecting the terms $x'\theta_3$, $y'\theta_1$, as we now know that they are of the order ω^2 (Art. 618), the equations of motion become

$$y'' - 2\theta_1 z' = -\kappa y', \qquad z'' = -g - \kappa z'.$$

The vertical motion is sensibly the same as if the earth were at rest. Substituting $z' = -gt$ in the first equation,

$$y'' + \kappa y' = -2g\theta_1 t, \qquad \therefore\ y' + \kappa y = -g\theta_1 t^2;$$

$$\therefore\ y = -\frac{1}{\delta+\kappa} g\theta_1 t^2 = -\frac{1}{\delta}\left(1 - \frac{\kappa}{\delta} + \frac{\kappa^2}{\delta^2} - \&c.\right) g\theta_1 t^2,$$

where $\delta = d/dt$. This leads at once to $y = -\frac{1}{3} g\theta_1 t^3 \left(1 - \frac{3\kappa}{4} t\right)$. *The easterly deviation is therefore slightly diminished by the resistance of the air.*

Ex. 4. Prove that, if the attraction of the earth on the falling particle were represented by $X = -gx/a$, $Y = -gy/a$, $Z = -g(1 - 2z/a)$, the time of falling from rest at a height h, as deduced from the equations of Art. 614, would be increased by the inappreciable fraction $5h/6a$ of itself. Thence show that the easterly deviation is not perceptibly altered.

Ex. 5. *The southern deviation.* A particle falls from relative rest at a point A situated on the vertical at a point O on the surface of the earth. Let the southern horizontal component of the attraction of the earth be represented by

$$X = \sin\lambda\cos\lambda\,(Ax + Cz),$$

where A and C are very small functions of the ellipticity and the angular velocity of the earth, the point O having been reduced to rest. Prove that the southern deviation measured on the tangent plane at O is $\sin\lambda\cos\lambda\, gt^4 \left(\frac{3}{8}\omega^2 + \frac{5}{24} C\right)$.

This result is obtained by substituting the approximate values of y and z obtained in Art. 618 in the small terms given in Art. 617. Expressions for the components of the attraction of the earth are to be found in treatises on the "figure of the earth" (see Stokes' *Mathematical and Physical Papers*, vol. II. p. 142). These give approximately (after some reduction) $C = (2m - \epsilon)\, 2g/a$, where $m = \omega^2 a/g$ and $\epsilon = 1/300$, hence $C = 2\omega^2$ nearly.

620. Two cases of motion. Two special cases of the motion of a particle deserve attention; (1) when the particle in its motion does not deviate far from the vertical and (2) when the motion is nearly horizontal.

Supposing the axis of z to be vertical, the horizontal velocities dx/dt and dy/dt are small compared with the vertical velocity dz/dt in the first case. The products of the horizontal velocities by ω are therefore of a higher order of small quantities than the product of the vertical velocity by ω and should be neglected in a first approximation.

In the second case, on the contrary, dz/dt is small and we neglect its product by ω. The two sets of equations are therefore as follows (Art. 614):

$$\left.\begin{aligned}\frac{d^2x}{dt^2}+2\frac{dz}{dt}\theta_2&=X,\\ \frac{d^2y}{dt^2}-2\frac{dz}{dt}\theta_1&=Y,\\ \frac{d^2z}{dt^2}&=-g+Z.\end{aligned}\right\}\qquad \left.\begin{aligned}\frac{d^2x}{dt^2}-2\frac{dy}{dt}\theta_3&=X,\\ \frac{d^2y}{dt^2}+2\frac{dx}{dt}\theta_3&=Y,\\ \frac{d^2z}{dt^2}-2\frac{dx}{dt}\theta_2+2\frac{dy}{dt}\theta_1&=-g+Z.\end{aligned}\right\}$$

We notice that when the motion is nearly vertical the components θ_1, θ_2 enter into the equations, while θ_3 does not appear until we proceed to higher approximations. It is therefore the component of the angular velocity about a tangent to the earth which affects the motion.

On the other hand when the motion of the particle is nearly horizontal it is the component of the earth's rotation about the vertical, viz. θ_3, which plays the principal part.

If we compare the x and y equations for the case in which the motion is nearly horizontal with those given in Art. 614, when the square of ω is neglected we see that they express the motion of a particle moving freely in space but referred to axes which turn round the vertical with an angular velocity θ_3. If, as is generally the case, the forces X, Y are either zero or independent of the changes of the nearly constant quantity z, *we can thus obtain these equations in an elementary way.* The particle moves freely in space, unaffected by the rotation of the earth, but the axes of reference move round the vertical and leave the particle behind. This geometrical interpretation of the equations may be made more evident by considering some simple cases.

621. As an example *consider the case of a pendulum.* When the bob makes small oscillations the motion is nearly horizontal. To construct the motion we suppose the pendulum to oscillate freely in space (with the proper initial conditions). This oscillation is left behind by the earth, and the effect is that the plane of oscillation appears to revolve about the vertical with an angular velocity equal and opposite to the vertical component of the earth's angular velocity. The plane of oscillation therefore turns from west to south with an angular velocity $\omega \sin \lambda$. This problem is more fully considered in Art. 624.

622. Flat trajectories. A bullet is projected from a gun, situated at the point O, with a great velocity V, in a direction making a small angle α with the horizon so that *the trajectory is nearly flat*. It is required to find the motion.

The initial velocity of the bullet in space (after O has been reduced to rest) is V. After leaving the gun the bullet describes a parabolic path in space, while the axes of reference turn with the earth round the vertical at O, and the bullet is left behind by the axes (Art. 620). Supposing that the initial plane of xz contains the direction of projection, the coordinates of the bullet at the time t are evidently

$$x = Vt\cos\alpha, \qquad y = -x\theta_3 t \text{ where } \theta_3 = -\omega\sin\lambda.$$

The deviation y is therefore always to the right of the plane of firing in the northern hemisphere, and to the left in the southern hemisphere. If R be the range the whole deviation is $Rt\omega\sin\lambda$. We notice also that the deviation y is independent of the azimuth of the plane of firing, and that the time of describing a given distance x is independent of the rotation of the earth.

The third equation of motion (Arts. 614, 615) gives

$$\frac{d^2z}{dt^2} = -g + 2\theta_2\frac{dx}{dt}, \qquad \therefore z = Vt\sin\alpha - \tfrac{1}{2}gt^2 - V\omega t^2\cos\alpha\cos\lambda\sin\beta,$$

where $\theta_2 = -\omega\cos\lambda\sin\beta$ and β is the angle the plane of firing makes with the meridian. The vertical deviation of the bullet from its parabolic path at the moment of reaching a target distant x from the gun is therefore $-xt\omega\cos\lambda\sin\beta$.

623. Deviation of a projectile. *Ex.* A particle is projected with a velocity V in a direction making an angle α with the horizontal plane, and the vertical plane through the direction of projection makes an angle β with the plane of the meridian, the angle β being measured from the south towards the west. If x is measured horizontally in the plane of projection, y horizontally in a direction making an angle $\beta + \frac{1}{2}\pi$ with the meridian, and z vertically upwards from the point of projection, prove that

$$x = V\cos\alpha t + (V\sin\alpha t^2 - \tfrac{1}{3}gt^3)\,\omega\cos\lambda\sin\beta,$$
$$y = (V\sin\alpha t^2 - \tfrac{1}{3}gt^3)\,\omega\cos\lambda\cos\beta + V\cos\alpha t^2\,\omega\sin\lambda,$$
$$z = V\sin\alpha t - \tfrac{1}{2}gt^2 - V\cos\alpha t^2\,\omega\cos\lambda\sin\beta,$$

where λ is the latitude of the place, and ω the angular velocity of the earth.

Prove also (1) that the increase of range on the horizontal plane through the point of projection is $4\omega\sin\beta\cos\lambda\sin\alpha\,(\frac{1}{3}\sin^2\alpha - \cos^2\alpha)\,V^3/g^2$,

(2) that the deviation to the right of the plane of projection is

$$4\omega\sin^2\alpha\,(\tfrac{1}{3}\cos\lambda\cos\beta\sin\alpha + \sin\lambda\cos\alpha)\,V^3/g^2,$$

and (3) that the time T of flight is decreased by $2T\cos\alpha\cos\lambda\sin\beta\,V\omega/g$.

It is not usual in practical gunnery to take account of the rotation of the earth except when V is very great, and then only the terms containing V are perceptible.

624. Disturbance of a pendulum. A particle of mass m is suspended by a fine wire of length l from a point O fixed relatively to the earth, and being drawn aside, so that the wire makes a small angle α with the vertical at O, is let go. It is required to find the motion; see Art. 621.

The equations of motion are those given in Art. 614. Taking the axis of z vertical and the origin at the position of equilibrium of the mass m we see that the ordinate z is less than $l(1-\cos\alpha)$, and the terms of the form $\theta\, dz/dt$ are of the order $l\omega\alpha^2$: these we shall reject. Let us also make the axes of x, y turn slowly round the vertical with such an angular velocity p relatively to the earth that $\theta_3 = -\omega\sin\lambda + p$ becomes zero, as explained in Art. 616. The equations of motion are now

$$\left.\begin{aligned} \frac{d^2x}{dt^2} &= -\frac{T}{m}\frac{x}{l}, \quad \frac{d^2y}{dt^2} = -\frac{T}{m}\frac{y}{l}, \\ \frac{d^2z}{dt^2} - 2\frac{dx}{dt}\theta_2 + 2\frac{dy}{dt}\theta_1 &= -g + \frac{T}{m}\frac{l-z}{l} \end{aligned}\right\} \text{........(1)},$$

where T is the tension of the string, and θ_1, θ_2 have the values given in Art. 616.

The third equation proves that the tension T differs from mg by quantities of the order $l\omega\alpha$ at least. Since x/l and y/l are of the order α, and we have agreed to reject terms of the order $\omega\alpha^2$, we must put $T = mg$ in the two first equations.

Since the two first equations are independent of ω, the motion of a real pendulum when affected by the rotation of the earth is the same as that of an ideal pendulum, unaffected by the rotation, but whose path, viewed by a spectator moving with the earth, appears to turn round the vertical with an angular velocity $p = \omega\sin\lambda$ in a direction south to west.

If $ln^2 = g$, the solutions of the equation are clearly

$$x = A\cos(nt + C), \quad y = B\sin(nt + D) \text{........(2)}.$$

It appears that the time of oscillation, viz. $2\pi/n$, is unaffected by the rotation of the earth. To determine the constants of integration, we notice that when the particle is drawn aside from the vertical and not yet liberated, it partakes of the velocity of the earth and has therefore a small velocity relative to the axes. This is equal to $-l\alpha\omega\sin\lambda$ and is transverse to the plane of displacement. Taking the plane of displacement as the plane of xz at the time $t=0$, the initial conditions are

$$x = l\alpha, \quad y = 0, \quad dx/dt = 0, \quad dy/dt = -l\alpha\omega\sin\lambda.$$

It is then easy to see that

$$A = l\alpha, \quad Bn = -l\alpha\omega \sin\lambda, \quad C = 0, \quad D = 0.$$

The particle therefore describes an ellipse whose semi-axes are A and $-B$. Since the ratio of the axes, viz. $\omega \sin\lambda \sqrt{(l/g)}$ is very small, the ellipse is very elongated and the particle appears to oscillate in a vertical plane. The effect of the rotation of the earth is to make this plane appear to turn round the vertical with an angular velocity $\omega \sin\lambda$.

625. It is known that, independently of all considerations of the rotation of the earth, the path of the bob of a pendulum is approximately an ellipse whose axes have a small nearly uniform motion round the vertical. This progression of the apses vanishes when the angle subtended at the point of suspension by either axis of the ellipse is zero; see Art. 566. As the presence of this progression will complicate the experiment, it is important (1) that the angle of displacement should be small, (2) that the pendulum when drawn aside should be liberated without giving the bob more transverse velocity than is necessary. This is usually effected by fastening the bob when displaced to some point fixed in earth by a thread, and when the mass has come to apparent rest it is set free by burning the thread. The progression of the apses due to the angular magnitude of the displacement is in the opposite direction to that caused by the rotation of the earth.

The advantage of using a long pendulum is that the linear displacement of the bob may be considerable though the angular displacement of the wire is very small. The bob should also be of some weight, for otherwise its motion would be soon destroyed by the resistance of the air; Art. 113.

626. As we have rejected some small terms it is interesting to examine if these could rise into importance on proceeding to solve the equations (1) to a second approximation. To determine this we substitute the first approximation of Art. 624 (2) in the differential equations. The third equation shows that $T/m - g$ has two sets of terms. First, there are terms independent of ω which lead to the solution already obtained in Art. 555, and need not be again considered here. Next, there are terms which contain ω as a factor and have the form $\sin(nt \pm \beta)$ where $\beta = pt$, Art. 616. These when multiplied by x/l or y/l give no terms of the form $\sin nt$ or $\cos nt$. None of the terms which contain ω can rise into importance (Art. 303).

627. The idea of proving the rotation of the earth by making experiments on falling bodies originated with Newton. But more than a hundred years elapsed before any observations of value were made. In 1791 Guglielmini of Bologna made some experiments in a tower 300 feet high. The liberation of the balls was effected by burning the thread by which they were suspended, and this was not done until they had entirely ceased to vibrate as observed by a microscope. The vertical was determined by a plumb line, but he had to wait several months before it came to rest. The results were disappointing for they showed a deviation towards the south nearly as great as that towards the east. This discrepancy was due to two causes, (1) the numerous apertures in the walls of the tower caused slight winds, (2) the vertical was not ascertained until a change in the seasons had altered its position. Other experiments were made by Benzenberg about 1802 in Hamburg, but Reich's experiments in 1831—3 in the mines of Freiberg are generally considered to be the most important. The height of the fall was $158\frac{1}{2}$ metres and the mean of 106 experiments gave a deviation to the east of $28\frac{1}{3}$ millimetres, the deviation to the south being about a twentieth of that towards the east. These were the experiments that Poisson selected to test the theory; he showed that the observed easterly deviation was within a thirtieth of that given by calculation. Poisson also investigates the general equations of motion of a particle relative to the earth and obtains equations equivalent to those given in Art. 617. He then applies them to a variety of problems. *Journal de l'école polytechnique*, 1838.

The defect of experiments on falling bodies is the smallness of the quantities to be measured. In 1851 Foucault invented a new method; he showed that the plane of oscillation of a simple pendulum appeared to rotate round the vertical with an angular velocity equal and opposite to the component of the earth's angular velocity. The advantage of this method is that the experiment can be continued through several hours, so that the slow deviation of the pendulum can be (as it were) integrated through a time long enough to make the whole displacement very large. Foucault's experiment was widely repeated with many improvements. Among English experiments we may mention those by Worms in 1859 at King's College, London, in Dublin by Galbraith and Haughton, at Bristol, at Aberdeen, at Waterford in 1895. The accuracy of the method is such that it is possible to deduce the time of rotation of the earth. Foucault's observations gave $23^h, 33^m, 57^s$, while the repetition of the experiment at Waterford led to $24^h, 7^m, 30^s$, the true time lying between the two (see *Engineering*, July 5, 1895). Though the experiment can be easily tried when only the general result is required, yet many difficulties arise when the deviation has to be found with accuracy. Indeed Foucault admitted that it was only after a long series of trials that he made the experiment succeed (see *Bulletin de la Société Astronomique de France*, Dec. 1896).

Inversion and Conjugate functions.

628. Inversion*. Let a point P of unit mass move under the action of forces whose potential in polar coordinates is $U=f(r, \theta, \phi)$. Produce any radius vector OP of the path to Q, where $OP \,.\, OQ=k^2$; the locus of Q is called the inverse path of that of P and any two points thus related are called inverse points. Let $OP=r$, $OQ=\rho$.

Let P', Q' be two other inverse points near the former, then since $OP \,.\, OQ = OP' \,.\, OQ'$, a circle can be described about the quadrilateral $PQP'Q'$. The elementary arcs PP', QQ' are therefore ultimately in the ratio $r : \rho$. If the points P, Q move so as

* The reader may consult a paper by Larmor in *The Proceedings of the London Mathematical Society*, vol. xv. 1884. The principle of least action is there applied to both the method of Inversion and that of Conjugate functions.

to be always inverse points, their velocities u, u_1, are connected by the equation $u/u_1 = r/\rho$.

The position of the point P in space is determined either by the quantities (ρ, θ, ϕ) or (r, θ, ϕ). Choosing the former as the coordinates, the Lagrangian equations of the motion of P are deduced from

$$T = \tfrac{1}{2}u^2 = \tfrac{1}{2}\frac{k^4}{\rho^4}(\rho'^2 + \rho^2\theta'^2 + \rho^2 \sin^2\theta\phi'^2),$$

$$U + C = f\left(\frac{k^2}{\rho}, \theta, \phi\right) + C.$$

These equations contain only the polar coordinates of Q. They primarily give the motion of a point Q describing the inverse path in such a manner that P and Q are always at inverse points.

Let us now transpose the factor k^4/ρ^4 from T to U. We then have (Art. 524)

$$T_2 = \tfrac{1}{2}(\rho'^2 + \&c.), \quad U_2 = \frac{k^4}{\rho^4}\left\{f\left(\frac{k^2}{\rho}, \theta, \phi\right) + C\right\}.$$

The Lagrangian equations derived from these give the motion of a particle which describes the same path as that of Q, but in a different time. Let the particle be called Π. The form of T_2 shows that Π moves as a free particle, acted on by forces whose potential is U_2. We see also that the masses of the particles P and Π are equal. See also Art. 650, Ex. 2.

The path of either particle may be inferred from that of the other. *If the path of the particle P described with a work function $f(r, \theta, \phi) + C$ is known, then the other particle Π, if properly projected, will describe the inverse path, with a work function*

$$U_2 = \frac{k^4}{\rho^4}\left\{f\left(\frac{k^2}{\rho}, \theta, \phi\right) + C\right\}.$$

629. To find the relation between the velocities u, v of the particles P, Π, when passing through any inverse points P, Q, we notice that by the principle of vis viva $\tfrac{1}{2}u^2 = U + C$, $\tfrac{1}{2}v^2 = U_2$. It follows immediately that $v = uk^2/\rho^2$, and therefore that $ur = v\rho$. Since the planes of motion OPP', OQQ' coincide, *the angular momenta of the particles, when at inverse points of their paths, about every axis through the centre of inversion are equal.*

The constant C is determined by the consideration that the known velocity u in the given path must satisfy the equation $\frac{1}{2}u^2 = U + C$.

The particles P, Π do not necessarily pass through inverse points of their respective paths at the same instant. Let t, τ be the times at which they pass through any pair P, Q, of inverse points; $t+dt$, $\tau+d\tau$ the times at which they pass through a neighbouring pair P', Q' of inverse points. Since the elementary arcs PP', QQ' are in the ratio $r : \rho$ while the velocities of P, Π are in the ratio $1/r : 1/\rho$, it follows by division that the elementary times dt, $d\tau$ are in the ratio $r^2 : \rho^2$. The relation between t and τ is found by integration from $\dfrac{dt}{d\tau} = \dfrac{r^2}{\rho^2}$. This agrees with the ratio given in Art. 524.

Supposing that the particles P, Π are projected from inverse points on their respective paths, their initial velocities must be inversely as their distances from the centre O of inversion. The initial directions of motion must be in the same plane and make supplementary angles with the radius vector which passes through both the initial positions.

630. If the particle P is constrained to move on a surface the argument needs but a slight alteration. The inverse point Q describes a curve which lies on the inverse surface. Let (ρ, θ, ϕ) be the polar coordinates of Q; then these may also be taken as the Lagrangian coordinates of P. Using the equation of the inverse surface, we have $\rho' = \dfrac{d\rho}{d\theta}\theta' + \dfrac{d\rho}{d\phi}\phi'$. Substituting the values of ρ, ρ' in the expressions for T and $U+C$ given in Art. 628, we proceed as before and arrive at similar results.

631. The Pressures. When the particles P, Π are constrained to move on a surface and the inverse surface respectively, *the pressures R_1, R_2, at any pair of inverse points are such that $R_1 r^3 = R_2 \rho^3$.*

To prove this we take any axis of z and resolve the forces on the particles perpendicularly to the meridian plane $zOPQ$, Art. 491. We then have

$$\frac{1}{r\sin\theta}\frac{dA}{dt} = \frac{1}{r\sin\theta}\frac{dU}{d\phi} + R_1\cos\alpha_1,$$

$$\frac{1}{\rho\sin\theta}\frac{dA}{d\tau} = \frac{1}{\rho\sin\theta}\frac{dU_2}{d\phi} + R_2\cos\alpha_2,$$

where A is the angular momentum of either particle about the axis of z, Art. 629, and dt, $d\tau$ are the times respectively occupied by the particles in passing from any pair of inverse points to an adjoining pair.

The forces R_1, R_2 act along the normals to the two surfaces. To understand the geometrical relations, we describe a sphere passing through P, Q and touching one surface. Then since the sphere has the property that for every chord the product $OP \,.\, OQ$ is the same, the sphere will touch the inverse surface also. The normals therefore meet in the centre of the sphere and will make equal angles with every straight line perpendicular to the radius vector OPQ. The angles a_1, a_2 of resolution are therefore equal, if the reactions are taken positively towards the centre of the sphere.

Since $\rho^2 dt = r^2 d\tau$ and $\rho^2 U_2 = r^2 (U + C)$, we see at once that $r^3 R_1 = \rho^3 R_2$. Since $ur = v\rho$, we have $R_1/u^3 = R_2/v^3$, i.e. the pressures at inverse points are also as the cubes of the velocities.

Ex. Deduce from the relations $\rho^2 U_2 = r^2 (U + C)$, $\quad \frac{1}{2}u^2 = U + C$,

(1) that the parallel components G, G' of the impressed forces on the particles P, Π in any direction perpendicular to the radius vector are connected by the equation $\rho^3 G' = r^3 G$.

(2) that the radial components F, F', are connected by $\rho^3 F' + r^3 F = -4r^2 (U + C)$.

632. *Ex.* 1. The path of a free particle under the action of no forces is a straight line; in this case we have $u^2 = 2U = 2C$. By inversion the path of a free particle, when $v^2 = u^2 \dfrac{r^2}{\rho^2} = 2U_2$, is the inverse of a straight line, i.e. a circle passing through the origin. This gives $U_2 = Ck^4/\rho^4$, and the central force $F = 4Ck^4/\rho^5$. This is Newton's theorem that a circle can be described freely about a centre of force on the circumference whose attraction varies as the inverse fifth power of the distance.

Ex. 2. Show that a particle can describe the curve $\rho^2 = a^2 \cos^2 \theta + b^2 \sin^2 \theta$ under the action of a force F in the origin which varies as $\dfrac{1}{\rho^5} \left\{ \dfrac{1}{a^2} + \dfrac{1}{b^2} - \dfrac{3}{2} \dfrac{1}{\rho^2} \right\}$.

When the axes a, b of the curve are so unequal that their ratio is greater than $\sqrt{2}$, the force F changes from attraction to repulsion as the particle proceeds from the extremity of one axis to the other. Verify this by tracing the curve, and show that the curve is convex at the extremity of the lesser axis.

Ex. 3. Prove that the central forces F, F', under the action of which a curve and its inverse can be described about the centre of inversion are so related that $\dfrac{F'r'^3}{h'^2} + \dfrac{Fr^3}{h^2} = 2\dfrac{r^2}{p^2}$; show also that the velocities v, v' at inverse points are connected by $vr = v'r'$. [This follows easily from the expression for F given in Art. 310. When $h = h'$, Art. 629, this agrees with Art. 631, Ex.]

Ex. 4. A particle P moves on a sphere under the action of a centre of attractive force situated at a point O on the surface, and the velocity v at any point is B/r^2 where $r = OP$. Prove that the path is a circle whose plane passes through O.

Inverting the sphere, we find that the stereographic projection is a straight line. The result follows at once, see Art. 609.

633. Conjugate functions. Let the Cartesian coordinates (x, y), (ξ, η) of two corresponding points P, Q be so related that

$$x + yi = f(\xi + \eta i) \qquad\qquad (1),$$

where f is any real function and $i = \sqrt{(-1)}$. Expanding the right-hand side we have

$$x + yi = \phi(\xi, \eta) + \psi(\xi, \eta)\, i \text{.................(2)},$$

where ϕ and ψ are real functions. The transformation is therefore effected by using the equations

$$x = \phi(\xi, \eta), \quad y = \psi(\xi, \eta) \text{.................(3)},$$

the motion of P following geometrically from that of Q. Differentiating (1) we find

$$x' + y'i = f'(\xi + \eta i) \,.\, \{\xi' + \eta' i\},$$
$$x' - y'i = f'(\xi - \eta i) \,.\, \{\xi' - \eta' i\};$$
$$\therefore\ x'^2 + y'^2 = \mu^2 \,.\, \{\xi'^2 + \eta'^2\} \text{.................(4)},$$

where μ^2 is a real positive quantity given by

$$\mu^2 = f'(\xi + \eta i) \,.\, f'(\xi - \eta i) \text{.................(5)}.$$

Let $U = F(x, y)$ be the work function of the forces which act on the particle P. The motions of P and Q may be deduced by the Lagrangian rule from

$$T = \tfrac{1}{2}\mu^2(\xi'^2 + \eta'^2), \quad U = F\{\phi(\xi, \eta), \psi(\xi, \eta)\},$$

the constant of U being included in F for the sake of brevity.

Transposing the factor μ^2 to the work function, the equations

$$T_2 = \tfrac{1}{2}(\xi_1^2 + \eta_1^2), \quad U_2 = \mu^2 F(\phi, \psi),$$

give by the same rule the motion of a particle Π, whose mass is equal to that of P, which (when properly projected) will describe the same path as the point Q, but in a different time, Art. 524.

To find the relation between the velocities u, v of the particles P, Π at corresponding points of their paths, we observe that since $\tfrac{1}{2}u^2 = U$, $\tfrac{1}{2}v^2 = U_2$, the velocities are such that $v = \mu u$.

To find the ratio of the times dt, $d\tau$ we notice that, by (4), the corresponding arcs ds, $d\sigma$ are such at $ds = \mu d\sigma$, while $\mu u = v$. It follows by division that $dt = \mu^2 d\tau$.

634. *Ex.* It is known that a particle can describe the ellipse $x^2/a^2 + y^2/b^2 = 1$, with a force tending to the centre equal to κr. It is required to find the conjugate path and law of force when we use the transformation $x \pm yi = (\xi \pm \eta i)^n / c^{n-1}$.

Let $x = r\cos\theta$, $y = r\sin\theta$; $\xi = \rho\cos\phi$, $\eta = \rho\sin\phi$; the equation of transformation then gives

$$r = \rho^n / c^{n-1}, \qquad \theta = n\phi.$$

The equation of the path is therefore

$$\frac{\cos^2 n\phi}{a^2} + \frac{\sin^2 n\phi}{b^2} = \frac{c^{2n-2}}{\rho^{2n}}.$$

Also, $$\mu^2=f'(\xi+\eta i)f'(\xi-\eta i)=n^2(\xi^2+\eta^2)^{n-1}/c^{2n-2};$$
$$\therefore\ \mu=n\rho^{n-1}/c^{n-1}.$$
Again in the elliptic orbit,
$$u^2=2(U+C)=\kappa(a^2+b^2-r^2).$$
Hence since $v=\mu u$,
$$v^2=2U_2=\frac{n^2\kappa}{c^{2n-2}}\rho^{2n-2}\left(a^2+b^2-\frac{\rho^{2n}}{c^{2n-2}}\right);$$
$$\therefore\ -F=\frac{dU_2}{d\rho}=\frac{n^2\kappa}{c^{2n-2}}\rho^{2n-3}\left\{(n-1)(a^2+b^2)-\frac{(2n-1)\rho^{2n}}{c^{2n-2}}\right\}.$$
The ratio of the angular momenta, viz. $v\rho/ur$, is easily seen to be equal to n.

When $n=-1$, this transformation becomes $r=c^2/\rho$, $\theta=-\phi$. The transformation reduces to a simple inversion, except that ϕ is measured positively in the opposite direction to θ.

635. *Ex.* If the particle P is constrained to move on any given curve with a work function U, while the equal particle Π is constrained to move on the conjugate curve, with a work function $U_2=\mu^2U$, the pressures R_1, R_2 on the two curves are in the ratio of the cubes of the velocities, i.e. $R_1/u^3=R_2/v^3$. This gives also $R_2=\mu^3R_1$.

The grouping of trajectories and Jacobi's solution.

636. The Cartesian equations of the motion of a free particle of unit mass are
$$x''=\frac{dU}{dx},\quad y''=\frac{dU}{dy},\quad z''=\frac{dU}{dz}\ \ \ldots\ldots\ldots\ldots(1),$$
and to these we join the equation of energy
$$v^2=x'^2+y'^2+z'^2=2U+2C\ \ldots\ldots\ldots\ldots\ldots(2).$$
When the equations (1) have been integrated we have x, y, z expressed by three functions of t with six constants whose values become known when the initial values a, b, c of the coordinates and the initial velocities a', b', c' are given.

Since t enters into the equations (1) only in the form dt, the differential equations are not altered by writing $t+\epsilon$ for t. One of the constants of integration therefore enters into the solution as a mere addition to the time. When we eliminate the time we arrive at two equations which are the equations of all the possible trajectories in space. The constant ϵ disappears with t, and the equations of the possible trajectories contain five constants, of which the energy C may be regarded as one. To understand the relations of these trajectories to each other it becomes necessary to group them into systems.

We first group the trajectories according to the values of the energy C. Taking any one group, having any given energy, the four remaining constants are determined for any special trajectory when the coordinates of some two points A, B arbitrarily chosen on it are given.

637. Action. If ds be an element of the arc of the trajectory, the integral $V=\int mvds$ is called *the action* as the particle passes from A to B. If mv^2 be the vis viva of the particle in any position we also have $V=\int mv^2dt$, the limits being the times t_1 and t of passing through A and B. When we are only concerned with the motion of a single particle, it is convenient to suppose its mass to be taken as unity.

Considering a single particle, let s be measured from A to B along the trajectory of least action and let the length AB be l. Let $A'B'$ be a neighbouring trajectory (Art. 590) from some point A' near A to a point B' near B. Proceeding as in Art. 591, writing v for ϕ, we find

$$\delta V=\left[v\frac{dx}{ds}\delta x+\&\text{c.}\right]+\int\left[\left\{\frac{dv}{dx}-\frac{d}{ds}\left(v\frac{dx}{ds}\right)\right\}\delta x+\&\text{c.}+\frac{dv}{dC}\delta C\right]ds\ldots(3),$$

where the part outside the integral is to be taken between the limits A and B and the energy C has been varied for the sake of generality. It is easy to deduce from the equations of motion (as in Art. 599) that the coefficients of δx, δy, δz inside the integral are zero. Also since $\frac{1}{2}v^2=U+C$, we have $vdv/dC=1$. Since vdx/ds is the x component of the velocity we thus have

$$\delta V=x'\delta x+y'\delta y+z'\delta z-a'\delta a-b'\delta b-c'\delta c+(t-t_1)\,\delta C\ldots(4).$$

When we consider the motion of a system of particles, either constrained or free, and all taking different paths, it is more convenient to take t as the independent variable. Let us imagine the system to be moving in some manner which we will call the actual course. Let the work function of the field be U and let L be the Lagrangian function, then $L=T+U$ (Art. 506). Let θ_1, θ_2, &c. be any independent coordinates of the system, a_1, a_2, &c. their values in some position A occupied by the system at a time t_1. Then θ_1, θ_2, &c. are functions of t, whose forms it is our object to discover.

Let us next suppose the system to move in some varied manner, i.e. let the coordinates be functions of t slightly different from those in the actual course. By

the fundamental theorem* in the calculus of variations, we have

$$\delta\int Ldt=\left[L\delta t+\Sigma\frac{dL}{d\theta'}\omega\right]_{t_1}^{t}+\int\Sigma\left(\frac{dL}{d\theta}-\frac{d}{dt}\frac{dL}{d\theta'}\right)\omega dt,$$

where $\omega=\delta\theta-\theta'\delta t$, Σ implies summation for all the coordinates θ_1, θ_2, &c. and the limits of integration are t_1 and t. Since each separate term inside the integral vanishes by Lagrange's equations (Art. 506), we have

$$\delta\int Ldt=\left[(T+U)\delta t+\Sigma\frac{dT}{d\theta'}(\delta\theta-\theta'\delta t)\right]_{t_1}^{t}.$$

If the geometrical conditions do not contain the time explicitly T will be a homogeneous function of θ_1', θ_2', &c. (Art. 510) and therefore $\Sigma\frac{dT}{d\theta'}\theta'=2T$. We also suppose that for each varied course the velocities are so arranged that the principle of energy holds, i.e. $T-U=C$, though C may be different for each course. Hence $L=2T-C$, and $\delta\int Cdt=\delta\{C(t-t_1)\}$. We now have the two equations

$$\delta\int Ldt=-C(\delta t-\delta t_1)+\Sigma\left(\frac{dT}{d\theta'}\delta\theta\right)-\Sigma\left(\frac{dT}{da'}\delta a\right)\quad\text{.................(A)}$$

$$\delta\int 2Tdt=(t-t_1)\,\delta C+\Sigma\left(\frac{dT}{d\theta'}\delta\theta\right)-\Sigma\left(\frac{dT}{da'}\delta a\right)\text{.....................(B).}$$

The action V of the system is the sum of the actions of the several particles. We therefore have $V=\int 2Tdt$. When the system reduces to a single particle of unit mass $2T=x'^2+y'^2+z'^2$, and the equation (B) becomes the same as (4)

638. Let us consider the motion of a single free particle and *let the energy C be given*, therefore $\delta C=0$. Let v_1, v_2 be the velocities at A, B; $\delta\sigma_1$, $\delta\sigma_2$ the displacements AA', BB'; θ_1, θ_2 the angles these displacements make with the positive directions of the tangents at A, B; then, as in Art. 592, (4) becomes

$$\delta V=v_2\cos\theta_2\delta\sigma_2-v_1\cos\theta_1\delta\sigma_1\quad\text{...........(IV).}$$

* The proof of this theorem is as follows. We have

$$\delta\int Ldt=\int(\delta Ldt+Ld\delta t)=[L\delta t]+\int(\delta Ldt-dL\delta t).$$

Now L is a function of the letters typified by θ, θ',

$$\therefore\ \delta L=\Sigma(L_\theta\delta\theta+L_{\theta'}\delta\theta'),\qquad dL=\Sigma(L_\theta d\theta+L_{\theta'}\theta''dt),$$

where suffixes imply partial differential coefficients. Since

$$\delta\frac{d\theta}{dt}=\frac{d\theta+d\delta\theta}{dt+d\delta t}-\frac{d\theta}{dt}=\frac{d\delta\theta}{dt}-\frac{d\theta}{dt}\frac{d\delta t}{dt},$$

$$\therefore\ \delta\theta'-\theta''\delta t=\frac{d}{dt}(\delta\theta-\theta'\delta t)=\omega',$$

substituting we find

$$\delta\int Ldt=[L\delta t]+\int\Sigma(L_\theta\omega+L_{\theta'}\omega')\,dt.$$

Integrating the last term by parts we immediately obtain the theorem in the text.

Introducing the mass m, this may be read, *the change of the action in passing from one trajectory AB to a neighbouring one is the difference of the virtual moments of the momenta at the two ends.*

Taking any arbitrary surface which we may call S_1, let us group together all the trajectories which cut S_1 orthogonally; then $\cos\theta_1 = 0$. On each of these trajectories let us take the point B so that the action from the surface S_1 to B is some given quantity. As we pass from one trajectory to a neighbouring one, B traces out a second surface which we may call S_2, and at every point of S_2 we have $\delta V = 0$. It follows that for this surface (supposing it to be of finite extent) $\cos\theta_2$ is also zero. The trajectories therefore intersect the surface S_2 at right angles.

Considering all possible trajectories we first group them according to the value of the energy. We classify them again by selecting all those at right angles to some given surface. We have now a congruence of trajectories. The theorem just proved asserts that all these trajectories can be cut orthogonally by a system of surfaces. These orthogonal surfaces are such that, when any two are given, the action from one to the other is the same for all the trajectories. See Thomson and Tait, *Treatise on Natural Philosophy*, 1879, vol. I. Art. 332.

All possible trajectories may be grouped together in the manner just described in many different ways. One method is to select a surface intersecting all the trajectories. Each point of this surface may be regarded as the centre of an infinitely small sphere which all the trajectories intersect at right angles. The surface S_1 is then reduced to a collection of points occupying an arbitrary surface. This is the method of grouping adopted in Arts. 159, 330, 339, &c. By a different grouping we obtain different orthogonal surfaces.

639. These considerations lead us to a rule which is a special case of that given by Jacobi for the solution of dynamical problems. When this method is applied to the dynamics of a particle the orthogonal surfaces are investigated first and the trajectories are afterwards deduced. In the general case of a system of rigid bodies the interpretation is not so simple.

640. *Let the action V be expressed as a function of the energy C and of the coordinates (x, y, z), (a, b, c) of the particle in the two arbitrary positions B and A.* Then by the principles of the differential calculus,

$$dV = \frac{\delta V}{dx}\delta x + \frac{dV}{dy}\delta y + \frac{dV}{dz}\delta z + \frac{dV}{da}\delta a + \frac{dV}{db}\delta b + \frac{dV}{dc}\delta c + \frac{dV}{dC}\delta C \ldots(5),$$

the energy being varied for the sake of generality. Comparing this with the expression (4) (Art. 637) we see that

$$x' = \frac{dV}{dx}, \text{ \&c., \&c.,} \quad a' = -\frac{dV}{da}, \text{ \&c., \&c.,} \quad t - t_1 = \frac{dV}{dC} \ldots(6).$$

Substituting in the equation (2) of energy, we find

$$\left(\frac{dV}{dx}\right)^2 + \left(\frac{dV}{dy}\right)^2 + \left(\frac{dV}{dz}\right)^2 = 2U + 2C, \quad \left(\frac{dV}{da}\right)^2 + \left(\frac{dV}{db}\right)^2 + \left(\frac{dV}{dc}\right)^2 = 2U_0 + 2C \ldots(7),$$

where U_0 is the value of U when we write for x, y, z their initial values a, b, c. These are called *the Hamiltonian equations of motion.*

It is obvious that if we can deduce from the equations (7) the proper form for the function V, the first set of (6) will give the component velocities of the particle and the second set will give the relations between the coordinates x, y, z and their initial values. The last equation will give the time.

Jacobi proved that it is not necessary to obtain the general integral of either differential equation. It is sufficient to discover one solution of the form

$$V = f(x, y, z, C, \alpha, \beta) + \gamma \ldots\ldots\ldots\ldots\ldots\ldots(8),$$

containing three new constants α, β, γ. He also proved that the introduction of the initial coordinates a, b, c into the expression for V is unnecessary. Instead of these he uses the two constants of integration here called α, β.

641. In the first differential equation (7) and in the complete integral (8), the quantities x, y, z are the independent variables. Jacobi's rule asserts that *if we establish the following relations between x, y, z and a new variable t, the equations of motion* (1) *will be satisfied.* These assumed relations are

$$\frac{df}{d\alpha} = -\alpha_1, \quad \frac{df}{d\beta} = -\beta_1, \quad \frac{df}{dC} = t + \epsilon \ldots\ldots\ldots\ldots(9),$$

where α_1, β_1, and ϵ are three new constants. These new relations make x, y, z functions of t, C and the five constants $\alpha, \beta, \alpha_1, \beta_1$, and ϵ.

To prove these relations we differentiate (9) with regard to t and thus arrive at three equations of the form

$$x' \frac{d^2 f}{dx\,d\alpha} + y' \frac{d^2 f}{dy\,d\alpha} + z' \frac{d^2 f}{dz\,d\alpha} = 0 \text{..............(10)}.$$

The other equations have β and C written for α, but in the third the zero on the right-hand side is replaced by unity. These equations determine x', y', z'.

Also since (8) is a solution of the first of the differential equations (7), it must satisfy that equation identically. We may therefore differentiate (7) *after substitution* with regard to each of the constants α, β, C. We thus arrive at three equations of the form

$$\frac{df}{dx}\frac{d^2 f}{dx\,d\alpha} + \frac{df}{dy}\frac{d^2 f}{dy\,d\alpha} + \frac{df}{dz}\frac{d^2 f}{dz\,d\alpha} = 0 \text{...........(11)}.$$

The other equations have β and C written for α, but in the third the zero is replaced by unity.

Comparing the three equations (10) with the three (11), we see at once that

$$x' = \frac{df}{dx}, \quad y' = \frac{df}{dy}, \quad z' = \frac{df}{dz} \text{..............(12)}.$$

It also follows that

$$x'' = \frac{d^2 f}{dx^2} x' + \frac{d^2 f}{dx\,dy} y' + \frac{d^2 f}{dx\,dz} z' \text{..............(13)},$$

with similar expressions for y'', z''.

We may also differentiate (7) after substitution from (8) partially with respect to any one of the three variables x, y, z;

$$\therefore \frac{df}{dx}\frac{d^2 f}{dx^2} + \frac{df}{dy}\frac{d^2 f}{dx\,dy} + \frac{df}{dz}\frac{d^2 f}{dx\,dz} = \frac{dU}{dx}.$$

Substituting from (12), the left-hand side becomes by (13) equal to x''. We therefore have

$$x'' = \frac{dU}{dx}, \quad y'' = \frac{dU}{dy}, \quad z'' = \frac{dU}{dz},$$

which are the equations of motion (1).

642. Consider the system of surfaces defined by

$$f(x, y, z, C, \alpha, \beta) = K \text{......................(14)},$$

where C, α, β are constants and K the parameter. The equations (12) prove that the direction of motion at any point is normal to that surface of the system which passes through the point. Thus *the surfaces* (14) *cut the trajectories at right angles.* These trajectories (with their parameters α_1, β_1) may be deduced from (14) by the rules given in the theory of differential equations or more easily by Jacobi's equations (9).

The trajectories in Jacobi's method are thus grouped together according to their orthogonal surfaces. By taking different complete integrals for (8), we group the same trajectories in different ways. Art. 638.

643. As an example which requires no long algebraical process, let us discuss the trajectories when the forces are absent. The Hamiltonian equation is

$$\left(\frac{dV}{dx}\right)^2 + \left(\frac{dV}{dy}\right)^2 + \left(\frac{dV}{dz}\right)^2 = 2C \text{ .... (15)}.$$

One complete integral, suggested by the rules for solving differential equations, is

$$V = \{\alpha x + \beta y + \sqrt{(1 - \alpha^2 - \beta^2)}\, z\} \sqrt{(2C)} + \gamma \text{ .... (16)},$$

another complete integral is

$$V = \{(x - \alpha)^2 + (y - \beta)^2 + z^2\}^{\frac{1}{2}} \sqrt{(2C)} \text{........................(17)}.$$

If we choose the first integral the surfaces $V = K$ are planes and the trajectories are grouped into systems of parallel lines, the lines taking all directions. If we choose the second integral, the surfaces $V = K$ are spheres having their centres on the plane of xy. The trajectories are grouped into systems of straight lines diverging from points on that plane.

To illustrate the use of equations (9) let us substitute in them the second integral. We have at once

$$\frac{x - \alpha}{r} = -\alpha_1, \quad \frac{y - \beta}{r} = -\beta_1, \quad \frac{r}{\sqrt{(2C)}} = t + \epsilon \text{ .... (18)},$$

where $r^2 = (x - \alpha)^2 + (y - \beta)^2 + z^2$. These evidently give a system of straight lines diverging from the point $x = \alpha$, $y = \beta$, $z = 0$, described with a velocity $\sqrt{(2C)}$.

644. When the coordinates chosen are not Cartesian the expression for the kinetic energy does not take the simple form given in (2). Let the kinetic energy T be given by

$$2T = P\theta'^2 + Q\phi'^2 + R\psi'^2 \text{..................(19)},$$

where P, Q, R are functions of the coordinates θ, ϕ, ψ. Let us now take as the Hamiltonian equation

$$\frac{1}{P}\left(\frac{dV}{d\theta}\right)^2 + \frac{1}{Q}\left(\frac{dV}{d\phi}\right)^2 + \frac{1}{R}\left(\frac{dV}{d\psi}\right)^2 = 2U + 2C \text{.....(20)}.$$

Proceeding exactly in the same way as before, we prove that if

$$V = f(\theta, \phi, \psi, C, \alpha, \beta) + \gamma \text{ ....(21)},$$

be an integral of (20), the first integrals of the Lagrangian equations of motion (Art. 506), are

$$P\theta' = \frac{df}{d\theta}, \quad Q\phi' = \frac{df}{d\phi}, \quad R\psi' = \frac{df}{d\psi} \text{ .... (22)}.$$

The trajectories, &c. are given by

$$\frac{df}{d\alpha} = -\alpha_1, \quad \frac{df}{d\beta} = -\beta_1, \quad \frac{df}{dC} = t + \epsilon \text{ ....(23)},$$

where α_1, β_1, and ϵ are new constants.

This enunciation includes the most useful cases of Jacobi's rule. But his method applies also to any dynamical system, in which T is a quadratic function of the velocities. For these generalizations we refer the reader to treatises on Rigid Dynamics.

645. *Ex.* 1. Apply Jacobi's rule to find the path of a projectile.

The Hamiltonian equation is

$$\left(\frac{dV}{dx}\right)^2 + \left(\frac{dV}{dy}\right)^2 = -2gy + 2C.$$

Separating the variables, we find that one complete integral is

$$V = \sqrt{(2\alpha)}\, x - \frac{1}{3g}(2C - 2\alpha - 2gy)^{\frac{3}{2}} + \gamma.$$

Ex. 2. Apply Jacobi's method to find the path of a particle in three dimensions about a fixed centre of force which attracts according to the Newtonian law.

Taking polar coordinates we have

$$2T = r'^2 + r^2\theta'^2 + r^2 \sin^2\theta\phi'^2, \qquad U = \frac{\mu}{r}.$$

The Hamiltonian equation (Art. 644) may be put into the form

$$\left\{r^2\left(\frac{dV}{dr}\right)^2 - 2\mu r - 2Cr^2\right\} + \left(\frac{dV}{d\theta}\right)^2 + \frac{1}{\sin^2\theta}\left(\frac{dV}{d\phi}\right)^2 = 0.$$

If we equate these three expressions respectively to α, $-\alpha + \beta \operatorname{cosec}^2\theta$ and $-\beta \operatorname{cosec}^2\theta$, we obtain three differential equations in which the variables are separated and whose solutions satisfy the Hamiltonian equation. Let the integrals of these be $V = f_1(r, \alpha)$, $V = f_2(\theta, \alpha, \beta)$, $V = f_3(\phi, \beta)$. It is obvious that $V = f_1 + f_2 + f_3 + \gamma$ is a complete integral from which all the trajectories may be deduced.

Ex. 3. Apply Jacobi's method to find the motion of a particle in elliptic coordinates (λ, μ, ν) when the work function is

$$U = \frac{(\mu^2 - \nu^2) f_1(\lambda) + (\nu^2 - \lambda^2) f_2(\mu) + (\lambda^2 - \mu^2) f_3(\nu)}{(\lambda^2 - \mu^2)(\mu^2 - \nu^2)(\nu^2 - \lambda^2)}.$$

Taking the expression for T given in Art. 577, the Hamiltonian equation (Art. 644) after a slight reduction becomes

$$(\lambda^2-h^2)(\lambda^2-k^2)(\mu^2-\nu^2)\left(\frac{dV}{d\lambda}\right)^2+(\mu^2-h^2)(\mu^2-k^2)(\nu^2-\lambda^2)\left(\frac{dV}{d\mu}\right)^2+(\nu^2-h^2)(\nu^2-k^2)(\lambda^2-\mu^2)\left(\frac{dV}{d\nu}\right)^2$$

$$=-2\{(\mu^2-\nu^2)f_1(\lambda)+(\nu^2-\lambda^2)f_2(\mu)+(\lambda^2-\mu^2)f_3(\nu)\}-2CD,$$

where $D=(\lambda^2-\mu^2)(\mu^2-\nu^2)(\nu^2-\lambda^2)$. Since

$$(\mu^2-\nu^2)+(\nu^2-\lambda^2)+(\lambda^2-\mu^2)=0,$$

$$\lambda^2(\mu^2-\nu^2)+\mu^2(\nu^2-\lambda^2)+\nu^2(\lambda^2-\mu^2)=0,$$

$$\lambda^4(\mu^2-\nu^2)+\mu^4(\nu^2-\lambda^2)+\nu^4(\lambda^2-\mu^2)=-D,$$

the differential equation is satisfied by assuming

$$(\lambda^2-h^2)(\lambda^2-k^2)\left(\frac{dV}{d\lambda}\right)^2=-2f_1(\lambda)+\alpha+\beta\lambda^2+2C\lambda^4,$$

with similar expressions for $dV/d\mu$ and $dV/d\nu$. In these trial solutions the variables λ, μ, ν have been separated, the first containing λ, the second μ, and the third ν. Supposing the integrals to be $V=F_1(\lambda, \alpha, \beta, C)$, $V=F_2(\mu, \&c.)$, $V=F_3(\nu, \&c.)$, the required complete integral is then $V=F_1+F_2+F_3+\gamma$. The solution then follows by simple differentiations with regard to the constants α, β, C.

This expression for U is given by Liouville in his *Journal*, vol. XII. 1847. He uses it in conjunction with Jacobi's solution.

We may also write the expression in a different form. Let p_1, p_2, p_3 be the perpendiculars from the origin on the tangent planes to the three confocals which intersect in any point, and let λ, μ, ν be as before the semi-major axes. We find by using the expressions for these perpendiculars in elliptic coordinates (Art. 577)

$$U=p_1^2F_1(\lambda)+p_2^2F_2(\mu)+p_3^2F_3(\nu).$$

Taking $U=p^2F(\lambda)$, (omitting the suffixes) we see at once that the level surfaces intersect the ellipsoids in the polhodes. The direction of the force at any point P is therefore normal to the polhode which passes through P. It may be shown by differentiation that the components, T and N, of the force, tangential and normal to the ellipsoid which passes through P, are

$$T=-2p^4F(\lambda)\{S_8-p^2S_6^2\}^{\frac{1}{2}}=-2p^6F(\lambda)\left\{\Sigma\frac{x^2y^2(\lambda^2-b^2)^2}{\lambda^8b^8}\right\}^{\frac{1}{2}},$$

$$N=2p^5F(\lambda)S_6+\frac{p^3}{\lambda}F'(\lambda),$$

where $S_n=\dfrac{x^2}{\lambda^n}+\dfrac{y^2}{b^n}+\dfrac{z^2}{c^n}$. The Cartesian components X, Y, Z are

$$X=\frac{2p^4x}{\lambda^2}\left\{-\frac{1}{\lambda^2}+2p^2S_6\right\}F(\lambda)+\frac{p^4x}{\lambda^2}\frac{F'(\lambda)}{\lambda},$$

with similar expressions for Y and Z.

We may obtain simpler expressions by combining the three terms of U. Putting $f_1(\lambda)=-\lambda^{2n+4}$, $f_2(\mu)=-\mu^{2n+4}$, $f_3(\nu)=-\nu^{2n+4}$, we see that U is equal to the sum of the different homogeneous products of λ^2, μ^2, ν^2 of n dimensions, each product being taken with a coefficient unity. This symmetrical function of the roots of the cubic in Art. 576 may be expressed as a rational function of the coefficients. We thus find possible forms for U in Cartesian coordinates. For example, putting $f_1(\lambda)=-\lambda^6$ &c., we find

$$U=\lambda^2+\mu^2+\nu^2=(x^2+y^2+z^2)+A.$$

As another example, put $f_1(\lambda) = -\lambda^8$ &c., we then have

$$U = \lambda^4 + \mu^4 + \nu^4 + \lambda^2\mu^2 + \mu^2\nu^2 + \nu^2\lambda^2$$
$$= (x^2+y^2+z^2)^2 + (x^2+y^2+z^2)(h^2+k^2) + h^2y^2 + k^2z^2 + B,$$

where A and B are two constants.

646. Principle of least action. Let the extremities A, B of the trajectories be given and let the particle be constrained to move from one point to the other along a smooth wire, the energy being given, Art. 636. Of all the different methods of conducting the particle from A to B there may be one which is the trajectory the particle would take if unconstrained. We see by Art. 637 that for this course the value of δV is given by equation (4). But since the points A, B are fixed, δx, δy, δz vanish at each end. We therefore have $\delta V = 0$. It follows therefore that the free trajectory is such that the change of action in passing from it to any neighbouring constrained course is zero. *The action for a free trajectory with given energy is either a maximum, a minimum, or is stationary.*

Conversely, if the path from A to B is required which makes the action a max-min, the principles of the Calculus of Variations require that the coefficients of δx, δy, δz inside the integral (3) in Art. 637 should be zero, provided the geometrical conditions of the problem permit δx, δy, δz to have arbitrary signs. Assuming this, the vanishing of the coefficients leads, as already explained, to the equations of motion. The result is that *the free trajectory from A to B is then the path of max-min action given by the calculus of variations.*

A similar theorem holds for the motion of a system either free or connected by geometrical relations. Let any two configurations or positions A, B be given. If we conduct the system from A to B by any varied paths as described in Art. 637 we have (since the variations of the coordinates of these positions are zero)

$$\delta \int L dt = -C(\delta t - \delta t_1) \ldots\ldots (A), \qquad \delta \int 2T dt = (t - t_1)\,\delta C \ldots\ldots (B).$$

Let us now suppose that in these varied paths the particles, without violating the geometrical relations, are conducted with such velocities that *the energy $C = T - U$ has a given value*, (the same as in the actual course,) then $\delta C = 0$, and the equation (B) shows that *the action $\int 2T dt$ is a max-min or is stationary in the actual path.*

The equation (A) gives a companion theorem. Let us suppose that in the varied paths the particles are so conducted that *the time $t - t_1$ is equal to a given quantity, then $\int L dt$ is a max-min or is stationary.*

647. The action from one given point to another cannot be a real maximum if the velocity is always the same function of the position of the particle. Every element of either of the integrals $\int v^2 dt$ or $\int v ds$ is positive and therefore, whatever path from A to B may be taken, we can increase the whole action by conducting the particle along a sufficiently circuitous but neighbouring path. Thus, if C be any point on the free course AB we can conduct the particle along that course to C, then compel it to make a circuit, and after returning to the neighbourhood of C conduct it along the remainder CB of the free path. Additional positive terms are thus given to the integral and the action is increased. The energy of the motion is unaltered, but the time of transit is longer.

Since every element of the integral is positive, there must be some path joining A and B which makes the action a true minimum. If the theory of max-min in the Calculus of Variations gives only one path, that path must be a minimum.

648. It may be that there are several free paths by which the particle could travel from A to B. Selecting one of these, say ADB, we may ask if the action along it is a true minimum. Let a neighbouring free path starting from A (the energy being the same) intersect ADB in C. To simplify matters let no other free path intersect ADB nearer to A than C. If B lie between A and C there is only one free path from A to B which is in accordance with the principles of mechanics, and that path makes the action a true minimum; Art. 647. If B is beyond C, there are two neighbouring free paths from A to C. It may be proved that the action from A to B is not in general a true minimum, the action for some neighbouring courses being greater and for others less than for the free path AB (Art. 653).

649. It may be that there is no free path from A to B, yet there must be a path of minimum action. For example, a heavy particle projected from A with a given velocity can by a free path arrive only at such points as lie within a certain paraboloid whose focus is at A, Art. 159. The path of minimum action from A to a point B beyond the paraboloidal boundary is not a free path. When deduced from the Calculus of Variations it falls under the case mentioned in Art. 646. Its position is such that it cannot be varied arbitrarily on all sides, i.e. the signs of the variations δx, δy, δz are not arbitrary along the whole length of the course.

Such limitations exist when the path runs along the boundary of the field of motion (Art. 299). We therefore draw verticals from A and B to intersect the level of zero velocity (which in this case is the directrix) in C and D. Let us conduct the particle from A along AC to a point as near C as we please, and thence along a course coinciding indefinitely nearly with the directrix to a point as near D as we please. The particle is finally conducted along the vertical DB to the given point B. Throughout this course the velocity is always supposed to be $\sqrt{(2gz)}$ where z is the depth below the directrix. The velocity being ultimately zero along the directrix the whole action from A to B is reduced to the sum of the actions along the vertical paths AC, DB. The path close to the directrix cannot be varied arbitrarily, because the particle cannot be conducted above that level without making the velocity imaginary. This minimum path is therefore not given by the ordinary rules of the Calculus of Variations.

A similar anomaly occurs in the case of brachistochrones. The parabola is a brachistochrone when the force acts parallel to the axis and is such that the velocity is inversely proportional to the square root of the distance from the directrix; Art. 605. The directrix being given in position, the initial and final points A, B of the course may be so far apart that no such parabola can be drawn. In this case the brachistochrone is found by conducting the particle along the vertical straight line AC in accordance with the given law of velocity, thence with an infinite velocity along the directrix CD, and finally along the vertical line DB to B.

The further discussion of these points is a part of the Calculus of Variations. Some remarks on the dynamics of the problem may be found in the author's *Rigid Dynamics*, vol. II. chap. X.

650. *Ex.* 1. Prove that the same path is a brachistochrone for $v^2=f(x, y, z)$ and a path of least action for $v'^2=A/f(x, y, z)$; Art. 599.

The brachistochrone is deduced from the calculus of variations by making $\int ds/v$ a minimum; the path of least action by making $\int v'ds$ a minimum. These must give the same curve if $v'=k^2/v$; (Jellett and Tait).

Ex. 2. Prove that, if a path be described by a particle P with such a work function that $v^2=f(r, \theta, \phi)$, the inverse path can be described by a particle Π with a velocity v', such that $v'^2=\dfrac{k^4}{\rho^4}f\left(\dfrac{k^2}{\rho}, \theta, \phi\right)$, where $r\rho=k^2$; Art. 628.

To find the first path we make $\int v ds$ a minimum. Since $ds'/ds=\rho/r$, the second path is found by making $\int v'ds\,\rho/r$ a minimum. These are the same integrals. This mode of proof applies equally whether the particle is free or constrained to move on a surface.

651. *Ex.* 1. Prove that in an elliptic orbit described about the focus S, the time is measured by the area described about the focus S and the action by the time described about the empty focus H.

If p, p' be the perpendiculars on the tangent from S and H, we know that $pp'=b^2$. Since $v=h/p$, the action $\int v ds$ becomes $\int p'ds\,.\,h/b^2$; the area described about H being $\frac{1}{2}\int p'ds$, the result follows at once. [Tait, *Dynamics of a particle.*]

Ex. 2. In an ellipse described about the centre C, perpendiculars PM, PN are drawn from P on the major and minor axes CA, CB, and A, B represent the elliptic areas PMA, $PNCA$ respectively. Prove that the action from A to P is

$$(a^2A+b^2B)\sqrt{\mu}/ab.$$

Ex. 3. Prove that the action in describing an arc of a central orbit is $\displaystyle\int\frac{h}{p}\left(1-\frac{p^2}{r^2}\right)^{-\frac{1}{2}}dr$. When the central force is $F=\mu/r^n$ and the initial velocity is that from infinity, prove also that the action is $\dfrac{2h}{n-3}\tan\dfrac{n-3}{2}\theta$, where θ is measured from the maximum or minimum radius vector; Art. 360.

Ex. 4. A heavy particle describes a parabola. Prove that the action from any point A to another B is κ times the sectorial area ASB, where S is the focus, $\kappa^2=16g/l$ and l is the semi-latus rectum.

Prove also that, if the chord AB pass through the focus, the action along the parabolic path is greater than that along the course AC, CD, DB where AC, BD are perpendiculars on the directrix. Arts. 159, 649.

652. *Ex.* 1. When a heavy particle is projected from a point A with a given velocity to pass through a point B, there are in general two possible parabolic paths. Prove that the action is a minimum along that parabola in which the arc AB is less than the arc AC where C is the other extremity of the chord drawn from A through the focus.

The action is a minimum when B is not beyond the intersection with the neighbouring parabola drawn from A; Art. 648. Since the chord of intersection ultimately passes through the focus of either of these neighbouring parabolas, Art. 159, the result given follows at once.

Ex. 2. When the force is central and varies according to the Newtonian law, there are in general two elliptic paths which a particle could take when projected from A with a given velocity to pass through B. Prove that the action is a minimum along that ellipse in which the arc AB is less than AC, where C is the other extremity of the chord drawn from A through the empty focus: Art. 339.

653. *Ex. A particle describes a circular orbit about a centre of force represented by $F=\mu/r^n$, situated in the centre O. It is required to find the change in the action when the particle is conducted with the same energy from a given point A to another B on the circle by some neighbouring path lying in the plane of the circle.*

Let a be the radius, then taking the normal resolution, the velocity $v_0=\sqrt{(\mu/a^{n-1})}$. The principle of energy for the varied path gives

$$\frac{v^2}{2}=\frac{\mu}{n-1}\frac{1}{r^{n-1}}+C.$$

Also $C=\frac{1}{2}\frac{n-3}{n-1}\frac{\mu}{a^{n-1}}$, since the energy C is the same for both paths.

Let the equation of the varied path be $r=a(1+\rho)$ where ρ is some function of θ. Substituting we find

$$v=v_0\{1-\rho+\tfrac{1}{2}(n-1)\rho^2+\dots\} \quad\dots\dots\dots\dots \quad (1).$$

Here ρ is equivalent to the δr of the Calculus of Variations.

Since $(ds)^2=r^2(d\theta)^2+(dr)^2$, we find by the same substitution

$$\frac{ds}{d\theta}=a\left\{1+\rho+\frac{1}{2}\left(\frac{d\rho}{d\theta}\right)^2+\dots\right\} \quad\dots\dots\dots\dots \quad (2).$$

The action therefore when θ increases from 0 to θ is

$$\int v\,ds=av_0\left\{\theta+\frac{1}{2}\int\left\{\left(\frac{d\rho}{d\theta}\right)^2-p^2\rho^2\right\}d\theta+\dots\right\} \quad\dots\dots\dots\dots \quad (3),$$

where $p^2=3-n$ as in Art. 367, and the limits are $\theta=0$ to θ. *By substituting for ρ the value corresponding to any assumed variation of the path, the change in the action follows immediately.*

If the particle starting from A were to describe a neighbouring free path with the same energy, we know by Art. 367 that the first intersection of the new path with the circle is at a point given by $\theta=\pi/p$ nearly.

We may easily deduce from the expression (3) that *the action from A to B is a* *true minimum if the angle* $AOB < \pi/p$; see Art. 594, 648. To prove this we use an artifice due to Lagrange*. Since

$$\frac{d}{d\theta}(\lambda\rho^2) = 2\lambda\rho\frac{d\rho}{d\theta} + \rho^2\frac{d\lambda}{d\theta} \quad\text{.............................(4)},$$

where λ is an arbitrary function of θ, we may write the integral on the right-hand side of (3) in the form

$$I = -[\lambda\rho^2] + \int\left\{\left(\frac{d\rho}{d\theta}\right)^2 + 2\lambda\rho\frac{d\rho}{d\theta} + \left(\frac{d\lambda}{d\theta} - p^2\right)\rho^2\right\}d\theta.$$

The term $\lambda\rho^2$ taken between the limits is zero, since both paths begin at A and end at B. Let us choose the function λ so that

$$\lambda^2 = \frac{d\lambda}{d\theta} - p^2, \quad \therefore \lambda = p\tan p(\theta - a) \quad\text{........................ (5)},$$

then

$$I = \int\left(\frac{d\rho}{d\theta} + \lambda\rho\right)^2 d\theta \quad\text{.............................(6)}.$$

Since this integral is essentially positive it follows from (3) that the action along every varied path from A to B is greater than that along the circle.

This argument requires that λ should not be infinite within the limits of integration. By taking $pa = \frac{1}{2}\pi - \epsilon$ where ϵ is a quantity as small as we please the values of λ given by (5) can be made finite from $\theta = 0$ to $\theta = \pi/p - \epsilon'$ where ϵ' is a quantity as small as we please. *The argument therefore requires that the point B should not make the angle* $AOB > \pi/p$.

When the angle AOB is greater than π/p *we can prove that the action along some varied curves extending from A to B is less, and along others is greater, than that in the circle.*

To prove this let us conduct the particle from A to B along the varied path whose equation is $\rho = L\sin g\theta$. Let β be the angle AOB, then since ρ vanishes at each end, g is arbitrary except that $g\beta$ is a multiple of π. Since $p\beta > \pi$ one value at least of g is less than p and the others are greater than p. Substituting in (3), we find that the integral is

$$I = \int\left\{\left(\frac{d\rho}{d\theta}\right)^2 - p^2\rho^2\right\}d\theta = \frac{L^2\beta}{2}(g^2 - p^2) \quad\text{.................... (7)},$$

the limits being $\theta = 0$ to $\theta = \beta$. The smaller values of g make I negative, while the greater values (which correspond to the more circuitous routes) make I positive. The conclusion is that *when the angle* $AOB > \pi/p$, *the action along the circle is not a true minimum.*

654. *Ex. A particle moves in a plane with a velocity* $v = \phi(x, y)$ *beginning at a given point A and ending at B. The path taken being that of minimum action, it is required to find in Cartesian coordinates the equation of the path and the change of action when the path is varied in an arbitrary manner.*

Let the elementary action $v\,ds = \phi\sqrt{(1 + y'^2)}\,dx$ be represented by $f(x, y, p)\,dx$, where p has been written for $y' = dy/dx$. Then writing $y + \delta y$, $p + \delta p$ for y and p,

---

* Lagrange *Théorie des fonctions Analytiques* 1797. He refers to Legendre, *Memoirs of the Academy of Sciences* 1786, and adds that it must be shown that λ does not become infinite between the limits of integration. Not being able to settle this question, he just missed Jacobi's discovery. See also Todhunter's *History of the Calculus of Variations*, page 4.

(but not varying x) the whole increase of action on the varied curve is by Taylor's theorem,

$$\delta A = \int [f_y \delta y + f_p \delta p + \tfrac{1}{2} \{f_{yy} (\delta y)^2 + 2 f_{yp} \delta y \delta p + f_{pp} (\delta p)^2\} + \&c.]\, dx,$$

where suffixes as usual represent partial differential coefficients. Integrating the second term by parts, as in Art. 591, we have

$$\delta A = [f_p \delta y] + \int \{(f_y - f_p') \delta y + \&c.\}\, dx,$$

where the part outside the integral, being taken between fixed limits, is zero, and accents denote total differentiation with regard to x. The path of minimum action is found by equating the coefficient of δy to zero, Art. 591. This path is therefore given by

$$f_y - f_p' = 0 \quad\text{..................................}\quad (1),$$

and the change of action in any varied path by

$$\delta A = \tfrac{1}{2} \int [f_{yy} (\delta y)^2 + 2 f_{yp} \delta y\, \delta p + f_{pp} (\delta p)^2]\, dx \quad\text{....................}\quad (2).$$

To find the path in Cartesian coordinates we integrate the equation (1). This can only be effected when the form of the function ϕ is given. The integration presents only those difficulties which are discussed in treatises on differential equations. We now proceed to find the change in the action given by (2).

To determine the sign of δA, we write (2) in the form

$$\delta A = [\lambda (\delta y)^2] + \tfrac{1}{2} \int [(f_{yy} - 2\lambda') (\delta y)^2 + 2 (f_{yp} - 2\lambda)\, \delta y \delta p + f_{pp} (\delta p)^2]\, dx \quad\text{......}\quad (3),$$

where the term outside the integral is zero, provided λ does not become infinite between the limits of integration.

Let $y = F(x, c_1, c_2)$ be the integral of (1), then changing the constants into $c_1 + \alpha$, $c_2 + \beta$ where α, β are indefinitely small,

$$y + \delta y = F + \frac{dF}{dc_1} \alpha + \frac{dF}{dc_2} \beta \quad\text{............................}\quad (4),$$

is also a solution of (1). We choose the constants c_1, c_2 so that the curve $y = F$ passes through the limiting points A and B. Making the varied curve (4) also pass through A, we have an equation to find β/α. Hence

$$\delta y = \alpha \left(\frac{dF}{dc_1} + \frac{dF}{dc_2} \frac{\beta}{\alpha} \right) = u \quad\text{...........................}\quad (5),$$

is the equation of a neighbouring path of minimum action beginning at A and making a small arbitrary angle with the path AB, the magnitude of the angle depending on that of α. If C is the *first point of intersection* of these two paths, then u is not zero between A and C.

Differentiating (1) we see that $\delta y = u$ satisfies the equation

$$f_{yy} \delta y + f_{yp} \delta p - \frac{d}{dx} (f_{yp} \delta y + f_{pp} \delta p) = 0;$$

$$\therefore \left(f_{yy} - \frac{d}{dx} f_{yp} \right) u = \frac{d}{dx} (f_{pp} u') \quad\text{..........................}\quad (6).$$

Returning to the integral (3) let us choose λ so that

$$(f_{yp} - 2\lambda)\, u = -f_{pp} u' \quad\text{................................}\quad (7).$$

Substituting in (6) we find

$$\left(f_{yy} - \frac{d}{dx} f_{yp} \right) u = -\frac{d}{dx} (f_{yp} - 2\lambda)\, u$$

$$= -\left(\frac{d}{dx} f_{yp} - 2 \frac{d\lambda}{dx} \right) u + \frac{(f_{yp} - 2\lambda)^2 u}{f_{pp}},$$

the last term being obtained by substituting for u' from (7). This becomes

$$\left(f_{yy} - 2\frac{d\lambda}{dx}\right) f_{pp} = (f_{yp} - 2\lambda)^2 \text{..........................(8).}$$

The quantity under the integral sign in (3) is therefore a perfect square. Remembering (7) we see that

$$\delta A = \frac{1}{2}\int f_{pp}\left\{\delta p - \frac{u'}{u}\delta y\right\}^2 dx \text{..........................(9).}$$

The value of λ is by (7)

$$2\lambda = \left(\frac{dv}{dy}p + \frac{v}{1+p^2}\frac{u'}{u}\right)\frac{1}{\sqrt{(1+p^2)}} \text{......................(10).}$$

Hence in order that both λ and the subject of integration in (9) may be finite *it is necessary that u should not vanish between the limits of integration.* The second limiting point B must therefore not be beyond C. It is supposed that v and dv/dy are finite between the same limits. See Art. 648.

Supposing this condition to be satisfied, every term of the integral (9) is positive if f_{pp} is positive from A to B. Since $f_{pp} = v(1+p^2)^{-\frac{3}{2}}$, and the velocity v is supposed to keep one sign throughout the motion, this condition also is satisfied. *The change of action caused by a variation of path is therefore always positive and its amount is determined by* (2) or (9).

This investigation can be applied to brachistochrones and may also be extended to any cases in which the subject of integration, viz. $f(x, y, p)$, is a function only of the coordinates y, x, and the first differential coefficient. In order that the course AB given by (1) should be a true minimum, no variation must exist which can make δA negative. The conditions for this are (1) the point B must not be beyond C, as explained in Arts. 594, 648, (2) the differential coefficient d^2f/dp^2 must be positive throughout the whole course AB.

If d^2f/dp^2 were negative for any portion PQ of the course given by (1), let us vary the remaining portions AP, QB so that δy is as nearly equal to u as we please, the portion PQ being varied in some other manner. In this variation such prominence is given to the negative elements of the integral (9) that δA is made negative. It is also evident from (7) that λ is finite if d^2f/dp^2, $d^2f/dp\,dy$ are finite.

A SWARM OF PARTICLES.

Note on Art. 414.

THE argument will be made more complete if we suppose that the boundary of the swarm is an ellipsoid instead of a sphere. Owing to the manner in which the forces of attraction depend on the shape of the swarm, the results for an ellipsoid are not altogether the same as those for a sphere.

Taking the same axes as before, the coordinates of the projection of any particle P on the plane of motion of the centre are $r+\xi$, η, while ζ is the distance of P from that plane. Treating the ellipsoid as homogeneous and of density D, the component attractions of the swarm at any internal point are $A\xi$, $B\eta$, $C\zeta$, where A, B, C are functions of the ratios of the axes of the bounding ellipsoid and their sum is $4\pi D$.

The equations (1) of Art. 414 are slightly modified by having their last terms replaced by $-A\xi$, $-B\eta$; and instead of (3) we have

$$\left.\begin{aligned}\frac{d^2\xi}{dt^2}-2n\frac{d\eta}{dt}+(A-3n^2)\,\xi&=0\\ \frac{d^2\eta}{dt^2}+2n\frac{d\xi}{dt}+B\eta\qquad\quad&=0\end{aligned}\right\}\qquad\text{(I)}.$$

The equation for ζ is evidently

$$\frac{d^2\zeta}{dt^2}=-\frac{M\zeta}{r^3}-C\zeta=-(n^2+C)\,\zeta\qquad\text{(II)}.$$

Putting $\xi=a\cos(pt+\alpha)$, $\eta=b\sin(pt+\alpha)$, and $\zeta=c\sin(qt+\gamma)$ we find by proceeding as in Art. 414,

$$\{p^2-(A-3n^2)\}\{p^2-B\}-4p^2n^2=0,\quad q^2=n^2+C\qquad\text{(III)}.$$

The condition for stability is therefore $A>3n^2$.

In an ellipsoid $A>B$ if the axis in the direction of ξ is less than that in the direction of η. It follows that if the axis of ξ is the least axis, A is greater for an ellipsoid than for a sphere. The swarm is therefore more stable for an ellipsoidal than for a spherical swarm provided the least axis of the ellipsoid is placed along the radius vector from the sun.

Let us suppose that all the particles are describing the same principal oscillation. The projections of their paths on the plane $\xi\eta$ are therefore given by $\xi=a\cos\theta$, $\eta=b\sin\theta$, where $\theta=pt+\alpha$. These paths are coaxial ellipses described in the same periodic time $2\pi/p$, the semi-axes of any ellipse being a, b. By substituting these values of ξ, η in the second of equations (I), we find $\dfrac{a}{b}=\dfrac{p^2-B}{-2np}$; it follows that all the ellipses are similar to each other. There will therefore be no collisions between the particles.

The ratio of the axes of the ellipses is not altogether arbitrary. By using (III) we find

$$\left(\frac{a}{b}\right)^2 = \frac{p^2 - B}{p^2 - (A - 3n^2)}, \quad \therefore p^2(a^2 - b^2) - Aa^2 + Bb^2 = -3a^2n^2,$$

where A, B and therefore p^2 are known functions of the ratios of the axes of the ellipsoid. We may deduce from the values of A, B given in the theory of Attractions that Aa^2 is less or greater than Bb^2 according as a^2 is greater or less than b^2. It then follows from this equation that in both the principal oscillations the axis of the ellipsoid in the direction of the radius vector from the sun is less than the axis of the ellipsoid in the direction of motion of the centre.

If P, Q, R be any three particles describing similar co-axial ellipses in the same time with an acceleration tending to their common centre, it is not difficult to prove that the area of the triangle PQR is constant throughout the motion. Let us apply this theorem to the motion of the projections of the particles on the plane of $\xi\eta$. Joining adjacent triads of particles, we divide the whole area into elementary triangles. If the swarm is homogeneous, the areas of these triangles are initially equal and we see that they will remain equal throughout the motion. The swarm will therefore remain homogeneous.

Consider next the motions of the particles perpendicular to the plane of $\xi\eta$. These are harmonic oscillations and are all described in the same time $2\pi/q$. The amplitude of each oscillation is the ordinate of the ellipsoid corresponding to the ellipse described by the projection and this is constant for the same particle. The distance between two adjacent particles moving in the same ordinate in the same direction is increasing or decreasing according as they are approaching or receding from the plane of $\xi\eta$. As there are as many particles approaching as receding, the uniformity of the density is not affected by this motion.

When both the principal oscillations are being described simultaneously the state of the motion becomes more complicated. The outer boundary is not strictly ellipsoidal, being dependent on both the states of motion. Since also the rotations in the principal oscillations are in opposite directions, we can no longer neglect the collisions between the particles.

To take account of the collisions we must have recourse to a statistical theory analogous to the kinetic theory of gases. But this would lead us too far from the methods of this treatise.

For an example of the application of the kinetic theory the reader is referred to a memoir by G. H. Darwin, *On the mechanical conditions of a swarm of meteorites, &c.*, *Phil. Trans.* 1889. He supposes a number of meteorites to be falling together from a condition of wide dispersion and to have not yet coalesced into a system of a sun and planets. No account is taken of the rotation of the system.

Callandreau has discussed the case in which a comet, regarded as a spherical swarm of particles, is heterogeneous, the density being a function of the distance from the centre. The effect of a passage near Jupiter has also been taken into account. See his *Étude sur la théorie des comètes périodiques.* He considers it probable that the periodic comets are undergoing a gradual disintegration and he points out that according to this hypothesis a few comets captured by the action of Jupiter could by repeated subdivisions produce all those known to exist. See *The Observatory*, Feb. 1898.

LAGRANGE'S EQUATIONS.

Note on Art. 524.

THIS rule may be put into another form. We know that if $L=T+U+C$ be the Lagrangian function and θ, ϕ, &c. the coordinates, the equations of motion are

$$\frac{d}{dt}\frac{dL}{d\theta'}=\frac{dL}{d\theta},\quad \frac{d}{dt}\frac{dL}{d\phi'}=\frac{dL}{d\phi},\ \text{\&c.} \qquad (1).$$

We now see that we may use the same equations, if we substitute

$$L=\frac{T_2}{M}+M(U+C) \qquad (2),$$

where M is any arbitrary function of the coordinates θ, ϕ, &c. which we may find suitable when solving the equations.

The expression for T_2 differs from T only in the fact that the differential coefficients are taken with regard to a different independent variable, which has been represented by τ. Thus

$$T=\tfrac{1}{2}A_{11}\left(\frac{d\theta}{dt}\right)^2+A_{12}\frac{d\theta}{dt}\frac{d\phi}{dt}+\text{\&c.};\qquad T_2=\tfrac{1}{2}A_{11}\left(\frac{d\theta}{d\tau}\right)^2+A_{12}\frac{d\theta}{d\tau}\frac{d\phi}{d\tau}+\text{\&c.} \ldots (3).$$

When the equations have been solved the paths of the particles are found by eliminating τ without enquiry into its meaning.

The equation of energy is supposed to be $T-U=C$; the constant C is therefore known when the initial values of θ, ϕ, &c., θ', ϕ', &c. are given.

We notice that one solution must be analogous to that given by the principle of vis viva. We therefore have $\frac{T_2}{M}=M(U+C)$. Since this must agree with the equation $T=U+C$, it immediately follows that $T=T_2\left(\frac{d\tau}{dt}\right)^2$, $T_2=M^2T$. The relation between τ and t is therefore $Md\tau=dt$.

When the paths of the particles are alone required, we may eliminate the time from the Lagrangian equations by using a new function instead of the Lagrangian function.

In this method we choose some one coordinate θ to be the independent variable and regard the others ϕ, ψ, &c. as unknown functions of θ whose forms are to be determined by the altered equations of motion. Let

$$T=\tfrac{1}{2}A_{11}\theta'^2+A_{12}\theta'\phi'+\tfrac{1}{2}A_{22}\phi'^2+A_{23}\phi'\psi'+\ldots \qquad (4),$$

where accents denote differential coefficients with regard to the time. Let also

$$T'=\tfrac{1}{2}A_{11}+A_{12}\phi_1+\tfrac{1}{2}A_{22}\phi_1^2+A_{23}\phi_1\psi_1+\ldots \qquad (5),$$

where the suffixes of ϕ, ψ, &c. here denote differentiations with regard to the new independent variable θ.

$$\therefore\ \frac{dT}{d\phi'}=\frac{dT'}{d\phi_1}\theta';\qquad \frac{dT}{d\phi}=\frac{dT'}{d\phi}\theta'^2 \qquad (6).$$

The equation of energy gives

$$T'\theta'^2 = U + C, \qquad \therefore\ \theta' = \left(\frac{U+C}{T'}\right)^{\frac{1}{2}} \text{ .... (7).}$$

The Lagrangian equation $\dfrac{d}{dt}\dfrac{dT}{d\phi'} - \dfrac{dT}{d\phi} = \dfrac{dU}{d\phi}$ becomes

$$\left(\frac{U+C}{T'}\right)^{\frac{1}{2}} \frac{d}{d\theta}\left\{\left(\frac{U+C}{T'}\right)^{\frac{1}{2}} \frac{dT'}{d\phi_1}\right\} = \frac{dT'}{d\phi}\,\frac{U+C}{T'} + \frac{dU}{d\phi},$$

where all the differential coefficients are partial except the $d/d\theta$.

Remembering that U is not a function of ϕ_1, this becomes

$$\frac{d}{d\theta}\frac{d}{d\phi_1}\{(U+C)\,T'\}^{\frac{1}{2}} = \frac{d}{d\phi}\{(U+C)\,T'\}^{\frac{1}{2}} \text{ .... (8).}$$

If then we use $Q = \{(U+C)\,T'\}^{\frac{1}{2}}$ *as if it were the Lagrangian function and regard* θ *as the independent variable, we have the equations*

$$\frac{d}{d\theta}\frac{dQ}{d\phi_1} = \frac{dQ}{d\phi}, \qquad \frac{d}{d\theta}\frac{dQ}{d\psi_1} = \frac{dQ}{d\psi}, \text{ \&c. .... (9),}$$

from which the paths may be found.

This result follows easily from the theorem of Art. 524 by putting $d\tau = d\theta$, and we have here reproduced so much of that article as is required for our present purpose. If $d\tau = d\theta$, we have $Md\theta = dt$ and therefore by (7) of this note $M = \left(\dfrac{T'}{U+C}\right)^{\frac{1}{2}}$. Substituting in (2) the Lagrangian function becomes

$$L = 2\,\{(U+C)\,T'\}^{\frac{1}{2}}.$$

We notice that however the expressions for the vis viva and the work function may be different in different problems, yet *so long as the product* $(U+C)\,T'$ *remains unchanged, the paths are determined by the same relations between the coordinates* θ, ϕ, *&c.*

Since in the Lagrangian equations, the letters θ, ϕ, &c. represent arbitrary functions of the quantities or coordinates which determine the position of the system, it is evident that we have here taken as the independent variable any arbitrary function of the coordinates.

If some one coordinate, say ϕ, *is absent from the product* $(U+C)\,T'$ (though T contains the differential coefficients of ϕ), *we see that one solution of the equations of motion is*

$$\frac{dQ}{d\phi_1} = a, \qquad \therefore\ (U+C)^{\frac{1}{2}}\frac{dT'^{\frac{1}{2}}}{d\phi_1} = a \text{ .... (10),}$$

where a *is an arbitrary constant.* If C is arbitrary, the product Q cannot be independent of ϕ unless T' and U are separately independent of ϕ. But when C is given by the initial conditions this limitation is not necessary. If we substitute for $dT'/d\phi_1$ and T' the values given by (6) and (7) this integral becomes $dT/d\phi' = 2a$, which is the same as that obtained in Art. 521.

We may deduce this extension directly from the Lagrangian equations. Suppose

$$T = M\,\{\tfrac{1}{2}A_{11}\theta'^2 + \text{\&c.}\}, \qquad U + C = \frac{1}{M} f(\theta, \psi, \text{\&c.}),$$

where M is a function of θ, ϕ, &c. while A_{11}, &c. are not functions of ϕ. In this case the product $T(U+C)$ is not a function of ϕ. The Lagrangian equation for ϕ gives

$$\frac{d}{dt}\frac{dT}{d\phi'} - \frac{dM}{d\phi}(\tfrac{1}{2}A_{11}\theta'^2 + \&c.) = -\frac{1}{M^2}\frac{dM}{d\phi}f(\theta, \psi, \&c.);$$

$$\therefore \frac{d}{dt}\frac{dT}{d\phi'} = \frac{dM}{d\phi}\frac{1}{M}(T - U - C) \quad\text{.........................(11).}$$

If then the initial circumstances are such that the equation of energy is $T = U + C$, we have $\dfrac{dT}{d\phi'} = a$.

As a simple example, consider the case of a projectile moving under the action of gravity. We have $T = \frac{1}{2}(x'^2 + y'^2)$, $U = -gy$. Since the product of these is independent of x we choose some other coordinate as the independent variable. Writing $x_1 = dx/dy$ we have

$$Q = \{(1 + x_1^2)(y + C)\}^{\frac{1}{2}}; \qquad \therefore \frac{dQ}{dx_1} = \frac{x_1(y + C)^{\frac{1}{2}}}{\sqrt{(1 + x_1^2)}} = a.$$

This by an easy integration leads to the parabola $(x - \beta)^2 = 4a^2(y + C - a^2)$.

The elimination of the time from the Lagrangian equations is given by Painlevé in his *Leçons sur l'intégration des équations différentielles de la Mécanique*, 1895. By an application of the principle of least action he obtains the function here called Q and writes the equations in the typical form $\dfrac{d}{dq_1}\dfrac{dQ}{dq'_n} = \dfrac{dQ}{dq_n}$. From these he deduces (page 239) that the Lagrangian equations may be written in the two forms

$$\frac{d}{dt}\frac{dT}{dq'} - \frac{dT}{dq} = \frac{dU}{dq}, \qquad \frac{d}{d\tau}\frac{dT'}{dq'} - \frac{dT'}{dq} = 0,$$

where $T' = T(U + C)$ and $d\tau = (U + C)\,dt$. This special result follows from that given at the beginning of this note by putting $1/M = U + C$. *Its importance lies in the fact that by this change the motion is made to depend on that of a system moving under no forces.*

The elimination of the time from Lagrange's equations is also given by Darboux in his *Leçons sur la théorie générale des surfaces*, Art. 571, 1889. He expresses his results in the same form as Painlevé.

We may obtain an extension of the theorem (2). In such problems as those discussed in Art. 255 the Lagrangian function takes the form

$$L = L_2 + L_1 + L_0 \quad\text{.................................(12),}$$

where L_n is a homogeneous function of θ', ϕ', &c. of the order n, the coefficients being functions of θ, ϕ, &c. but not of t. We then find as in Art. 512, Ex. 3, that the equation of energy becomes

$$L_2 - L_0 = C \quad\text{......................................(13).}$$

Proceeding as in Art. 524, we change dt into $d\tau$ and write

$$L = \frac{L_2}{M} + L_1 + M(L_0 + C) \quad\text{............................ (14).}$$

We may now use this as the Lagrangian function.

INDEX.

The numbers refer to the articles.

Cambridge:
PRINTED BY J. AND C. F. CLAY,
AT THE UNIVERSITY PRESS.

Printed in the United States
By Bookmasters